THE CHEMICAL BASIS OF LIFE

Readings from

**SCIENTIFIC
AMERICAN**

THE CHEMICAL BASIS OF LIFE

An Introduction to Molecular and Cell Biology

With introductions by

Philip C. Hanawalt
Stanford University

Robert H. Haynes
York University, Toronto

W. H. Freeman and Company
San Francisco

Library of Congress Cataloging in Publication Data

Hanawalt, Philip C   1931–   comp.
    The chemical basis of life.

    1. Molecular biology—Addresses, essays,
lectures.   2. Cytology—Addresses, essays, lectures.
I. Haynes, Robert H., 1931–   joint comp.
II. Scientific American.   III. Title.
[DNLM: 1. Biochemistry—Collected works.
2. Molecular biology—Collected works.   QH506
H233c 1973]
QH501.H35   574.8′8   73–8899
ISBN 0–7167–0882–5
ISBN 0–7167–0881–7 (pbk.)

Most of the SCIENTIFIC AMERICAN articles in
THE CHEMICAL BASIS OF LIFE: AN INTRO-
DUCTION TO MOLECULAR AND CELL BIOL-
OGY are available as separate Offprints. For a
complete list of more than 950 articles now avail-
able as Offprints, write to W. H. Freeman and
Company, 660 Market Street, San Francisco,
California 94104.

Printed in the United States of America

9 8 7 6 5 4 3 2 1

PREFACE

Our first collection of readings from *Scientific American*, published in 1968 and entitled *The Molecular Basis of Life*, has been widely used in biology courses as a text and as a source of supplementary readings. The clear exposition and vivid illustrations of *Scientific American* articles are often more stimulating to students than are those found in traditional textbooks. The authors of the articles convey the excitement of molecular biological research, and their detailed descriptions of experimental strategy enliven the presentation of ideas. Our earlier volume had an historical flavor, and it included many of the now classic *Scientific American* articles that did so much to excite widespread interest in the field. The earliest article in that previous reader was by George W. Beadle (September 1948); the latest was by Brian F. C. Clark and Kjeld A. Marcker (January 1968). In the present collection the earliest article is by Gunther S. Stent (May 1953), and twenty-eight of the thirty-eight articles were first published during the period 1966–1973. In the most recent contribution to this collection (March 1973) by O. L. Miller, Jr., the central dogma of molecular biology is vividly brought to life in electron micrographs of individual genes in action.

In our introductory essay in *The Molecular Basis of Life*, we closed on a rather funereal note—the excitement of the field seemed to have peaked, and avant-garde molecular biologists were moving on to what appeared to be the greener pastures of neurobiology and the problems of embryonic development. However, the gloomy prophecies made five years ago have proven to be premature since molecular biology is still very much alive and is growing as vigorously as the political estate will permit. Although the genetic code has been cracked and we now have a much more detailed understanding of the elements of protein synthesis, a wealth of new problems has been uncovered, and surprises are occurring year by year, among them the discovery that RNA can direct the synthesis of DNA, a finding that may well be of key importance in solving the cancer problem.

The present selection of articles is based upon the experience of ourselves and others with the use of *Scientific American* Offprints in teaching. We have retained half (seventeen) of the articles from our previous collection and have added twenty-one new ones that update and provide more comprehensive coverage of the field. Notable is the inclusion of a series of recent articles on membranes and cell surface phenomena—an exciting new area of biology now that we appreciate how much of the chemistry of the living cell is carried out on membranes rather than in an aqueous "cell soup." A more detailed treatment of regulation includes a description of the recent isolation of repressor proteins, a molecular species that had been postulated some years before on the basis of purely genetic experiments. New elements of regulation such as cyclic AMP are discussed, and the student is introduced to frontier studies on eucaryotic chromosomes. Separate articles specifically treat the structure, function, and assembly of the most important subcellular organelles; the ribosomes, the mitochondria, and the chloroplasts. Overall, there is an increased emphasis on the contributions from biochemistry to the development of cell biology.

The volume is divided into four parts. The first deals with energy capture, transfer and utilization in living systems, and the origin of life itself. The second part on molecular architecture includes an article illustrating each of the major functional types of proteins: structural proteins, enzymes, hormones, antibodies, oxygen carriers, and electron transport proteins. The story of the discovery of DNA is followed by an account of the isolation of the first DNA polymerase enzyme by Arthur Kornberg and the use of this enzyme to synthesize biologically active DNA in the test tube. The third part deals with supramolecular complexes and the operation of principles of self-assembly and efficient design in the formation of cells from relatively few types of building blocks. Six articles in the collection deal with the structures, the life cycles, and the assembly of those ultimate parasites, the viruses. The final section, on information transfer and control, contains articles on the regulation of protein synthesis and introduces the reader to some of the most exciting frontier areas of molecular biology, such as the visualization of the processes of transcription and translation in the electron microscope.

A concluding essay by Gunther S. Stent places the entire field of molecular biology in the context of human creative endeavor and suggests how our scientific perceptions and world view might be significantly influenced by the neural structure of the human mind.

Philip C. Hanawalt
April 1973   Robert H. Haynes

CONTENTS

IV INFORMATION TRANSFER AND CONTROL

Note on cross-references: References to articles included in this book are noted by the title of the article and the page on which it begins; references to articles that are available as Offprints, but are not included here, are noted by the article's title and Offprint number; references to articles published by SCIENTIFIC AMERICAN, but which are not available as Offprints, are noted by the title of the article and the month and year of its publication.

THE CHEMICAL BASIS OF LIFE

I

ENERGY FLOW IN LIVING SYSTEMS

ENERGY FLOW IN LIVING SYSTEMS

INTRODUCTION

Matter and energy are two fundamental and interconvertible components of the physical world, and the history of this world as understood by scientists today is an account of the gradual organization of matter into structures such as atoms, molecules, stars and planets under the influence of nuclear, electromagnetic, and gravitational forces. Astrophysical and geochemical evidence further suggests that we inhabit an expanding universe with an *evolutionary history.* Following the burst of a primordial "fireball" of matter-energy, some 11 billion years ago, stars and galactic systems began to form in which changing conditions of temperature and pressure permitted the step-wise build-up of the ninety-two elements and the formation of the simplest gaseous molecules. Thermonuclear fusion and neutron capture of subatomic particles in hot stellar interiors are basic mechanisms of *atomic evolution,* and these processes continue today in stars throughout the universe.

Matter ejected by stars in the course of their development and debris from the explosions of supernovae give rise to vast clouds of interstellar gas and dust. Under the influence of stellar gravitational fields, some of this material has accreted into planetary systems. Radioactive dating of meteorites indicates that our solar system was formed about 4.5 billion years ago although the oldest exposed rock of the earth's crust is about 3.5 billion years old. In the cooling atmosphere and seas of the primitive earth, conditions were appropriate for further stages of chemical or *molecular evolution.* During the first 1 or 2 billion years of earth history these latter processes gradually gave rise to the complex organic molecules necessary for the assembly of the first primitive microorganisms and the onset of Darwinian or *biological evolution.* (The oldest known microfossils, found in the eastern Transvaal region of South Africa, are about 3.1 billion years old.)

Evolutionary processes depend upon a supply of the appropriate sorts of matter and a suitable flow of energy to achieve the formation of atoms, molecules, or organisms. The flow of energy responsible for the origin and continuing existence of life on earth derives from the constant flux of light from the sun. The materials that are constantly recycled through living organisms as a result of the steady flow of solar energy came originally from the earth's crust, the oceans, and the atmosphere.

The nature and properties of the entities that can be produced in atomic, molecular, and biological evolution are governed by quantum-mechanical, physicochemical, and biological laws. These laws permit the existence of only certain types of atoms, the formation of only certain types of chemical compounds, or the evolution of only certain categories of organisms out of the almost infinite variety that might be imagined. However, it is interesting to note that as one ascends through the hierarchy of atoms, molecules, macromolecules, and organisms, the "laws" seem to become less restrictive, and the potential variety, diversity and complexity of entities becomes astonishingly large; thus, there are a few hundred types of atomic nuclei, a few hundred thousand discrete, naturally occurring chemical compounds (although over 2 million have been synthesized by man), and an estimated 12 million species of plants, animals, and microorganisms. Furthermore, because the macromolecular constituents of living cells are long, chain-like molecules (linear polymers), the scope for variety and diversity in the biological world is vastly greater than even these figures would suggest. Proteins are the predominant type of macromolecule found in cells; a typical one would consist of a chain of about 200 amino acid units. Each amino acid link in this chain could be any one of the twenty amino acids that normally occur in living organisms. Thus, it is evident that $20^{200} = 10^{260}$ different variants of such a protein could exist. It is difficult to appreciate the significance of such a number, as it is what mathematicians call a "combinatorial" number and is vastly greater than other large numbers associated with familiar objects in nature. For example, it might be noted that the mass of the earth is equivalent to the mass of 10^{55} electrons, and the mass of the entire solar system (99.8 percent of which is in the sun) is equivalent to the mass of "only" 10^{60} electrons. Clearly, the scope for biological variation among individual organisms and, hence, for evolutionary experimentation, is enormously greater than what has ever been realized even amongst all the organisms that have ever existed on earth.

The "laws" of physics, chemistry, and biology can be thought of as ultimate sources of "information" necessary for the formation of the resulting atomic, molecular, and biological structures. Indeed, three fundamental ingredients, *matter, energy,* and *information* are required for the evolution of natural objects, just as building materials, the energy of workers and architectural information, are necessary for the construction of buildings or other artifacts of man. We have, accordingly, arranged the articles that constitute this introduction in the order discussed in the preface. The picture of an evolutionary world in which ever more complex and highly ordered systems emerge hierarchically in the natural course of events might strike some readers as being strangely inconsistent with the second law of thermodynamics. That law asserts that isolated physical systems tend to decay spontaneously and inexorably toward an equilibrium state which maximizes entropy, or disorder within the system. Put more picturesquely, the universe is decaying toward a "heat death" or "entropic doom." Evolutionary processes oppose or diminish this tendency to disorder, at least for some systems and over restricted periods of time. However, there is no real incompatibility between evolution and the second law because the latter applies to closed systems approaching thermal equilibrium whereas the surface of the earth is an open system maintained far from equilibrium by virtue of the fact that it constantly receives solar energy and in turn emits longer wavelength radiation to outer space. Thus, the earth's surface is part of an open, steady-state system through which there is a constant flow of energy from a source (the sun) to a sink (outer space). Harold J. Morowitz, a biophysicist at Yale University, has shown that such systems are subject to the following physical principle: The flow of energy through a steady-state system from a source to a sink will lead to at least one cycle of material in the system. However, the nature of the cycles and of any organized physicochemical structures that might arise in the system will depend on the initial composition of the system and any other constraints that might emerge as the system evolves. Arguing from this principle, one can show that it is thermodynamically possible for evolutionary processes to take place in open, steady-state systems and that the dynamic integrity of any resulting structures is buttressed against entropic decay by virtue of the continuing flux of energy through the system.

In order for material cycles to arise in steady-state systems, it is of course necessary for some of the incident energy to be *absorbed* by molecules within the system. Energy that is reflected or merely transmitted through the system can have no effect upon it. Light and other forms of electromagnetic radiation are absorbed by atoms and molecules in discrete packets called "quanta." The amount of energy associated with these quanta depends on the wavelength of the radiation: The shorter the wavelength, the higher the "quantum energy" of the radiation. Thus, quanta of ultraviolet or blue light contain more energy than quanta of red or infrared radiation so that a molecule absorbing a "blue" quantum receives more energy than one absorbing a "red" quantum. When a molecule absorbs a quantum of electromagnetic energy, an orbital (or valence) electron is promoted into one of the many "excited states" it can occupy. The energy of these excited states might be reemitted by the original absorbing molecule, (i.e., fluorescence), transferred to other molecules or stored in the form of new chemical bonds. Energy stored in chemical bonds can subsequently be released to do useful physical, chemical, or biological work. In the course of these processes the original high energy quanta are, in effect, degraded and ultimately reemitted as quanta of lower energy. In steady-state systems such as that involving the sun, the earth, and outer space, an energy balance is maintained and the total *amount* of energy incident upon the planet is equal to the amount lost to space. However, the quanta are degraded (and work is done) during their passage through the atmosphere. Since the energy degradation is irreversible, the second law of thermodynamics is not violated for the "sun-earth-outer space" system as a whole.

Most of the solar energy incident upon the earth's atmosphere and surface drives the cycles of wind and rain and ocean currents, and only a small, but significant, fraction (0.1 percent) is responsible for the turnover of water, carbon dioxide, oxygen, and other materials in "biogeochemical cycles" of the global ecosystem (or biosphere). Meteorological cycles arise primarily from temperature gradients caused by solar heating of the atmosphere, oceans, and land areas of the globe, and the basic processes involved are physical rather than chemical in nature. On the other hand, most of the energy that enters the biosphere is absorbed specifically by chlorophyll molecules in green plants, and chemical rather than physical work is done. Both weather and life are material cycles maintained by the steady flow of energy through the earth's atmosphere from sun to outer space; the fantastic qualitative differences between them arise from the chemical differences between the molecules that absorb and process the incident solar energy in the atmosphere and in green plants.

In addition to being the ultimate energy source for the persistence of weather and life today, it is now realized that solar radiation was chiefly responsible for the initiation of molecular evolution on the primitive earth. Indeed, the origin of life can be viewed as one aspect of the physical development of the global ecosystem.

It seems appropriate to begin our consideration of energy flow in living systems with Nobel laureate George Wald's account of traditional and modern ideas on the origin of life. There are only two possible ways life could have arisen: creation by some supernatural power, or spontaneous generation from the materials and energy sources available on the primitive earth. (The hypothesis that the first organisms on earth arose from the arrival of extraterrestrial spores or bacteria does not solve the general problem; it merely removes it to another planet!) Most scientists (and even some theologians) viewed spontaneous generation as nothing more than a matter of common sense observation until Louis Pasteur apparently disproved it in 1860. However, what Pasteur showed was that microbial growth would not occur in boiled organic infusions from which spores or airborne microorganisms were excluded; he did not prove the impossibility of spontaneous generation under other conditions. Nonetheless, as a result of Pasteur's work many scientists came to regard the origin of life as an impossible problem and so lost interest in it. It is one of the nicer ironies of scientific history that revival of the idea of spontaneous generation in modern dress by the British geneticist J. B. S. Haldane and the Russian biochemist A. I. Oparin has today given rise to a whole new field of experimental enquiry.

Attempts to deduce the causes of past events, whether the extinction of the dinosaur or the appearance of the star over Bethlehem, are bedeviled by the fact that one cannot go back in time to make observations or carry out experiments. While it is always easy and, therefore, tempting to invoke divine intervention or natural catastrophes to account for historical phenomena, the only satisfactory approach for resolving scientific problems is provided by the principle of uniformitarianism. This principle, proposed in 1785 by the Scottish geologist James Hutton, asserts that the present is the key to the past, that the laws of nature remain unchanged through time, and that past events may be understood and sometimes simulated by processes that take place and can be studied today. Uniformitarianism has been of the greatest significance in geology, although the experiments of Harold Urey and Stanley Miller with simulated primitive earth atmospheres provide a striking example of the effectiveness of this approach in molecular biology. In his article, Wald describes the Miller-Urey experiments which showed that many of the organic molecules essential for the origin of life could have been formed spontaneously under the influence of solar radiation in the primitive atmosphere and seas. And, as Wald shows, although there are still serious gaps in our understanding, it is now possible to construct a plausible scenario for the origin of life by spontaneous generation some 3 billion years ago!

In 1925, Alfred Lotka of Johns Hopkins University dramatically illustrated the energy flow principle by comparing the global ecosystem to "a great world engine in which plants and animals act as *coupled transformers of energy* in the mill-wheel of life that is driven by solar radiation." Today, in this era of ecological consciousness, it has become commonplace to emphasize the interdependence of organisms, which is intrinsic to the flow of energy and exchange of nutrients among them and with their inorganic environment. At the ecological level, animals are viewed as "consumer organisms" while green plants are seen as producers that capture solar energy and convert it, in the process of photosynthesis, into the chemical bond energy of carbohydrates. This energy can then be released in the stepwise breakdown of carbohydrates that constitutes the processes of fermentation and respiration. The energy can be used by cells to do the chemical work of synthesizing further cells, the mechanical work of moving or contracting, or the osmotic work associated with absorbing nutrients, expelling waste products and maintaining ionic gradients across cell membranes. However, plants use for their own maintenance and growth only a portion of the energy they transform in photosynthesis. They accumulate a surplus of organic material that can be used subsequently by animal consumers for *their* growth and proliferation. Ecological cycles are closed by microorganisms, the "decomposers," that degrade the carcasses and wastes of plants and animals, thereby returning carbon dioxide and water to the inorganic environment for further use by the producers. Thus, we see that solar radiation drives a vast cycle in which the processes of photosynthesis and respiration are coupled together, and if we are to understand the motive force of life, we must understand the molecular bases of these processes.

Photosynthesis is but one of several biological phenomena that depend upon the absorption of light. In the second article in this section, Sterling B. Hendricks describes the initial molecular events associated with the absorption and transduction of electromagnetic radiation in photosynthesis, vision, and plant and animal responses to the daily and seasonal cycles of light and darkness. All photobiological processes are based on the absorption of light by specific molecules called "chromophores." The initial absorption events are appropriately called "light reactions" whereas the subsequent transfer and utilization of the absorbed energy takes place through various "dark reactions."

In his article on the mechanism of photosynthesis R. P. Levine describes both the light and dark reactions and shows that the light absorbed by chlorophyll serves to promote electrons to excited states and that this energy of excitation is subsequently converted into the chemical bond energy of adenosine triphosphate (ATP), the universal carrier of energy in biological systems. The energy associated with the attachment of the terminal phosphate groups in ATP can be used to drive almost any energy-requiring process in biology, including the synthesis of carbohydrates in the dark reactions of photosynthesis.

Photosynthesis and respiration take place in two subcellular organelles, the chloroplast and the mitochondrion, respectively. Biologists have often remarked on the fact that extensively folded membranes are a prominent feature of both organelles. Enzymes associated with photosynthesis and respiration are organized in ordered arrays on these membranes, and this specific organization is apparently essential for the efficient operation of the closely coupled biochemical reactions involved. In his article on the membrane of the mitochondrion, Efraim Racker describes how he has been able to take mitochondrial membranes apart and put them back together again. This is one of the most difficult yet rewarding of all experimental strategies in molecular biology because it gives one confidence that the subunits obtained upon dismemberment of the organelle correspond to those that are used by the cell in the original assembly of the structure. As we shall see in several further papers in this collection, the assembly in the test tube of relatively complex subcellular structures is becoming increasingly practicable in molecular biology today. Dare we hope that some day man will also achieve the synthesis of living cells?

BIBLICAL ACCOUNT of the origin of life is part of the Creation, here illustrated in a 16th-century Bible printed in Lyons. On the first day (*die primo*) God created heaven and the earth. On the second day (*die secundo*) He separated the firmament and the waters. On the third day (*die tertio*) He made the dry land and plants. On the fourth day (*die quarto*) He made the sun, the moon and the stars. On the fifth day (*die quinto*) He made the birds and the fishes. On the sixth day (*die sexto*) He made the land animals and man. In this account there is no theological conflict with spontaneous generation. According to *Genesis* God, rather than creating the animals and plants directly, bade the earth and waters bring them forth. One theological view is that they retain this capacity.

1

THE ORIGIN OF LIFE

GEORGE WALD

August 1954

How did living matter first arise on the earth? As natural scientists learn more about nature they are returning to a hypothesis their predecessors gave up almost a century ago: spontaneous generation

About a century ago the question, How did life begin?, which has interested men throughout their history, reached an impasse. Up to that time two answers had been offered: one that life had been created supernaturally, the other that it arises continually from the nonliving. The first explanation lay outside science; the second was now shown to be untenable. For a time scientists felt some discomfort in having no answer at all. Then they stopped asking the question.

Recently ways have been found again to consider the origin of life as a scientific problem—as an event within the order of nature. In part this is the result of new information. But a theory never rises of itself, however rich and secure the facts. It is an act of creation. Our present ideas in this realm were first brought together in a clear and defensible argument by the Russian biochemist A. I. Oparin in a book called *The Origin of Life*, published in 1936. Much can be added now to Oparin's discussion, yet it provides the foundation upon which all of us who are interested in this subject have built.

The attempt to understand how life originated raises a wide variety of scientific questions, which lead in many and diverse directions and should end by casting light into many obscure corners. At the center of the enterprise lies the hope not only of explaining a great past event—important as that should be—but of showing that the explanation is workable. If we can indeed come to understand how a living organism arises from the nonliving, we should be able to construct one—only of the simplest description, to be sure, but still recognizably alive. This is so remote a possibility now that one scarcely dares to acknowledge it; but it is there nevertheless.

One answer to the problem of how life originated is that it was created. This is an understandable confusion of nature with technology. Men are used to making things; it is a ready thought that those things not made by men were made by a superhuman being. Most of the cultures we know contain mythical accounts of a supernatural creation of life. Our own tradition provides such an account in the opening chapters of *Genesis*. There we are told that beginning on the third day of the Creation, God brought forth living creatures—first plants, then fishes and birds, then land animals and finally man.

Spontaneous Generation

The more rational elements of society, however, tended to take a more naturalistic view of the matter. One had only to accept the evidence of one's senses to know that life arises regularly from the nonliving: worms from mud, maggots from decaying meat, mice from refuse of various kinds. This is the view that came to be called spontaneous generation. Few scientists doubted it. Aristotle, Newton, William Harvey, Descartes, van Helmont, all accepted spontaneous generation without serious question. Indeed, even the theologians—witness the English Jesuit John Turberville Needham—could subscribe to this view, for *Genesis* tells us, not that God created plants and most animals directly, but that He bade the earth and waters to bring them forth; since this directive was never rescinded, there is nothing heretical in believing that the process has continued.

But step by step, in a great controversy that spread over two centuries, this belief was whittled away until nothing remained of it. First the Italian Francesco Redi showed in the 17th century that meat placed under a screen, so that flies cannot lay their eggs on it, never develops maggots. Then in the following century the Italian abbé Lazzaro Spallanzani showed that a nutritive broth, sealed off from the air while boiling, never develops microorganisms, and hence never rots. Needham objected that by too much boiling Spallanzani had rendered the broth, and still more the air above it, incompatible with life. Spallanzani could defend his broth; when he broke the seal of his flasks, allowing new air to rush in, the broth promptly began to rot. He could find no way, however, to show that the air in the sealed flask had not been vitiated. This problem finally was solved by Louis Pasteur in 1860, with a simple modification of Spallanzani's experiment. Pasteur too used a flask containing boiling broth, but instead of sealing off the neck he drew it out in a long, S-shaped curve with its end open to the air. While molecules of air could pass back and forth freely, the heavier particles of dust, bacteria and molds in the atmosphere were trapped on the walls of the curved neck and only rarely reached the broth. In such a flask the broth seldom was contaminated; usually it remained clear and sterile indefinitely.

This was only one of Pasteur's experiments. It is no easy matter to deal with so deeply ingrained and common-sense a belief as that in spontaneous generation. One can ask for nothing better in such a pass than a noisy and stubborn opponent, and this Pasteur had in the naturalist Félix Pouchet, whose arguments before the French Academy of Sciences drove Pasteur to more and more rigorous experiments. When he had finished, nothing remained of the belief in spontaneous generation.

We tell this story to beginning students of biology as though it represents a triumph of reason over mysticism. In fact it is very nearly the opposite. The reasonable view was to believe in spontaneous generation; the only alternative, to believe in a single, primary act of supernatural creation. There is no third position. For this reason many scientists a century ago chose to regard the belief in spontaneous generation as a "philosophical necessity." It is a symptom of the philosophical poverty of our time that this necessity is no longer appreciated. Most modern biologists, having reviewed with satisfaction the downfall of the spontaneous generation hypothesis, yet unwilling to accept the alternative belief in special creation, are left with nothing.

I think a scientist has no choice but to approach the origin of life through a hypothesis of spontaneous generation. What the controversy reviewed above showed to be untenable is only the belief that living organisms arise spontaneously under present conditions. We have now to face a somewhat different problem: how organisms may have arisen spontaneously under different conditions in some former period, granted that they do so no longer.

The Task

To make an organism demands the right substances in the right proportions and in the right arrangement. We do not think that anything more is needed—but that is problem enough.

The substances are water, certain salts—as it happens, those found in the ocean—and carbon compounds. The latter are called *organic* compounds because they scarcely occur except as products of living organisms.

Organic compounds consist for the most part of four types of atoms: carbon, oxygen, nitrogen and hydrogen. These four atoms together constitute about 99 per cent of living material, for hydrogen and oxygen also form water. The organic compounds found in organisms fall mainly into four great classes: carbohydrates, fats, proteins and nucleic acids. The illustrations on this and the next three pages give some notion of their composition and degrees of complexity. The fats are simplest, each consisting of three fatty acids joined to glycerol. The starches and glycogens are made of sugar units strung together to form long straight and branched chains. In general only one type of sugar appears in a single starch or glycogen; these molecules are large, but still relatively simple. The principal function of carbohydrates and fats in the organism is to serve as fuel—as a source of energy.

The nucleic acids introduce a further level of complexity. They are very large structures, composed of aggregates of at least four types of unit—the nucleotides—brought together in a great variety of proportions and sequences. An almost endless variety of different nucleic acids is possible, and specific differences among them are believed to be of the highest importance. Indeed, these structures are thought by many to be the main constituents of the genes, the bearers of hereditary constitution.

Variety and specificity, however, are most characteristic of the proteins, which include the largest and most complex molecules known. The units of

which their structure is built are about 25 different amino acids. These are strung together in chains hundreds to thousands of units long, in different proportions, in all types of sequence, and with the greatest variety of branching and folding. A virtually infinite number of different proteins is possible. Organisms seem to exploit this potentiality, for no two species of living organism, animal or plant, possess the same proteins.

Organic molecules therefore form a large and formidable array, endless in variety and of the most bewildering complexity. One cannot think of having organisms without them. This is precisely the trouble, for to understand how organisms originated we must first of all explain how such complicated molecules could come into being. And that is only the beginning. To make an organism requires not only a tremendous variety of these substances, in adequate amounts and proper proportions, but also just the right arrangement of them. Structure here is as important as composition—and what a complication of structure! The most complex machine man has devised—say an electronic brain—is child's play compared with the simplest of living organisms. The especially trying thing is that complexity here involves such small dimensions. It is on the molecular level; it consists of a detailed fitting of molecule to molecule such as no chemist can attempt.

The Possible and Impossible

One has only to contemplate the magnitude of this task to concede that the spontaneous generation of a living organism is impossible. Yet here we are—as a result, I believe, of spontaneous generation. It will help to digress for a mo-

CARBOHYDRATES comprise one of the four principal kinds of carbon compound found in living matter. This structural formula represents part of a characteristic carbohydrate. It is a polysaccharide consisting of six-carbon sugar units, three of which are shown.

ment to ask what one means by "impossible."

With every event one can associate a probability—the chance that it will occur. This is always a fraction, the proportion of times the event occurs in a large number of trials. Sometimes the probability is apparent even without trial. A coin has two faces; the probability of tossing a head is therefore 1/2. A die has six faces; the probability of throwing a deuce is 1/6. When one has no means of estimating the probability beforehand, it must be determined by counting the fraction of successes in a large number of trials.

Our everyday concept of what is impossible, possible or certain derives from our experience: the number of trials that may be encompassed within the space of a human lifetime, or at most within recorded human history. In this colloquial, practical sense I concede the spontaneous origin of life to be "impossible." It is impossible as we judge events in the scale of human experience.

We shall see that this is not a very meaningful concession. For one thing, the time with which our problem is concerned is geological time, and the whole extent of human history is trivial in the balance. We shall have more to say of this later.

But even within the bounds of our own time there is a serious flaw in our judgment of what is possible. It sounds impressive to say that an event has never been observed in the whole of human history. We should tend to regard such an event as at least "practically" impossible, whatever probability is assigned to it on abstract grounds. When we look a little further into such a statement, however, it proves to be almost meaningless. For men are apt to reject reports of very improbable occurrences. Persons of good

judgment think it safer to distrust the alleged observer of such an event than to believe him. The result is that events which are merely very extraordinary acquire the reputation of never having occurred at all. Thus the highly improbable is made to appear impossible.

To give an example: Every physicist knows that there is a very small probability, which is easily computed, that the table upon which I am writing will suddenly and spontaneously rise into the air. The event requires no more than that the molecules of which the table is composed, ordinarily in random motion in all directions, should happen by chance to move in the same direction. Every physicist concedes this possibility; but try telling one that you have seen it happen. Recently I asked a friend, a Nobel laureate in physics, what he would say if I told him that. He laughed and said that he would regard it as more probable that I was mistaken than that the event had actually occurred.

We see therefore that it does not mean much to say that a very improbable event has never been observed. There is a conspiracy to suppress such observations, not among scientists alone, but among all judicious persons, who have learned to be skeptical even of what they see, let alone of what they are told. If one group is more skeptical than others, it is perhaps lawyers, who have the harshest experience of the unreliability of human evidence. Least skeptical of all are the scientists, who, cautious as they are, know very well what strange things are possible.

A final aspect of our problem is very important. When we consider the spontaneous origin of a living organism, this is not an event that need happen again and again. It is perhaps enough for it to happen once. The probability with

which we are concerned is of a special kind; it is the probability that an event occur *at least once*. To this type of probability a fundamentally important thing happens as one increases the number of trials. However improbable the event in a single trial, it becomes increasingly probable as the trials are multiplied. Eventually the event becomes virtually inevitable. For instance, the chance that a coin will not fall head up in a single toss is 1/2. The chance that no head will appear in a series of tosses is $1/2 \times 1/2 \times 1/2$ . . . as many times over as the number of tosses. In 10 tosses the chance that no head will appear is therefore 1/2 multiplied by itself 10 times, or 1/1,000. Consequently the chance that a head will appear at least once in 10 tosses is 999/1,000. Ten trials have converted what started as a modest probability to a near certainty.

The same effect can be achieved with any probability, however small, by multiplying sufficiently the number of trials. Consider a reasonably improbable event, the chance of which is 1/1,000. The chance that this will not occur in one trial is 999/1,000. The chance that it won't occur in 1,000 trials is 999/1,000 multiplied together 1,000 times. This fraction comes out to be 37/100. The chance that it will happen at least once in 1,000 trials is therefore one minus this number—63/100—a little better than three chances out of five. One thousand trials have transformed this from a highly improbable to a highly probable event. In 10,000 trials the chance that this event will occur at least once comes out to be 19,999/20,000. It is now almost inevitable.

It makes no important change in the argument if we assess the probability that an event occur at least two, three, four or some other small number of

FATS are a second kind of carbon compound found in living matter. This formula represents the whole molecule of palmitin, one of the commonest fats. The molecule consists of glycerol (*11 atoms at the far left*) and fatty acids (*hydrocarbon chains at the right*).

times rather than at least once. It simply means that more trials are needed to achieve any degree of certainty we wish. Otherwise everything is the same.

In such a problem as the spontaneous origin of life we have no way of assessing probabilities beforehand, or even of deciding what we mean by a trial. The origin of a living organism is undoubtedly a stepwise phenomenon, each step with its own probability and its own conditions of trial. Of one thing we can be sure, however: whatever constitutes a trial, more such trials occur the longer the interval of time.

The important point is that since the origin of life belongs in the category of at-least-once phenomena, time is on its side. However improbable we regard this event, or any of the steps which it involves, given enough time it will almost certainly happen at least once. And for life as we know it, with its capacity for growth and reproduction, once may be enough.

Time is in fact the hero of the plot. The time with which we have to deal is of the order of two billion years. What we regard as impossible on the basis of human experience is meaningless here. Given so much time, the "impossible" becomes possible, the possible probable, and the probable virtually certain. One has only to wait: time itself performs the miracles.

Organic Molecules

This brings the argument back to its first stage: the origin of organic compounds. Until a century and a quarter ago the only known source of these substances was the stuff of living organisms. Students of chemistry are usually told that when, in 1828, Friedrich Wöhler synthesized the first organic compound, urea, he proved that organic compounds do not require living organisms to make

them. Of course it showed nothing of the kind. Organic chemists are alive; Wöhler merely showed that they can make organic compounds externally as well as internally. It is still true that with almost negligible exceptions all the organic matter we know is the product of living organisms.

The almost negligible exceptions, however, are very important for our argument. It is now recognized that a constant, slow production of organic molecules occurs without the agency of living things. Certain geological phenomena yield simple organic compounds. So, for example, volcanic eruptions bring metal carbides to the surface of the earth, where they react with water vapor to yield simple compounds of carbon and hydrogen. The familiar type of such a reaction is the process used in old-style bicycle lamps in which acetylene is made by mixing iron carbide with water.

Recently Harold Urey, Nobel laureate in chemistry, has become interested in the degree to which electrical discharges in the upper atmosphere may promote the formation of organic compounds. One of his students, S. L. Miller, performed the simple experiment of circulating a mixture of water vapor, methane (CH_4), ammonia (NH_3) and hydrogen—all gases believed to have been present in the early atmosphere of the earth—continuously for a week over an electric spark. The circulation was maintained by boiling the water in one limb of the apparatus and condensing it in the other. At the end of the week the water was analyzed by the delicate method of paper chromatography. It was found to have acquired a mixture of amino acids! Glycine and alanine, the simplest amino acids and the most prevalent in proteins, were definitely identified in the solution, and there were indications it contained aspartic acid and two others. The yield was surprisingly

high. This amazing result changes at a stroke our ideas of the probability of the spontaneous formation of amino acids.

A final consideration, however, seems to me more important than all the special processes to which one might appeal for organic syntheses in inanimate nature.

It has already been said that to have organic molecules one ordinarily needs organisms. The synthesis of organic substances, like almost everything else that happens in organisms, is governed by the special class of proteins called enzymes—the organic catalysts which greatly accelerate chemical reactions in the body. Since an enzyme is not used up but is returned at the end of the process, a small amount of enzyme can promote an enormous transformation of material.

Enzymes play such a dominant role in the chemistry of life that it is exceedingly difficult to imagine the synthesis of living material without their help. This poses a dilemma, for enzymes themselves are proteins, and hence among the most complex organic components of the cell. One is asking, in effect, for an apparatus which is the unique property of cells in order to form the first cell.

This is not, however, an insuperable difficulty. An enzyme, after all, is only a catalyst; it can do no more than change the *rate* of a chemical reaction. It cannot make anything happen that would not have happened, though more slowly, in its absence. Every process that is catalyzed by an enzyme, and every product of such a process, would occur without the enzyme. The only difference is one of rate.

Once again the essence of the argument is time. What takes only a few moments in the presence of an enzyme or other catalyst may take days, months or years in its absence; but given time, the end result is the same.

NUCLEIC ACIDS are a third kind of carbon compound. This is part of desoxyribonucleic acid, the backbone of which is five-carbon sugars alternating with phosphoric acid. The letter R is any one of four nitrogenous bases, two purines and two pyrimidines.

Indeed, this great difficulty in conceiving of the spontaneous generation of organic compounds has its positive side. In a sense, organisms demonstrate to us what organic reactions and products are *possible*. We can be certain that, given time, all these things must occur. Every substance that has ever been found in an organism displays thereby the finite probability of its occurrence. Hence, given time, it should arise spontaneously. One has only to wait.

It will be objected at once that this is just what one cannot do. Everyone knows that these substances are highly perishable. Granted that, within long spaces of time, now a sugar molecule, now a fat, now even a protein might form spontaneously, each of these molecules should have only a transitory existence. How are they ever to accumulate; and, unless they do so, how form an organism?

We must turn the question around. What, in our experience, is known to destroy organic compounds? Primarily two agencies: decay and the attack of oxygen. But decay is the work of living organisms, and we are talking of a time before life existed. As for oxygen, this introduces a further and fundamental section of our argument.

It is generally conceded at present that the early atmosphere of our planet contained virtually no free oxygen. Almost all the earth's oxygen was bound in the form of water and metal oxides. If this were not so, it would be very difficult to imagine how organic matter could accumulate over the long stretches of time that alone might make possible the spontaneous origin of life. This is a crucial point, therefore, and the statement that the early atmosphere of the planet was virtually oxygen-free comes forward so opportunely as to raise a suspicion of special pleading. I have for this reason taken care to consult a number of

geologists and astronomers on this point, and am relieved to find that it is well defended. I gather that there is a widespread though not universal consensus that this condition did exist. Apparently something similar was true also for another common component of our atmosphere—carbon dioxide. It is believed that most of the carbon on the earth during its early geological history existed as the element or in metal carbides and hydrocarbons; very little was combined with oxygen.

This situation is not without its irony. We tend usually to think that the environment plays the tune to which the organism must dance. The environment is given; the organism's problem is to adapt to it or die. It has become apparent lately, however, that some of the most important features of the physical environment are themselves the work of living organisms. Two such features have just been named. The atmosphere of our planet seems to have contained no oxygen until organisms placed it there by the process of plant photosynthesis. It is estimated that at present all the oxygen of our atmosphere is renewed by photosynthesis once in every 2,000 years, and that all the carbon dioxide passes through the process of photosynthesis once in every 300 years. In the scale of geological time, these intervals are very small indeed. We are left with the realization that all the oxygen and carbon dioxide of our planet are the products of living organisms, and have passed through living organisms over and over again.

Forces of Dissolution

In the early history of our planet, when there were no organisms or any free oxygen, organic compounds should have been stable over very long periods. This is the crucial difference between

the period before life existed and our own. If one were to specify a single reason why the spontaneous generation of living organisms was possible once and is so no longer, this is the reason.

We must still reckon, however, with another destructive force which is disposed of less easily. This can be called spontaneous dissolution—the counterpart of spontaneous generation. We have noted that any process catalyzed by an enzyme can occur in time without the enzyme. The trouble is that the processes which synthesize an organic substance are reversible: any chemical reaction which an enzyme may catalyze will go backward as well as forward. We have spoken as though one has only to wait to achieve syntheses of all kinds; it is truer to say that what one achieves by waiting is *equilibria* of all kinds—equilibria in which the synthesis and dissolution of substances come into balance.

In the vast majority of the processes in which we are interested the point of equilibrium lies far over toward the side of dissolution. That is to say, spontaneous dissolution is much more probable, and hence proceeds much more rapidly, than spontaneous synthesis. For example, the spontaneous union, step by step, of amino acid units to form a protein has a certain small probability, and hence might occur over a long stretch of time. But the dissolution of the protein or of an intermediate product into its component amino acids is much more probable, and hence will go ever so much more rapidly. The situation we must face is that of patient Penelope waiting for Odysseus, yet much worse: each night she undid the weaving of the preceding day, but here a night could readily undo the work of a year or a century.

How do present-day organisms manage to synthesize organic compounds against the forces of dissolution? They do so by a continuous expenditure of

PROTEINS are a fourth kind of carbon compound found in living matter. This formula represents part of a polypeptide chain, the backbone of a protein molecule. The chain is made up of amino acids. Here the letter R represents the side chains of these acids.

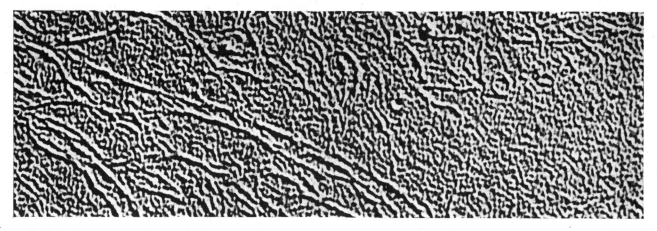

FILAMENTS OF COLLAGEN, a protein which is usually found in long fibrils, were dispersed by placing them in dilute acetic acid. This electron micrograph, which enlarges the filaments 75,000 times, was made by Jerome Gross of the Harvard Medical School.

energy. Indeed, living organisms commonly do better than oppose the forces of dissolution; they grow in spite of them. They do so, however, only at enormous expense to their surroundings. They need a constant supply of material and energy merely to maintain themselves, and much more of both to grow and reproduce. A living organism is an intricate machine for performing exactly this function. When, for want of fuel or through some internal failure in its mechanism, an organism stops actively synthesizing itself in opposition to the processes which continuously decompose it, it dies and rapidly disintegrates.

What we ask here is to synthesize organic molecules without such a machine. I believe this to be the most stubborn problem that confronts us—the weakest link at present in our argument. I do not think it by any means disastrous, but it calls for phenomena and forces some of which are as yet only partly understood and some probably still to be discovered.

Forces of Integration

At present we can make only a beginning with this problem. We know that it is possible on occasion to protect molecules from dissolution by precipitation or by attachment to other molecules. A wide variety of such precipitation and "trapping" reactions is used in modern chemistry and biochemistry to promote syntheses. Some molecules appear to acquire a degree of resistance to disintegration simply through their size. So, for example, the larger molecules composed of amino acids—polypeptides and proteins—seem to display much less tendency to disintegrate into their units than do smaller compounds of two or three amino acids.

Again, many organic molecules dis-

play still another type of integrating force—a spontaneous impulse toward structure formation. Certain types of fatty molecules—lecithins and cephalins—spin themselves out in water to form highly oriented and well-shaped structures—the so-called myelin figures. Proteins sometimes orient even in solution, and also may aggregate in the solid state in highly organized formations. Such spontaneous architectonic tendencies are still largely unexplored, particularly as they may occur in complex mixtures of substances, and they involve forces the strength of which has not yet been estimated.

What we are saying is that possibilities exist for opposing *intra*molecular dissolution by *inter*molecular aggregations of various kinds. The equilibrium between union and disunion of the amino acids that make up a protein is all to the advantage of disunion, but the aggregation of the protein with itself or other molecules might swing the equilibrium in the opposite direction: perhaps by removing the protein from access to the water which would be required to disintegrate it or by providing some particularly stable type of molecular association.

In such a scheme the protein appears only as a transient intermediate, an unstable way-station, which can either fall back to a mixture of its constituent amino acids or enter into the formation of a complex structural aggregate: amino acids $\leftrightarrows$ protein $\rightarrow$ aggregate.

Such molecular aggregates, of various degrees of material and architectural complexity, are indispensable intermediates between molecules and organisms. We have no need to try to imagine the spontaneous formation of an organism by one grand collision of its component molecules. The whole process must

be gradual. The molecules form aggregates, small and large. The aggregates add further molecules, thus growing in size and complexity. Aggregates of various kinds interact with one another to form still larger and more complex structures. In this way we imagine the ascent, not by jumps or master strokes, but gradually, piecemeal, to the first living organisms.

First Organisms

Where may this have happened? It is easiest to suppose that life first arose in the sea. Here were the necessary salts and the water. The latter is not only the principal component of organisms, but prior to their formation provided a medium which could dissolve molecules of the widest variety and ceaselessly mix and circulate them. It is this constant mixture and collision of organic molecules of every sort that constituted in large part the "trials" of our earlier discussion of probabilities.

The sea in fact gradually turned into a dilute broth, sterile and oxygen-free. In this broth molecules came together in increasing number and variety, sometimes merely to collide and separate, sometimes to react with one another to produce new combinations, sometimes to aggregate into multimolecular formations of increasing size and complexity.

What brought order into such complexes? For order is as essential here as composition. To form an organism, molecules must enter into intricate designs and connections; they must eventually form a self-repairing, self-constructing dynamic machine. For a time this problem of molecular arrangement seemed to present an almost insuperable obstacle in the way of imagining a spontaneous origin of life, or indeed the laboratory

FIBRILS OF COLLAGEN formed spontaneously out of filaments such as those shown on the opposite page when 1 per cent of sodium chloride was added to the dilute acetic acid. These long fibrils are identical in appearance with those of collagen before dispersion.

synthesis of a living organism. It is still a large and mysterious problem, but it no longer seems insuperable. The change in view has come about because we now realize that it is not altogether necessary to *bring* order into this situation; a great deal of order is implicit in the molecules themselves.

The epitome of molecular order is a crystal. In a perfect crystal the molecules display complete regularity of position and orientation in all planes of space. At the other extreme are fluids—liquids or gases—in which the molecules are in ceaseless motion and in wholly random orientations and positions.

Lately it has become clear that very little of a living cell is truly fluid. Most of it consists of molecules which have taken up various degrees of orientation with regard to one another. That is, most of the cell represents various degrees of approach to crystallinity—often, however, with very important differences from the crystals most familiar to us. Much of the cell's crystallinity involves molecules which are still in solution— so-called liquid crystals—and much of the dynamic, plastic quality of cellular structure, the capacity for constant change of shape and interchange of material, derives from this condition. Our familiar crystals, furthermore, involve only one or a very few types of molecule, while in the cell a great variety of different molecules come together in some degree of regular spacing and orientation—*i.e.*, some degree of crystallinity. We are dealing in the cell with highly mixed crystals and near-crystals, solid and liquid. The laboratory study of this type of formation has scarcely begun. Its further exploration is of the highest importance for our problem.

In a fluid such as water the molecules are in very rapid motion. Any molecules dissolved in such a medium are under a constant barrage of collisions with water molecules. This keeps small and moderately sized molecules in a constant turmoil; they are knocked about at random, colliding again and again, never holding any position or orientation for more than an instant. The larger a molecule is relative to water, the less it is disturbed by such collisions. Many protein and nucleic acid molecules are so large that even in solution their motions are very sluggish, and since they carry large numbers of electric charges distributed about their surfaces, they tend even in solution to align with respect to one another. It is so that they tend to form liquid crystals.

We have spoken above of architectonic tendencies even among some of the relatively small molecules: the lecithins and cephalins. Such molecules are insoluble in water yet possess special groups which have a high affinity for water. As a result they tend to form surface layers, in which their water-seeking groups project into the water phase, while their water-repelling portions project into the air, or into an oil phase, or unite to form an oil phase. The result is that quite spontaneously such molecules, when exposed to water, take up highly oriented positions to form surface membranes, myelin figures and other quasicrystalline structures.

Recently several particularly striking examples have been reported of the spontaneous production of familiar types of biological structure by protein molecules. Cartilage and muscle offer some of the most intricate and regular patterns of structure to be found in organisms. A fiber from either type of tissue presents under the electron microscope a beautiful pattern of cross striations of various widths and densities, very regularly spaced. The proteins that form these structures can be coaxed into free solution and stirred into completely random orientation. Yet on precipitating, under proper conditions, the molecules realign with regard to one another to regenerate with extraordinary fidelity the original patterns of the tissues [*see illustration above*].

We have therefore a genuine basis for the view that the molecules of our oceanic broth will not only come together spontaneously to form aggregates but in doing so will spontaneously achieve various types and degrees of order. This greatly simplifies our problem. What it means is that, given the right molecules, one does not have to do everything for them; they do a great deal for themselves.

Oparin has made the ingenious suggestion that natural selection, which Darwin proposed to be the driving force of organic evolution, begins to operate at this level. He suggests that as the molecules come together to form colloidal aggregates, the latter begin to compete with one another for material. Some aggregates, by virtue of especially favorable composition or internal arrangement, acquire new molecules more rapidly than others. They eventually emerge as the dominant types. Oparin suggests further that considerations of optimal size enter at this level. A growing colloidal particle may reach a point at which it becomes unstable and breaks down into smaller particles, each of which grows and redivides. All these phenomena lie within the bounds of known processes in nonliving systems.

The Sources of Energy

We suppose that all these forces and factors, and others perhaps yet to be revealed, together give us eventually the first living organism. That achieved, how does the organism continue to live?

We have already noted that a living organism is a dynamic structure. It is the site of a continuous influx and outflow of matter and energy. This is the very sign of life, its cessation the best evidence of death. What is the primal organism to use as food, and how derive the energy it needs to maintain itself and grow?

For the primal organism, generated under the conditions we have described, only one answer is possible. Having arisen in an oceanic broth of organic molecules, its only recourse is to live upon them. There is only one way of doing that in the absence of oxygen. It is called fermentation: the process by which organisms derive energy by breaking organic molecules and rearranging their parts. The most familiar example of such a process is the fermentation of sugar by yeast, which yields alcohol as one of the products. Animal cells also ferment sugar, not to alcohol but to lactic acid. These are two examples from a host of known fermentations.

The yeast fermentation has the following over-all equation: $C_6H_{12}O_6 \rightarrow 2\ CO_2 + 2\ C_2H_5OH$ + energy. The result of fragmenting 180 grams of sugar into 88 grams of carbon dioxide and 92 grams of alcohol is to make available about 20,000 calories of energy for the use of the cell. The energy is all that the cell derives by this transaction; the carbon dioxide and alcohol are waste products which must be got rid of somehow if the cell is to survive.

The cell, having arisen in a broth of organic compounds accumulated over the ages, must consume these molecules by fermentation in order to acquire the energy it needs to live, grow and reproduce. In doing so, it and its descendants are living on borrowed time. They are consuming their heritage, just as we in our time have nearly consumed our heritage of coal and oil. Eventually such a process must come to an end, and with that life also should have ended. It would have been necessary to start the entire development again.

Fortunately, however, the waste product carbon dioxide saved this situation. This gas entered the ocean and the atmosphere in ever-increasing quantity. Some time before the cell exhausted the supply of organic molecules, it succeeded in inventing the process of photosynthesis. This enabled it, with the energy of sunlight, to make its own organic molecules: first sugar from carbon dioxide and water, then, with ammonia and nitrates as sources of nitrogen, the entire array of organic compounds which it requires. The sugar synthesis equation is: $6\ CO_2 + 6\ H_2O$ + sunlight $\rightarrow C_6H_{12}O_6 + 6\ O_2$. Here 264 grams of carbon dioxide plus 108 grams of water plus about 700,000 calories of sunlight yield 180 grams of sugar and 192 grams of oxygen.

This is an enormous step forward. Living organisms no longer needed to depend upon the accumulation of organic matter from past ages; they could make their own. With the energy of sunlight they could accomplish the fundamental organic syntheses that provide their substance, and by fermentation they could produce what energy they needed.

Fermentation, however, is an extraordinarily inefficient source of energy. It leaves most of the energy potential of organic compounds unexploited; consequently huge amounts of organic material must be fermented to provide a modicum of energy. It produces also various poisonous waste products—alcohol, lactic acid, acetic acid, formic acid and so on. In the sea such products are readily washed away, but if organisms were ever to penetrate to the air and land, these products must prove a serious embarrassment.

One of the by-products of photosynthesis, however, is oxygen. Once this was available, organisms could invent a new way to acquire energy, many times as efficient as fermentation. This is the

EXPERIMENT of S. L. Miller made amino acids by circulating methane (CH_4), ammonia (NH_3), water vapor (H_2O) and hydrogen (H_2) past an electrical discharge. The amino acids collected at the bottom of apparatus and were detected by paper chromatography.

process of cold combustion called respiration: $C_6H_{12}O_6 + 6 O_2 \rightarrow 6 CO_2 + 6 H_2O$ + energy. The burning of 180 grams of sugar in cellular respiration yields about 700,000 calories, as compared with the approximately 20,000 calories produced by fermentation of the same quantity of sugar. This process of combustion extracts all the energy that can possibly be derived from the molecules which it consumes. With this process at its disposal, the cell can meet its energy requirements with a minimum expenditure of substance. It is a further advantage that the products of respiration—water and carbon dioxide—are innocuous and easily disposed of in any environment.

Life's Capital

It is difficult to overestimate the degree to which the invention of cellular respiration released the forces of living organisms. No organism that relies wholly upon fermentation has ever amounted to much. Even after the advent of photosynthesis, organisms could have led only a marginal existence. They could indeed produce their own organic materials, but only in quantities sufficient to survive. Fermentation is so profligate a way of life that photosynthesis could do little more than keep up with it. Respiration used the material of organisms with such enormously greater efficiency as for the first time to leave something over. Coupled with fermentation, photosynthesis made organisms self-sustaining; coupled with respiration, it provided a surplus. To use an economic analogy, photosynthesis brought organisms to the subsistence level; respiration provided them with capital. It is mainly this capital that they invested in the great enterprise of organic evolution.

The entry of oxygen into the atmosphere also liberated organisms in another sense. The sun's radiation contains ultraviolet components which no living cell can tolerate. We are sometimes told that if this radiation were to reach the earth's surface, life must cease. That is not quite true. Water absorbs ultraviolet radiation very effectively, and one must conclude that as long as these rays penetrated in quantity to the surface of the earth, life had to remain under water. With the appearance of oxygen, however, a layer of ozone formed high in the atmosphere and absorbed this radiation. Now organisms could for the first time emerge from the water and begin to populate the earth and air. Oxygen provided not only the means of obtaining adequate energy for evolution but the protective blanket of ozone which alone made possible terrestrial life.

This is really the end of our story. Yet not quite the end. Our entire concern in this argument has been to bring the origin of life within the compass of natural phenomena. It is of the essence of such phenomena to be repetitive, and hence, given time, to be inevitable.

This is by far our most significant conclusion—that life, as an orderly natural event on such a planet as ours, was inevitable. The same can be said of the whole of organic evolution. All of it lies within the order of nature, and apart from details all of it was inevitable.

Astronomers have reason to believe that a planet such as ours—of about the earth's size and temperature, and about as well-lighted—is a rare event in the universe. Indeed, filled as our story is with improbable phenomena, one of the least probable is to have had such a body as the earth to begin with. Yet though this probability is small, the universe is so large that it is conservatively estimated at least 100,000 planets like the earth exist in our galaxy alone. Some 100 million galaxies lie within the range of our most powerful telescopes, so that throughout observable space we can count apparently on the existence of at least 10 million million planets like our own.

What it means to bring the origin of life within the realm of natural phenomena is to imply that in all these places life probably exists—life as we know it. Indeed, I am convinced that there can be no way of composing and constructing living organisms which is fundamentally different from the one we know—though this is another argument, and must await another occasion. Wherever life is possible, given time, it should arise. It should then ramify into a wide array of forms, differing in detail from those we now observe (as did earlier organisms on the earth) yet including many which should look familiar to us—perhaps even men.

We are not alone in the universe, and do not bear alone the whole burden of life and what comes of it. Life is a cosmic event—so far as we know the most complex state of organization that matter has achieved in our cosmos. It has come many times, in many places—places closed off from us by impenetrable distances, probably never to be crossed even with a signal. As men we can attempt to understand it, and even somewhat to control and guide its local manifestations. On this planet that is our home, we have every reason to wish it well. Yet should we fail, all is not lost. Our kind will try again elsewhere.

2

HOW LIGHT INTERACTS WITH LIVING MATTER

STERLING B. HENDRICKS

September 1968

Light activates three key processes of life: photosynthesis, vision, and photoperiodism (the response of plants and animals to the cycle of night and day). Such activation is mediated by specific pigments

Life is believed to have arisen in a primordial broth formed by sunlight acting on simple molecules at the surface of the cooling earth. It could have been sustained by the broth for aeons, but eventually, with the arrival of photosynthesis, some living things came to use sunlight more directly. So it remains today, with photosynthesis by plants serving to capture sunlight for the energy needs of all forms of life.

As various kinds of animals evolved, the ones that were best able to sense their surroundings were favored to survive. Because light acts over considerable distances it is well suited to sensing. To exploit light the animals needed some kind of detector: a tissue, an eyespot or an eye. The detector had to be coupled to a responding system: a ganglion or a brain. Signals from the system controlled locomotion toward food or away from danger.

Photosynthesis and vision do not exhaust the potential of the luminous environment. Both plants and animals have evolved mechanisms to respond to the changing daily cycle of light and dark. It is this photoperiodism that provides the seasonal schedule for, among other things, the flowering of plants, the pupation of insects and the nesting of birds.

To understand these phenomena one must ask how light acts in life. Part of the answer is very simple: it acts by exciting certain absorbing molecules. What happens to the molecules in the course of absorption is more difficult to describe, but many details of the processes are now reasonably well known. On the other hand, our ideas about how the molecular events are coupled to the responses of plants and animals are still quite tentative.

In discussing the present state of knowledge about light and life I shall treat vision first because this phenomenon has some features in common with both photosynthesis and photoperiodism. In all three processes light acts through absorption by a small, colored molecule—a chromophore—that is associated with a large molecule of protein. In the case of vision the light-sensitive molecules are responsible for the pink and purplish color of the retina. In the retina of the human eye there are some 100 million thin rod-shaped cells and five million slightly cone-shaped ones. Each is connected through a synapse, or junction, to a nerve fiber leading to the brain. Electron micrographs show that the outer end of both rods and cones is packed with thin membranous sacs, and with these sacs are associated the light-absorbing chromophores. (Vision and photosynthesis share this association of a chromophore with a membrane.) Excitation of the chromophore by light causes some kind of change in the membrane, and this change gives rise to a signal in the nerve fiber.

In vision the nature of the receiving chromophore and the manner of its excitation by light are well understood. Both have much in common with light reception in photoperiodism. As George Wald of Harvard University established, the receiving chromophore is vitamin-A aldehyde (in structural terms 11-*cis* retinal). The chromophore is found in association with a protein, opsin. The opsins are fatty proteins; thus they have an affinity for the sac membranes, which consist largely of lipid—that is, fatty—material. There are four types of opsin, one in the rods and three in the cones. Combined with 11-*cis* retinal, they respectively form rhodopsin and three kinds of iodopsin. On excitation by light all four opsins change in the same way.

In vision, photosynthesis and photoperiodism alike the chromophore molecule is notable for its alternating single and double chemical bonds. Known to chemists as conjugated systems, molecules of this kind are structurally quite stable because the groups of atoms attached by double bonds cannot rotate around the bonds. Each conjugated system, if it is adequately extended, has a rather low energy state that can be excited by visible light. When the system is excited, its double-bond character is somewhat relaxed, so that a *cis* configuration can change to a *trans* one [*see top illustration on next page*]. This ability to change form is a key element in vision and photoperiodism. In photosynthesis, however, no change of form takes place because the change is constrained by the ring structures of the chlorophyll chromophores.

The effects of light in vision and photoperiodism are determined by measur-

LIGHT-SENSITIVE PIGMENT that triggers photoperiodic responses in plants is shown in its two states in the photograph on the opposite page. Called phytochrome, the pigment, which is seen here in a .2 percent solution, is instrumental in a number of seasonal occurrences such as plants flowering and seeds germinating. In one state (*left*) phytochrome is excitable by far-red light, in the other (*right*) by red light. Alternating exposures to these colors change the pigment from one state to the other and back again. The phytochrome shown here was extracted from oat seedlings in the laboratory of F. E. Mumford and E. L. Jenner in the Central Research Department of E. I. du Pont de Nemours & Co. It is contained in square quartz cells designed for studies of light absorption. The faint numerals near the top indicates the length of the light path through the cell: 1.000 centimeter.

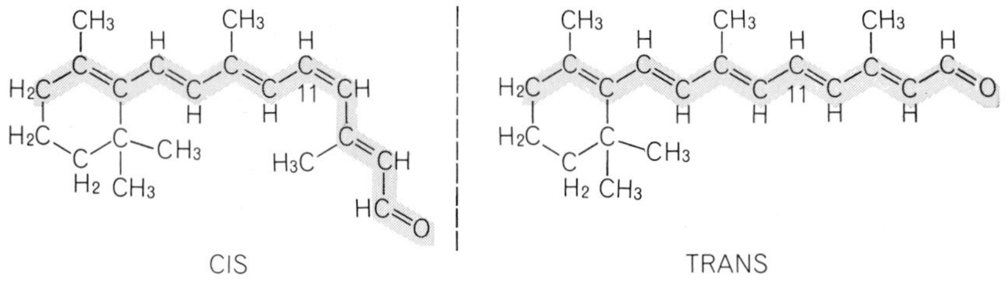

VISION depends on a light-sensitive chromophore molecule, 11-*cis* retinal (*left*), which has alternating single and double bonds (*color*). When light excites the molecule, its configuration changes from *cis* to the *trans* form (*right*), thereby setting in train a series of complex changes in the structure of the proteins with which the retinal chromophore is associated.

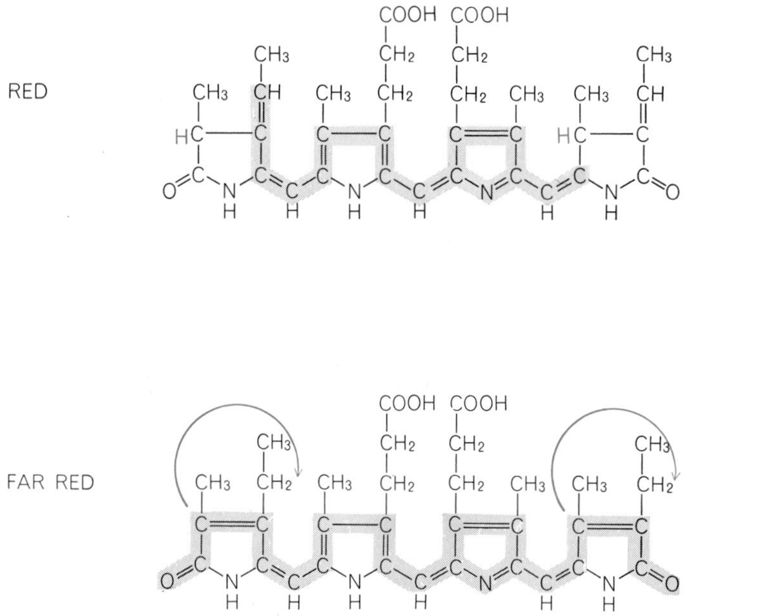

PHOTOPERIODISM in plants depends on phytochrome, another molecule that is sensitive to light. Like the retinal molecule, phytochrome has alternating single and double bonds (*color*). When excited by light, it changes from a configuration sensitive to red light (*top*) to one sensitive to far-red light, probably because two hydrogen atoms shift (*bottom*).

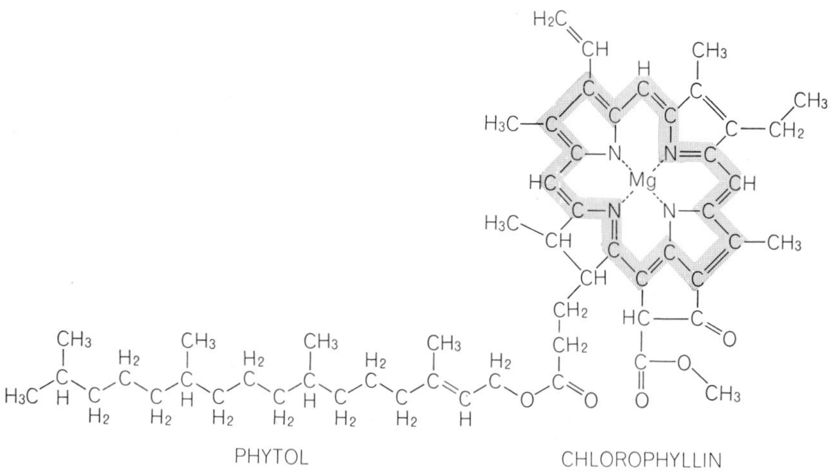

PHOTOSYNTHESIS depends on the light-sensitive chromophore molecules of several kinds of chlorophylls that have differing side groups. The molecule shown here is chlorophyll *a*. Like the vision and photoperiodism chromophores, chlorophyll molecules include singly and doubly bonded atoms, but these form a closed loop within the chlorophyllin portion of the molecule (*color*). When excited by light, chlorophylls forward the energy they receive to centers where it induces chemical changes (*see illustrations on page 27*).

ing the molecular changes produced by light excitation. Some of these changes are very rapid: they may occur in less than a millionth of a second. Changes as fast as this can be followed only if they are excited in an even shorter time, for instance by a very brief but intense flash of light. The measurement of the change also must be made quite rapidly. The method employed is flash excitation at room temperature or lower, followed by photoanalysis—the technique for which George Porter and R. G. W. Norrish shared (with Manfred Eigen) the 1967 Nobel prize in chemistry [see "The Chemical Effects of Light," by Gerald Oster; SCIENTIFIC AMERICAN, September, 1968]. Low temperatures slow down the molecular changes and make them more amenable to measurement.

When the vision chromophore is excited by light, it changes from the *cis* to the *trans* form. The result is the conversion of rhodopsin into prelumirhodopsin with an all-*trans* chromophore. The production of this single change is the one and only role that light plays in vision. The change is followed by several rapid shifts in the structure of the opsin and also changes in the relation of the chromophore to the opsin. To judge by the time it takes for a retinal-cell signal to arrive at a nerve ending, the signal is induced by the shifts that take place in the first thousandth of a second.

The course of the molecular changes can be traced by studying rhodopsin in solution. Prelumirhodopsin, identifiable by its maximum absorption of light at a wavelength of 543 nanometers, can be held at temperatures below −140 degrees Celsius and reversibly changed back into rhodopsin. When the prelumirhodopsin is warmed to −40 degrees C., it is converted into lumirhodopsin. The same conversion probably occurs at body temperature but much more rapidly. This change and subsequent ones, including the formation of metarhodopsins, involve shifts in the molecular configuration of opsin. Among vertebrates the changes finally lead to the dissociation of the chromophore from the protein. The released all-*trans* retinal then has to be reduced to the alcohol form and oxidized back to the aldehyde form to regenerate 11-*cis* retinal. Once the *cis* retinal is regenerated it spontaneously recombines with opsin to form rhodopsin again.

Analysis of the changes in electric potential that occur simultaneously with these molecular changes shows that a potential appears within 25 millionths of a second after a flash of light. The potential is positive with respect to the cornea in a circuit that includes neighboring tissues and the retina. The positive potential is followed in a thousandth of a second by a growing signal of the opposite sign. These events take place during the period when prelumirhodopsin and lumirhodopsin are present. The first potential probably accompanies the change of rhodopsin to prelumirhodopsin. The second depends markedly on the temperature at some distance from the place of light action, probably in the outer membrane of the rod or cone. Currently there is much interest in the possible identification of these changes in

potential as early steps in the eventual excitation of the nerve fiber. Another view is that nerve excitation is associated with the transitions involving metarhodopsin I and metarhodopsin II [see upper illustration below].

Color vision depends on the three opsins, each found in a different cone cell. Their absorption spectra have been measured, and curves were found with peaks at wavelengths of 450, 525 and 555 nanometers (respectively in the blue, green and yellow regions of the visible spectrum). Activation by light leads to the same sequence of molecular changes described for rhodopsin. The singularity of the nerve associations with the rod and

cone cells preserves the retinal detail, or register, in the transmission of the visual signal; the differences in absorption among the three kinds of cone retain the color pattern of the image.

The responses of plants to variations in the length of day and night involve light-induced molecular changes that closely parallel those involved in vision. Because photoperiodic responses are not as well known as visual ones I shall present some illustrative examples. Chrysanthemums and many other plants flower in response to the increasing length of the nights as fall approaches. If the long nights are experimentally inter-

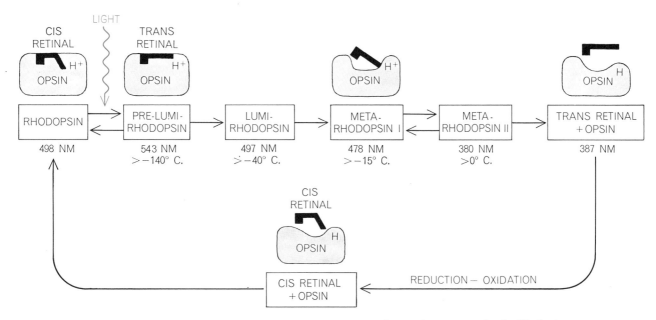

CHANGES IN RHODOPSIN, the visual pigment contained in the rod cells of the vertebrate eye, can be traced in the laboratory at low temperatures. Four successive forms of the pigment appear (rectangles) as the retinal molecule (black) first attains its *trans* configuration and then dissociates from the protein opsin. When it becomes *cis* retinal again, it recombines and completes the cycle.

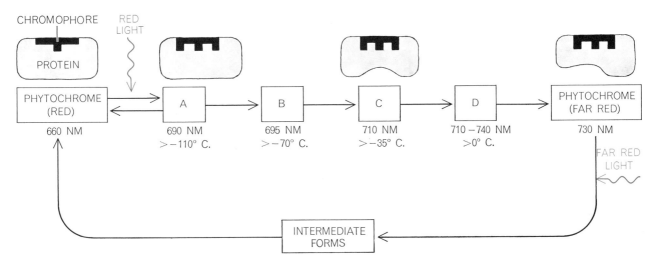

CHANGES IN PHYTOCHROME, the plant-photoperiodism chromophore, can also be traced at low temperature. As with rhodopsin, intermediate forms with characteristic light-absorption peaks appear before the initial red-absorbing form of the pigment is turned into the far-red-absorbing form. Unlike retinal, the chromophore (black) remains associated with its protein throughout the cycle.

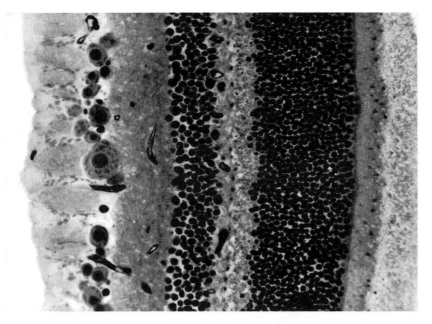

CAT RETINA is seen in cross section, enlarged 670 times. The nerve-fiber layer (*left*) is the part of the retina that lies in contact with the eye's vitreous body. The entering light must penetrate this and five additional layers of retinal tissue before reaching rods and cones (*right*). The micrograph was made by A. J. Ladman of the University of New Mexico.

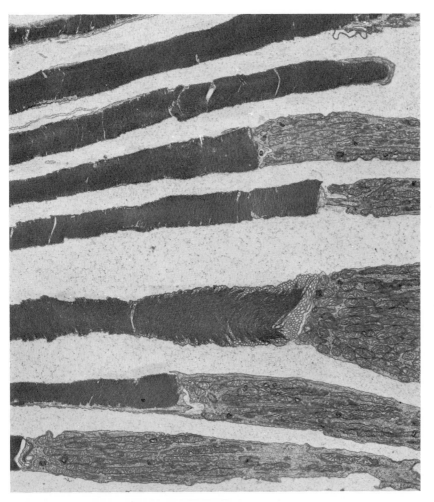

HUMAN RODS AND CONES, enlarged 7,200 times, are seen in an electron micrograph of retina. The visual pigments are concentrated in platelike layers of membrane called lamellae. The micrograph was made by Toichiro Kuwabara of the Harvard Medical School.

rupted by exposing the plants to short periods of light near midnight, the plants will not flower. Red light with an absorption maximum at a wavelength of 660 nanometers is most effective in preventing flower formation. Thus we anticipate that the light-receiving pigment in the plant is blue—the complementary color to the absorbed red. If shortly after exposure to red light the plants are exposed to light near the limit of vision in the far red (730 nanometers), they will flower.

The ornamental plant kalanchoe clearly illustrates the reversible response. The red light evidently converts the photoreversible pigment to a far-red-absorbing form. This changes the plant from the flowering state to the nonflowering one. The far-red light returns the pigment to its red-absorbing form, which enables flowering to proceed. Control of flowering by length of night is a very important factor in determining what varieties of soybean, wheat and other commercial crops are best suited for being grown in various latitudes with different periods of light and darkness.

Many kinds of seeds will germinate only if the photoreversible pigment has been activated. The seeds of some pine and lettuce species, for example, will not germinate in the laboratory unless they are briefly exposed to red light (or, to be sure, light containing red light). If the red-light activation of the seeds is followed by a short exposure to far-red light before the seeds are returned to darkness, the seeds remain dormant. The activation-reversal cycle can be repeated many times; germination or continued dormancy depends on the last exposure in the sequence.

The requirement of light for seed germination is a major cause of the persistence of weeds in cultivated crops. A seed that is dormant when it first falls to the ground is usually covered by soil in the course of the winter. As the seed lies buried the pigment that controls its germination changes into the red-absorbing form; now the seed will not germinate until it is again exposed to sunlight by cultivation or some other disturbance of the soil. When it is exposed, the sunlight converts some of the red-absorbing pigment back to the far-red-absorbing form, and germination begins. Seeds of one common weed, lamb's-quarters, are known to have lain buried for 1,700 years and then to have germinated on exposure to light.

The activation of the photoreversible pigment also controls the growth of trees and many common flowering plants. If

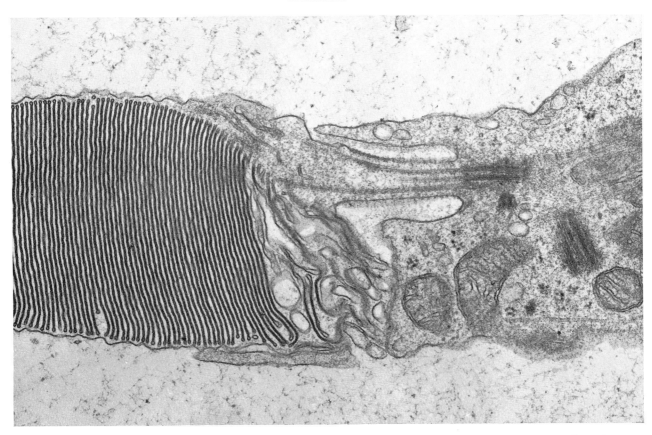

PART OF HUMAN ROD is magnified 44,000 times in electron micrograph. The outer segment of the rod (*left*) is filled with the membrane of the lamellae. The inner segment (*right*) is less complex in structure. This micrograph was also made by Kuwabara.

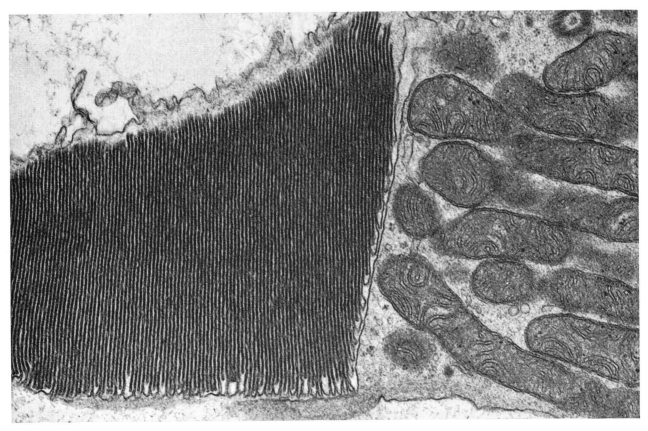

PART OF HUMAN CONE is magnified 44,000 times in another micrograph by Kuwabara that emphasizes the area connecting the structure's inner and outer segments. The lamellae differ from rod lamellae in being "packaged," some singly and some in groups.

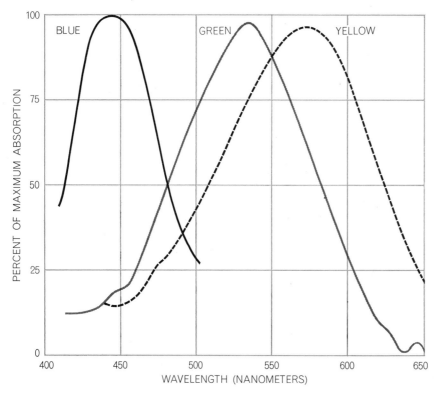

COLOR PERCEPTION in humans arises from the combination of retinal with three dissimilar opsins in the cones of the retina. The three different iodopsin pigments formed thereby absorb the greatest amount of visible light at three different wavelengths. The differences between the signals from each group of cones reflect the color pattern of the image.

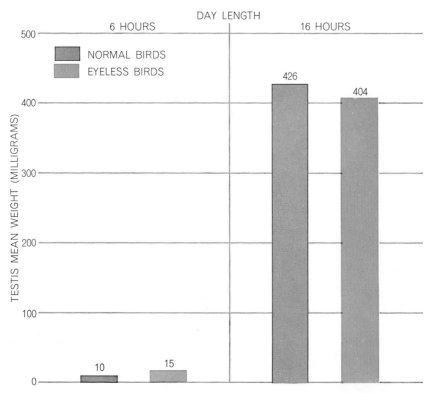

PHOTOPERIODISM IN SPARROWS has been shown to involve some light receptor other than the eye. The testis weight of both eyeless and normal sparrows remained low when their cages were lighted to simulate short days and long nights over a two-month period (*left*). When eyeless and normal birds instead underwent two months of long days and short nights, their testis weight showed a nearly identical increase (*right*). The experiment was conducted by Michael Menaker and Henry Keatts of the University of Texas.

such plants are to continue growing, they must have long periods of daylight. As the days become short growth stops and the plants' buds go into a dormant state that protects them against the low temperatures of winter.

The photoreversible pigment of plants has been named phytochrome. It is invisible in plant tissue because of its low concentration. It was isolated by methods widely used in the preparation of enzymes and other proteins. The pigment is indeed blue [*see illustration on page 18*]. Its photoreversibility is exactly what was expected on the basis of plant responses to light.

The chemical structure of the phytochrome molecule shows that it is related to the greenish-yellow pigments of human bile and the blue pigments of blue-green algae. The molecule comprises an open group of atoms that is closely related to the rings in the chlorophyll molecule. It has two side groups that can change from the *cis* form to the *trans* when they are excited by light. A more probable excitation change, however, is a shift in the position of the molecule's hydrogen atoms.

The changes in the phytochrome molecule following excitation by a flash of light are similar to those in rhodopsin. The first excitation response takes place in a few millionths of a second and gives rise to a form of the molecule that is analogous to prelumirhodopsin. The change stops at this point if the temperature is below −110 degrees C. At these low temperatures the molecule can be reconverted into its initial red-absorbing form by the action of light. At temperatures higher than −110 degrees several more intermediate phytochromes are formed before the final far-red-absorbing molecule appears. These intermediate stages also involve alterations in the molecular form of the protein associated with phytochrome, just as there are alterations in the form of opsin, the protein of rhodopsin. In its final form phytochrome differs from rhodopsin in that the molecule of phytochrome remains linked to the protein rather than being dissociated from it. Far-red light will reverse the process and convert the final form of phytochrome back to its initial red-absorbing form, although a different series of intermediate molecular forms is involved.

Flowering, seed germination and most other plant responses follow slowly on the excitation of phytochrome. Unlike vision, in which the response follows the rapid appearance of intermediate molecules, the photoperiodic response of plants depends on the presence of the final, far-red-absorbing form of phytochrome. Little is known about how the far-red-absorbing molecule does its work. One view is that it regulates enzyme production by controlling the genetic material in cell nuclei. Another view is that the molecule's lipid solubility results in its being attached to membranes in the cell, such as the cell wall and the membrane of the nucleus. Changes in the form of the phytochrome molecule would then affect the permeability of the membranes and therefore the functioning of the cell.

The continuous exposure of plants to blue and far-red wavelengths in the visible spectrum opposes the action of the far-red-absorbing form of the phytochrome molecule. It may be that excitation by far-red light causes a continuous displacement of the far-red-absorbing molecules from cell membranes. Continuous excitation of this kind is what happens, for example, during the long light periods that so markedly influence the growth of Douglas firs. If the trees are exposed to 12-hour days and 12-hour nights, they remain dormant. If the length of the day increases, however, they grow continuously.

Photoperiodism is not confined to the plant kingdom: animals also respond to changes in the length of the day. The migration and reproduction of many birds, the activity cycles of numerous mammals and the diapause (suspended animation) of insects are controlled in

LONG LIGHT PERIODS markedly influence the growth of Douglas fir. When exposed to short days, or days and nights of equal length, the tree will remain dormant (*left*). Excitation by additional light produces continuous growth. One tree (*center*) received an hour of dim illumination during its 12-hour night; the other (*right*) had its 12-hour day extended by eight hours of dim light.

this way. These examples of photoperiodism (and some less clear-cut responses in man) depend on the action of several hormones working in sequence. Such sequences of hormone action can have a regular rhythm. They provide a basis for the circadian (meaning about one day) rhythms of "biological clocks." The 24-hour cycle of such clocks is established by light.

The diapause of insects illustrates one form of the interplay of hormone action and light. Some silkworms and the larva of the codling moth, for example, go into a dormant form when the days are short. In this state, which helps the insects to survive the winter, the release of a hormone from a group of cells in the central part of the brain is suspended. The unreleased hormone is the first in a series that leads to a final hormone, ecdysone, that controls the metamorphosis of the pupa into an adult moth. When the brain cells of the dormant pupa are exposed to light for long days, the brain hormone is released and triggers the metamorphosis. Ecdysone injected into a resting pupa brings on metamorphosis even when the days are short.

Here the sole action of the light is the release of a brain hormone. The pigment in the brain cells that absorbs the light has not yet been identified, but current work in the U.S. Department of Agriculture on the response to light by codling moths in diapause promises an answer. The blue-green part of the spectrum (between 500 and 560 nanometers), and probably shorter wavelengths as well, appears to be most effective for breaking diapause. The pigment is possibly of the porphyrin type, with a central structure resembling the ring system of chlorophyll.

Man's dependence on a biological clock is apparent in the unease he feels when the relation of his circadian rhythm to the actual cycle of day and night is quickly disrupted, as it is when he travels by air for distances measured in many degrees of longitude. His hormonal controls are disturbed or out of phase. Deer mice and other small mammals also display cyclic periods of activity such as running that seem to be regulated by light.

Involved at an early stage in the release of the hormones that trigger activity cycles is the region of the brain known as the hypothalamus. Whether this region contains a pigment receptor for the small amount of light that might penetrate the skull or whether it is stimulated by a signal from the eye, or the region of the eye, is unknown. The hypothalamus also controls the pituitary, hormones from which affect the reproductive organs, the cortex of the adrenal gland and other target organs. At present, however, the existence of a pigment responsible for vertebrate photoperiodism, its physical location and the nature of its action on the molecular level remain to be established.

By exciting a chromophore light acts as a trigger both in vision and in photoperiodism, initiating processes that depend for their energy on the organism's own metabolism. In the third major area of light's interaction with living matter—photosynthesis—the opposite is true: the energy of light is utilized to manufacture the fuels that support life. For this to happen there must be (1) a system to receive the light, (2) an arrangement to transfer energy between molecules and (3) some means of coupling light energy to chemical change. Chlorophyll molecules (or rather the molecules of several chlorophylls that differ in their side groups) constitute the principal receiving system. An electron in the chlorophyll molecule is excited from its normal energy level to a higher level by the impact of visible light. The excited electron reverts to its normal state in less than a hundred-millionth of a second. This reversion might be accomplished by the reemission of visible light, but that, of course, would not advance the photosynthetic process. Instead the reversion proceeds in several steps during which the energy necessary for photosynthesis is transferred along a chain of molecules. A small part of the energy released by the reversion is reemitted as light of a longer wavelength, and hence of lower energy, than the light that was absorbed. (This is the dark red light characteristically emitted by chlorophyll when it is excited to fluorescence.) The remainder of the energy is transferred by way of other chlorophyll molecules to an ultimate recipient, a molecule that receives the energy and effects the chemi-

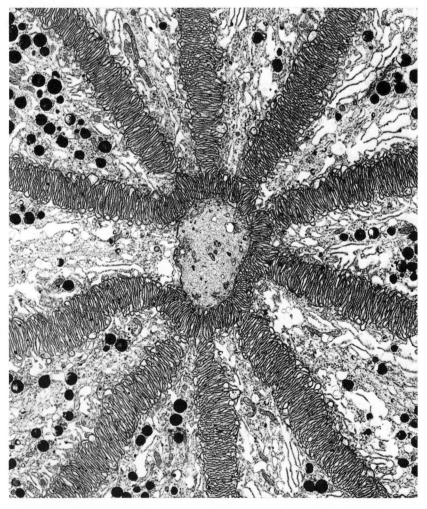

LIGHT-RECEPTOR ORGAN of an invertebrate is enlarged 8,000 times in an electron micrograph. Called an ommatidium, it is one of about 1,000 such units that comprise a horseshoe crab's compound eye. The spokelike arrays and the central ring are photosensitive. William H. Miller of the Yale University School of Medicine made the micrograph.

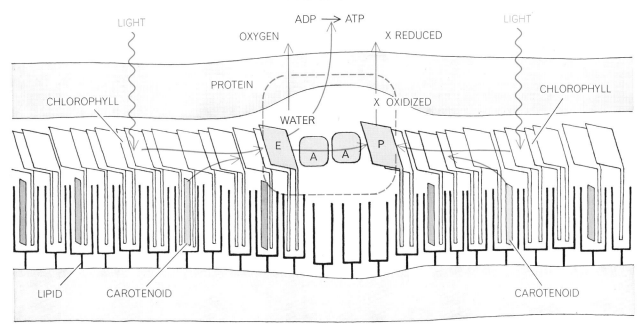

PLANT PHOTORECEPTORS contain molecular arrays, shown here in schematic form. Within the chloroplast, chlorophyll molecules are held together both by their mutual attraction and by the affinity of each molecule's phytol "tail" for lipids and its main body for proteins. Other molecules of pigment, such as carotenoids, are also embedded in the array. For each 500 or so chlorophyll molecules is found a specialized energy-transfer center, comprising two energy sinks, *E* and *P* (*color, center*), linked by a system for transferring electrons, represented here by units labeled *A*. The electron-cascade system by means of which this array of molecules turns energy from light (*colored inward arrows*) into chemical energy (*outward arrows*) is seen in detail in the illustration below.

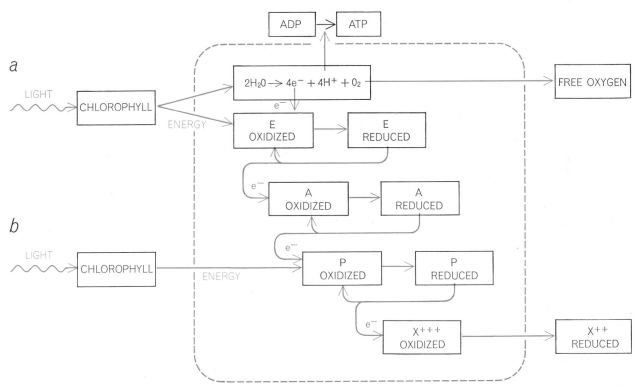

ELECTRON CASCADE that is responsible for the main action of photosynthesis, the use of energy from light to reduce carbon dioxide to sugar, is shown schematically here. Within a zone that contains two energy sinks, *E* and *P*, water and its components (*top rectangle*) are in a state of equilibrium until several chlorophyll molecules pass the energy received from light (*a*) to the first energy sink, *E*. This event starts the cascade; *E*, driven to a higher state of excitation by the energy it receives, seizes an electron (*colored arrow*) from a water component. Next *E* falls back to its lower state; the seized electron is released and cascades onward via the transfer system *A*. It arrives at the second energy sink, *P*, soon after that sink has received energy from other light-excited chlorophyll molecules (*b*). The cascade ends when *P* falls back to its lower state, passing both electron and energy to *X*, an electron-rich compound. The event energizes *X* sufficiently to let it power the carbon dioxide reduction process (*arrow, lower right*); this is the main photosynthetic action. Two other events, however, are also consequences of the cascade. Hydrogen ions (*top arrow*) provide the energy gradient needed to transform adenosine diphosphate (ADP) into adenosine triphosphate (ATP). Similarly, the ion neutralized by the electron loss that initiated the cascade joins with other components to form water, thereby freeing oxygen (*arrow, upper right*).

cal synthesis. For these energy transfers to be efficient the molecules in the chain must meet two criteria. First, they must be physically close together. Second, there must be a close match between the amount of energy available from the donor molecule and the amount acceptable by the recipient.

The plastids of plant cells, the microscopic bodies that contain the chlorophyll pigments, are made up of layered structures known as lamellae that have a high content of protein and lipid [*see top illustration on preceding page*]. The chlorophyll molecule has one end (the phytol end) that is soluble in lipid and a main body (the chlorophyllin end) that has an affinity for protein. These affinities give rise to a structural system in which the chlorophyll molecules are closely packed.

The lamellae also contain other molecules with conjugated bonds. These include carotenoids that are similar in structural arrangement to retinal, and phycocyanin, which has a chromophore closely related to the chromophore of phytochrome. These accessory molecules also absorb radiant energy and transfer it to the chlorophyll molecules.

The accumulated energy is finally transferred from the chlorophyll molecules to a relatively few molecules that act as energy-trapping "sinks." In each lamella there is about one sink for every 500 chlorophyll molecules. This small number, whereas it effects a desirable parsimony in the systems required for the chemical steps of photosynthesis, constitutes a bottleneck insofar as energy transfer is concerned. When light reach-

es a level of intensity about a fifth the intensity of full sunlight, energy arrives at the sinks faster than it can be utilized. Saturation at this level of intensity is nonetheless a good compromise because the average plant leaf is somewhat shaded and seldom receives energy much above the one-fifth level.

The energy accumulated by the sink molecules is ultimately applied to split water molecules into hydrogen and oxygen and to yield an electron-rich compound, which here I shall call "X," that acts as a final electron-acceptor. An oxidized material (that is, one that has given up electrons) is formed as a waste product. In green plants this material is oxygen. It is as a result of this aspect of photosynthesis that the earth's atmosphere contains the oxygen essential to all animal life.

Measured in terms of its products, the effectiveness of the photosynthetic chemical system decreases as the wavelength of the light being absorbed becomes longer. Absorption in the far-red region of the visible spectrum can be made effective, however, if supplementary light of shorter wavelength is also present. This suggests that two steps are involved in electron transfer rather than one; perhaps two energy sinks work together in some kind of booster action. The processes associated with each type of sink are a subject of current investigation, as is the manner in which electron flow might be coordinated between the two steps.

The electron-transport system, as it is now conceived, can be represented schematically [*see illustrations on page 27*]. Trapping centers are indicated as points E and P. An electron is thought to be transferred from P to X by one act of light absorption (*b*): the electron loss leaves P oxidized, whereupon X, the electron-acceptor, becomes an electron-rich, or reduced, compound. In close order a second act of light absorption (*a*) transfers an electron from water to point E, leaving E electron-rich and leading eventually to free oxygen as the oxidized substance in green plants. The scheme is completed by electron transfer from reduced E to oxidized P. Functioning of the electron transport steps from water to X again requires close association of the necessary parts in the lipid-rich lamellae of the plastids.

The bare skeleton of the scheme serves the purpose of exposition as far as the "photo" part of photosynthesis is concerned, but it leaves much to be told about the "synthesis" part. "Synthesis" implies an output that can be used in a

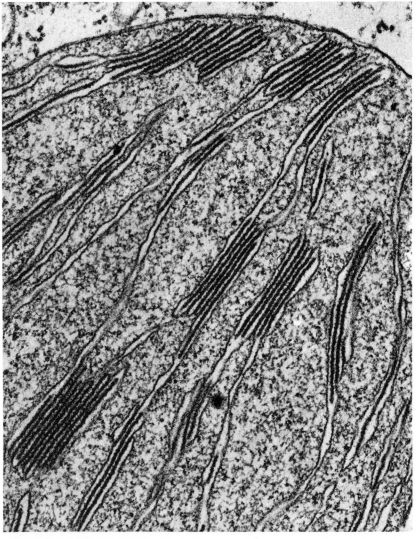

PART OF PHOTOSYNTHETIC ORGAN, a chloroplast of the alga *Nitella*, is seen enlarged 133,000 times in an electron micrograph made by Myron C. Ledbetter of the Brookhaven National Laboratory. Chlorophylls and other pigments involved in the process of photosynthesis are associated with the many lamellae scattered throughout the chloroplast.

life process. Oxygen, although it is an oxidized waste product of the scheme, eventually closes back on the electron-rich compound X through the process of respiration. X is an immediately useful product for reactions outside the lamellae. It serves as the chief energy-transferring agent in the reduction of carbon dioxide to sugar. A further reaction transforms X through intermediates, along with carbon dioxide and water, into electron-poor, or oxidized, X and phosphorus-containing sugars. The reaction needs more energy than can be supplied by X alone. This energy, as well as the needed phosphate, comes from adenosine triphosphate (ATP). ATP is formed by the removal of water from adenosine diphosphate (ADP) and the addition of a phosphate molecule. Energy for the transformation of X is available when the added phosphate of ATP

is transferred to some other molecule or is split from ATP by water.

Returning to our illustration of the scheme, it appears that the energy difference in the electron transfer from E to P is adequate to make ATP. This until recently was thought to be the most likely way at least part of a plant's supply of ATP was produced. A view that is now being vigorously debated suggests that hydrogen ions appear inside the lamella in the electron transfer that follows light absorption. The enhanced acidity with respect to the outside of the retaining membrane that the ions inside the membrane provide would give an energy gradient adequate for the formation of ATP. With regard to the first action of light (b), electrons excited by that event can also be transferred through X back to the starting point P with coupling to ATP along the

way—a process known as cyclic phosphorylation.

This broad outline of energy transfer in photosynthesis has been developed chiefly during the past 10 years. There is still much to be learned about the molecular details of oxygen liberation, the formation of ATP and the coordination of electron flow in various parts of the process. New discoveries may well alter some of today's concepts of photosynthesis at the most basic level. The situation is much the same with regard to our present understanding of vision and photoperiodism. Examination of the immediate changes after light absorption has proved to be a more fruitful realm of study than the search for the ensuing steps that lead to the responses of sight, growth and biological rhythm.

3

THE MECHANISM OF PHOTOSYNTHESIS

R. P. LEVINE
December 1969

Light of two wavelengths is required to activate two photochemical systems. Together they provide the electrons, protons and energy-rich molecules needed to convert carbon dioxide and water into food

Light interacts with living organisms in such processes as vision, bioluminescence and photosynthesis, but it is apparently only during photosynthesis that light energy is converted into useful forms of chemical energy. That energy, in turn, is used to build up complex molecules—notably carbohydrates—that animals require as food. Photosynthetic organisms also provide most of the oxygen in the atmosphere, and the evolution of animals was certainly dependent on the existence of oxygen-evolving microorganisms in the primitive seas.

The study of photosynthesis spans the disciplines of photophysics, photochemistry, biochemistry and physiology. In recent decades such studies have revealed many remarkable aspects of the photosynthetic process. The simple equation that summarizes the process has been known since the 19th century: water (H_2O) plus carbon dioxide (CO_2) yields some form of carbohydrate (represented by CH_2O) and oxygen (O_2). The reaction is driven by light energy. Photons are first absorbed by chlorophyll and other photosynthetic pigments. The red and blue-green algae, for example, contain pigments called phycobilins in addition to chlorophyll [*see illustration on page 32*].

Once the light energy is absorbed, it is used for two purposes. First, it is used to generate what a chemist calls "reducing power." Reduction involves the addition of electrons or the removal of protons, or both. Molecules that are rich in reducing power can transfer electrons to more oxidized molecules. The reducing agent produced by photosynthesis is NADPH, the reduced form of nicotinamide adenine dinucleotide diphosphate (NADP). Second, the light energy becomes converted into the energy-rich phosphate compound adenosine triphosphate (ATP). ATP and NADPH have certain structural elements in common [*see illustrations on page 33*]. Both are needed to reduce CO_2, a relatively oxidized molecule, into carbohydrate. The overall balance sheet for photosynthesis shows that three molecules of ATP and two molecules of NADPH are required for each molecule of CO_2 reduced.

In most algae and in higher plants photosynthesis occurs in the intricate, membrane-filled structure known as the chloroplast [*see illustration on opposite page*]. Within the chloroplast light energy is trapped and the rapid photophysical and photochemical reactions take place that generate the ATP and NADPH to be used in the more leisurely biochemical process of reducing carbon dioxide to carbohydrate. Once ATP and NADPH have been formed they are released into the nonmembranous, or soluble, phase of the chloroplast; there the fixation of carbon dioxide can proceed in the absence of light with the assistance of a number of soluble enzymes.

The Absorption of Light Energy

When chlorophyll or one of the other photosynthetic pigments absorbs photons, the pigment passes from its lowest energy state, or ground state, to a higher energy state. The excited state is not stable, and the pigment can return to the ground state within 10^{-9} second. If in that brief period the energy is not used for the generation of ATP or NADPH, it can be dissipated as fluorescent light. Since 100 percent efficiency is never achieved in biological systems, fluorescence is always observed during photosynthesis; the fluorescence is at a longer —hence less energetic—wavelength than the wavelength originally absorbed. As we shall see, fluorescence has been useful to the investigator of photosynthesis.

Each of the photosynthetic pigments has its characteristic absorption spectrum: it absorbs more or less light at different wavelengths depending on its molecular structure. For example, the two chlorophylls designated *a* and *b* have major and distinctive absorption bands in the blue and red regions of the spectrum [*see top illustration on page 34*]. When studied in isolation outside the chloroplast, each pigment also has a characteristic fluorescence spectrum. Inside the chloroplast, however, chlorophyll *b* fluorescence is never detected, even when the incident light is of a wavelength known to be absorbed by that chlorophyll. Similarly, fluorescence is never observed from the carotenoid pigments and the phycobilins. Only one of the pigments fluoresces naturally inside the chloroplast: chlorophyll *a*.

This surprising phenomenon has now been explained, largely through the work of Louis N. M. Duysens of the Netherlands. He showed that chlorophyll *b*, the carotenoids and the phycobilins do not participate directly in photosynthesis but rather act only as "antennas" to help gather light energy. When they absorb energy and become excited, they transfer their excitation energy to chlorophyll *a*. Only chlorophyll *a* is actively involved in the subsequent reactions of photosynthesis; when its energy cannot be used for photosynthesis, it dissipates its excitation energy as fluorescence.

The mechanism for the transfer of excitation energy between pigment molecules in photosynthesis is not clearly understood, but a process called inductive resonance is one possibility. If an excited molecule is close enough to an unexcited one (say within 30 angstroms), it can dissipate its energy by inducing an excited state in the neighboring molecule. In this way energy can pass from chlorophyll *b* to chlorophyll *a*. The reverse process is not possible, however, because to become excited chlorophyll *b* requires more excitation energy than an excited chlorophyll *a* can provide.

Ultimately the energy of excitation reaches a photosynthetic reaction center, where it is transferred to a special long-wavelength form of chlorophyll *a*. Because this pigment absorbs at a longer wavelength, and hence at a lower energy, than the surrounding pigment molecule, it can be considered a kind of energy sink. The transfer of excitation energy from an excited molecule of normal chlorophyll *a* to such a special chlorophyll *a* molecule probably takes place within 10^{-12} second, which is 1,000 times

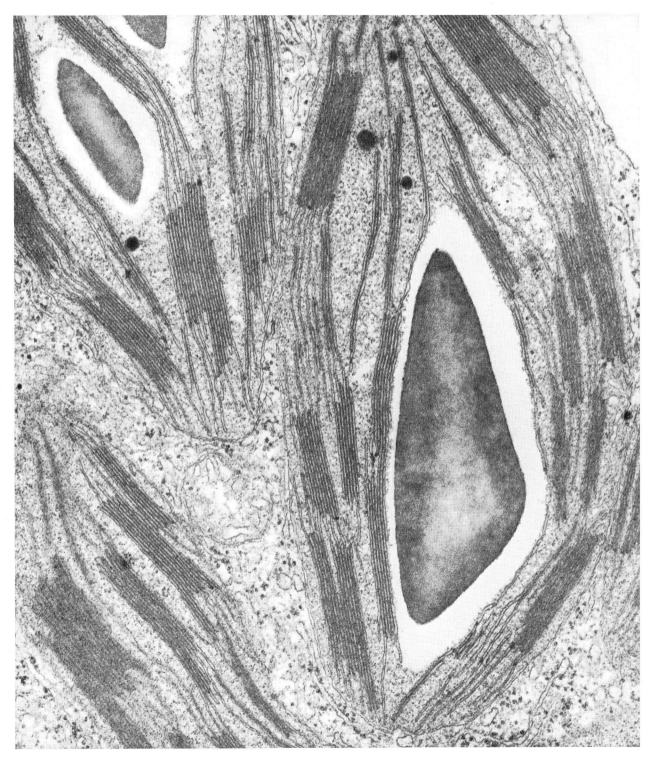

SITE OF PHOTOSYNTHESIS is the organelle known as a chloroplast, present in the cells of all higher plants and most algae. This electron micrograph made by Peter Hepler of Harvard University shows portions of three chloroplasts. Photosynthesis takes place inside the dark membranes that lie in long parallel bundles. The large triangular object in the chloroplast at the right is a kernel of starch, produced by photosynthesis. The chloroplast at the upper left contains two kernels. The magnification is 58,000 diameters.

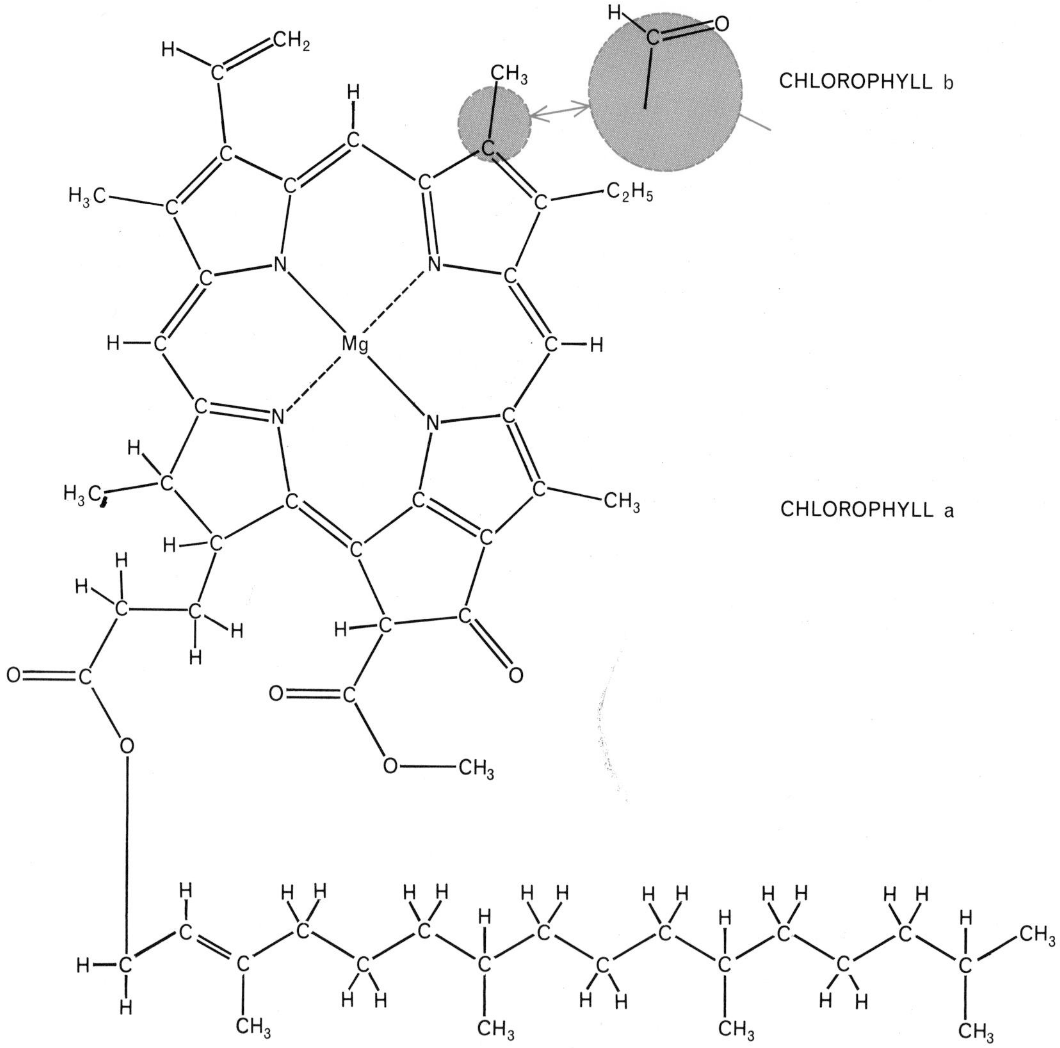

CH₂

CHLOROPHYLL b

CHLOROPHYLL a

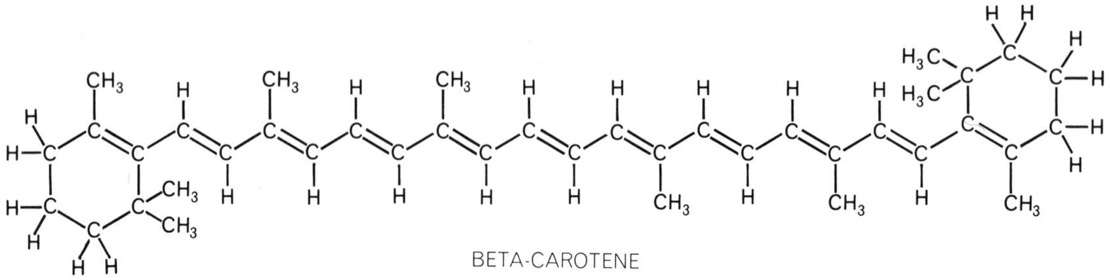

BETA-CAROTENE

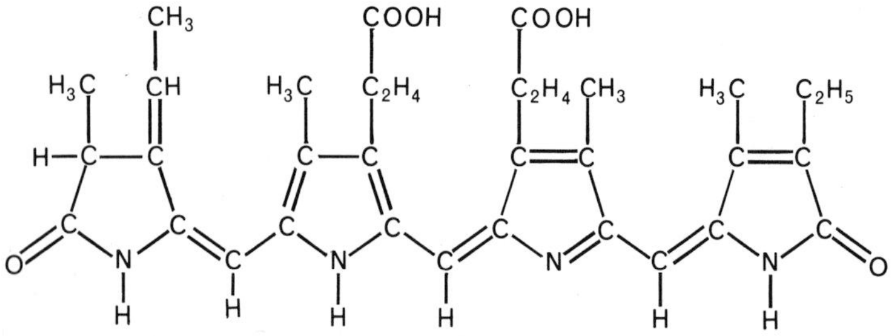

PHYCOCYANOBILIN

PHOTOSYNTHETIC PIGMENTS have the ability to capture photons and convert their energy into molecular excitation energy. Chlorophyll *a* is the pigment found in algae and in the leaves of higher plants. Chlorophyll *b* has the same structure except that a –CHO group replaces a –CH₃ group in one corner of the porphyrin ring. Beta-carotene is another photosynthetic pigment present in many higher plants. Red and blue-green algae contain still a third class of pigments, the phycobilins, of which phycocyanobilin is one.

faster than the time taken for the "waste" energy of chlorophyll *a* to emerge as fluorescence. Thus there is ample time for an excited chlorophyll molecule to disperse its energy in a chemically useful way.

The First Chemical Steps

Once light energy has been relayed to the special chlorophyll *a* molecule, the energy sink in the reaction center, the chemistry begins: the excitation energy must be used to form an oxidant and a reductant. The oxidant must be capable of oxidizing water, that is, capable of splitting the water molecule into free oxygen, protons and electrons. (Actually two molecules of H_2O are split into one molecule of O_2 plus four electrons and four protons.) The reductant must accept the reducing equivalents (electrons and protons) that arise from the oxidation of water. Ultimately these equivalents will be used in the reduction of carbon dioxide. The oxidant and the reductant must be formed within the very short lifetime of the excited state of chlorophyll *a*. How this comes about constitutes one of the biggest gaps in our knowledge of photosynthesis, and a great deal of what follows must be conjecture.

One simple way to visualize the initiation of the first chemical steps of photosynthesis is to imagine the existence of an electron-donor molecule *D* and an electron-acceptor molecule *A*. The donor in the oxidized form (D^+) will oxidize water and the acceptor in reduced form (A^-) will ultimately transfer its reducing equivalents to NADP, converting it to NADPH.

A simple model of the primary reaction sequence involving the donor, the acceptor and an excited molecule of chlorophyll *a* is shown in the top illustration on page 35. In this sequence the chlorophyll *a* in the reaction center is raised to an excited state by receiving excitation energy from surrounding pigment molecules. Each reaction-center chlorophyll molecule is in close association with the donor and acceptor molecules in the membrane of the chloroplast. When the chlorophyll returns to the ground state, the release of the excitation energy is sufficient to extract an electron from the donor molecule, thereby oxidizing it to D^+, and to transfer this electron to the acceptor, thus reducing it to A^-. Such charge-transfer processes are known to operate in nonbiological systems where the organic molecules involved have properties similar to those involved in photosynthesis.

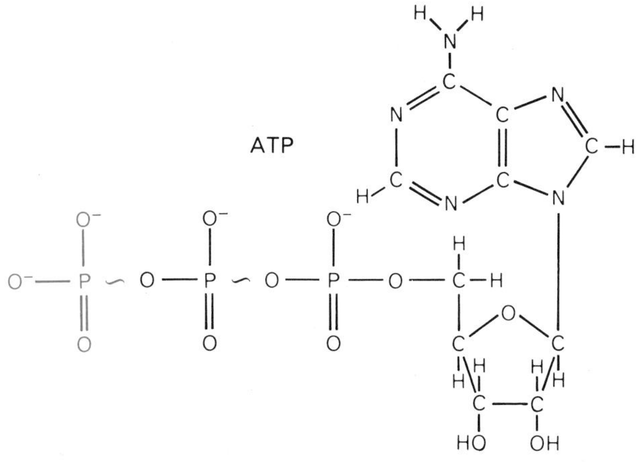

ADENOSINE TRIPHOSPHATE, or ATP, is produced from adenosine diphosphate (ADP), with the energy collected by the photosynthetic pigments. The wavy lines are links to energy-rich phosphate groups. If the last group (*color*) is removed, ATP becomes ADP. In the process ATP supplies energy for converting carbon dioxide into carbohydrates.

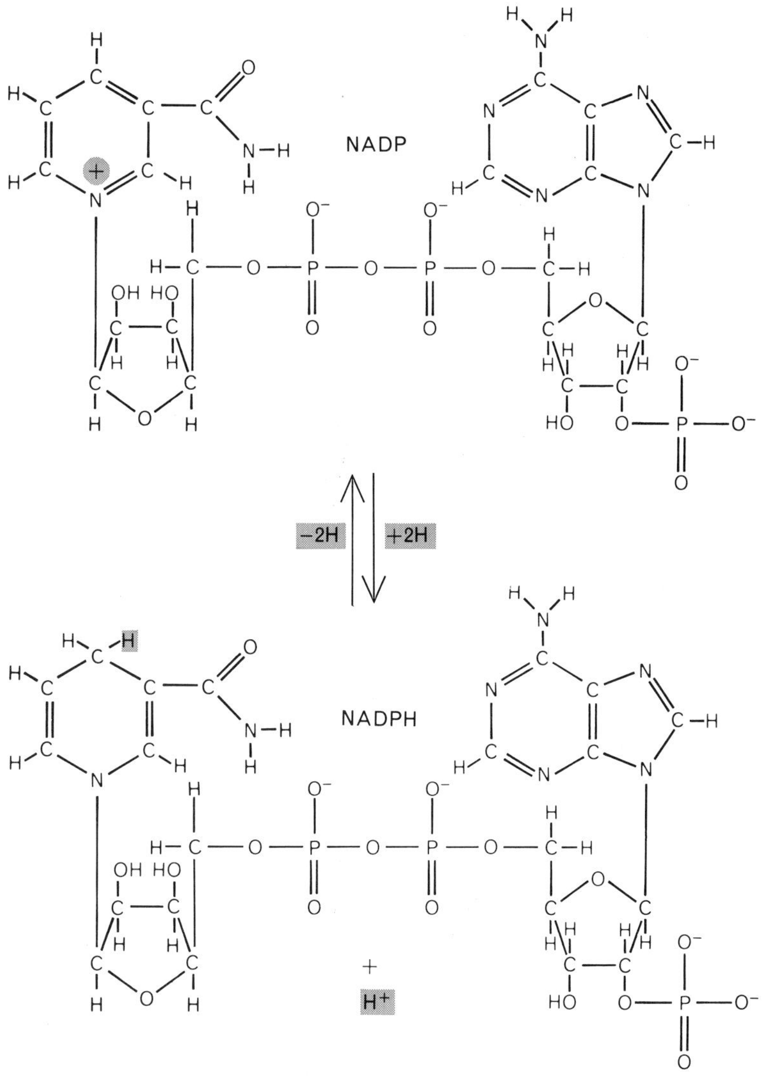

NICOTINAMIDE ADENINE DINUCLEOTIDE DIPHOSPHATE, or NADP, is reduced to NADPH during photosynthesis. NADP becomes NADPH by the addition of two hydrogen atoms. One binds directly to the molecule while the other loses its electron and is released as a proton (H^+). NADPH supplies "reducing power" for fixation of carbon dioxide.

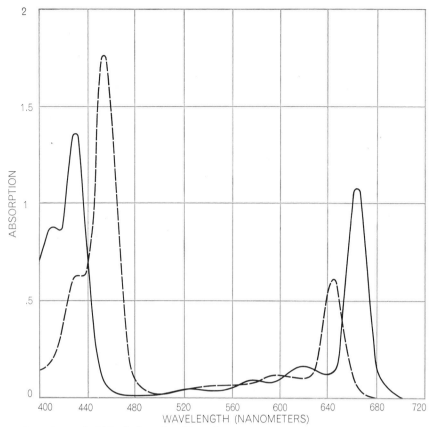

ABSORPTION SPECTRA show that chlorophyll *a* (*solid line*) and chlorophyll *b* (*broken line*) strongly absorb blue and far-red light. The green, yellow and orange wavelengths lying between the peaks are reflected and give both pigments their familiar green color.

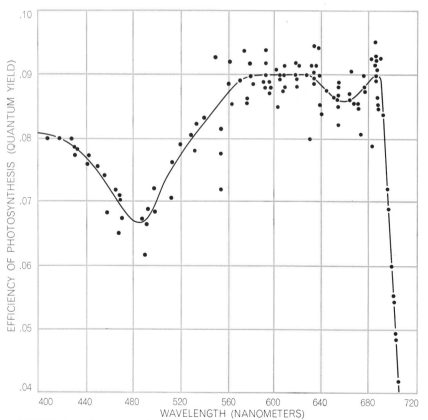

EFFICIENCY OF LIGHT ABSORPTION in the alga chlorella was determined by Robert Emerson of the University of Illinois. Efficiency of photosynthesis falls off sharply in the far-red region beyond 680 nanometers even though chlorophyll *a* still absorbs at that wavelength. If light of shorter wavelength is added to far-red light, efficiency rises sharply.

Regardless of how the charge transfer may be accomplished in the chloroplast, its net effect is to separate oxidizing and reducing equivalents. Very little is known about how the hypothetical D^+ participates in the oxidation of water with the concomitant evolution of oxygen. Much more is known about the transfer of reducing equivalents from the hypothetical A^- to NADP. With this step the mechanism of photosynthesis moves from a high-speed photochemical phase to a slower biochemical phase in which electrons are transported through a series of reactions, ultimately to yield NADPH and ATP.

The Biochemical Phase

Our understanding of the biochemical phase of photosynthesis owes much to investigations showing that two light reactions (and not one, as has been tacitly assumed so far) take place in the photosynthetic process used by algae and higher plants. Two experiments set the stage for this discovery. The first provided measurements of the rate of photosynthesis at different wavelengths of light over the range absorbed by the photosynthetic pigments. The result is a curve showing how the quantum efficiency of the process varies at different wavelengths [*see bottom illustration at left*]. The curve reveals a curious fact: in the far-red region, beyond a wavelength of 680 nanometers, the efficiency of photosynthesis falls rapidly to zero even though the pigments still absorb light.

This surprising result led to the second set of experiments, reported in 1956 by Robert Emerson and his colleagues at the University of Illinois. They found that although photosynthesis is very inefficient at wavelengths greater than 680 nanometers, it can be enhanced by adding light of a shorter wavelength, 650 nanometers for example. Moreover, the rate of photosynthesis in the presence of both wavelengths is greater than the sum of the rates obtained when the two wavelengths are supplied separately. This phenomenon, now known as the Emerson enhancement effect, can be explained if photosynthesis is assumed to require two light-driven reactions, both of which can be driven by light of less than 680 nanometers but only one by light of longer wavelength.

These two sets of experiments marked the beginning of an exciting period in the effort to understand the mechanism of photosynthesis. They gave rise to a provocative hypothesis and to several revealing lines of research. The hypothesis was provided by Robert Hill and Fay Bendall of the University of Cambridge. They proposed a scheme showing how electrons could be transported along a biochemical chain in which two separate reactions are triggered by light. Before describing the Hill-Bendall scheme I should briefly touch on some characteristics of electron transport.

One must distinguish first between a transfer in which electrons go *with* an electrochemical gradient (the easy direction) and a transfer in which the electrons go *against* that gradient (the hard direction). Electron-donor and electron-acceptor molecules can be characterized by the quantity called oxidation-reduction potential, which can be positive or negative and is usually expressed in volts. Electrons can be transferred from donors that have a more negative potential to acceptors that have a more positive potential without any input of energy. In fact, when electron transfer takes place along this gradient, energy is released; the greater the gap in potential between donor and acceptor, the greater the yield of energy. To transfer electrons against the electrochemical gradient, on the other hand, requires an input of energy. The greater the gap between the donor and the acceptor, the greater the energy required.

In the mitochondria of both plant and animal cells energy, in the form of ATP, is generated as a consequence of electron transport down an electrochemical gradient between a series of electron-donors and -acceptors called cytochromes. At least some of the electron-transport steps in photosynthesis, however, must go against an electrochemical gradient because the oxidation-reduction potential of water (the primary electron-donor) is +.8 volt whereas that of NADP (the terminal acceptor) is −.3 volt.

The Two Photochemical Systems

Hill and his co-workers had earlier identified and characterized a number of cytochromes found in the chloroplast; Hill and Bendall saw that ATP might be generated in photosynthesis if advantage were taken of the difference in oxidation-reduction potential between two of these cytochromes. One of them, a *b*-type cytochrome, has a potential close to zero and the other, a *c*-type cytochrome, has a potential of about +.35 volt. In the Hill-Bendall scheme, therefore, two light reactions provide the energy to go *against* the electrochemical

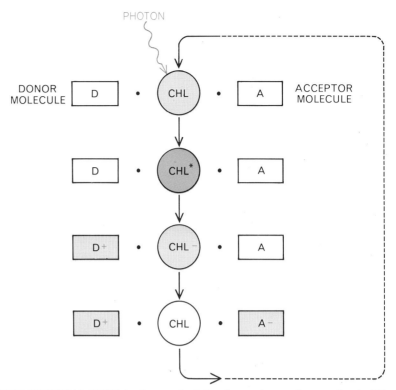

FIRST CHEMICAL STEPS in photosynthesis involve an electron-donor molecule (D) and an electron-acceptor molecule (A) in close association with a special chlorophyll (Chl) in the reaction center. An incoming photon can raise the chlorophyll to an excited state (Chl^*). When the excited chlorophyll returns to the ground state in less than 10^{-9} second, the energy released extracts an electron from D, oxidizing it to D^+, and transfers the electron to A, reducing it to A^-. Later D^+ oxidizes water and A^- reduces NADP to NADPH.

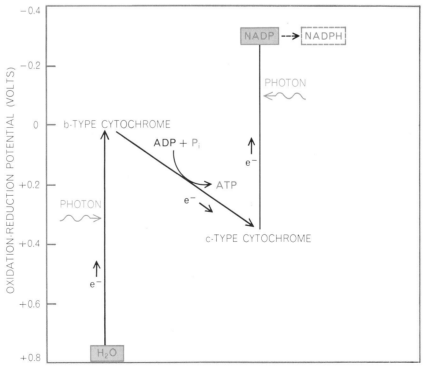

HILL-BENDALL MODEL of electron transport in photosynthesis suggests how electrons (e^-) removed from water can be boosted against an electrochemical gradient, finally reaching NADP. With the aid of protons, also provided by water, NADP is converted to NADPH. A change in an upward direction requires an input of energy, supplied by photons; a change in a downward direction yields energy. ADP is presumably converted to ATP on the downslope between the two cytochromes, which act as electron-acceptors and -donors.

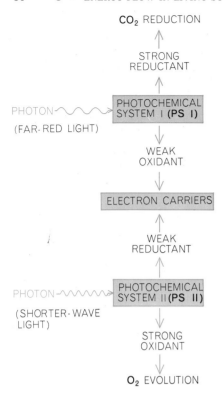

CO₂ REDUCTION

STRONG REDUCTANT

PHOTON〰〰 PHOTOCHEMICAL SYSTEM I (PS I)
(FAR-RED LIGHT)

WEAK OXIDANT

ELECTRON CARRIERS

WEAK REDUCTANT

PHOTON〰〰 PHOTOCHEMICAL SYSTEM II (PS II)
(SHORTER-WAVE LIGHT)

STRONG OXIDANT

O₂ EVOLUTION

PS I AND PS II, two photochemical systems, cooperate in the fixation of carbon dioxide in algae and higher plants. Each system has its own reaction center containing a photosynthetic pigment. The pigment in PS I is a species of chlorophyll *a* known as P-700 because its maximum absorption is at a wavelength of 700 nanometers. The strong oxidant of PS II is able to oxidize water. The strong reductant of PS I has the power to reduce NADP to NADPH. The reactions driven by the photochemical systems are shown in the equations at the bottom of the page.

gradient while electron transport between the two cytochromes goes *with* the gradient [*see bottom illustration on preceding page*].

The Hill-Bendall formulation indicated that the two light reactions occur in two different photochemical systems [*see illustration above*]. Each system has a reaction center within which an oxi-

dant and a reductant are formed. Photochemical system II (PS II) sensitizes a reaction that results in the oxidation of water and in the formation of a weak reductant. The chlorophyll in the reaction center of PS II has not yet been identified, but it is presumed to be some form of chlorophyll *a*. Photochemical system I (PS I) sensitizes a reaction that yields a weak oxidant and a strong reductant. The chlorophyll in the reaction center of PS I has been identified as a species of chlorophyll *a* whose absorption peak is at 700 nanometers and is therefore known as P-700. The two photochemical systems are linked in series by electron-carriers, so that the weak reductant produced in PS II is oxidized by the weak oxidant produced in PS I.

Duysens and his co-workers provided some of the early evidence for this model of two photochemical systems acting in series when they showed that the *c*-type cytochrome in the chloroplast is reduced by the shorter-wavelength light absorbed by PS II and oxidized by the longer-wavelength light absorbed by PS I. Such "antagonistic" effects indicate that the cytochrome lies in the path of electron flow between the two systems. Other investigators have since demonstrated similar antagonistic effects on the *b*-type cytochrome of the chloroplast and on P-700.

Additional evidence for the series model has involved the use of the potent weed killer DCMU (dichlorophenyldimethyl urea), which owes its effectiveness to its ability to inhibit the flow of electrons from water to NADP. In its presence both the *c*-type and the *b*-type cytochromes can be oxidized by PS I but they cannot be reduced by PS II. One can therefore assume that the DCMU acts at a site somewhere between PS II and the cytochromes [*see illustration on page 37*]. The photoreduction of NADP is thus blocked by DCMU, but it can be restored if an artificial electron-donor (such as a reduced

indophenol dye) is introduced into the chloroplast. In the presence of the dye NADP can once again be photoreduced but now light absorbed by PS I alone is sufficient. The effect of DCMU indicates the existence of two light reactions coupled by a system of electron-donors and -acceptors.

Electron Path from PS II

The portion of the photosynthetic electron-transport chain that carries electrons from water to PS II is known as the oxidizing "side" of PS II. As mentioned above, the oxidation of water is effected by the oxidized form of a hypothetical donor molecule, D^+. Experimental evidence for electron transport between water and PS II has recently been provided by Takashi Yamashita and Warren L. Butler of the University of California at San Diego, but the nature of the electron-carrier (or carriers) involved, and its relation to D, has not yet been determined.

The reducing "side" of PS II is the portion of the electron-transport chain between PS II and its electron-acceptor A. The fluorescence properties of the chloroplast have provided information on the nature of A. If chloroplasts are irradiated with short-wavelength light, the yield of fluorescence is high, but if they are illuminated with the longer wavelength of light that can be absorbed by PS I, the fluorescence yield decreases. From these observations Duysens and his associates have inferred that when A is reduced by PS II, fluorescence is high, but when it is oxidized by PS I, fluorescence is low. They called the acceptor component Q rather than A, Q standing for quencher of fluorescence. Q in the oxidized form quenches fluorescence, whereas Q in the reduced form does not and therefore the fluorescence yield increases. The yield increases even more in the presence of DCMU, suggesting that the weed killer acts at a site between Q and PS I in the photosynthetic electron-transport chain.

The chemical nature of Q has not been determined with certainty. Norman I. Bishop of Oregon State University has obtained evidence suggesting that it may be a compound known as plastoquinone. Regardless of its chemical nature, Q is probably the electron-acceptor of PS II.

Electron Path from Q to P-700

Proceeding along the electron-transport chain from PS II to PS I, one finds

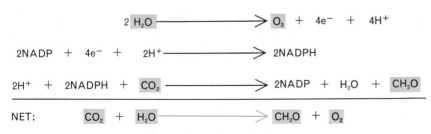

$$2\,H_2O \longrightarrow O_2 + 4e^- + 4H^+$$

$$2NADP + 4e^- + 2H^+ \longrightarrow 2NADPH$$

$$2H^+ + 2NADPH + CO_2 \longrightarrow 2NADP + H_2O + CH_2O$$

$$NET: \quad CO_2 + H_2O \longrightarrow CH_2O + O_2$$

CHEMISTRY OF PHOTOSYNTHESIS is summarized by these four equations. Water is oxidized, releasing free oxygen, electrons and protons. The electrons and two protons reduce NADP to NADPH. NADPH plus two protons and carbon dioxide yield NADP, water and carbohydrate (CH_2O). Thus carbon dioxide and water yield carbohydrate and oxygen.

that there is a cytochrome on the downhill slope from Q to PS I. This is a b-type cytochrome. It is followed by a c-type cytochrome. The reduction of both cytochromes is sensitized by PS II; their oxidation is sensitized by PS I. As mentioned above, this differential oxidation and reduction pattern localizes the cytochromes between the two photochemical systems.

Between the b-type and the c-type cytochromes there is at least one more component, which is not yet identified. Evidence for its existence comes from experiments that Donald Gorman and I have conducted in the Biological Laboratories of Harvard University, using mutant strains of a unicellular green alga that had lost the capacity to carry out normal photosynthetic electron transport. Of the mutant strains, one lacked the b-type cytochrome, another lacked the c-type cytochrome and a third lacked an unknown component.

The third mutant strain proved to possess both cytochromes (b-type and c-type), but when it was illuminated with long-wavelength light of the kind absorbed by PS I, only the c-type cytochrome was oxidized. When light of the kind absorbed by PS II was used, only the b-type cytochrome was reduced. Since the first kind of light normally oxidizes both cytochromes and the second kind of light normally reduces both cytochromes, it was clear that some component was missing from the mutant strain that ordinarily acts as an electron-donor and -acceptor between the two cytochromes. For want of a specific identification we have designated it M.

Another component in the electron-transport chain is the copper-containing protein plastocyanin. Although this protein can act as an electron-acceptor and -donor, its role is not clearly understood. At present some experiments indicate that plastocyanin lies between the c-type cytochrome and PS I, and other experiments give equally convincing evidence that it lies on the uphill side of the c-type cytochrome.

We now come to P-700, the chlorophyll that absorbs far-red light in the reaction center of PS I. Its discoverers, Bessel Kok of the Research Institute for Advanced Study in Baltimore and George E. Hoch of the University of Rochester, showed that P-700 is oxidized by light absorbed by PS I and reduced by light absorbed by PS II. On being oxidized it transfers its electron to its electron-acceptor. It can then be reduced again by electrons coming from water and passed along the transport

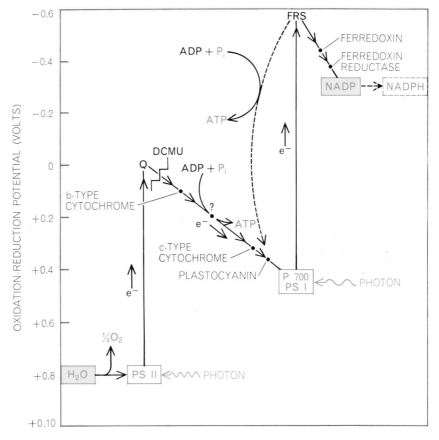

ELECTRON-TRANSPORT CHAIN in photosynthesis follows the Hill-Bendall model (*see bottom illustration on page 35*). It shows how electrons removed from water are passed along by various acceptors and donors until they finally reach NADP and participate in its reduction to NADPH. Along the chain, at one or more places not yet clearly identified, energy is extracted to form ATP from ADP and inorganic phosphate (P_i). The energy for boosting electrons against the electrochemical gradient is supplied by photons that excite chlorophyll molecules in the two photochemical systems, PS I and PS II. Electrons leaving PS II are evidently accepted directly by a substance called Q (for quencher of fluorescence) and electrons leaving PS I by the ferredoxin-reducing substance (FRS). The downhill path from Q to PS I contains a number of donors and acceptors, which are discussed in the text. The path that transports electrons from FRS to NADP appears to be less complicated; the end result is NADPH. Many details of the electron-transport chain have been clarified with the help of a weed killer called DCMU, which interrupts electron flow to the right of Q.

chain from PS II. The immediate electron-donor to P-700 is either the c-type cytochrome or plastocyanin or both.

The Path between P-700 and NADP

Moving along the electron-transport chain, we see that the electron donated by PS I, which has a potential of about +.45 volt, must travel against a large electrochemical gradient before it can reach NADP, which has a potential of −.3 volt. The electron-acceptor of PS I has been much debated ever since it was proposed a few years ago that the acceptor might be ferredoxin, an electron-acceptor and -donor molecule whose negative potential (−.43 volt) is even higher than that of NADP. It was clear that if light energy could boost an electron against the potential gradient from

P-700 to ferredoxin, the final step to NADP would be downhill. The difficulty was that several investigators found that PS I can boost electrons against potentials even more negative than that of ferredoxin. It did not seem economical of nature to provide a greater boosting capacity than is actually required, and this cast doubt on ferredoxin's being the primary acceptor of electrons from PS I.

Recently Charles Yocum and Anthony San Pietro of Indiana University and Achim Trebst of the University of Göttingen have discovered what is apparently the true acceptor. A substance with a potential of about −.6 volt, it has been given the tentative name ferredoxin-reducing substance, or FRS. Its chemical nature is now under study. Its absorption spectrum suggests that it will turn out to have a complex structure consisting of more than one molecular species.

We have now nearly reached the end of the photosynthetic electron-transport chain. The FRS transfers its electron to ferredoxin, and NADP is reduced to NADPH in the presence of an enzyme called ferredoxin-NADP reductase.

ATP Formation and CO₂ Fixation

Much current research is focused on how the production of ATP is coupled to photosynthetic electron transport. Theoretically, as Hill and Bendall originally suggested, sufficient energy is available in the downhill flow of electrons between the *b*-type and the *c*-type cytochromes to phosphorylate a molecule of ADP, converting it into a molecule of ATP. And indeed there is evidence for a site of ATP formation between the two cytochromes. There is also evidence for a cyclic flow of electrons around PS I (most likely involving FRS), and ATP formation is coupled to this electron flow.

In spite of extensive investigation there is uncertainty regarding the mechanism of the coupling of ATP formation and electron flow not only in the chloro-plast but also in mitochondria. To review current opinions regarding the mechanism would require an article in itself.

We have now reached the last phase of the photosynthetic process: the reduction of carbon dioxide to carbohydrate. Much of our knowledge of this final phase is due to the work of Melvin Calvin, James A. Bassham and Andrew A. Benson of the University of California [see "The Path of Carbon in Photosynthesis," by J. A. Bassham; SCIENTIFIC AMERICAN Offprint 122]. In this cycle one molecule of ribulose diphosphate and one molecule of carbon dioxide react, with the aid of suitable enzymes, to form two molecules of phosphoglyceric acid (PGA). The two molecules of PGA are converted to two molecules of glyceraldehyde phosphate in a reaction that requires two molecules of NADPH and two of ATP. One other step requires ATP (the production of ribulose diphosphate from the monophosphate), so that the overall requirement is three molecules of ATP and two of NADPH for each molecule of carbon dioxide reduced to carbohydrate [see illustration below]. This sequence is thought to represent the pathway of carbon dioxide fixation in higher plants, algae and photosynthetic bacteria.

Quite recently, however, M. D. Hatch and C. R. Slack in Australia have shown that there is a different kind of pathway in certain species of tropical grasses. The first step of CO₂ fixation in these grasses involves the carboxylation of phospho-pyruvic acid (rather than of ribulose diphosphate), yielding oxaloacetic acid, which then serves as a precursor of PGA.

We have now followed the mechanism of photosynthesis from the initial trapping of the electromagnetic energy of light, through the conversion of energy into chemical energy, then through the electron-transport steps that lead to the generation of NADPH and ATP and finally to the terminal events of carbon dioxide fixation. We have seen that some parts of the process are much better understood than others. The most enigmatic part is the one associated with events at photochemical system II. The means by which four electrons and four protons are extracted from water with the concomitant evolution of a molecule of oxygen is one of the most fascinating problems still to be solved.

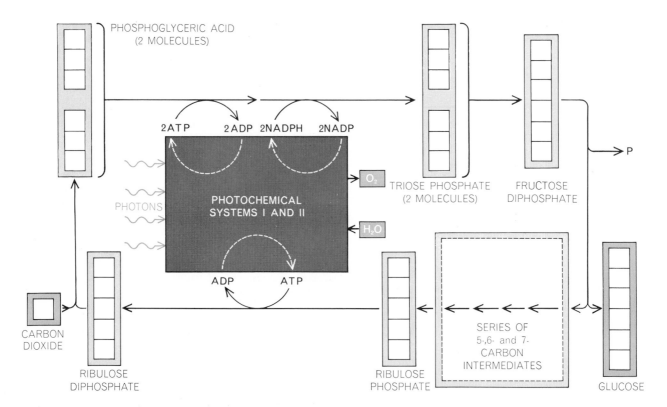

FIXATION OF CARBON DIOXIDE is achieved by a cycle of chemical reactions powered by photons that are trapped by the two photochemical systems. These systems package part of the energy in the form of ATP and remove electrons and protons from water, releasing oxygen. The electrons and protons enter the cycle in the form of NADPH. Two molecules of NADPH and three of ATP are required to fix one molecule of carbon dioxide, shown entering the cycle at the lower left. In the cycle each white square represents a carbon atom. The carbon atoms from CO₂ can be incorporated into a variety of compounds and removed at various points in the cycle. Here six atoms of carbon supplied by CO₂ are shown leaving the cycle as glucose, $C_6H_{12}O_6$, a simple carbohydrate.

4

THE MEMBRANE OF THE MITOCHONDRION

EFRAIM RACKER

February 1968

The folded inner membrane of this intracellular body is the site of the major process of energy metabolism in the living cell. It is studied by taking it apart and attempting to put it together again

The seat of oxidative phosphorylation, the process by which most plant and animal cells produce the energy required to sustain life, is the inner membrane of the intracellular particles called mitochondria [*see bottom illustration on page 40*]. Associated with this membrane are the enzymes of oxidative phosphorylation, embedded in a complex matrix that binds them tenaciously in an ordered array. The mechanism of energy production in mitochondria has long defied analysis, since a complex chemical pathway in a living organism cannot really be understood until its intermediate products have been identified and the enzymes that catalyze each step of the process have been individually resolved as soluble components. A decade ago my colleagues and I set out to attack the problem by trying to take the inner membrane of the mitochondrion apart and put it back together. We have been partially successful in the attempt, and along the way we have made some exciting discoveries and developed new methods of studying enzymes bound in membranes.

The universal energy carrier of the cell is adenosine triphosphate (ATP). This molecule functions by transferring its energetic terminal phosphate group to another molecule. In so doing it is converted to adenosine diphosphate (ADP), which in turn can be transformed into ATP by energy-generating systems in the cell. This regeneration of ATP occurs at several stages in the course of the breakdown and oxidation of foodstuffs. Some ATP is formed during glycolysis, a well-understood metabolic pathway utilizing soluble enzymes that break carbohydrates down to simpler compounds.

Most of the ATP is formed, however,

during the course of oxidative phosphorylation in mitochondria. Pyruvate, the end product of glycolysis, is delivered to the mitochondria, where it is oxidized to carbon dioxide and water by the enzymes of the Krebs cycle [*see top illustration on pages 40 and 41*]. As hydrogen is removed from the successive intermediate products, it is captured by the coenzyme diphosphopyridine nucleotide (DPN), which contains the vitamin nicotinamide. The electrons of hydrogen are passed along a series of respiratory enzymes, notably yellow flavoproteins and red cytochromes, ultimately combining with protons and oxygen to form water. The energy of this oxidation process is utilized at three sites to regenerate ATP from ADP and inorganic phosphate. Under physiological conditions such "coupling" of oxidation with phosphorylation is compulsory, and respiration takes place only when ADP and phosphate are available, which is to say when ATP is being utilized. This "tight coupling" represents an ingenious control mechanism through which energy production is regulated by the rate of energy consumption.

Two chemicals that affect oxidative phosphorylation serve as tools with which to analyze the process. One is dinitrophenol (DNP), which uncouples oxidation from phosphorylation so that respiration proceeds but produces heat instead of ATP. The other is the antibiotic oligomycin, which acts differently. It interferes with the production of ATP, thereby inhibiting respiration as long as the system is tightly coupled. When dinitrophenol is added, the inhibition by oligomycin is overcome and respiration returns to its original rate, although it produces no ATP.

The enzymes that catalyze electron transport had been isolated and char-

acterized, but only after the disruption of mitochondria with detergents. This process left the oxidation enzymes able to function but damaged the phosphorylation system severely. It was accordingly believed for a long time that the intact mitochondrial structure was essential for oxidative phosphorylation and that the component parts could not be separated without destroying them.

In 1956 the system did begin to yield to fractionation. Independent experiments reported almost simultaneously from the laboratories of Albert L. Lehninger, Henry A. Lardy, David E. Green and W. W. Kielley showed that chemicals such as digitonin and physical methods such as sonic oscillation would break mitochondria into "submitochondrial particles" much smaller than mitochondria and yet able to catalyze oxidative phosphorylation. This accomplishment was an important step forward, and yet there was still no indication that a true resolution—a separation of soluble components—would ever be possible. Such resolution was the task we undertook at the Public Health Research Institute of the City of New York.

In 1957 the first successful resolution of the system of oxidative phosphorylation was achieved when Harvey Penefsky, Maynard E. Pullman and I fragmented beef-heart mitochondria by agitating them with glass beads in a powerful device called a Nossal shaker. We removed the heavier unbroken mitochondria by centrifuging the mixture at low speed and then respun the lighter fraction at high speed [*see top illustration on page 42*]. The resulting sediment—the submitochondrial particles—still contained the respiratory enzymes but could not produce much ATP; the remaining fraction contained a soluble component that was necessary for the coupling of phosphorylation to oxidation. We called this soluble component a coupling factor, F_1. In time various treatments of mitochondria separated other coupling factors that were also required for phosphorylation, and these we called F_2, F_3 and F_4.

The experiments demonstrating the resolution of F_1 were difficult to reproduce. In laboratory jargon our data were "in the right direction"—and that is always a sign of trouble. F_1 as much as doubled phosphorylation, but that was not enough stimulation to provide a reliable assay of its coupling activity. And a reliable assay was required if we were to purify F_1 and characterize it. Now, it was known that mitochondria could catalyze the splitting of ATP into ADP and inorganic phosphorus. In fact, as early as 1945 Lardy, working at the University of Wisconsin, had suggested that this enzymatic ("ATP-ase") activity might be the inverse of some step in oxidative phosphorylation. We discovered that partially purified F_1 did in fact exhibit ATP-ase activity.

Since this was the first time ATP-ase had been extracted as a soluble component from mitochondria, we decided to go after this enzyme. We realized it was a gamble that might not shed any light on oxidative phosphorylation, but the ATP-ase assay was simple and accurate and at least we had had experience in purifying soluble enzymes. We felt that it should not take long to establish whether or not the ATP-ase and the coupling-factor activity were related.

Yet sometimes experience gets in one's way. Working in a "cold room," as one ordinarily does in enzyme research, we found the ATP-ase to be quite unstable (in contrast to the ATP-ase activity of submitochondrial particles, which was quite stable) and we made little progress.

One day we discovered that this enzyme was "cold labile": at 0 degrees centigrade it lost all activity in a few hours, but at room temperature it was stable for days. That was a turning point in our investigations; from then on purification was simple. Furthermore, we had a decisive tool for determining the relationship between ATP-ase and coupling factor. The fact that both activities decayed at the same rate at 0 degrees indicated that the same protein was responsible for both.

Chemical fractionation of F_1 gave us a pure enzyme in good yield. In fact, at first the yield seemed to be too good: often our final preparation had more units of ATP-ase than we had estimated were present in the crude preparation. Moreover, the ratio of coupling activity to ATP-ase activity was not constant during purification. An examination of these discrepancies by Pullman revealed that the crude mitochondrial extract contained a protein that inhibited ATP-ase activity but not coupling activity, and that the removal of this inhibitor during purification explained the unexpected increase in total ATP-ase activity.

The purified F_1 had one puzzling property. Whereas Lardy and his collaborators had shown that both oxidative phosphorylation and the ATP-ase activity of mitochondria were very sensitive to oligomycin, our soluble enzyme was completely insensitive. This apparent discrepancy caused some of our colleagues to challenge the significance of our observations with the soluble enzyme. I had heard that the late Oswald T. Avery had once said: "It doesn't matter if you fall down, as long as you pick up something from the floor when you get up." And so we accepted the challenge and embarked on a project to find

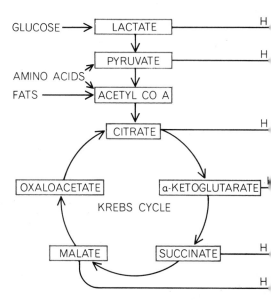

OXIDATIVE PHOSPHORYLATION is the process whereby energy from the oxidation of foodstuffs is harnessed to produce ATP, the energy carrier of the cell. Sugars, fats

out why oligomycin inhibited the enzyme in mitochondrial particles but not the soluble enzyme.

We started with the working hypothesis that there must be a component in mitochondria that confers oligomycin sensitivity on the enzyme. To show this we first had to prepare submitochondrial particles from which all the bound, oligomycin-sensitive ATP-ase had been removed, then add F_1 to them and observe what happened. We were able to eliminate the ATP-ase activity from particles by treating them with urea at 0 degrees, but to our surprise oligomycin-sensitive ATP-ase activity kept reappearing on dilution or aging. It developed that most of the ATP-ase in submitochondrial particles was latent—masked, apparently, by Pullman's inhibitor—and was more resistant to urea than the manifest enzyme was. We had to learn how

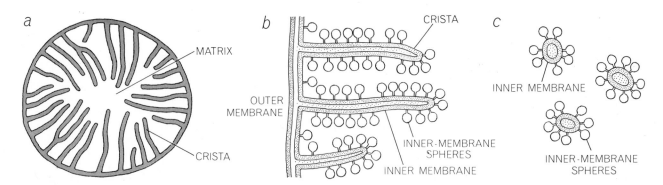

MITOCHONDRION, seen in a schematic cross section (a), has two membranes, each about 60 angstrom units (six millionths of a millimeter) thick. The inner membrane is deeply folded into "cristae" covered with the inner-membrane spheres, each about 85 ang- **stroms in diameter (b). The inner membrane, with its spheres, is the site of oxidative phosphorylation. Mitochondria exposed to sonic oscillation become fragmented into small submitochondrial particles (c), which are still capable of oxidative phosphorylation.**

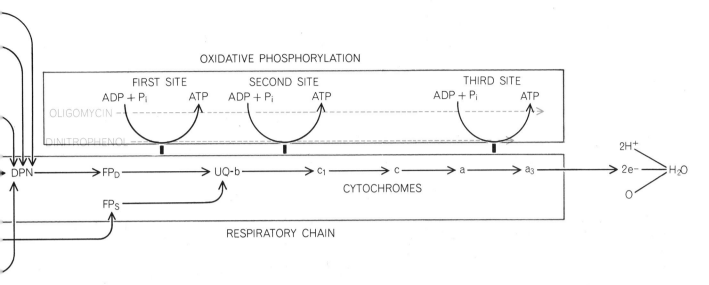

and proteins are partially metabolized and then, in mitochondria, enter the Krebs cycle, in which they are broken down to carbon dioxide. In the process hydrogen atoms are accepted by the coenzyme diphosphopyridine nucleotide (*DPN*). The chain of respiratory enzymes, including flavoproteins (*FP*) and cytochromes b, c_1, c, a and a_3, catalyze a stepwise transfer of electrons to form water. At three sites phosphorylation is "coupled" to electron transfer. It can be uncoupled by dinitrophenol and inhibited by oligomycin.

to unmask this ATP-ase by removing the inhibitor before urea treatment. We found that if submitochondrial particles were first treated with trypsin, a digestive enzyme, and only then with urea, the resulting "*TU* particles" were depleted of virtually all ATP-ase activity [*see middle illustration on next page*]. More recently, when we found that the trypsin was damaging the mitochondrial membrane, Lawrence Horstman of our laboratory discovered that the inhibitor could be removed more gently by passing the submitochondrial particles through a column of Sephadex, a molecular sieve that separates small bodies from large ones. When this procedure is followed by treatment with urea, the resulting "*SU* particles" are analogous to *TU* particles but are much more effective in reconstitution experiments.

When we added F_1 to *TU* or *SU* particles, the enzyme was bound to the particles and the ATP-ase activity became not only sensitive to oligomycin but also stable at 0 degrees. Thus our working hypothesis was confirmed: mitochondria contain a component or components that alter the properties of F_1. We have become increasingly aware that this phenomenon is not unusual. Enzymes bound to membranes almost invariably have some properties that are different from those of the same enzymes in solution. Gottfried Schatz of our laboratory suggested the word "allotopy" (from the Greek for "other" and "position") to designate this phenomenon. We observed, furthermore, that the properties not only of the enzyme but also of the

membrane to which it is attached are changed depending on whether they are separate or bound to one another [*see top illustration on page 46*]. An allotopic property of an enzyme can be used to devise a quantitative assay to serve during the purification of the membrane, since one can test successively purer membrane preparations to see if they are still capable of changing the properties of the added enzyme.

The *TU* particles that conferred oligomycin sensitivity on F_1 still contained the entire electron-transport chain, and we went on, with the allotopic property of F_1 as the tool, in an attempt to further resolve this membrane system. One day I subjected *TU* particles to sonic oscillation without including the usual salt buffer. Centrifugation of the resulting mixture at high speed yielded a soluble extract that conferred oligomycin sensitivity on F_1. We called the factor responsible for this property F_0. The discovery seemed even more exciting when the soluble preparation turned out to contain the entire electron-transport chain and even some residual phosphorylating activity: it appeared that we had actually rendered the entire system soluble. Then the addition of salt solution made the preparation turbid, which meant that particles had formed from the soluble system. In other words, in the presence of salt buffer—which must be added in biological experiments to keep the medium constant—F_0 was still particulate. At the time this was disappointing, but the observation led us

into new investigations of the relation between membrane structure and function.

In collaboration with Donald F. Parsons of the University of Toronto Faculty of Medicine and Britton Chance of the University of Pennsylvania School of Medicine, we examined all our membranous preparations of F_0 by negative staining in the electron microscope. We saw, first, that the submitochondrial particles we had started with were similar to those prepared by earlier investigators: sac-shaped structures outlined by a membrane that was covered with the characteristic "inner-membrane spheres" that had been discovered by Humberto Fernandez-Moran of the University of Chicago. The treatment with trypsin caused little change in structure. Subsequent treatment with urea, however, had a dramatic effect: although it left the membrane intact, it removed the inner-membrane spheres [*see bottom illustration on next page*]. This was unexpected, since David Green had once maintained that these spheres, which he called "elementary particles," represented groups of enzymes of the electron-transport chain [see "The Mitochondrion," by David E. Green; SCIENTIFIC AMERICAN, January, 1964]. We had found, on the contrary, that the *TU* particles (which lacked spheres) contained the entire electron-transport chain!

If the spheres did not contain respiratory enzymes, what did they contain? We calculated that most of the protein removed by urea treatment could be accounted for by the removal of ATP-ase,

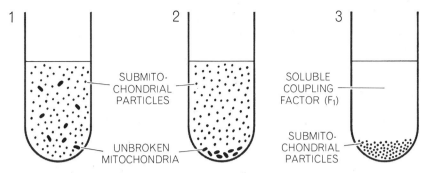

COUPLING FACTOR F_1 is separated by centrifugation. Mitochondria subjected to sonic oscillation (*1*) are centrifuged at low speed to separate particles from intact mitochondria (*2*). Then the particles are spun at high speed. The resulting light fraction (*3*) contains a soluble component (F_1) that is required for ATP production in the course of oxidation.

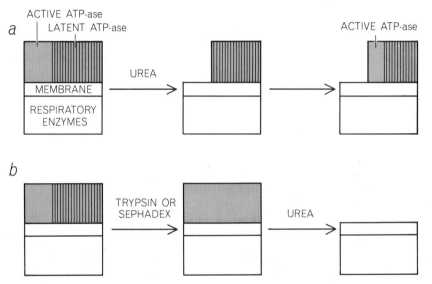

ENZYMATIC ACTIVITY (ATP-ase activity) of soluble F_1 was found resistant to oligomycin, unlike that of the intact membrane. To see if the membrane conferred this sensitivity on soluble F_1, it was first necessary to remove all native F_1 from the membrane. Most of the F_1 is masked by an inhibitor (*bars*), however; treatment with urea removed only exposed F_1, and ATP-ase activity reappeared (*a*). The destruction of inhibitor by trypsin (*b*) exposed the latent F_1 to removal by urea. Later Sephadex was substituted for trypsin.

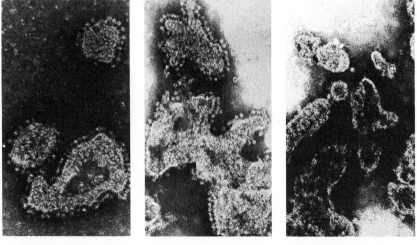

ELECTRON MICROGRAPHS trace the procedure diagrammed in the preceding illustration. The membrane of submitochondrial particles, enlarged about 100,000 diameters, is lined with inner-membrane spheres (*left*). Trypsin has little effect on appearance (*center*). Urea removes the spheres from the particles, leaving "*TU particles*" without spheres (*right*).

and so we suspected that the spheres were identical with F_1. We were encouraged in this belief when a preparation of pure F_1 turned out to have the characteristic appearance of the 85-angstrom-unit inner-membrane spheres [*see micrograph at bottom left on page 43*].

One further experiment was needed to identify F_1 unambiguously with the spheres: the reconstitution of a depleted particle by the addition of F_1, resulting in the restoration of the submitochondrial particle's typical shape and function. This was accomplished only recently, after the development of the *SU* particles. The addition of F_1 to these particles yielded a preparation that was indistinguishable in structure from fully functional submitochondrial particles [*see illustration on page 43*] and confirmed that coupling factor F_1 is identical with the inner-membrane spheres.

The morphological reconstitution was not paralleled by restoration of function, however. In an effort to regain oxidative phosphorylation we added three more coupling factors, F_2, F_3 and F_4—proteins that had been obtained from mitochondria by various extraction procedures. *SU* particles reconstituted with all four coupling factors oxidized succinate, a compound of the Krebs cycle, with a high efficiency of ATP synthesis: for each molecule of oxygen consumed, up to 1.8 molecules of ATP were formed. That is very close to the best value—two molecules—that can be achieved with intact mitochondria.

With these experiments one of the aims of our investigation had been achieved: a resolution of soluble components and a reconstitution of structure and function. Another aim has been to get some insight into the mode of action of the coupling factors. How do they fit into the mechanism of oxidative phosphorylation?

There are currently two views of the general nature of that mechanism. One is a chemically oriented hypothesis originally suggested by E. C. Slater of the University of Amsterdam in 1953, in analogy to the mechanism of ATP formation in glycolysis. It proposes that during electron transport high-energy intermediate compounds ($A{\sim}x$, $B{\sim}x$, $C{\sim}x$) are formed at each coupling site, composed of a member of the respiratory chain (A, B, C) and an unknown (x). These compounds are transformed into a common intermediate by interaction with another unknown (y) to form $x{\sim}y$. This intermediate in turn combines with inorganic phosphate to yield $x{\sim}P$, which

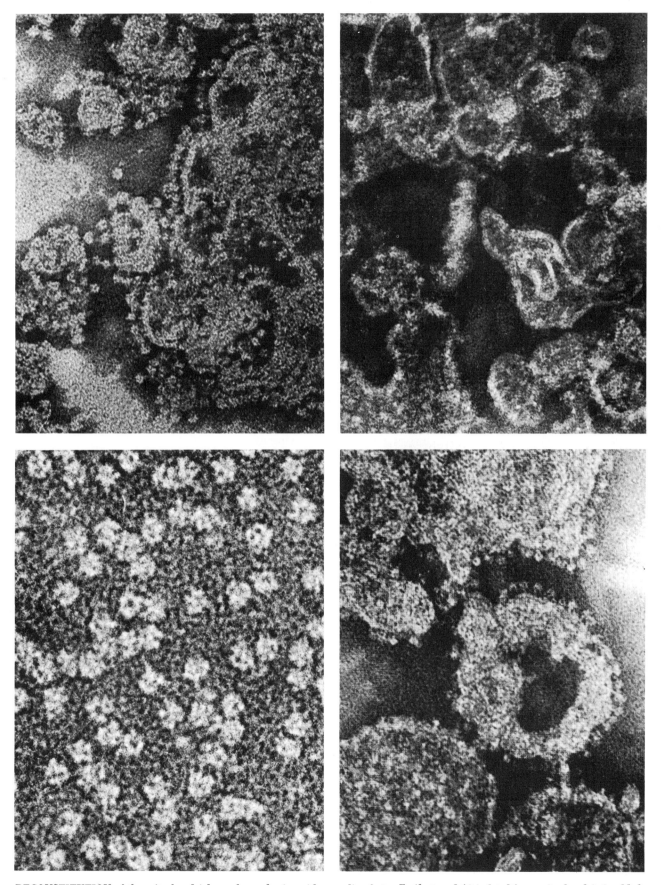

RECONSTITUTION of the mitochondrial membrane begins with submitochondrial particles lined with inner-membrane spheres (*top left*). Treatment with a molecular sieve (Sephadex) and urea produces "SU particles" without spheres (*top right*). When coupling factor F_1 (*bottom left*) isolated from mitochondria is added, the characteristic shape of submitochondrial particles is restored (*bottom right*). F_1 spheres are enlarged about 600,000 diameters, other preparations about 300,000 in these electron micrographs.

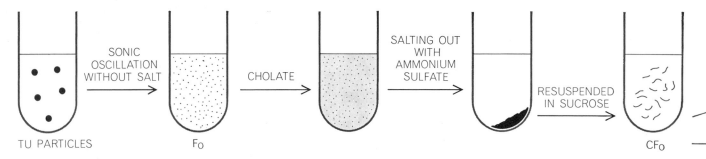

MEMBRANE OF MITOCHONDRION was isolated as shown here. *TU* particles were subjected to sonic oscillation, producing F_O, which had the capacity to bind F_1. When F_O was dissolved in cho-late and fractionated by "salting out," the colorless precipitate CF_O was obtained. It lacked respiratory enzymes and lipids; added to F_1, it inhibited ATP-ase activity. Addition of phospholipid

ultimately transfers its energetic phosphate group to ADP to form ATP.

Recently Peter Mitchell of the Glynn Research Ltd. laboratories in England has challenged this chemical hypothesis with some new and provocative ideas. Instead of a high-energy intermediate compound of the respiratory chain, he proposes that an electrical potential develops during respiration that provides the energy for ATP production: The positively charged hydrogen ions (protons) are moved to one side of the membrane while the negatively charged electrons are channeled to the other side. The separation of charges is utilized by a complex mechanism to give rise to the high-energy intermediate $x\sim y$, which powers the formation of ATP. At the core of this hypothesis is an ATP-ase located in the inner membrane.

In some respects the two hypotheses are not much different: both include a high-energy intermediate, $x\sim y$, to generate ATP from ADP and phosphate. In the Mitchell hypothesis, however, $x\sim y$ is formed by means of an electrical membrane potential. This requires a much higher integrity of the membrane structure than is required by the chemical hypothesis. Indeed, Mitchell considers that uncouplers such as dinitrophenol act by making the membrane "leaky" to protons, thus preventing a separation of charges. It is apparent, therefore, that further studies of the inner membrane are of utmost importance for the evaluation of the two hypotheses.

What is the role of coupling factor according to these two formulations? Mitchell proposes that F_1, together with F_O, represents the reversible ATP-ase

that utilizes the electrical potential to generate ATP. According to the chemical hypothesis, F_1 catalyzes the last step in ATP formation, the "transphosphorylation" from $x\sim P$ to ADP. Indeed, every reaction associated with oxidative phosphorylation that requires ATP can be shown to be dependent on F_1. June Fessenden-Raden in our laboratory has prepared an antibody against F_1 and has found that these ATP-dependent reactions are inhibited by the antibody.

In collaboration with Mrs. Fessenden-Raden, Richard McCarty and Gottfried Schatz, I have recently found that a coupling factor may have, in addition to a catalytic function that is inhibited by its antibody, a second, "structural" function that is not impaired by the antibody. Our first example was the stimulation by

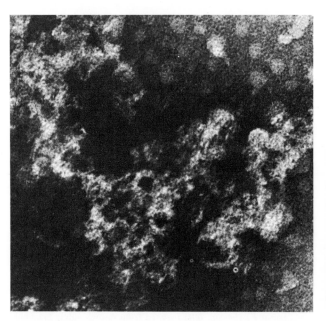 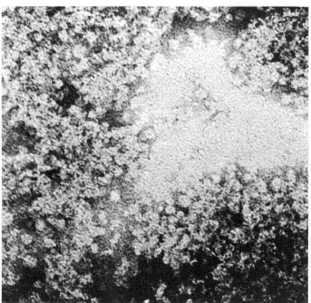

CF_O FRACTION, enlarged about 250,000 diameters in an electron micrograph, appears amorphous (*left*). When F_1 is added to the CF_O preparation, it appears that the F_1 spheres attach themselves to the CF_O, but no distinct structure is seen (*center*). When phospholipids are added, the distinct structure that emerges (*right*) resembles that of submitochondrial particles. In other words, CF_O

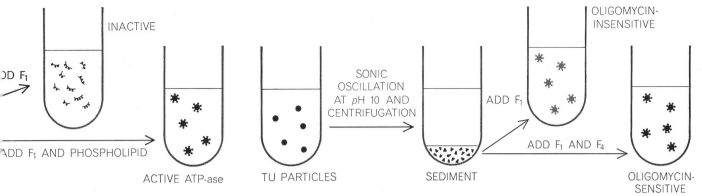

INACTIVE

OLIGOMYCIN-INSENSITIVE

SONIC OSCILLATION AT *p*H 10 AND CENTRIFUGATION

ADD F_1

ADD F_1 AND PHOSPHOLIPID

ADD F_1 AND F_4

OD F_1

ACTIVE ATP-ase   TU PARTICLES   SEDIMENT   OLIGOMYCIN-SENSITIVE

restored ATP-ase activity, which was now oligomycin-sensitive. CF_O and phospholipid may thus comprise the membrane proper.

COUPLING FACTOR F_4 is apparently required for oligomycin sensitivity. When *TU* particles are broken down under alkaline conditions, the sedimented particles cannot confer oligomycin sensitivity on F_1. Addition of F_4 in the presence of salt restores this capacity.

F_1 of a reaction that is catalyzed by mitochondria but does not involve ATP. The second example was the observation that in chloroplasts, the energy-generating particles of plant cells, a coupling factor (chloroplast F_1) is required not only for all reactions that involve ADP or ATP but also for a "proton pump" that is driven by light energy without ATP. In contrast to the ATP-dependent reactions, however, this proton pump was not inhibited by an antibody against the chloroplast coupling factor. The factor therefore appears to contribute to the integrity of the chloroplast membrane, which is required for proton transport.

A third example of the "structural" role of a coupling factor was observed with a preparation of F_1 from yeast mitochondria, which stimulated phosphorylation in beef-heart particles that still contained some residual beef F_1. This stimulation was apparently due to a structural effect of yeast F_1, since it was not inhibited by an antibody against yeast F_1. In beef-heart particles (*SU* particles) that were completely devoid of native F_1, the yeast factor had no effect; apparently it could not fulfill the catalytic functions of native coupling factor.

We are beginning to suspect that the dual role played by F_1 is representative of a common occurrence in the interaction of enzymes and membranes and is an important expression of the allotopy phenomenon. The contribution of F_1 to the integrity of the mitochondrial membrane and the fact that it is required for the operation of the proton pump of the chloroplast have obvious bearing on the Mitchell hypothesis, and may lead to a clarification of the role played by the

membrane in oxidative phosphorylation.

While the work with F_1 was going forward as described above, we therefore also pursued the problem of resolving the inner mitochondrial membrane. F_O had turned out to be a yellow-brown complex of many components including the entire electron-transport chain. By chemical fractionation in the presence of a bile salt, Yasuo Kagawa isolated a virtually colorless fraction (CF_O) with some interesting properties. It lacked respiratory activity, having lost almost all the flavoproteins and cytochromes present in the original submitochondrial particles, and it contained only traces of phospholipids, the fat constituents of the membrane.

When F_1 was added to CF_O, the ATP-ase activity of F_1 was almost completely inhibited. The subsequent addition of

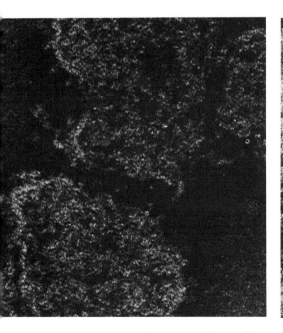

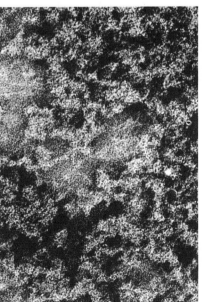

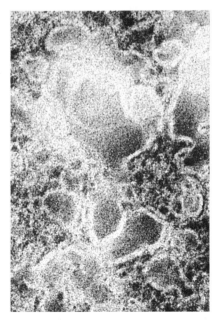

and phospholipid seem to suffice, without respiratory enzymes, to bind F_1 and reconstitute the shape of submitochondrial particles.

F_4 PREPARATION, seen in an electron micrograph at a magnification of 250,000 diameters, appears to be amorphous (*left*). The addition of phospholipid to soluble F_4 yields particles with a sac-shaped structure similar to that of submitochondrial particles (*right*).

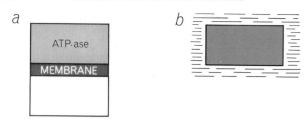

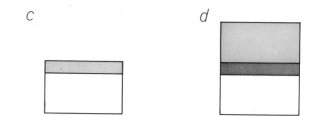

ALLOTOPIC PROPERTIES of purified F_1 and the mitochondrial membrane are indicated. The ATP-ase activity of particles (*a*) was known to be sensitive to oligomycin (*light color*). When soluble F_1 was discovered (*b*), it was found to be resistant to oligomycin (*dark color*), and membrane from which F_1 had been removed (*c*) was sensitive to trypsin (*light gray*). When the enzyme and membrane were bound (*d*), each was changed: the F_1 became sensitive to oligomycin, the membrane resistant to trypsin (*dark gray*).

phospholipid to this inactive complex fully restored ATP-ase activity, which was now sensitive to oligomycin! Equally striking results were observed in the electron microscope. CF_O was amorphous. After the addition of F_1 numerous inner-membrane spheres became attached to CF_O, which remained amorphous. Then, with the addition of phospholipid, the characteristic saclike membranous structures covered with spheres became apparent [*see illustration, page 44*]. They could not be distinguished from functional submitochondrial particles, even though they lacked major components of such particles, the respiratory enzymes. These enzymes had always been assumed to be an integral part of the inner membrane—and now they were found not to be present in what appears to be the isolated membrane. How, then, are the respiratory enzymes associated with the membrane? What are the constituents of the inner membrane itself? How are they organized?

To answer some of these questions we proceeded to disrupt the membrane further to see which constituents were necessary for the interaction with F_1. Several years ago Thomas E. Conover and Richard L. Prairie in our laboratory had separated a coupling factor, F_4, that was necessary for phosphorylation in particles obtained by sonic oscillation of

mitochondria under highly alkaline conditions. When Kagawa exposed *TU* particles to sonic oscillation under alkaline conditions, after high-speed centrifugation he obtained a sediment ("*TUA* particles") that no longer made F_1 sensitive to oligomycin. The addition of F_4 to *TUA* particles restored oligomycin sensitivity to the complex. Recent experiments by Bernard Bulos in our laboratory at Cornell University revealed that F_1 is bound by *TUA* particles in the presence of salt but is nevertheless not inhibited by oligomycin. On addition of small amounts of a highly purified preparation of F_4, he observed a time-dependent restoration of oligomycin sensitivity, suggesting that an enzymatic process may be taking place. This is the first clue to the mode of action of F_4.

Several years ago Richard S. Criddle, Stephen H. Richardson and their collaborators at the University of Wisconsin isolated an insoluble "structural protein" from mitochondria with the help of detergents and solvents. Our crude preparation of F_4 was similar to that protein in its capacity to combine with some flavoproteins and cytochromes of the respiratory chain but, not having been exposed to damaging chemicals, it remained soluble. In the electron microscope it appeared quite amorphous. When phospholipids were added to soluble F_4,

however, a precipitate formed that appeared to be membranous and shaped into sacs [*see illustration on page 45*]. We have therefore proposed that F_4 may have a function as an organizational protein in the mitochondrial membrane—a kind of backbone for the association of the respiratory enzymes and the coupling factors involved in the transformation of oxidative energy into ATP.

With this concept as a working hypothesis we have embarked on what promises to be a long and venturesome journey. Taking the membrane-like complex of F_4 and phospholipid as starting material, we are adding isolated soluble flavoproteins and cytochromes of the respiratory chain step by step, checking at each stage to see if some of the allotopic properties of the respiratory chain are restored. In our laboratory Alessandro Bruni and Satoshi Yamashita have constituted sections of the respiratory chain with the appropriate allotopic properties (such as sensitivity to the respiratory poison antimycin). These experiments have given us confidence that we shall eventually achieve a complete reconstitution of the respiratory chain from soluble components. Then we shall turn to the final task: the reconstitution of the system of oxidative phosphorylation from its individual components.

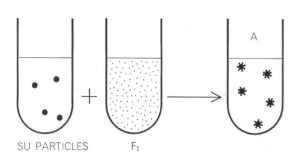

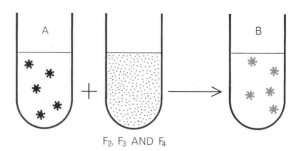

SU PARTICLES, from which coupling factor F_1 had been removed, were reconstituted to the shape of submitochondrial particles (A) by the addition of F_1 (*see micrographs on page 43*). To restore function as well as structure it was necessary also to add F_2, F_3 and F_4. When this was done, the reconstituted particles (B) were capable of generating ATP almost as well as intact mitochondria.

II

MOLECULAR ARCHITECTURE

The Structures and Functional Roles of Proteins

MOLECULAR ARCHITECTURE

Hydrogen, the simplest of all atoms, is by far the most abundant substance in the universe, and so it is not surprising that geochemists are generally agreed that in the primitive atmosphere of the earth the elements carbon, nitrogen, oxygen, and sulphur were present in the highly reduced forms of methane (CH_4), ammonia (NH_3), water (H_2O) and hydrogen sulphide (H_2S); small amounts of carbon dioxide and sulphur dioxide, derived from volcanic gases, probably were also present, but there was essentially no free oxygen. What is surprising is that if mixtures of these gases, simulating the earth's early atmosphere, are exposed to ultraviolet light, heat, or other energy sources, the principal products in aqueous condensates are all biologically important molecules, whereas many other conceivable substances are not formed. Amino acids were produced in the first experiment of this type, carried out at the University of Chicago by Harold Urey and Stanley Miller (described by George Wald in the first article in this collection). Others have subsequently demonstrated in similar experiments the synthesis of fatty acids, carbohydrates, purines, pyrimidines, porphyrin—indeed, essentially all of the small molecular building blocks used in the formation of living cells. It is hard to avoid the conclusion that the occurrence of these molecules in cells was to a large extent predetermined by spontaneous photochemical processes in the earth's early atmosphere and seas.

Discrete molecules much larger or more complex than amino acids, carbohydrates, fatty acids, or nucleotides (a purine or pyrimidine base plus a sugar and a phosphate) are not formed under simulated primitive earth conditions. The route to larger and more complex molecular structures seems to have involved the linking of these simpler molecular units together to form long chain-like "polymers." The unit molecules linked together in polymers are called "monomers." So again it is not surprising that the principal macromolecular constituents of living cells are polymers of molecules capable of being formed photochemically in the early atmosphere and seas. Polymers of amino acids are called "polypeptides"; proteins consist of one or more polypeptide chains. The nucleic acids DNA and RNA are nucleotide polymers (polynucleotides), and the polysaccharides are polymers of simple carbohydrates.

Chemists have recently demonstrated that simple polypeptides and polynucleotides can be formed under abiotic conditions by the condensation of amino acids and nucleotides, respectively. Similar processes could have occurred in the primitive seas, but ultimately an equilibrium between polymer and monomer would have been established, thereby blocking the evolution of more complex biological structures. Further evolutionary advance seems to have involved the dynamic coupling of protein and nucleic acid chemistry in cycles of macromolecular synthesis and degradation. The maintenance of this coupling probably depended on a degree of isolation of the primitive proteins and nucleic acids from their aqueous environment. This could have been achieved through the formation of simple macromolecular complexes involving nucleic acids, proteins, and lipids in which fatty material (complexes of lipid and protein) acted as a kind of primitive membrane. However, one of the fundamental unanswered questions in molecular biology today is how this necessary coupling of protein and nucleic acid chemistry occurred; the answer might also provide an explanation for the origin of the genetic code. Unfortunately, all we can say now is that protein and nucleic acid syntheses are universally coupled in nature today and that plausible biochemical arguments make it unlikely that life could have arisen on the basis of either proteins or nucleic acids alone.

In this section we will be concerned primarily with proteins and nucleic acids, as these polymers are of obvious central importance in the genetic control of cellular activities. However, we shall see, especially in Part III that the walls and membranes that preserve the integrity of cells by separating them from their aqueous environment, and other important features of cell and tissue architecture, are critically dependent on the physical and chemical properties of lipids and polysaccharides.

The protein content of whole tissues varies widely (1.2 percent in lettuce, 37 percent in soybeans), but within cells protein is the most abundant material, constituting more than 50 percent of the cell's dry weight, and they are involved in almost all cellular activities. One major class of proteins, the enzymes, serves as specific catalysts for virtually all biochemical reactions in cells; other important classes are structural proteins, transport proteins, antibodies, hormones, storage proteins, contractile proteins, and toxins. Representatives of the first five of these eight major classes are considered in detail in the articles in this section. Simple proteins are comprised only of amino acids, but others are conjugated with nucleic acids (nucleoproteins), lipids (lipoproteins), carbohydrates (glycoproteins), or other molecules (e.g., heme in hemoglobin). Both the number of amino acids and the number of polypeptide chains in proteins are found to vary over a considerable range. The enzyme ribonuclease is a relatively small protein, consisting of one polypeptide chain of 124 amino acids whereas the enzyme complex "fatty acid synthetase" from yeast contains about 20,000 amino acids divided among twenty-one discrete chains.

One of the more inspired insights of the founders of molecular biology was that from among the 150 distinct amino acids found in various cells and tissues, only twenty should be considered as essential building blocks of proteins. It is now known that these twenty standard amino acids are the only ones coded in the cell's genetic material. The others either do not occur in proteins, or, if they do, they are produced by chemical modification of standard amino acids subsequent to polypeptide synthesis. (For example, the hydroxyproline and hydroxylysine residues in collagen are derived in this way from the two standard amino acids, proline and lysine.)

Amino acids can be thought of as having tripartite structure. Attached to a "central" carbon atom is a carboxyl group, an amino group, and a specific "side group." In polypeptide chains amino acids are linked in "head-to-tail" fashion through the formation of a covalent "peptide" bond between the amino group (head) of the second with the carboxyl group (tail) of the first. Thus, the amino and carboxyl groups, common to all amino acids, are involved in the formation of the polypeptide backbone whereas the various side groups are ultimately responsible for the chemical and biological specificity of proteins.

Amino acid side groups range in complexity from a simple hydrogen atom in glycine to complex aromatic rings as in tryptophan. The nature of the interaction between water and these side groups is an important determinant of the final three-dimensional shape and, hence, of many of the properties of proteins. Eight of the twenty standard amino acids have hydrophobic (or nonpolar) side groups which tend to force the protein chain into three-dimensional configurations such that they are as far removed as possible from neighboring water molecules. The remaining twelve amino acids have hydrophilic (polar) side groups, and, in a complementary way, these tend to force the protein into configurations such that they achieve maximal contact with water. The twelve polar amino acids can in turn be subdivided into three categories, those with uncharged side groups and those whose side groups possess net negative or positive charges at neutral pH. The charge state of the polar side groups can also affect the three-dimensional conformation and chemical properties of proteins.

It is convenient to think of protein structure at four levels: primary, secondary, tertiary, and quaternary. The precise sequence of amino acids in a single polypeptide chain constitutes the primary structure of the molecule. As we shall see in subsequent sections of this book, especially in the article by Charles Yanofsky, information specifying the amino acid sequences of proteins is encoded in the genes forming a kind of "genetic blueprint" that is transmitted from one generation to the next. Secondary structure refers to the various ways in which polypeptide chains become twisted in space to form fibrous structures. Extended polypeptides are relatively flexible molecules and under certain conditions in solution can take up configurations as random as pieces of thread blown in the wind. Such a state is appropriately called an "extended random coil." Under other conditions they can take up stable fibrous or globular configurations, or some combination of these. There are three principal stable configurations associated with fibrous proteins such as those found in hair, silk, or connective tissue. One of these is the alpha helix, first found in the α-keratins of wool and hair in which successive peptide bonds trace out a helical configuration around a central axis. The alpha helix is stabilized by the formation of hydrogen bonds between every fourth peptide bond group in the chain. If the hydrogen bonds are broken and the fibers extended, as occurs after steaming and stretching wool, the polypeptide chain assumes the extended "β-conformation." Fibroin, the protein of silk, consists of antiparallel polypeptide chains in the β-conformation, linked together by *interchain* hydrogen bonds to form the so-called "pleated sheet" structure. On the other hand, the protein collagen, found in cartilage and other connective tissues, consists of three polypeptide chains twisted around one another to form a triple helix. Tertiary structure refers to the further folding of helical (or randomly coiled) chains that is found in the compact globular proteins; and, finally, the specific way in which two or more globular units are arranged in space, to form the so-called oligomeric proteins, is called quaternary structure.

Each level of protein structure depends on the physicochemical properties of the level below, and so ultimately the complex three-dimensional shape of proteins depends on the particular sequence of amino acids in the polypeptide chains that constitute them. Since the specific biological functions of proteins depend on the arrangement in three-dimensional space of their constituent amino acids, it is clear how genes, by controlling amino acid sequences, exert their indirect but nonetheless effective control over so many biological activities. It should also be clear that the technical problems involved in unraveling the full three-dimensional structures of proteins remain among the most formidable in science today.

The features of protein structure summarized briefly above are described in greater detail in the first seven articles in this section. Paul Doty presents the structure of the twenty standard amino acids and of the peptide bond that links them into polypeptide chains. He also describes the three basic types of secondary structure found in fibrous proteins and some of the important physical approaches that have been used in the elucidation of these structures. The protein collagen, described by Jerome Gross in the second article, is the single most abundant protein in the animal kingdom. It is interesting in that it contains two "nonstandard" amino acids that have a significant effect on its structure. In an elegant diagram on p. 00 Gross shows how the structure of a tissue (the cornea of the eye) is built up stepwise from the level of small molecules.

The hormone insulin governs sugar metabolism in the body, and its absence gives rise to diabetes. It is one of the smallest proteins, containing only fifty-one amino acid residues in two polypeptide chains that are linked together by two disulphide bridges. Frederick Sanger of Cambridge University received the Nobel prize for his pioneering work in establishing the complete amino acid sequence of insulin. It was the first time such a feat of chemical virtuosity had been achieved, and as one of Sanger's co-workers, E. O. P. Thompson, predicted in the third article here, the year 1954 has gone down as a landmark in the history of protein chemistry. Sanger had to develop new and sophisticated techniques for his analysis; yet it is interesting to note that the rate of advance in this field has been so great that now the determination of amino sequences is virtually an automated procedure. Today the amino sequences of some hundreds of proteins are known, and this has even made possible quantitative investigations at the molecular level of the rates of evolutionary change among organisms. An excellent example of this latter type of study is provided next by Richard E. Dickerson in his article on the structure and history of an ancient protein. In this article Dickerson considers the changes in the amino acid sequence of the protein cytochrome *c* which plays a crucial role in respiration (oxidative metabolism). The evolutionary changes in cytochrome *c* have been relatively small in comparison with those that have occurred in other proteins. Dickerson argues that this can be attributed to the fact that in this small protein most amino acid substitions would have such a drastic effect on the folding pattern that the molecules function would be destroyed, and, as he says, "no cytochrome, no respiration; no respiration, no life." Here we see in operation at the molecular level two principles well known to evolutionary biologists: First, natural selection is essentially conservative in preserving well-adapted structures; and, second, the fact that selection is seen to act on the folded, functioning protein and not directly on its amino acid squence, demonstrates that even at the molecular level, selection acts on phenotype rather than genotype.

The fine art of determining the amino acid sequences of proteins has been carried to its highest level by Gerald M. Edelman who recently received the Nobel prize for sequencing an antibody, in this case an immunoglobulin produced in excessive quantities by patients with multiple myeloma, a malignant proliferation of antibody-producing cells in the bone marrow. Not only are antibodies about twenty-five times the size of insulin molecules, but they are also heterogeneous in structure. (One of the attractions of insulin for Frederick Sanger was that it is easily available in pure form.) In 1959 Edelman showed that antibodies consist of two types of protein subunits called light and heavy chains because of marked differences in their molecular weights. Both chains contain regions in which the amino acid sequence is constant for all antibodies together with regions in which the sequence varies, thereby conferring the particular antigen-binding specificity for each class of antibodies. Edelman was able to obtain large quantities of a particular myeloma protein by collecting blood plasma from one patient with the disease (each patient produces different myeloma proteins). He also established that the so-called "Bence Jones proteins" secreted in the patient's urine were homogeneous light chains of the antibody made by the tumor but not incorporated into whole molecules. By sequencing the light and heavy chains of the antibody and comparing these sequences with those of the Bence Jones protein and another heavy chain that was sequenced concurrently, Edelman was able to identify the variable and constant regions of the antibody molecule. Although the three-dimensional configuration of the antibody is not known, Edelman's work, here presented, stands as a major landmark in modern immunology.

Because the biochemical function of proteins is directly related to the specific way in which polypeptide chains fold upon themselves in three-dimensional space and because we lack a proper theory capable of predicting such folding patterns on the basis of known amino acid sequences, the tertiary and quaternary structures of proteins must be determined by direct physical measurements. Pure proteins can be crystallized, and because so much water is trapped within crystals, the solid environment does not radically alter the tertiary structure of proteins from that which holds within cells. Thus, it becomes possible to use the techniques of x-ray crystallography to deduce the exact three-dimensional shapes of proteins. This approach to protein structure was pioneered in the famous Cavendish physics laboratory at Cambridge University where John C. Kendrew and Max F. Perutz worked out the structures of myoglobin and hemoglobin, respectively, and for which both have been awarded the Nobel prize. In the article reprinted here, Perutz outlines the principles and problems of x-ray diffraction analysis and indicates how audacious he was, as a young Ph.D. student in 1937, to begin work on hemoglobin, a molecule containing 10,000 atoms of various types; for up to that time the most complex organic molecule whose structure had been determined by x-ray analysis contained only 58 atoms. Hemoglobin is a complex oligomeric protein containing four polypeptide chains, each of which is complexed to a heme group containing the iron atom which carries oxygen from the lungs to the tissues of the animal body. Although it is the nonprotein heme group that is directly involved in binding oxygen, it is important to realize that heme alone, in the absence of the protein globin, will not bind oxygen: The binding characteristics of heme are modified by the chemical environment in which it is located. Perutz again emphasizes how the biological function of proteins depends on their three-dimensional structures and points out the remarkable fact that the enormous hemoglobin molecule actually changes its conformation during the binding and release of a molecule as tiny as oxygen.

The notion that changes in the tertiary structure of proteins are associated with the biochemical tasks they perform recurs in David C. Phillips's article on the enzyme lysozyme, the first for which the precise structure of the active catalytic site was worked out in detail. Lysozyme contains 129 amino acid residues in a single polypeptide chain and so is about the same size as myoglobin or one of the four hemoglobin subunits. However, its tertiary structure is substantially more complex than myoglobin: It has less than one-half the alpha helical content of myglobin, and it also contains a region of antiparallel pleated sheet structure (β-configuration). Phillips observes that folding patterns in which hydrophobic side groups are kept out of contact with the surrounding water allow the protein and solvent molecules to reach a thermodynamically stable state in which free energy is minimized and entropy maximized. He also suggests that folding begins before the synthesis of the polypeptide chain is complete and that the initial folds near the amino terminal of the molecule may serve as a kind of "internal template" or center around which the rest of the chain is folded as synthesis proceeds. But perhaps the most remarkable feature of Phillips's work is his demonstration of the specific three-dimensional relation between enzyme and substrate in a surface cleft of the proton where the active catalytic site is located. It also becomes clear how the amino acid residues that constitute the active site of the enzyme can be some distance apart in the amino acid sequence but are still brought into functional proximity in the folded tertiary structure. The conformation of the protein changes when substrate is bound to it, and this perhaps makes it possible for the large enzyme molecule to exert stress on the substrate as well as producing the most favorable alignment of the chemical groups involved in the subsequent catalytic reaction. All of this is consistent with the "induced fit" theory of enzyme catalysis put forward by Daniel Koshland of the University of California, Berkeley, but Phillips's painstaking x-ray analysis has provided us with the most intimate view of enzyme action so far attained.

The two broad categories of nucleic acids, DNA (deoxyribonucleic acid) and RNA (ribonucleic acid), play central roles in the storage and transfer of genetic information in living cells and viruses. The genetic blueprint of cells is encoded in the genes they contain, and all genes, except for those in certain viruses, are made of DNA. RNA is involved in the biosynthetic processes that utilize this genetic information in the manufacture of proteins. In protein synthesis, a specific sequence of nucleotide bases in DNA is translated into a specific sequence of amino acids in a polypeptide chain.

All the biochemical reactions that are involved in the replication and expression of genetic information are catalysed by specific enzymes, and it is meaningless to ask whether nucleic acids or proteins are more important in cells. The necessary interdependence of nucleic acid and protein synthesis in the economy of the cell reflects the coupling of protein and nucleic acid chemistry, alluded to above, that must have occurred very early in the evolution of living systems. This coupling is not symmetrical in that genetic information flows only from nucleic acids to proteins, and the central dogma of molecular biology asserts that the direction of information flow cannot be reversed in any living cell. Nonetheless, from an evolutionary standpoint, we seem to be left with a "chicken-egg" problem. Is protein (and the cell) nothing more than a nucleic acid's way of reproducing itself, or is the central dogma an accident of evolutionary history in that the device of genetic coding arose as a protein's way of preserving and propagating its amino acid sequence? The latter possibility is not as wild as it sounds since enzymes alone cannot be used to synthesize specific proteins. This is because there would have to be specific "ordering enzymes" to add each new amino acid to the growing protein chain; and since enzymes are themselves proteins, still additional ordering enzymes would be needed to synthesize the first set of enzymes, and so on, in infinite regress. However, as indicated earlier, we do not yet know enough about the chemistry of protein-nucleic acid interactions to really come to grips with the coupling problem. And just as the evolution of any plant or animal species can only be understood in the broader context of the species' environment, so too an understanding of the evolution of genetic replicating systems might require much more knowledge of the early molecular environment and the development of the global ecosystem than we now possess.

Molecular biology is often viewed as the most significant and characteristic expression of biological thought in the twentieth century. It is therefore sobering to consider how deep are its roots in nineteenth-century science. Mendel's first paper, in which he postulated the existence of atomistic genetic factors or genes, was published in 1866, and Friedrich Miescher discovered nucleic acid in 1869, just ten years after the appearance of Darwin's *Origin of Species*. However, Mendel's paper had no impact on biology until the turn of the century when his results and ideas, rediscovered by Carl Correns and Hugo de Vries, could be incorporated into the mainstream of genetic thinking.

In the article on the discovery of DNA, Alfred E. Mirsky points out that Miescher was one of a small group who pioneered the application of physicochemical concepts and techniques to biological problems and at one point in his writing came tantalizingly close to the idea that nucleic acids are involved in genetic processes. However, Miescher later repudiated this notion despite the subsequent histochemical demonstration of the occurrence of nucleic acids in chromosomes and the assertion by such a luminary of the biological establishment as Oskar Hertwig that nucleic acids may be responsible both for fertilization and the transmission of hereditary characteristics. So again we have an example of a revolutionary idea that was ahead of its time, for it was not until 1952 that Alfred Hershey and Martha Chase finally settled the question of the chemical identity of the genetic material. Mirsky deals with the discovery of the existence of nucleic acids, and he carries the story up to the discovery of the role of DNA in the transformation of bacteria. In the concluding essay in this collection Gunther S. Stent argues that one of the reasons for the surprising frequency with which revolutionary ideas are overlooked or eclipsed for many years is that by their very nature they cannot be immediately incorporated into the accepted canon of orthodox scientific belief.

The double-helical structure of DNA was worked out in 1953 by James D. Watson and Francis H. C. Crick. Maurice H. F. Wilkins shared the Nobel Prize with Watson and Crick for providing the x-ray diffraction pattern that was the principal experimental basis from which the model was derived. This feat stands as perhaps the most dramatic single advance in 20th century biology. The structure and synthesis of DNA is presented in the article by Nobel laureate Arthur Kornberg, who first isolated from the bacterium *Escherichia coli* an enzyme, DNA polymerase, that was capable of catalysing the synthesis of high molecular weight DNA in the presence of an appropriate template of natural DNA and the four nucleotides that constitute the monomeric subunits of DNA.

The first synthetic DNA had no biological activity but after fourteen years of painstaking work Kornberg and his colleagues succeeded in replicating in vitro the DNA from a simple bacterial virus that was then infective and capable of directing the synthesis of new virus particles when taken up by host bacteria.

At the time Kornberg's article was written (1968) many believed that the mechanism of DNA replication in living cells must be essentially equivalent to that which had been characterized in the test tube. There were nagging problems, but even the most critical molecular biologists have been startled by the further dramatic development of this subject during the past four years: Two additional DNA polymerases have been discovered in *E. coli*; RNA synthesis has been found to play an important role in DNA synthesis; the original Kornberg enzyme has been shown to participate in DNA repair; and the whole question of the mechanism of DNA replication is again in active flux.

Doubts about the role of DNA polymerase I (as Kornberg's original enzyme is now called) in normal replication arose from several sources, one of which is revealed in the elegant autoradiographs of the bacterial chromosome in the article by John Cairns (p. 00). Here it will be seen that the *E. coli* chromosome consists of one long thread of double-stranded DNA with the ends joined to form a circle. Replication is initiated at a unique origin where the two strands of the parental DNA double helix separate, thereby allowing the two daughter strands to be synthesized using the complementary parental strands as templates (semi-conservative replication). When Cairns wrote this article, it was generally believed that there was a single point of replication that moved continuously in one direction around the chromosome; a "swivel" device at the origin was postulated to permit unwinding of the two parental strands during replication. However, it is now known that replication is normally bidirectional in *E. coli,* and this model is equally consistent with Cairn's autoradiographs: Two points of replication begin at the initiation site and proceed in opposite directions around the chromosome until they meet on the other side. The topology of unwinding still demands that there be swivel points; these can be provided by introducing repairable, single-strand breaks near each replication fork. Unidirectional DNA replication is characteristic of some other biological systems and of the *E. coli* system under specialized conditions (e.g., during the process of bacterial mating or conjugation).

Whether replication is bidirectional or unidirectional in overall topology, it is clear that the two complementary strands must be synthesized almost simultaneously at the growing points. Now single DNA strands have a polarity that arises from the fact that a phosphate group in the backbone of the chain is connected on one side to the 5' carbon atom of the sugar moiety but on the other side to the 3' carbon atom of the sugar (see diagram in Kornberg's article on p. 00). In the double helix the two complementary DNA strands run in opposite directions with respect to this polarity. All of the known DNA polymerases utilize nucleotides with the phosphate groups on the 5' side, and they add these nucleotides to the 3' ends of growing polynucleotide chains. How then can replication proceed on the daughter strand that ends with a 5' phosphate? A way out of this dilemma was provided by Reiji Okazaki in Japan who discovered that replication was discontinuous along at least one of the parental DNA strands. Thus, replication could proceed in short starts opposite to the over-all direction of replication, and these "Okazaki fragments," as they are called, could be later joined to form an intact daughter strand. Such details are of course not visible at the level of resolution possible in autoradiography, but this model has now been confirmed by high-resolution electron microscopy of replicating DNA molecules. Within the past year Okazaki has shown further that these fragments themselves are initiated by a short stretch of RNA, which then serves as a primer (by providing a 3' end) for their synthesis by a DNA polymerase. Thus, RNA synthesis plays a very crucial role in the process of DNA synthesis. Studies in other laboratories have shown that RNA synthesis is also involved in the initiation of the entire replication cycle in *E. coli* and in a number of bacteriophage systems.

E. coli contains several hundred molecules of DNA polymerase I. After a massive experimental search John Cairns discovered a mutant strain that contained only one or two functional molecules of this polymerase per cell. Yet this mutant was viable and quite capable of replicating its DNA. The existence of this mutant enabled other researchers to look more closely for new DNA polymerases in *E. coli* that had been previously overshadowed by the polymerase I activity. Soon Charles Richardson at Harvard and Tom Kornberg with Malcolm Gefter at Columbia reported two new DNA polymerases, II and III. Mutants deficient in DNA polymerase III have been isolated and shown to be unable to replicate DNA; thus, this enzyme appears to be the principal one involved in DNA chain elongation in *E. coli*. Meanwhile, it has been shown that the polymerase I deficient mutants are inefficient in the joining of Okazaki fragments, while also *in vitro* studies indicate that polymerase I is ideally suited to perform this step. The enzyme possesses a $5'$ exonuclease activity that can remove the RNA nucleotides from the initiation region of one Okazaki fragment as it replaces them with DNA nucleotides joined to the $3'$ end of the adjacent fragment or chain. Once the RNA has been removed, the final joining of the newly synthesized DNA fragment to the contiguous DNA strand utilizes another enzyme, polynucleotide ligase, discovered and characterized by Robert Lehman at Stanford and Martin Gellert at the National Institute of Health. (It has long been known that the final product daughter DNA molecules do not contain RNA segments.) Thus, DNA polymerase I does perform an essential role in the normal replication of DNA. As we shall learn in Part IV, it also performs an essential role in the repair of damaged DNA. Finally, it carries out an editing function in which it recognizes and removes mispaired bases from the $3'$ ends of growing daughter chains *before* it adds new nucleotides. Thus, it ensures that the mistake frequency in DNA replication is very low—in *E. coli* this frequency has been estimated at less than one error per 10^{10} nucleotides polymerized during replication. Evidence is accumulating that the various DNA polymerases in *E. coli* overlap somewhat in their capabilities and can substitute for one another in several of the critical functions in replication and repair.

Most classical geneticists viewed the gene as an atomistic, indivisible unit. Even at the time that Beadle and Tatum introduced the "one gene-one enzyme" hypothesis to account for the action of genes in controlling hereditary traits, chromosomes were envisioned as strings of unbreakable beads (the genes). Genetic rearrangement (recombination) was thought to occur primarily by breaking the "thread" between two genes on each of two homologous chromosomes and then rejoining the left-hand piece of one string to the right-hand piece of the other, and vice versa. It was Seymour Benzer who first "split the gene" and showed that recombination could take place at sites *within* individual genes. As described in his article on the fine structure of the gene, Benzer accumulated literally thousands of independent mutants in the rII locus of bacteriophage T4 and then proceeded to make pairwise crosses of these mutant strains. The frequency of recombinant individuals arising from these crosses enabled him to determine the genetic "distance" between the various mutant sites in the rII region and thereby construct a genetic "fine structure map" of the region. Not only did this work show that recombination could occur within genes, but it also demonstrated that recombination separates genetic sites corresponding to contiguous nucleotides in the phage chromosome.

Benzer's work clearly demonstrated that individual nucleotides in DNA constitute the letters of the genetic alphabet. However, if we are to "read" a genetic message, it is necessary both to identify the particular sequence of nucleotides in actual messages and to deduce the manner in which the nucleotide letters are combined into code words. Finally, we must decipher the meaning of the code words in amino acid language.

The determination of a nucleotide sequence is analogous to the problem of determining the amino acid sequence of a protein, but technically it is much more difficult. This very difficult experimental problem was first solved by Robert W. Holley, who received the Nobel prize for working out the complete nucleotide sequence of a transfer RNA molecule. Transfer RNAs serve as the "adaptors" in protein synthesis (see the diagram on p. 174); they have a unique "hairpin" structure and contain a number of nucleotides that are not found in either DNA or other forms of RNA. The significance of the special secondary structure of transfer RNA and the biological role of the unusual bases are problems of great current interest in molecular biology. Transfer RNA is the crucial chemical link between amino acids and coding nucleotides, and so it is possible that further detailed study of these molecules will provide the essential clues needed to understand the evolutionary coupling of protein and nucleic acid chemistry.

The physicist George Gamow was one of the first to introduce an explicit model for genetic coding into biology. Although the details of his original scheme proved to be incorrect, there is no question that Gamow's proposal was one of the most seminal ideas in molecular biology. The basic problem is to deduce how the twenty standard amino acids might be represented in a four-letter nucleotide language. Obviously, there cannot be a one-to-one correspondence between each of the amino acids and either individual nucleotides or pairs of nucleotides (a minimum of twenty words are needed in the genetic language, but pairwise combinations of four nucleotides would yield only sixteen words). On the other hand, four nucleotides can be permuted three at a time to yield sixty-four triplets, more than enough to specify the twenty amino acids. Obviously, more complex codes can be devised, especially if various possible modes of punctuation are taken into account. In his first article in this collection, on the genetic code, Francis H. C. Crick shows how the discovery of a new class of mutants in bacteriophage T4 (reading-frame shift mutants) led him and his colleagues to design an elegant series of mutational experiments from which they were able to deduce the general features of the code. It turns out to be one of the simplest of the many theoretical possibilities: a non-overlapping, unpunctuated language, in which the reading must, as the King of Hearts demanded, "begin at the beginning and go on till you come to the end; then stop." Indeed, if the reading does not begin at the correct starting nucleotide and then move on from one triplet codon to the next, the whole "reading frame" will be out of register so that a protein with a garbled amino acid sequence will result.

The actual deciphering of the triplet code words is described by Crick in the final article in this section. The tabular display of the full code on p. 194 marks one of the greatest achievements in modern biology, comparable to the elucidation of the periodic table of elements in chemistry. Crick describes many of the important general features of the code, including its redundancy (only two amino acids, methionine and tryptophan, are represented by only a single triplet), the indeterminacy or "wobble" associated with the third letter of many triplets, and the existence of three "nonsense codons" which do not specify any amino acid but rather serve as "termination codons" that stop the reading process. UAAUAGUGA!

5

PROTEINS

PAUL DOTY

September 1957

The principal substance of living cells, these giant molecules have identical backbones. Each is adapted to its specific task by a unique combination of side groups, size, folding and shape

Thousands of different proteins go into the make-up of a living cell. They perform thousands of different acts in the exact sequence that causes the cell to live. How the proteins manage this exquisitely subtle and enormously involved process will defy our understanding for a long time to come. But in recent years we have begun to make a closer acquaintance with proteins themselves. We know they are giant molecules of great size, complexity and diversity. Each appears to be designed with high specificity for its particular task. We are encouraged by all that we are learning to seek the explanation of the function of proteins in a clearer picture of their structure. For much of this new understanding we are indebted to our experience with the considerably simpler giant molecules synthesized by man. High-polymer chemistry is now coming forward with answers to some of the pressing questions of biology.

Proteins, like synthetic high polymers, are chains of repeating units. The units are peptide groups, made up of the monomers called amino acids [*see diagram below*]. There are more than 20 different amino acids. Each has a distinguishing cluster of atoms as a side group [*see next two pages*], but all amino acids have a certain identical group. The link-

ing of these groups forms the repeating peptide units in a "polypeptide" chain. Proteins are polypeptides of elaborate and very specific construction. Each kind of protein has a unique number and sequence of side groups which give it a particular size and chemical identity. Proteins seem to have a further distinction that sets them apart from other high polymers. The long chain of each protein is apparently folded in a unique configuration which it seems to maintain so long as it evidences biological activity.

We do not yet have a complete picture of the structure of any single protein. The entire sequence of amino acids has been worked out for insulin [*see the article "The Insulin Molecule," by E. O. P. Thompson, beginning on page 76*]; the determination of several more is nearing completion. But to locate each group and each atom in the configuration set up by the folded chain is intrinsically a more difficult task; it has resisted the Herculean labors of a generation of X-ray crystallographers and their collaborators. In the early 1930s W. T. Astbury of the University of Leeds succeeded in demonstrating that two X-ray diffraction patterns, which he called alpha and beta, were consistently associated with certain fibers, and he identified a third with collagen, the pro-

tein of skin, tendons and other structural tissues of the body. The beta pattern, found in the fibroin of silk, was soon shown to arise from bundles of nearly straight polypeptide chains held tightly to one another by hydrogen bonds. Nylon and some other synthetic fibers give a similar diffraction pattern. The alpha pattern resisted decoding until 1951, when Linus Pauling and R. B. Corey of the California Institute of Technology advanced the notion, since confirmed by further X-ray diffraction studies, that it is created by the twisting of the chain into a helix. Because it is set up so naturally by the hydrogen bonds available in the backbone of a polypeptide chain [*see top diagram on page 62*], the alpha helix was deduced to be a major structural element in the configuration of most proteins. More recently, in 1954, the Indian X-ray crystallographer G. N. Ramachandran showed that the collagen pattern comes from three polypeptide helixes twisted around one another. The resolution of these master plans was theoretically and esthetically gratifying, especially since the nucleic acids, the substance of genetic chemistry, were concurrently shown to have the structure of a double helix. For all their apparent general validity, however, the master plans did not give us the complete configuration in three dimensions

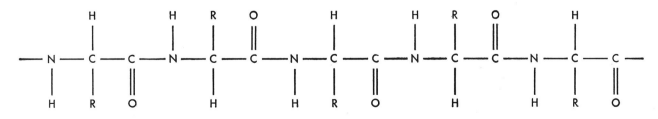

POLYPEPTIDE CHAIN is a repeating structure made up of identical peptide groups (CCONHC). The chain is formed by amino acids, each of which contributes an identical group to the backbone plus a distinguishing radical (R) as a side group.

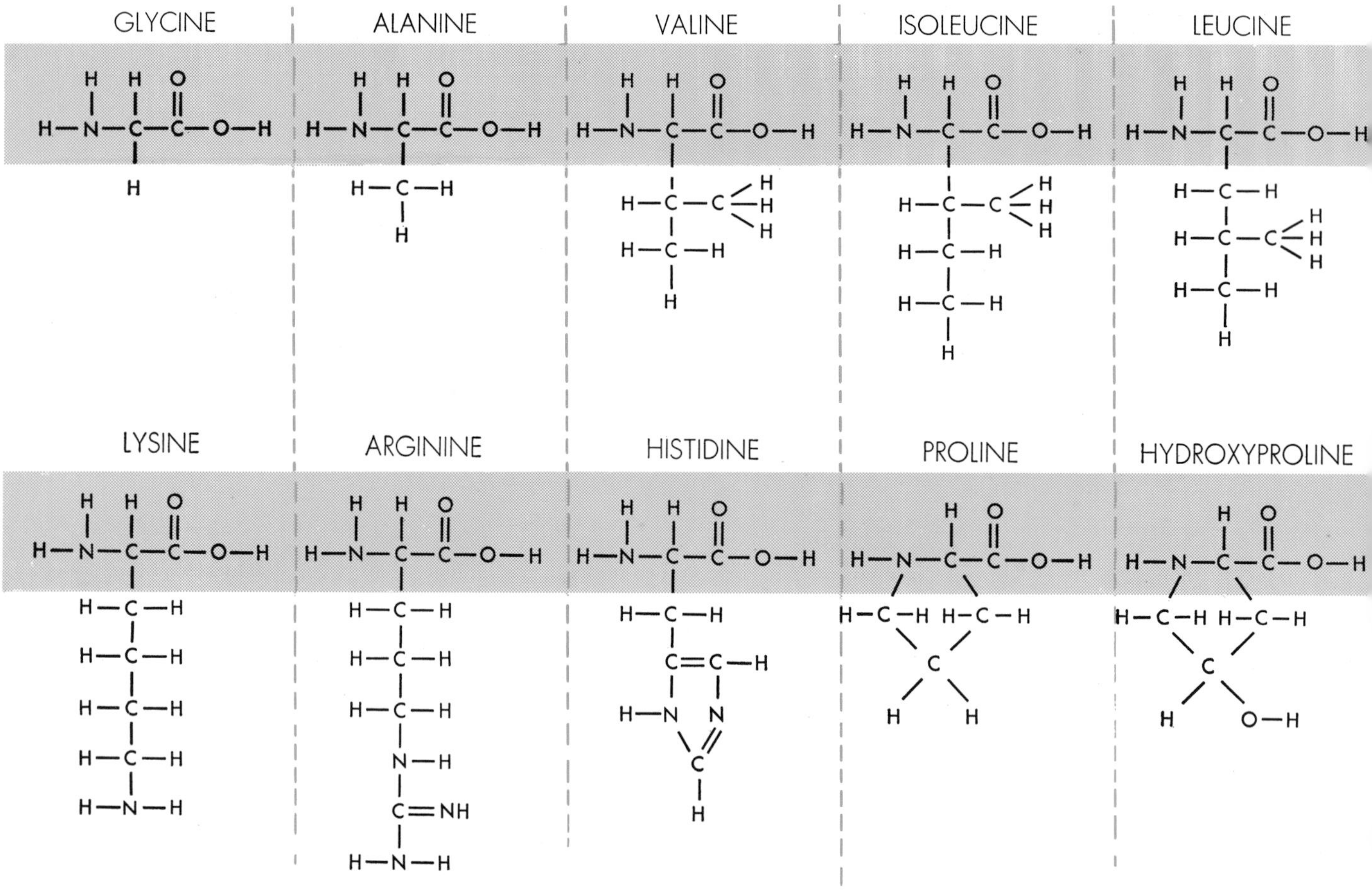

GLYCINE   ALANINE   VALINE   ISOLEUCINE   LEUCINE

LYSINE   ARGININE   HISTIDINE   PROLINE   HYDROXYPROLINE

AMINO ACIDS, the 20 commonest of which are shown in this chart, have identical atomic groups (*in colored bands*) which react to form polypeptide chains. They are distinguished by their unique side groups. In forming a chain, the amino group (NH₂) of one

of any single protein.

The X-ray diffraction work left a number of other questions up in the air. Since the alpha helix had been observed only in a few fibers, there was no solid experimental evidence for its existence elsewhere. There was even a suspicion that it could occur only in fibers, where it provides an economical way to pack polypeptides together in crystalline structures. Many proteins, especially chemically active ones such as the enzymes and antibodies, are globular, not linear like those involved in fibers and structural tissues. In the watery solutions which are the natural habitat of most proteins, it could be argued, the affinity of water molecules for hydrogen bonds would disrupt the alpha helix and reduce the chain to a random coil. These doubts and suppositions have prompted investigations by our group at Harvard University in collaboration with E. R. Blout of the Children's Cancer Research Foundation in Boston.

In these investigations we have employed synthetic polypeptides as laboratory models for the more complex and sensitive proteins. When Blout and co-workers had learned to polymerize them to sufficient length—100 to 1,000 amino acid units—we proceeded to observe their behavior in solution.

Almost at once we made the gratifying discovery that our synthetic polypeptides could keep their helical coils wound up in solutions. Moreover, we found that we could unwind the helix of some polypeptides by adjusting the acidity of our solutions. Finally, to complete the picture, we discovered that we could reverse the process and make the polypeptides wind up again from random coils into helixes.

The transition from the helix to the random coil occurs within a narrow range as the acidity is reduced; the hydrogen bonds, being equivalent, tend to let go all at once. It is not unlike the melting of an ice crystal, which takes place in a narrow temperature range. The reason is the same, for the ice crystal is held together by hydrogen bonds. To complete the analogy, the transition from the helix to the random coil can also be induced by heat. This is a true melting process, for the helix is a one-dimensional crystal which freezes the otherwise flexible chain into a rodlet.

From these experiments we conclude that polypeptides in solution have two natural configurations and make a reversible transition from one to the other, depending upon conditions. Polypeptides in the solid state appear to prefer the alpha helix, though this is subject to the presence of solvents, especially water. When the helix breaks down here, the transition is to the beta configuration, the hydrogen bonds now linking adjacent chains. Recently Blout and Henri Lenormant have found that fibers of polylysine can be made to undergo the alpha-beta transition reversibly by mere alteration of humidity. It is tempting to speculate that a reversible alpha-beta transition may underlie the process of muscle contraction and other types of

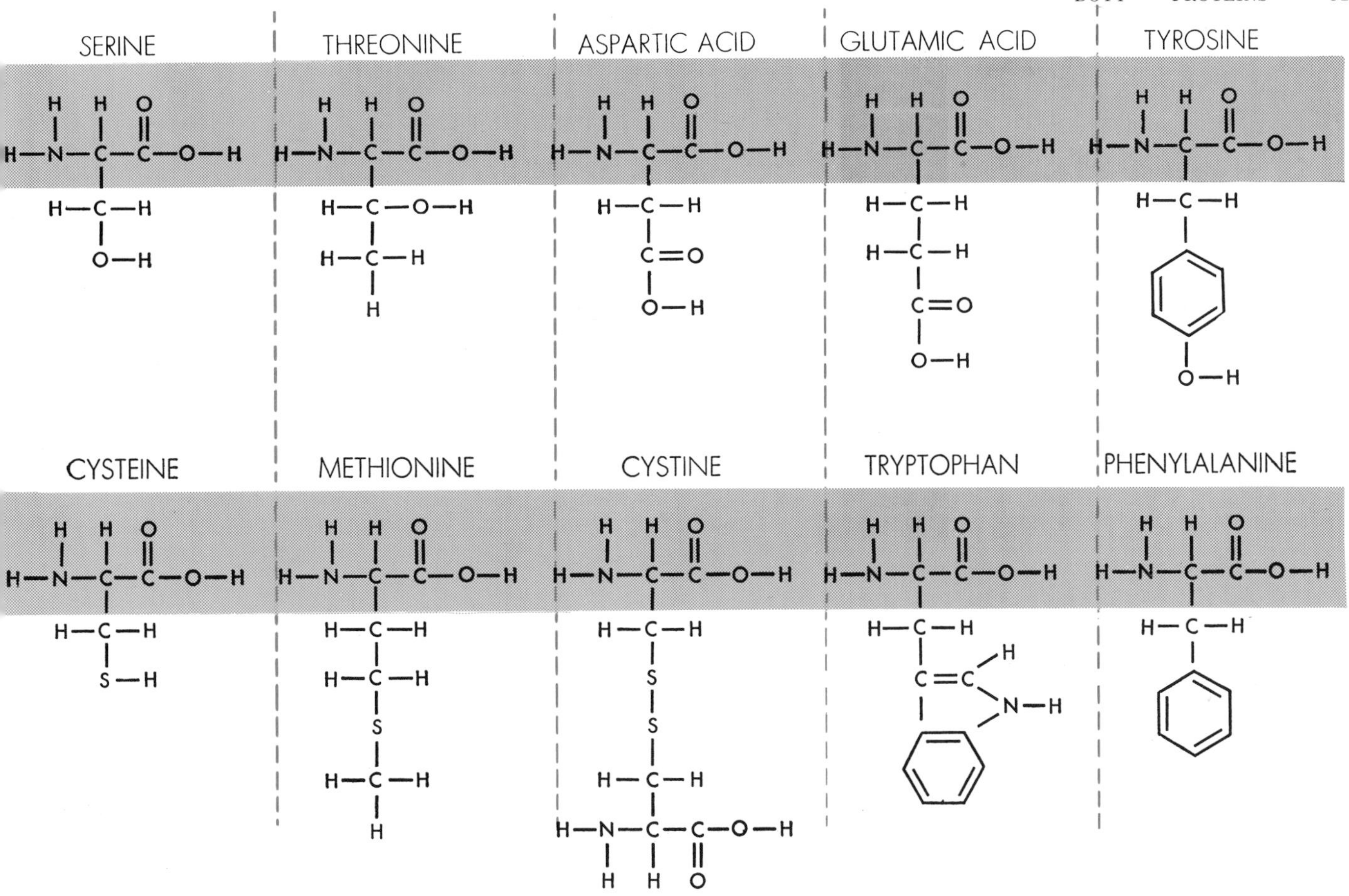

SERINE   THREONINE   ASPARTIC ACID   GLUTAMIC ACID   TYROSINE

CYSTEINE   METHIONINE   CYSTINE   TRYPTOPHAN   PHENYLALANINE

molecule reacts with the hydroxyl group (OH) of another. This reaction splits one of the amino hydrogens off with the hydroxyl group to form a molecule of water. The nitrogen of the first group then forms the peptide bond with the carbon of the second.

movement in living things.

Having learned to handle the polypeptides in solution we turned our attention to proteins. Two questions had to be answered first: Could we find the alpha helix in proteins in solution, and could we induce it to make the reversible transition to the random coil and back again? If the answer was yes in each case, then we could go on to a third and more interesting question: Could we show experimentally that biological activity depends upon configuration? On this question, our biologically neutral synthetic polypeptides could give no hint.

For the detection of the alpha helix in proteins the techniques which had worked so well on polypeptides were impotent. The polypeptides were either all helix or all random coil and the rodlets of the first could easily be distinguished from the globular forms of the second by use of the light-scattering technique. But we did not expect to find that any of the proteins we were going to investigate were 100 per cent helical in configura-

tion. The helix is invariably disrupted by the presence of one of two types of amino acid units. Proline lacks the hydrogen atom that forms the crucial hydrogen bond; the side groups form a distorting linkage to the chain instead. Cystine is really a double unit, and forms more or less distorting cross-links between chains. These units play an important part in the intricate coiling and folding of the polypeptide chains in globular proteins. But even in globular proteins, we thought, some lengths of the chains might prove to be helical. There was nothing, however, in the over-all shape of a globular protein to tell us whether it had more or less helix in its structure or none at all. We had to find a way to look inside the protein.

One possible way to do this was suggested by the fact that intact, biologically active proteins and denatured proteins give different readings when observed for an effect called optical rotation. In general, the molecules that exhibit this effect are asymmetrical in

atomic structure. The side groups give rise to such asymmetry in amino acids and polypeptide chains; they may be attached in either a "left-handed" or a "right-handed" manner. Optical rotation provides a way to distinguish one from the other. When a solution of amino acids is interposed in a beam of polarized light, it will rotate the plane of polarization either to the right or to the left [see diagrams at top of page 64]. Though amino acids may exist in both forms, only left-handed units, thanks to some accident in the chemical phase of evolution, are found in proteins. We used only the left-handed forms, of course, in the synthesis of our polypeptide chains.

Now what about the change in optical rotation that occurs when a protein is denatured? We knew that native protein rotates the plane of the light 30 to 60 degrees to the left, denatured protein 100 degrees or more to the left. If there was some helical structure in the protein, we surmised, this shift in rotation might be induced by the disappearance of the helical structure in the denaturation process. There was reason to believe that the helix, which has to be either left-handed or right-handed, would have optical activity. Further, although it appeared possible for the helix to be wound either way, there were grounds for assuming that nature had chosen to make all of its helixes one way or the other. If

it had not, the left-handed and right-handed helixes would mutually cancel out their respective optical rotations. The change in the optical rotation of proteins with denaturation would then have some other explanation entirely, and we would have to invent another way to look for helixes.

To test our surmise we measured the optical rotation of the synthetic poly-

peptides. In the random coil state the polypeptides made an excellent fit with the denatured proteins, rotating the light 100 degrees to the left. The rotations in both cases clearly arose from the same cause: the asymmetry of the amino acid units. In the alpha helix configuration the polypeptides showed almost no rotation or none at all. It was evident that the presence of the alpha helix caused a

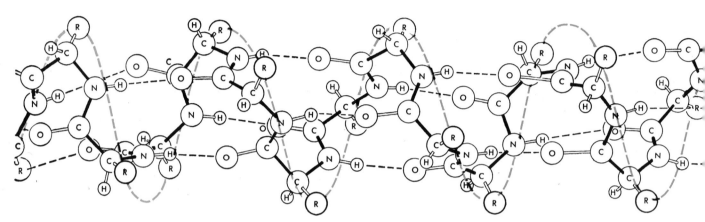

ALPHA HELIX gives a polypeptide chain a linear structure shown here in three-dimensional perspective. The atoms in the repeating unit (CCONHC) lie in a plane; the change in angle between one unit and the next occurs at the carbon to which the side group

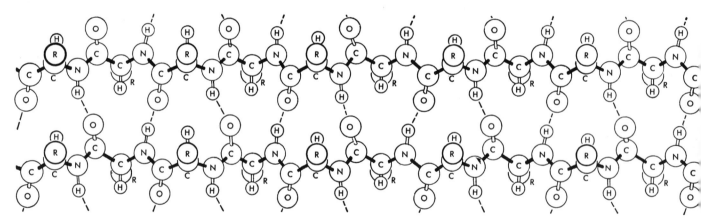

BETA CONFIGURATION ties two or more polypeptide chains to one another in crystalline structures. Here the hydrogen bonds do not contribute to the internal organization of the chain, as in the alpha helix, but link the hydrogen atoms of one chain to the oxygen

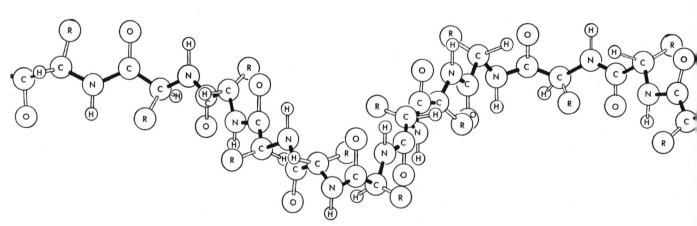

RANDOM CHAIN is the configuration assumed by the polypeptide molecule in solution, when hydrogen bonds are not formed. The flat configuration of the repeating unit remains, but the chain rotates about the carbon atoms to which the side groups are at-counter-rotation to the right which nearly canceled out the leftward rotation of the amino acid units. The native proteins also had shown evidence of such counter-rotation to the right. The alpha configuration did not completely cancel the leftward rotation of the amino acid units, but this was consistent with the expectation that the protein structures would be helical only in part. The experiment thus strongly indicated the presence of the alpha helix in the structure of globular proteins in solution. It also, incidentally, seemed to settle the question of nature's choice of symmetry in the alpha helix: it must be right-handed.

When so much hangs on the findings of one set of experiments, it is well to double check them by observations of another kind. We are indebted to William Moffitt, a theoretical chemist at Harvard, for conceiving of the experiment that provided the necessary confirmation. It is based upon another aspect of the optical rotation effect. For a given substance, rotation varies with the wavelength of the light; the rotations of most substances vary in the same way. Moffitt predicted that the presence of

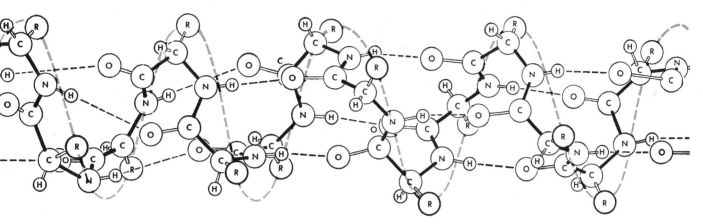

(R) is attached. The helix is held rigid by the hydrogen bond (*broken black lines*) between the hydrogen attached to the nitrogen in one group and the oxygen attached to a carbon three groups along the chain. The colored line traces the turns of the helix.

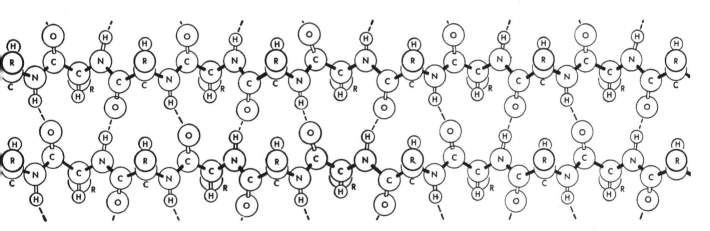

atoms in the adjoining chain. The beta configuration is found in silk and a few other fibers. It is also thought that polypeptide chains in muscle and other contractile fibers may make reversible transitions from alpha helix to beta configuration when in action.

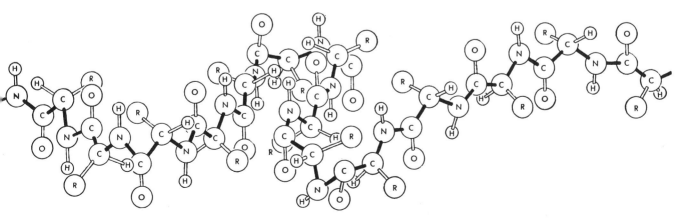

tached. The random chain may be formed from an alpha helix when hydrogen bonds are disrupted in solution. A polypeptide chain may make a reversible transition from alpha helix to random chain, depending upon the acid-base balance of the solution.

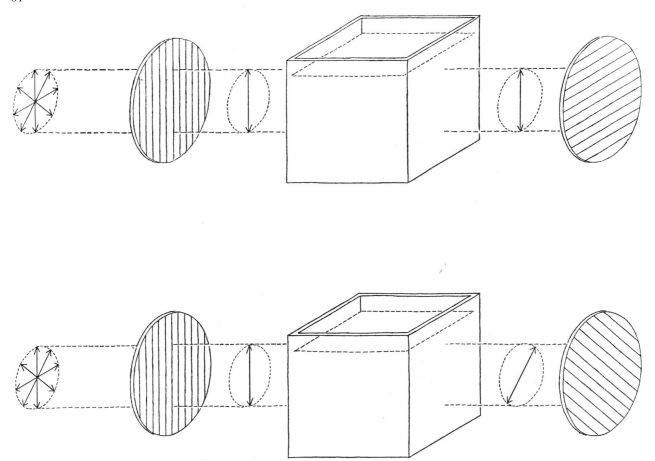

OPTICAL ROTATION is induced in a beam of polarized light by molecules having certain types of structural asymmetry. At top a beam of light is polarized in the vertical plane and transmitted unchanged through a neutral solution. At bottom asymmetrical molecules in the solution cause the beam to rotate from the vertical plane. The degree of rotation may be determined by turning the second polarizing filter (*right*) to the point at which it cuts off the beam. The alpha helix in a molecule causes such rotation.

the alpha helix in a substance would cause its rotation to vary in a different way. His prediction was sustained by observation: randomly coiled polypeptides showed a normal variation while the helical showed abnormal. Denatured and native proteins showed the same contrast. With the two sets of experiments in such good agreement, we could conclude with confidence that the alpha helix has a significant place in the structure of globular proteins. Those amino acid units that are not involved in helical configurations are weakly bonded to each other or to water molecules, probably in a unique but not regular or periodic fashion. Like synthetic high-polymers, proteins are partly crystalline and partly amorphous in structure.

The optical rotation experiments also provided a scale for estimating the helical content of protein. The measurements indicate that, in neutral solutions, the helical structure applies to 15 per cent of the amino acid units in ribonuclease, 50 per cent of the units in serum albumin and 85 per cent in tropomyosin. With the addition of denaturing agents to the solution, the helical content in each case can be reduced to zero. In

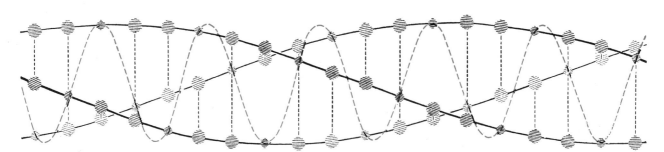

COLLAGEN MOLECULE is a triple helix. The colored broken line indicates hydrogen bonds between glycine units. The black broken lines indicate hydrogen bonds which link hydroxyproline units and give greater stability to collagens in which they are found.

some proteins the transition is abrupt, as it is in the synthetic polypeptides. On the other hand, by the use of certain solvents we have been able to increase the helical content of some proteins—in the case of ribonuclease from 15 to 70 per cent. As in the polypeptides, the transition from helix to random coil is reversible. The percentage of helical structure in proteins is thus clearly a variable. In their natural environment, it appears, the percentage at any given time represents the equilibrium between the inherent stability of the helix and the tendency of water to break it down.

In a number of enzymes we have been able to show that biological activity falls off and increases with helical content. Denaturation is now clearly identified with breakdown of configuration, certainly insofar as it involves the integrity of the alpha helix. This is not surprising. It is known that catalysts in general must have rigid geometrical configurations. The catalytic activity of an enzyme may well require that its structure meet similar specifications. If this is so, the rigidity that the alpha helix imposes on the otherwise flexible polypeptide chain must play a decisive part in establishing the biological activity of an enzyme. It seems also that adjustability of the stiffness of structure in larger or smaller regions of the polypeptide chain may modify the activity of proteins in response to their environment. Among other things, it could account for the versatility of the gamma globulins; without any apparent change in their amino acid make-up, they are able somehow to adapt themselves as antibodies to a succession of different infectious agents.

The next step toward a complete anatomy of the protein molecule is to determine which amino acid units are in the helical and which in the nonhelical regions. Beyond that we shall want to know which units are near one another as the result of folding and cross-linking, and a myriad of other details which will supply the hues and colorings appropriate to a portrait of an entity as intricate as protein. Many such details will undoubtedly be supplied by experiments that relate change in structure to change in function, like those described here.

In the course of our experiments with proteins in solution we have also looked into the triple-strand structure of collagen. That structure had not yet been resolved when we began our work, so we did not know how well it was designed for the function it serves in structural tissues. Collagen makes up one third of the proteins in the body and 5 per cent of its total weight; it occurs as tiny fibers or fibrils with bonds that repeat at intervals of about 700 Angstroms. It had been known for a long time that these fibrils could be dissolved in mild solvents such as acetic acid and then reconstituted, by simple precipitation, into their original form with their bandings restored. This remarkable capacity naturally suggested that the behavior of collagen in solution was a subject worth exploring.

Starting from the groundwork of other investigators, Helga Boedtker and I were able to demonstrate that the collagen molecule is an extremely long and thin rodlet, the most asymmetric molecule yet isolated. A lead pencil of comparable proportions would be a yard long. When a solution of collagen is just slightly warmed, these rodlets are irreversibly broken down. The solution will gel, but the product is gelatin, as is well known to French chefs and commercial producers of gelatin. The reason the dissolution cannot be reversed was made clear when we found that the molecules in the warmed-up solution had a weight about one third that of collagen. It appeared that the big molecule of collagen had broken down into three polypeptide chains.

At about the same time Ramachandran proposed the three-strand helix as the collagen structure. Not long afterward F. H. C. Crick and Alexander Rich at the University of Cambridge and Pauline M. Cowan and her collaborators at King's College, London, worked out the structure in detail. It consists of three polypeptide chains, each incorporating three different amino acid units—proline, hydroxyproline and glycine. The key to the design is the occurrence of glycine, the smallest amino acid unit, at every third position on each chain. This makes it possible for the bulky proline or hydroxyproline groups to fit into the links of the triple strand, two of these nesting in each link with the smaller glycine unit [see diagram on page 64].

One question, however, was left open in the original model. Hydroxyproline has surplus hydrogen bonds, which, the model showed, might be employed to reinforce the molecule itself or to tie it more firmly to neighboring molecules in a fibril. Independent evidence seemed to favor the second possibility. Collagen in the skin is irreversibly broken down in a first degree burn, for example, at a temperature of about 145 degrees Fahrenheit. This is about 60 degrees higher than the dissolution temperature of the collagen molecule in solution. The obvious inference was that hydroxyproline lends its additional bonding power to the tissue structure. Moreover, tissues with a high hydroxyproline content withstand higher temperatures than those with lower; the skin of codfish, with a low hydroxyproline content, shrivels up at about 100 degrees. Tomio Nishihara in

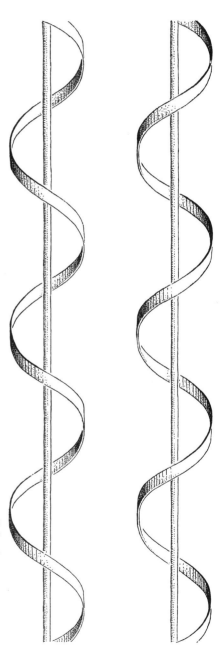

ASYMMETRY of a helix is either left-handed (*left*) or right-handed. Helix in proteins appears to be exclusively right-handed.

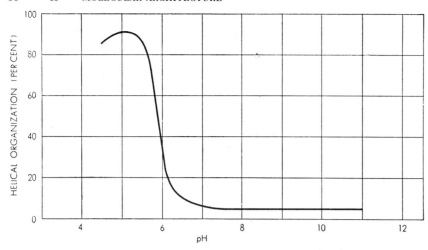

ALPHA HELIX BREAKDOWN is induced in solutions of some polypeptides when the pH (acidity or alkalinity) reaches a critical value at which hydrogen bonds are disrupted.

our laboratory has compared the breakdown temperatures of collagen molecules and tissues from various species and found that the tissue temperature is uniformly about 60 degrees higher. Thus we must conclude that the extra stability conferred by hydroxyproline goes directly to the molecule and not to the fibril.

The structure of collagen demonstrates three levels in the adaptation of polypeptide chains to fit the requirements of function. First there are the chains as found in gelatin, with their three amino acids lined up in just the right sequence. These randomly coiled and quite soluble molecules are transformed into relatively insoluble, girderlike building units when united into sets of three by hydrogen bonds. The subtly fashioned collagen molecules are still too fragile to withstand body temperatures. When arranged side by side, however, they form a crystalline structure which resists comparatively high temperatures and has fiber-like qualities with the vast range of strengths and textures required in the different types of tissues that are made of collagen.

The story of collagen, like that of other proteins, is still far from complete. But it now seems that it will rank among the first proteins whose molecular structure has been clearly discerned and related in detail to the functions it serves.

6

COLLAGEN

JEROME GROSS
May 1961

The main constituent of connective tissue, it accounts for a third of the protein in the human body. Its nature has been clarified by dissolving it and allowing its molecules to reassemble into fibers

Collagen is perhaps the most abundant protein in the animal kingdom. It is the major fibrous constituent of skin, tendon, ligament, cartilage and bone. Its properties are diverse and remarkable. In tendon it has a tensile strength equal to that of light steel wire; in the cornea it is as transparent as water. It accounts for the toughness of leather, the tenacity of glue and the viscousness of gelatin. It also underlies the development of crippling deformities associated with the rheumatic diseases. and with a number of congenital defects of the skeleton, blood vessels and other connective tissue.

By piecing together information derived from X-ray diffraction, chemical analysis, electron microscopy and many other techniques, it is now possible to present a reasonable account of the way collagen fibers are built up from long-chain molecules. Particularly instructive was the discovery that tiny collagen fibers, or fibrils, can be dissociated into their constituent molecules and then reaggregated, outside the living organism, into their original form. This disassembly-reassembly technique, which was perhaps first clearly demonstrated with collagen, has since been successfully applied to other giant molecules and to even more complex biological systems [see "Tissues from Dissociated Cells," by A. A. Moscona; SCIENTIFIC AMERICAN, May, 1959]. The procedure of taking biological materials apart and putting them together again provides much more insight into dynamic mechanisms than does the traditional biochemical approach of breaking things to bits and analyzing the pieces.

Collagen appeared very early in evolution, at least as far back as the coelenterates (a phylum including jellyfishes and sea anemones) and sponges,

and it seems to have changed very little in structure and composition since then. It usually appears as bundles of individual, nonbranching fibrils, varying greatly in diameter from tissue to tissue. In skin, under an ordinary light microscope, these bundles appear to be woven together at random, but a definite order emerges if larger areas of tissue are examined. In tendon, collagen fibers are arranged in long parallel bundles. In the cornea of the eye, transparency depends upon the orderly arrangement of collagen fibrils that probably have a refractive index identical to that of the substance in which they are embedded. (The same type of order appears in the translucent skin of the developing amphibian embryo, but after metamorphosis the skin acquires a somewhat random intermeshing structure.) In bone, the collagen fibrils are organized much like the struts and girders of a bridge; the mineralization of bone follows the detailed fine structure of the fibrils. In cartilage, which coats the inner surface of joints and which must have considerable elasticity and smoothness, the collagen fibrils are usually very thin, randomly oriented and embedded in a large volume of extracellular matrix.

Collagen is synthesized primarily by cells called fibroblasts. The basic collagen molecule is a group of three polypeptide chains each composed of about a thousand amino acid units linked together. In ordinary proteins the chains are assembled from the standard assortment of 22 amino acids, and once linked together end to end the acids are not further altered. In the synthesis of collagen, however, two unusual amino acids, hydroxyproline and hydroxylysine, seem to be formed *after* the molecular chain has been assembled; the new amino acids are created by addition of

hydroxyl (OH) groups to some of the proline and lysine units in the chain. This alteration of the primary molecular structure has not been observed in the synthesis of other proteins.

Proline and hydroxyproline, which together make up as much as 25 per cent of the links in the collagen molecule, prevent easy rotation of the regions in which they are located, thus imparting rigidity and stability to the collagen molecule. The higher the content of proline and hydroxyproline, the higher the resistance of the molecule to heat or chemical denaturation. Preliminary studies suggest that collagens have another distinctive feature: in long stretches of the molecular chain every fourth position seems to be occupied by glycine, which is followed immediately by proline or hydroxyproline. In any case, all collagens studied so far, regardless of their source, contain about 30 per cent glycine, with a variation of less than 5 per cent.

Collagen owes its properties not only to its chemical composition but also to the physical arrangement of its individual molecules. The basic molecular chain is twisted into a left-handed helix, and three such helices are wrapped around each other to form a right-handed superhelix [see illustration on page 69]. The three chains appear to be held together by hydrogen bonds established between the oxygen atoms, located where amino acids are joined by peptide linkages in one chain, and the nitrogen atoms, located at peptide linkages in an adjacent chain. This picture of the structure was first proposed in 1954 by the Indian workers G. N. Ramachandran and G. Kartha, and later refined by two British groups: Alexander Rich and F. H. C. Crick in Cambridge and Pauline

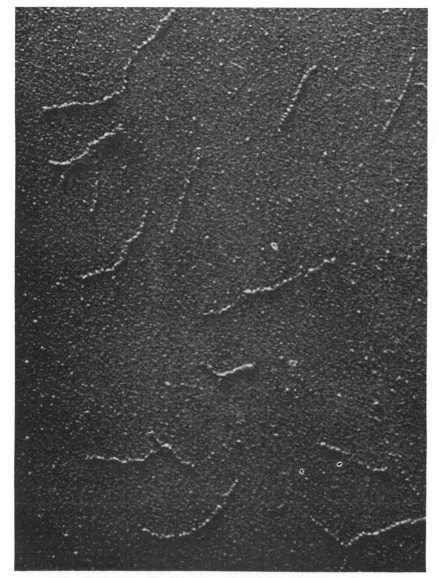

COLLAGEN MOLECULES, dissolved from the collagen of the fish swim bladder, are enlarged 140,000 diameters. They are 2,800 to 2,900 A. long and 14 to 15 A. wide. This electron micrograph was made by Cecil E. Hall of the Massachusetts Institute of Technology.

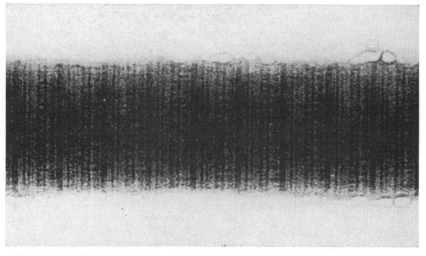

COLLAGEN FIBRIL is built up of collagen (or tropocollagen) molecules that overlap as shown on the opposite page. The intricate fine structure repeats about every 700 A.; the magnification is 250,000 diameters. Alan J. Hodge made electron micrograph at M.I.T.

M. Cowan and S. McGavin in London. The superhelix varies in cross section and electric charge along its length. There is convincing evidence from recent electron micrographs that these variations, or "bumps," are irregularly spaced. Alan J. Hodge and Francis O. Schmitt of the Massachusetts Institute of Technology have suggested that the molecule also has a short flexible "tail" at each end that participates in fibril formation.

For a long time it was believed that the collagen molecule was about 700 angstroms long (an angstrom is one ten-billionth of a meter). This length was inferred from early electron micrographs of collagen fibrils, which showed a series of regular bands with such a spacing. It was assumed that the bands marked the places where the molecules were joined end to end. As so often happens in science, things turned out to be more complex and more interesting than they had seemed at first.

The clue to the currently accepted value for length came from experimentally reconstituted collagen. It has been known since at least 1872 that if collagen fibers are dissolved in acid, reconstituted fibers will automatically appear when the acid is neutralized. The experiment remained little more than a curiosity until 1942, when it was repeated by Schmitt and his associates. It was this group's electron micrographs of natural collagen that had first shown the 700-angstrom periodicity. They were pleased but not greatly surprised to find that the same periodicity appeared in the reconstituted fibers [see top illustration on page 74]. The molecular length of 700 angstroms seemed to be confirmed.

Several years later V. N. Orekhovich and his associates in the U.S.S.R. found needle-like entities when they observed samples of reconstituted collagen under a light microscope. They believed that these entities, which they called procollagen, were newly formed collagen molecules capable of linking up to form the native fibril. Upon learning of this work Schmitt and his co-workers were naturally curious to examine the new material in the electron microscope. When Schmitt, John H. Highberger and I followed the Orekhovich method, we discovered a new type of reconstituted collagen fibril different in structure from any seen before. This new fibril showed bands spaced at about 2,800 angstroms, and a fine structure that surprised us by being symmetrical [see middle illustration on page 74]. We designated

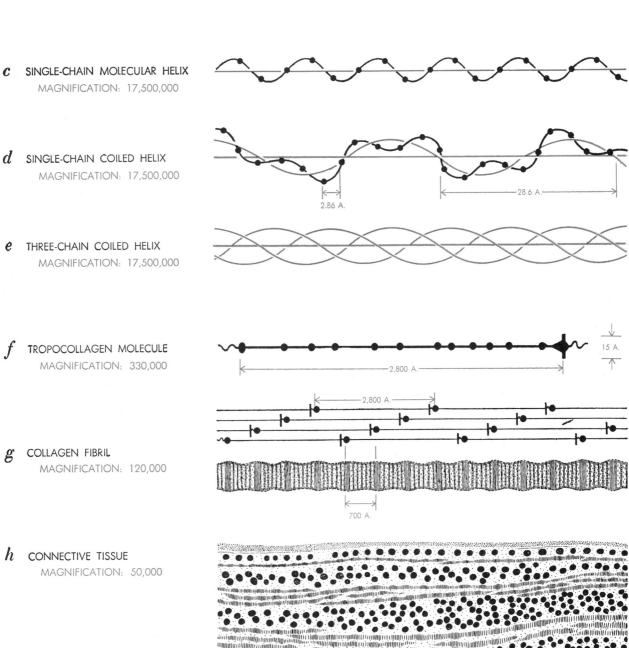

a   FREE AMINO ACIDS

GLYCINE   PROLINE   X

b   MOLECULAR CHAIN

—X—GLY—PRO—HYPRO—GLY—X—

c   SINGLE-CHAIN MOLECULAR HELIX
     MAGNIFICATION: 17,500,000

d   SINGLE-CHAIN COILED HELIX
     MAGNIFICATION: 17,500,000

2.86 A.   28.6 A.

e   THREE-CHAIN COILED HELIX
     MAGNIFICATION: 17,500,000

f   TROPOCOLLAGEN MOLECULE
     MAGNIFICATION: 330,000

2,800 A.   15 A.

g   COLLAGEN FIBRIL
     MAGNIFICATION: 120,000

2,800 A.   700 A.

h   CONNECTIVE TISSUE
     MAGNIFICATION: 50,000

FORMATION OF COLLAGEN can be visualized in seven steps. The starting materials (*a*) are amino acids; the letter "R" in amino acid X represents any of some 20 different side chains. "Hypro" stands for hydroxyproline, created from proline after the molecular chain (*b*) has been formed. The chain twists itself into a left-handed helix; three chains then intertwine to form a right-handed superhelix, which is the tropocollagen molecule. Many molecules line up in staggered fashion (*g*), overlapping by one-quarter of their length, to form a fibril. Fibrils in tissue (*h*) are often stacked in layers with fibrils aligned at right angles.

COLLAGEN FIBRILS, carefully pulled away from human skin, show bands spaced about 700 angstrom units (A.) apart. It was first believed that this represented the length of the underlying collagen molecule; actually the length is about four times greater (*see illustration on page 69*). This electron micrograph by the author is reproduced at a magnification of 42,000 diameters.

this new type of structure "fibrous long spacing," or FLS, simply to indicate that it was fibrous and had a long period. The FLS material can be produced in almost 100 per cent yield by adding a negatively charged large molecule (such as the alpha-one acid glycoprotein derived from blood serum) to a dilute acetic acid solution of purified collagen.

We soon discovered that collagen could be recrystallized into still a third type of structure, which lacked the characteristic fibrous or beltlike appearance of the two other forms. The new material looked superficially like isolated segments of FLS, but closer examination revealed that the numerous crossbands were asymmetrically spaced [*see bottom illustration on page 74*]. This structure is composed of threadlike units running perpendicularly to the bands. We called the arrangement "segment long spacing," or SLS, to indicate a nonfibrous material with a long period. SLS can be obtained by adding adenosine triphosphate (ATP) to acid solutions of collagen.

Any of the three structurally different forms can be dissolved and converted into either of the other two forms. We concluded that all three forms were being created from threadlike units about 2,800 angstroms long and less than 50 angstroms wide. Since this was evidently the fundamental unit of collagen structure, we named it "tropocollagen" (from the Greek meaning "turning into collagen"). The tropocollagen molecule, as we now know, is composed of the three helical chains.

A more refined estimate of the dimensions of tropocollagen was subsequently made by Helga Boedtker and Paul Doty after they had studied collagen in solution. They estimate that the molecule is 2,900 angstroms long by 14 angstroms wide. Cecil E. Hall of M.I.T. has confirmed these dimensions by electron microscopy of individual molecules [*see top illustration on page 68*]. On the basis of these and other studies we can postulate how the three basic forms of collagen are assembled from the tropocollagen molecule.

In the native collagen fibril the molecules are lined up facing in the same direction and overlapping by about one-quarter of their length [*see illustration at bottom of next two pages*]. It is this overlapping that creates the periodicity of about 700 angstroms.

In the FLS form the molecules again lie side by side, but they are not all facing in the same direction and they do not overlap. Since there is no overlapping, the major periodicity measures

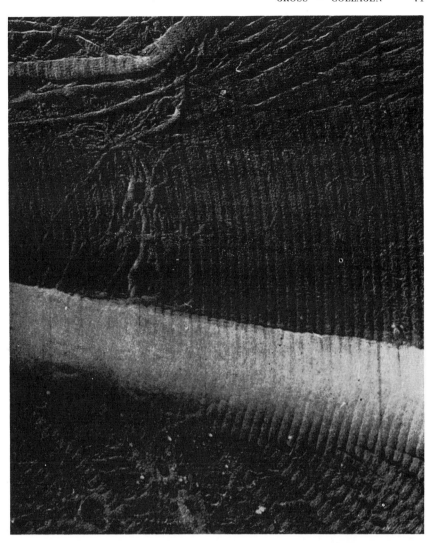

SHEET OF COLLAGEN precipitated from solution shows that tropocollagen molecules will lock themselves into an orderly two-dimensional array. Investigators are seeking conditions that will produce order in three-dimensions. Magnification is 40,000 diameters.

about 2,800 angstroms. The random positioning of "heads" and "tails" of the adjacent molecules accounts for the symmetrical fine structure, even though the discontinuities, or "bumps," along the length of the individual molecule are asymmetrically spaced. This can be understood by imagining two identical but nonsymmetrical molecules lying side by side and facing in opposite directions. If one were to take a blurred photograph, the two molecules would seem to blend into one and the two sets of irregularities would merge so that they would appear to be symmetrically spaced.

In the SLS form the molecules are also nonoverlapping but they all face the same way. As a result the discontinuities of the individual molecules are revealed in the electron microscope as an asymmetrical pattern of bands. Actually it is this pattern in electron micrographs that leads one to infer that the basic molecule contains irregularly spaced discontinuities.

Since these in vitro experiments employ acids in concentrations considered "unphysiological," they left us puzzled as to how fiber formation actually occurs in an animal. We began to gain insight into the natural process about eight years ago when we discovered that a certain fraction of collagen from young animals could be dissolved in cold neutral salt solution, and that simply by warming such solutions to body temperature we could make the dissolved collagen molecules polymerize spontaneously to form a typical cross-striated fibril with the native periodicity. The more rapidly the animal grows, the larger is the amount of collagen that can be extracted. If growth ceases as a result of starvation for a period as short as two days, this collagen fraction disappears from tissues. British investigators have demonstrated that the collagen extractable in cold salt solution is newly synthesized by the cells; it is still soluble because it is not yet tightly aggregated.

As the collagen becomes older, dilute acids are required to extract it; upon further aging it becomes insoluble even in acids.

A similar aging process can be demonstrated in vitro with collagen extracted in cold neutral salt solution. When the solution is warmed to body temperature, a gel composed of typical banded fibrils forms; if it is quickly cooled again, most of the fibrils redissolve. If, however, the gel is allowed to stand at body temperature for 24 hours, it no longer dissolves upon cooling. And if it is aged at body temperature for two weeks, it becomes completely insoluble in dilute acids as well. Since the collagen is highly purified, we do not believe that this time-dependent aging results from interaction with enzymes or other substances. The aging is probably a function of the highly specific structure of the molecule alone. It is likely that if two parallel collagen molecules overlap by a quarter of their length, the charge distribution and three-dimensional configuration of adjacent sections are complementary and therefore attract each other; the "bump" of one fits into the "groove" of the other. The decreasing solubility with time can be explained by the increasing perfection of fit between the molecules as they gradually pack together in a "lock and key" type of association along their length. The "glue" binding the molecules ever more tightly is fundamentally a secondary bond created by electric forces, which rise sharply in strength as surfaces are brought closer and closer together.

These studies suggest a reasonable picture for the first steps in the process by which the body produces collagenous or connective tissue. The fibroblast evidently synthesizes complete collagen molecules (the three-stranded tropocollagen) and extrudes them into the space outside the cells, where they polymerize into fibrils. Although polymerization may require nothing more than time and body heat, other substances in the extracellular environment may play a regulatory role.

We are having difficulty explaining what happens next. How do collagen fibrils become organized into the highly ordered patterns that can be seen in skin, bone and cornea? The cornea, in particular, has a plywood-like structure in which successive layers of fibrils are laid down at right angles to each other in near-crystalline array [see illustrations on page 75]. In studying sections of cornea, Marie A. Jakus of the Retina Foundation in Boston has observed that the fibrils of one layer lie along dark cross-striations in the fibrils below, which are oriented at right angles.

There is still no generally accepted mechanism to explain this intricate ordering. I am inclined to think that the tropocollagen molecules, after extrusion from the fibroblast, form a suspension of liquid crystals (loose semicrystalline aggregates). From such a suspension fibrils could be expected to condense in an orderly fashion. It has been shown that such suspensions of certain rod-shaped giant molecules will turn spontaneously into aggregates that display order in three dimensions. While we have not yet been able to duplicate this experiment with collagen, we have been able to precipitate collagen in broad sheets with a periodicity extending laterally over many square microns [see illustration on page 71]. A significant aspect of this pattern is the appearance of two degrees of order that are mutually perpendicular. Whereas the molecules themselves are laid down in parallel rows, other rows having like electric charge (visible as bands in the collagen fine structure) run at right angles to the molecular rows. One can imagine that with a suitable adjustment of conditions another layer of collagen molecules might precipitate on the first and follow not the molecular rows but the rows of electric charge.

There are a number of questions to answer: If collagen molecules are being secreted at a constant rate from the cells, why should they not form fibrils all

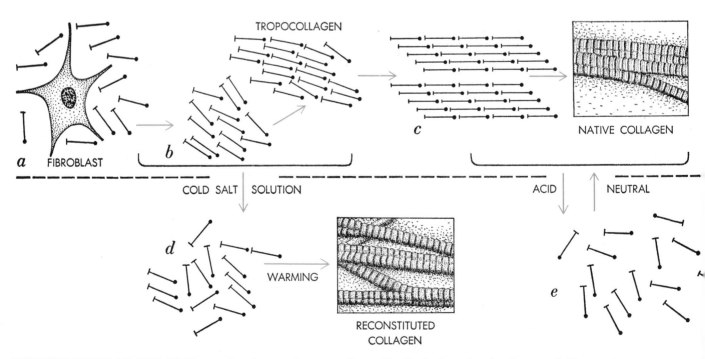

RECONSTITUTION OF COLLAGEN can take place in three basic ways. In an animal, tropocollagen molecules (b), manufactured by cells called fibroblasts (a), overlap to form native collagen (c). Newly formed molecules are soluble in cold salt solution (d); simple warming yields reconstituted fibrils duplicating the native form. Alternatively, native collagen can be disoriented in the same direction? And why should there be variations in the number of fibrils stacked one above the other within different layers? These puzzles might be explained if the secretion of collagen molecules from the cells were discontinuous in time, or "pulsed." In different sites at different times the duration of the pulse and the total amount of collagen secreted may differ. The thickness of the collagen layer would be determined by the duration of the "secretion pulse" and the amount of collagen secreted. The interval between pulses would allow time for fibrils to polymerize and would prevent intermingling of collagen in one layer with the next. The collagen molecules laid down during the single secretion pulse would be oriented in one direction only. Molecules secreted in the next pulse might then be oriented perpendicularly to the molecular rows of the preceding layer, being laid down along the rows of electric charge created by the periodicities lying in register.

We would also like to account for the remarkable uniformity in diameter of fibril bundles. Although the diameter varies from one type of tissue to another, it is often constant for each type. Again the idea of discontinuous secretion of collagen molecules seems useful. As a first step we can imagine that a single group of molecules is secreted at about the same time and that these molecules are rather evenly distributed as a loose liquid crystal in a particular layer of extracellular space. The next step calls for the appearance of nuclei of some sort around which the molecules can begin to condense. It seems reasonable that nuclei would appear at about the same time throughout a given volume of space. In the last step the fibrils grow in

BLOCK OF COLLAGEN (*right*) forms when a cold neutral salt solution of collagen (*left*) is warmed to body temperature for 10 minutes. Gel redissolves if cooled promptly.

size until the original supply of collagen molecules is exhausted. We need only specify that the fibrils grow at the same rate to explain how they all end up having about the same diameter. It is possible that the timing and the rate at which fibrils form and grow are determined by noncollagenous substances in the matrix; the viscosity and charge distribution of this matrix might control the freedom of movement of the individual collagen molecules. Attempts at explanation such as these prove nothing in themselves. Their principal value is that they suggest experiments that may lead to better understanding.

One incentive for studying collagen

so intensively is that it provides a valuable model for investigating the way in which the body assembles complex, reproducible structures from simple molecular building blocks. Another incentive is the medical one. In its growth and development the organism is continuously remodeling its tissues, a process of exquisite precision with regard to place, time and degree. It is possible that disturbances in the precise sequences of remodeling are responsible, at least in part, for some congenital malformations, for the crippling end results of rheumatic diseases and perhaps even for some of the changes in aging. Whether or not collagen itself is

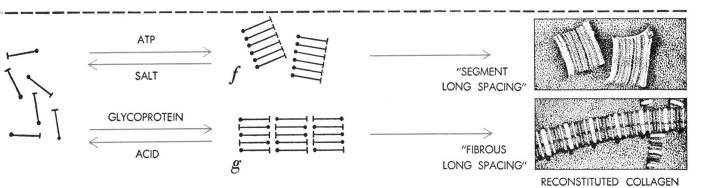

RECONSTITUTED COLLAGEN

solved in acetic acid (*e*). Treating the resulting solution with adenosine triphosphoric acid produces the nonoverlapping, segment-long-spacing form of collagen (*f*). Treating the solution with glycoprotein produces the fibrous-long-spacing form (*g*), in which molecules face randomly in addition to not overlapping. The fine structure reflects the asymmetry of the tropocollagen molecule.

RECONSTITUTED COLLAGEN fibrils form spontaneously when an acid solution of native collagen is neutralized. The reconstituted fibrils duplicate the native form. The three electron micrographs on this page are all at the same magnification: 70,000 diameters.

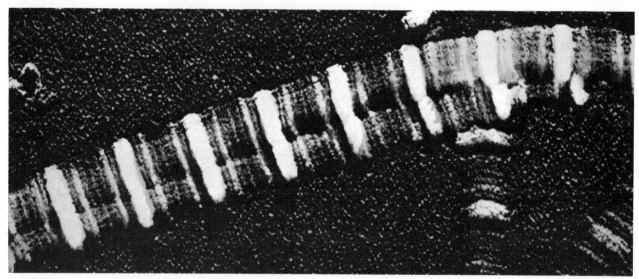

"FIBROUS LONG SPACING" form of collagen is produced by adding glycoprotein to an acid solution of native collagen. The chief feature of this form is its symmetrical intraperiod fine structure. The spacing of the period is about 2,800 angstroms.

"SEGMENT LONG SPACING" form of collagen is produced by adding adenosine triphosphoric acid to an acid solution of collagen. The fine structure is asymmetrical, reflecting the underlying asymmetry of the tropocollagen molecule. The molecular arrangements that produce these three forms of collagen are depicted in the illustration at the bottom of pages 72 and 73.

a target for the causative agent in diseases affecting connective tissue, such as rheumatoid arthritis, is still a matter of dispute. There is little doubt that severe crippling deformities of bones and joints, and the scarring of the heart, kidneys, blood vessels, lungs and other organs, are a manifestation of excessive production and aberrant arrangement of collagen in the affected tissues.

It is characteristic of medicine that frontal attacks on problems such as these seldom yield quick results. There is every reason to hope, however, that studying collagen at the fundamental level of molecular structure will lead ultimately to knowledge that can be applied in medicine.

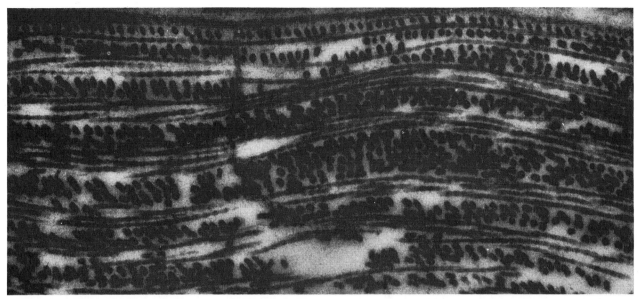

TADPOLE CORNEA consists of layers of collagen fibrils; the fibrils in one layer are at right angles to those in the next. The magnification is 56,000 diameters. This electron micrograph and the one below were made by Marie A. Jakus of the Retina Foundation.

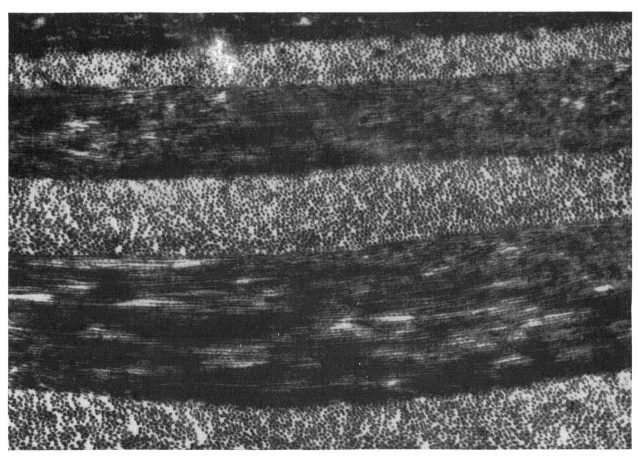

FISH CORNEA surpasses that of the tadpole in precision. In some unexplained manner a random collection of collagen molecules is converted into fibrils that are stacked neatly into layers lying at right angles to each other. The magnification is 28,000 diameters.

7

THE INSULIN MOLECULE

E. O. P. THOMPSON
May 1955

In 1954, after a full decade of intensive work at Cambridge University, Frederick Sanger and his colleagues for the first time totally described the chemical structure of a protein

Proteins, the keystone of life, are the most complex substances known to man, and their chemistry is one of the great challenges in modern science. For more than a century chemists and biochemists have labored to try to learn their composition and solve their labyrinthine structure [see "Proteins," by Joseph S. Fruton; SCIENTIFIC AMERICAN Offprint 10]. In the history of protein chemistry the year 1954 will go down as a landmark, for that year a group of investigators finally succeeded in achieving the first complete description of the structure of a protein molecule. The protein is insulin, the pancreatic hormone which governs sugar metabolism in the body.

Having learned the architecture of the insulin molecule, biochemists can now go on to attempt to synthesize it and to investigate the secret of the chemical activity of this vital hormone, so important in the treatment of diabetes. Furthermore, the success with insulin has paved the way toward unraveling the structure of other proteins with the same techniques, and work on some of them has already begun.

The insulin achievement was due largely to the efforts of the English biochemist Frederick Sanger and a small group of workers at Cambridge University. Sanger had spent 10 years of intensive study on this single molecule. When he commenced his investigation of protein structure in 1944, he chose insulin for several reasons. Firstly, it was one of the very few proteins available in reasonably pure form. Secondly, chemists had worked out a good estimate of its atomic composition (its relative numbers of carbon, hydrogen, nitrogen, oxygen and sulfur atoms). Thirdly, it appeared that the key to insulin's activity as a hormone lay in its

structure, for it contained no special components that might explain its specific behavior.

Insulin is one of the smallest proteins. Yet its formula is sufficiently formidable. The molecule of beef insulin (from cattle) is made up of 777 atoms, in the proportions 254 carbon, 377 hydrogen, 65 nitrogen, 75 oxygen and 6 sulfur. Certain general features of the organization of a protein molecule have been known for a long time, thanks to the pioneering work of the German chemist Emil Fischer and others. The atoms form building units called amino acids, which in turn are strung together in long chains to compose the molecule. Of the 24 amino acids, 17 are present in insulin. The total number of amino acid units in the molecule is 51.

Sanger's task was not only to discover the over-all chain configuration of the insulin molecule but also to learn the sequence of all the amino acids in the chains. The sequence is crucial: a change in the order of amino acids

changes the nature of the protein. The number of possible arrangements of the amino acids of course is almost infinite. One can get some notion of the complexity of the protein puzzle by remembering that the entire English language is derived from just 26 letters (two more than the number of amino acids) combined in various numbers and sequences.

Sanger followed the time-honored method used by chemists to investigate large molecules: namely, breaking them down into fragments and then attempting to put the pieces of the puzzle together. A complete breakdown into the amino acid units themselves makes it possible to identify and measure these components. But this gives no clue to how the units are combined and arranged. To investigate the structure a protein chemist shatters the molecule less violently and then examines these larger fragments, consisting of combinations of two, three or more amino acids. The procedure is somewhat like dropping a pile of plates on the floor. The

COMPLETE MOLECULE of insulin is depicted in this structural formula. Each amino acid in the molecule is presented by an abbreviation rather than its complete atomic structure. The key to these abbreviations is in the chart on page 79. The molecule consists first plate may break into 10 pieces; the second plate may also give 10 pieces but with fractures at different places; the next plate may break into only eight fragments, and so on. Since the sample of protein contains billions of molecules, the experiment amounts to dropping billions of plates. The chemist then pores through this awesome debris for recognizable pieces and other pieces that overlap the breaks to show how the broken sections may be combined.

An amino acid consists of an amino group (NH_3^+), a carboxyl group (COO^-) and a side chain attached to a carbon atom. All amino acids have the amino and carboxyl groups and differ only in their side chains. In a protein molecule they are linked by combination of the carboxyl group of one unit with the amino group of the next. In the process of combination two hydrogen atoms and an oxygen atom drop out in the form of a water molecule and the link becomes CO–NH. This linkage is called the peptide bond. Because of loss of the water molecule, the units linked in the chain are called amino acid "residues." A group of linked amino acids is known as a peptide: two units form a dipeptide, three a tripeptide and so on.

When a peptide or protein is hydrolyzed—treated chemically so that the elements of water are introduced at the peptide bonds—it breaks down into amino acids. The treatment consists in heating the peptide with acids or alkalis. To break every peptide bond and reduce a protein to its amino acids it must be heated for 24 hours or more. Less prolonged or drastic treatment, known as partial hydrolysis, yields a mixture of amino acids, peptides and some unbroken protein molecules. This is the plate-breaking process by which the de-

tailed structure of a protein is investigated.

One of the key inventions that enabled Sanger to solve the jigsaw puzzle was a method of labeling the end amino acid in a peptide. Consider a protein fragment, a peptide, which is composed of three amino acids. On hydrolysis it is found to consist of amino acids *A*, *B* and *C*. The question is: What was their sequence in the peptide? The first member of the three-part chain must have had a free (uncombined) amino (NH_3) group. Sanger succeeded in finding a chemical marker which could be attached to this end of the chain and would stay attached to the amino group after the peptide was hydrolyzed. The labeling material is known as DNP (for dinitrophenyl group). It gives the amino acid to which it is attached a distinctive yellow color. The analysis of the tripeptide sequence proceeds as follows. The tripeptide is treated with the labeling material and is then broken down into its three amino acids. The amino acid which occupied the end position, say *B*, is now identified by its yellow color. The process is repeated with a second sample of the tripeptide, but this time it is only partly hydrolyzed, so that two amino acids remain as a dipeptide derivative colored yellow. If *B* is partnered with, say, *A* in this fragment, one knows that the sequence must be BA, and the order in the original tripeptide therefore was BAC.

Another tool that played an indispensable part in the solution of the insulin jigsaw puzzle was the partition chromatography method for separating amino acids and peptides, invented by the British chemists A. J. P. Martin and R. L. M. Synge [see "Chromatography," by William H. Stein and Stanford Moore; SCIENTIFIC AMERICAN Offprint 81]. Ob-

viously Sanger's method of analysis required separation and identification of extremely small amounts of material. With paper chromatography, which isolates peptides or amino acids in spots on a piece of filter paper, it is possible to analyze a mixture of as little as a millionth of a gram of material with considerable accuracy in a matter of days. As many as 40 different peptides can be separated on a single sheet.

With the knowledge that the insulin molecule was made up of 51 amino acid units, Sanger began his attack on its structure by investigating whether the units were strung in a single long chain or formed more than one chain. Among the components of insulin were three molecules of the amino acid cystine. The cystine molecule is unusual in that it has an amino and a carboxyl group at each end [see its formula in table on page 79]. Since such a molecule could cross-link chains, its presence in insulin suggested that the protein might consist of more than one chain. Sanger succeeded in proving that there were indeed two chains, which he was able to separate intact by splitting the sulfur links in the cystine molecule. Using the DNP labeling technique, he also showed that one chain began with the amino acid glycine and the other with phenylalanine.

Sanger proceeded to break each chain into fragments and study the pieces —especially overlaps which would permit him to build up a sequence. Concentrating on the beginning of the glycine chain, Sanger labeled the glycine with DNP and examined the peptide fragments produced by partial hydrolysis. In the debris of the broken glycine chains he found these sequences attached to the labeled glycine molecules: glycine-isoleucine;   glycine-isoleucine-

of 51 amino acid units in two chains. One chain (*top*) has 21 amino acid units; it is called the glycyl chain because it begins with glycine (Gly). The other chain (*bottom*) has 30 amino acid units; it is called the phenylalanyl chain because it begins with phenylalanine (Phe). The chains are joined by sulfur atoms (S-S). The dotted lines indicate the fragments which located the bridges.

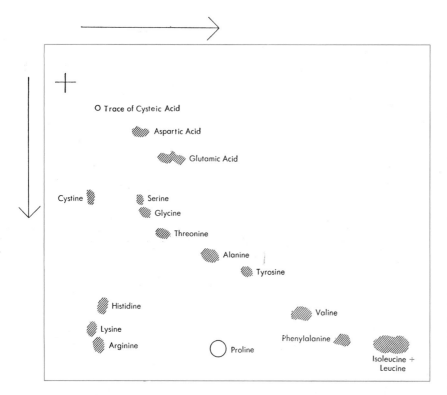

O Trace of Cysteic Acid

Aspartic Acid

Glutamic Acid

Cystine

Serine

Glycine

Threonine

Alanine

Tyrosine

Histidine

Valine

Lysine

Phenylalanine

Arginine

Proline

Isoleucine + Leucine

PAPER CHROMATOGRAPHY separates the 17 amino acids of insulin. In the chromatogram represented by this diagram insulin was broken down by hydrolysis and a sample of the mixture placed at the upper left on the sheet of paper. The sheet was hung from a trough filled with solvent which carried each amino acid a characteristic distance down the paper. The sheet was then turned 90 degrees and the process repeated. The amino acids, with the exception of proline, appear as purple spots when sprayed with ninhydrin.

valine; glycine-isoleucine-valine–glutamic acid; glycine-isoleucine-valine–glutamic acid–glutamic acid. Thus it was evident that the first five amino acids in the glycine chain were glycine, isoleucine, valine and two glutamic acids. Similar experiments on the phenylalanine chain established the first four amino acids in that sequence: phenylalanine, valine, aspartic acid and glutamic acid.

Sanger and a colleague, Hans Tuppy, then undertook the immense task of analyzing the structure of the entire phenylalanine chain. It meant breaking down the chain by partial hydrolysis, separating and identifying the many fragments and then attempting to put the pieces of the puzzle together in proper order. The chain, made up of 30 amino acids, was by far the most complex polypeptide on which such an analysis had ever been attempted.

The bewildering mixture of products from partial breakdown of the chain—amino acids, dipeptides, tripeptides, tetrapeptides and so on—was much too complicated to be sorted out solely by chromatography. Sanger and Tuppy first employed other separation methods (electrophoresis and adsorption on charcoal and ion-exchange resins) which divided the peptide fragments into groups. Then they analyzed these simpler mixtures by paper chromatography. They succeeded in isolating from the fractured chain 22 dipeptides, 14 tripeptides and 12 longer fragments [see chart on pages 80 and 81]. Although these were obtained only in microscopic amounts, they were identified by special techniques and the sequences of their amino acids were determined.

These were the jigsaw pieces that had to be reassembled. Just as in a jigsaw puzzle there are key pieces around which the picture grows, so in this case there were some key pieces as starting points. For instance, the chain was known to contain just one aspartic acid. Six peptides with this amino acid were found in the debris from partial breakdown of the chain [see chart]. The aspartic acid was attached to from one to four other amino acids in these pieces. Their sequences showed that in the original make-up of the chain the order must have been phenylalanine-valine–aspartic acid–glutamic acid–histidine.

Other sequences were pieced together in a similar way until five long sections of the chain were reconstructed. But this still left several gaps in the chain. Sanger and Tuppy now resorted to another method to find the missing links. They split the phenylalanine chain with enzymes instead of by acid hydrolysis. The enzyme splitting process yields longer fragments, and it leaves intact certain bonds that are sensitive to breakage by acid treatment. Thus the investigators obtained long chain fragments which bridged the gaps and revealed the missing links.

After about a year of intensive work Sanger and Tuppy were able to assemble the pieces and describe the structure of insulin's phenylalanine chain. Sanger then turned to the glycine chain and spent another year working out its structure, with the assistance of the author of this article. The glycine chain is shorter (21 amino acids) but it provided fewer clues: there were fewer key pieces that occurred only once, and two amino acids (glutamic acid and cystine) cropped up in so many of the fragments that it was difficult to place them unequivocally in the sequence.

One detail that remained to be decided before the structure could be completed was the actual composition of two amino acids in the chain. Certain amino acids may occur in two forms: e.g., glutamic acid and glutamine. Glutamic acid has two carboxyl (COO^-) groups, whereas glutamine has an amide ($CONH_2$) group in the place of one of the carboxyls [see opposite page]. The difference gives them completely different properties in the protein. Similarly there are aspartic acid and asparagine. Now acid hydrolysis changes glutamine to glutamic acid and asparagine to aspartic acid. Consequently after acid hydrolysis of a protein one cannot tell which form these amino acids had in the original chain. The question was resolved by indirect investigations, one of which involved comparing the products obtained when the same peptide was broken down by acid hydrolysis and by enzymes which do not destroy the amide groups.

By the end of 1952 the two chains were completely assembled. There remained only the problem of determining how the two chains were linked together to form the insulin molecule. But this was easier said than done. As so often happens, what looked simple in theory had complications in practice.

The bridges between the chains, as we have noted, must be cystine, because this amino acid has symmetrical bonds at both ends. The fact that insulin

FORMULA	NAME	ABBREVIATION	PHENYLALANYL	GLYCYL
$CH_2(NH_3{}^+) \cdot COO^-$	Glycine	Gly	3	1
$CH_3 - CH(NH_3 \cdot COO^-$	Alanine	Ala	2	1
$CH_2OH - CH(NH_3{}^+) \cdot COO^-$	Serine	Ser	1	2
$CH_3 \cdot CHOH - CH(NH_3{}^+) \cdot COO^-$	Threonine	Thr	1	0
$\begin{array}{l}CH_3\\ \quad\diagdown CH - CH(NH_3{}^+) \cdot COO^-\\ CH_3\end{array}$	Valine	Val	3	2
$\begin{array}{l}CH_3\\ \quad\diagdown CH \cdot CH_2 - CH(NH_3{}^+) \cdot COO^-\\ CH_3\end{array}$	Leucine	Leu	4	2
$\begin{array}{l}CH_3 \cdot CH_2\\ \qquad\qquad\diagdown CH - CH(NH_3{}^+) \cdot COO^-\\ \quad\quad CH_3\end{array}$	Isoleucine	Ileu	0	1
$\begin{array}{l}CH_2 - CH_2\\ \ \mid \qquad \mid\\ CH_2 \quad CH - COO^-\\ \quad\diagdown\ \diagup\\ \qquad NH^+\end{array}$	Proline	Pro	1	0
$\begin{array}{l}\quad CH = CH\\ CH\diagup\qquad\quad\diagdown C \cdot CH_2 - CH(NH_3{}^+) \cdot COO^-\\ \quad\diagdown CH - CH \diagup\end{array}$	Phenylalanine	Phe	3	0
$\begin{array}{l}\qquad\quad CH = CH\\ HO - C\diagup\qquad\quad\diagdown C \cdot CH_2 - CH(NH_3{}^+) \cdot COO^-\\ \qquad\diagdown CH - CH \diagup\end{array}$	Tyrosine	Tyr	2	2
$NH_2CO \cdot CH_2 - CH(NH_3{}^+) \cdot COO^-$	Asparagine	Asp (NH$_2$)	1	2
$COOH \cdot CH_2 \cdot CH_2 - CH(NH_3{}^+) \cdot COO^-$	Glutamic Acid	Glu	2	2
$NH_2 \cdot CO \cdot CH_2 \cdot CH_2 - CH(NH_3{}^+) \cdot COO^-$	Glutamine	Glu (NH$_2$)	1	2
$\begin{array}{l}NH_2 - C - NH \cdot CH_2 \cdot CH_2 \cdot CH_2 - CH(NH_3{}^+) \cdot \mathbf{COO^-}\\ \qquad\quad\parallel\\ \qquad\quad NH\end{array}$	Arginine	Arg	1	0
$\begin{array}{l}CH = C \cdot CH_2 - CH(NH_3{}^+) \cdot COO^-\\ \ \mid \qquad \mid\\ NH \quad N\\ \quad\diagdown\ \diagup\!\!/\\ \qquad CH\end{array}$	Histidine	His	2	0
$CH_2NH_2 \cdot CH_2 \cdot CH_2 \cdot CH_2 - CH(NH_3{}^+) \cdot COO^-$	Lysine	Lys	1	0
$\begin{array}{l}COO^-\qquad\qquad\qquad\qquad\qquad\qquad\diagup COO^-\\ \quad\diagdown CH - CH_2 - S - S - CH_2 - CH\\ NH_3{}^+ \qquad\qquad\qquad\qquad\qquad\qquad\ \diagdown NH_3{}^+\end{array}$	Cystine	CyS | CyS	2	4
			30	21

AMINO ACIDS, of insulin are listed in this chart. Their chemical formulas are at the left. The dots in the formulas represent chemical bonds other than those suggested by the atoms adjacent to each other. The number of amino acid units of each kind found in the phenylalanyl chain are listed in the fourth column of the chart. The fifth column comprises a similar listing for the glycyl chain.

	1	2	3	4	5	6	7	8	9	10	11	12	13	14	15	16
PEPTIDES FROM ACID HYDROLYZATES	Phe	Val		Glu	His		CySO₃H	Gly		His	Leu		Glu	Ala		
		Val	Asp		His	Leu					Leu	Val		Ala	Leu	
			Asp	Glu		Leu	CySO₃H		Ser	His		Val	Glu			
	Phe	Val	Asp			Leu	CySO₃H	Gly				Val	Glu	Ala		Tyr
				Glu	His	Leu			Ser	His	Leu					
		Val	Asp	Glu							Leu	Val	Glu			
					His	Leu	CySO₃H							Ala	Leu	Tyr
	Phe	Val	Asp	Glu					Ser	His	Leu	Val				Tyr
					His	Leu	CySO₃H	Gly			Leu	Val	Glu	Ala		
	Phe	Val	Asp	Glu	His				Ser	His	Leu	Val	Glu			
				Glu	His	Leu	CySO₃H			His	Leu	Val	Glu			
									Ser	His	Leu	Val	Glu	Ala		
SEQUENCES DEDUCED FROM THE ABOVE PEPTIDES	Phe	Val	Asp	Glu	His	Leu	CySO₃H	Gly								Tyr
									Ser	His	Leu	Val	Glu	Ala		
PEPTIDES FROM PEPSIN HYDROLYZATE	Phe	Val	Asp (NH₂)	Glu (NH₂)	His	Leu	CySO₃H	Gly	Ser	His	Leu					
												Val	Glu	Ala	Leu	
					His	Leu	CySO₃H	Gly	Ser	His	Leu					
PEPTIDES FROM CHYMOTRYPSIN HYDROLYZATE	Phe	Val	Asp (NH₂)	Glu (NH₂)	His	Leu	CySO₃H	Gly	Ser	His	Leu	Val	Glu	Ala	Leu	Tyr
PEPTIDES FROM TRYPSIN HYDROLYZATE																
STRUCTURE OF PHENYLALANYL CHAIN OF OXIDIZED INSULIN	Phe	Val	Asp (NH₂)	Glu (NH₂)	His	Leu	CySO₃H	Gly	Ser	His	Leu	Val	Glu	Ala	Leu	Tyr

STRUCTURE OF GLYCYL CHAIN OF OXIDIZED INSULIN

Gly · Ileu · Val · Glu · Glu(NH₂) · CySO₃H · CySO₃H · Ala · Ser · Val · CySO₃H · Ser · Leu · Tyr ·

SEQUENCE OF AMINO ACIDS in the phenylalanyl chain was deduced from fragments of the chain. The entire sequence is at the bottom above the dotted line. Each fragment is indicated by a horizontal sequence of amino acids joined by dots. The fragments are arranged so that each of their amino acids is in the vertical column above the corresponding amino acid in the entire chain.

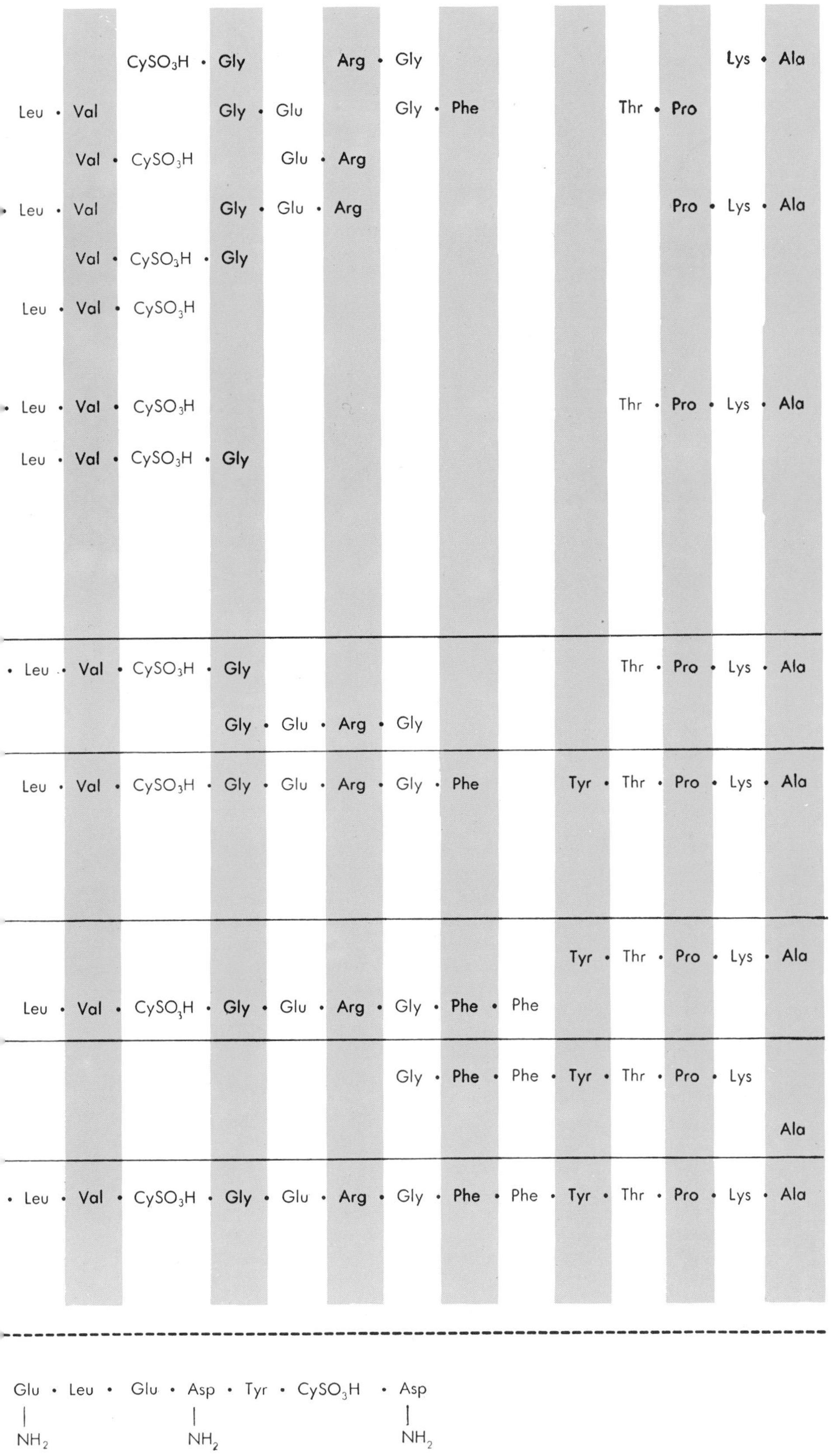

Leu · Val $CySO_3H$ · Gly   Arg · Gly   Lys · Ala

Val · $CySO_3H$ Gly · Glu   Gly · Phe   Thr · Pro

Leu · Val   Glu · Arg   Gly · Glu · Arg   Pro · Lys · Ala

Val · $CySO_3H$ · Gly

Leu · Val · $CySO_3H$

Leu · Val · $CySO_3H$ Thr · Pro · Lys · Ala

Leu · Val · $CySO_3H$ · Gly

Leu · Val · $CySO_3H$ · Gly   Thr · Pro · Lys · Ala

Gly · Glu · Arg · Gly

Leu · Val · $CySO_3H$ · Gly · Glu · Arg · Gly · Phe   Tyr · Thr · Pro · Lys · Ala

Tyr · Thr · Pro · Lys · Ala

Leu · Val · $CySO_3H$ · Gly · Glu · Arg · Gly · Phe · Phe

Gly · Phe · Phe · Tyr · Thr · Pro · Lys

Ala

Leu · Val · $CySO_3H$ · Gly · Glu · Arg · Gly · Phe · Phe · Tyr · Thr · Pro · Lys · Ala

Glu · Leu · Glu · Asp · Tyr · $CySO_3H$ · Asp
 | | |
NH_2 NH_2 NH_2

The shorter fragments (*group at the top*) were obtained by hydrolyzing insulin with acid. The longer fragments (*groups third, fourth and fifth from the top*) were obtained with enzymes. The same method was used to deduce the sequence in the glycyl chain (*bottom*).

contains three cystine units suggested that there might be three bridges, or cross-links, between the chains. It appeared that it should be a simple matter to locate the positions of the bridges by a partial breakdown of the insulin molecule which gave cystine-containing fragments with sections of the two chains still attached to the "bridge" ends.

When Sanger began this analysis, he was puzzled to find that the cystine-containing peptides in his broken-down mixtures showed no significant pattern whatever. Cystine was joined with other amino acids in many different combinations and arrangements, as if the chains were cross-linked in every conceivable way. Sanger soon discovered the explanation: during acid hydrolysis of the insulin molecule, cystine's sulfur bonds opened and all sorts of rearrangements took place within the peptides. Sanger and his associate A. P. Ryle then made a systematic study of these reactions and succeeded in finding chemical inhibitors to prevent them.

By complex analyses which employed both acid hydrolysis and enzyme breakdown, Sanger and his co-workers L. F. Smith and Ruth Kitai eventually fitted the bridges into their proper places and obtained a complete picture of the structure of insulin [*see diagram at bottom of pages 76 and 77*]. So for the first time the biochemist is able to look at the amino-acid arrangement in a protein molecule. The achievement seems astounding to those who were working in the field 10 years ago.

To learn how insulin's structure determines its activity as a hormone is still a long, hard road. It will be difficult to synthesize the molecule, but once that has been accomplished, it will be possible to test the effect of changes in the structure on the substance's physiological behavior. Evidently slight variations do not affect it much, for Sanger has shown that the insulins from pigs, sheep and steers, all equally potent, differ slightly in structure.

The methods that proved so successful with insulin, plus some newer ones, are already being applied to study larger proteins. Among the improvements are promising new techniques for splitting off the amino acids from a peptide chain one at a time—clearly a more efficient procedure than random hydrolysis. The rate of progress undoubtedly will be speeded up as more biochemists turn their attention to the intriguing problem of relating the structure of proteins to their physiological functions.

8

THE STRUCTURE AND HISTORY OF AN ANCIENT PROTEIN

RICHARD E. DICKERSON

April 1972

To oxidize food molecules all organisms from yeasts to man require a variant of cytochrome c. Differences in this protein from species to species provide a 1.2-billion-year record of molecular evolution

Between 1.5 and two billion years ago a profound change took place in some of the single-celled organisms then populating our planet, a change that in time would contribute to the rise of many-celled organisms. The machinery evolved for extracting far more energy from foods than before by combining food molecules with oxygen. One of the central components of the new metabolic machinery was cytochrome *c*, a protein whose descendants can be found today in every living cell that has a nucleus. By studying the cytochrome *c* extracted from various organisms it has been possible to determine how fast the protein has evolved since plants and animals diverged into two distinct kingdoms and in fact to provide an approximate date of 1.2 billion years ago for the event. For example, the cytochrome *c* molecules in men and chimpanzees are exactly the same: in the cells of both the molecule consists of 104 amino acid units strung together in exactly the same order and folded into the same three-dimensional structure. On the other hand, the cytochrome *c* in man has diverged from the cytochrome *c* in the red bread mold *Neurospora crassa* in 44 out of 104 places, yet the three-dimensional structures of the two cytochrome *c* molecules are essentially alike. We think we can now explain how it is that so many of the 104 amino acid units in cytochrome *c* are interchangeable and also why certain units cannot be changed at all without destroying the protein's activity.

Let us try to visualize the earth before cytochrome *c* first appeared. The first living organisms on the planet were little more than scavengers, extracting energy-rich organic compounds (includ-

ing their neighbors) from the water around them and releasing low-energy breakdown products. We still have the "fossils" of this life-style in the universal process of anaerobic (oxygenless) fermentation, as when a yeast extracts energy from sugar and releases ethyl alcohol, or when an athlete who exercises too rapidly converts glucose to lactic acid and gets muscle cramps. Anaerobic fermentation is part of the common biochemical heritage of all living things.

The upper limit on how much life the planet could support with only fermentation as an energy source was determined by the rate at which high-energy compounds were synthesized by nonbiological agencies: ultraviolet radiation, lightning discharges, radioactivity or heat. When some organisms developed the ability to tap sunlight for energy, photosynthesis was born and the life-carrying capacity of the earth increased enormously. This was the age of the bac-

SKELETON OF CYTOCHROME *c* MOLECULE is depicted in the illustration by Irving Geis on the opposite page. A variant of this protein molecule is found in the cells of every living organism that utilizes oxygen for respiration. The illustration shows in simplified form how 104 amino acid units are linked in a continuous chain that grips and surrounds a heme group, a complex rosette with an atom of iron (*Fe*) at its center. The picture is color-coded to indicate how much variation has been tolerated by evolution at each of the 104 amino acid sites in the molecule. Some species lack the 104th amino acid, and all species except vertebrates have as many as eight extra amino acids at the beginning of the chain (*see table on pages 84 and 85*). The amino acids that are most invariant throughout evolution, and presumably the most important, are shown in red and orange; the more variable sites appear in yellow-green, blue-green, blue and purple. The indispensable heme group is crimson. Each amino acid is represented only by its "alpha" carbon atom: the atom that carries a side chain unique for each of the 20 amino acids. The upper drawing at left below shows how two amino acids link up through an amide group (*colored panel*); the side chains connected to the alpha carbons (*color*) are represented by the balls labeled *R*. The lower drawing at left below shows the scheme used in the cytochrome *c* skeleton on the opposite page; all amide linkages (—CO—NH—) are omitted and the only side groups shown are those that are attached to the heme. The amino acids at the 35 invariant sites of the cytochrome *c* molecule (*red*) are designated in abbreviated form (*see key at right below*).

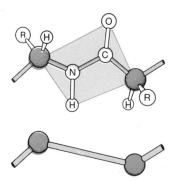

Ala	Alanine	Leu	Leucine
Asp	Aspartic acid	Lys	Lysine
Asn	Asparagine	Met	Methionine
Arg	Arginine	Phe	Phenylalanine
Cys	Cysteine	Pro	Proline
Gly	Glycine	Ser	Serine
Glu	Glutamic acid	Thr	Threonine
Gln	Glutamine	Trp	Tryptophan
His	Histidine	Tyr	Tyrosine
Ile	Isoleucine	Val	Valine

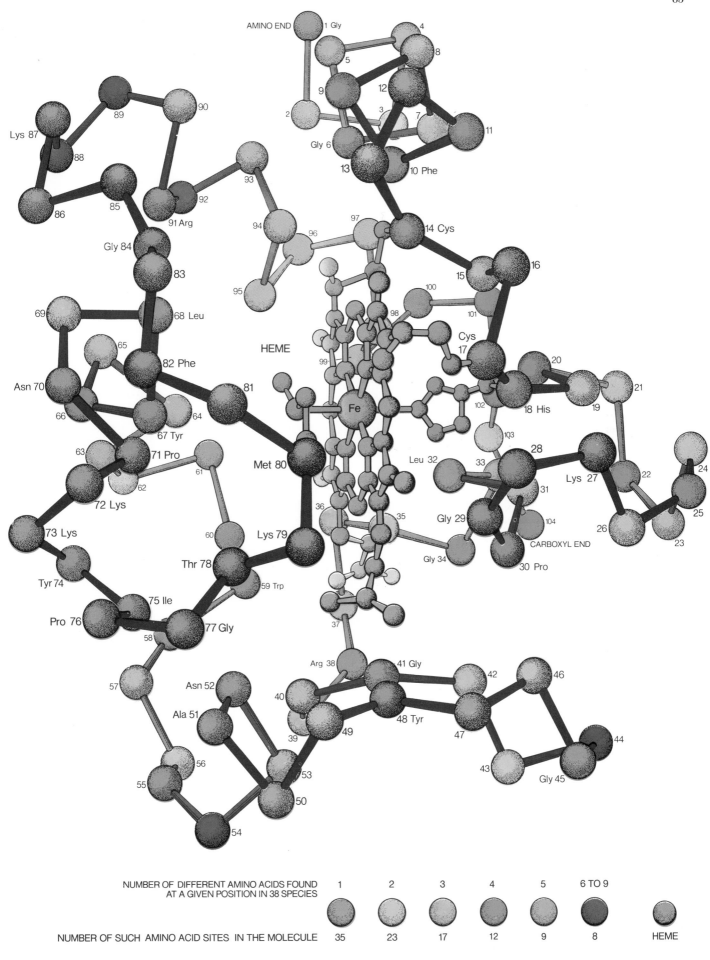

NUMBER OF DIFFERENT AMINO ACIDS FOUND
AT A GIVEN POSITION IN 38 SPECIES

1	2	3	4	5	6 TO 9	

NUMBER OF SUCH AMINO ACID SITES IN THE MOLECULE   35   23   17   12   9   8   HEME

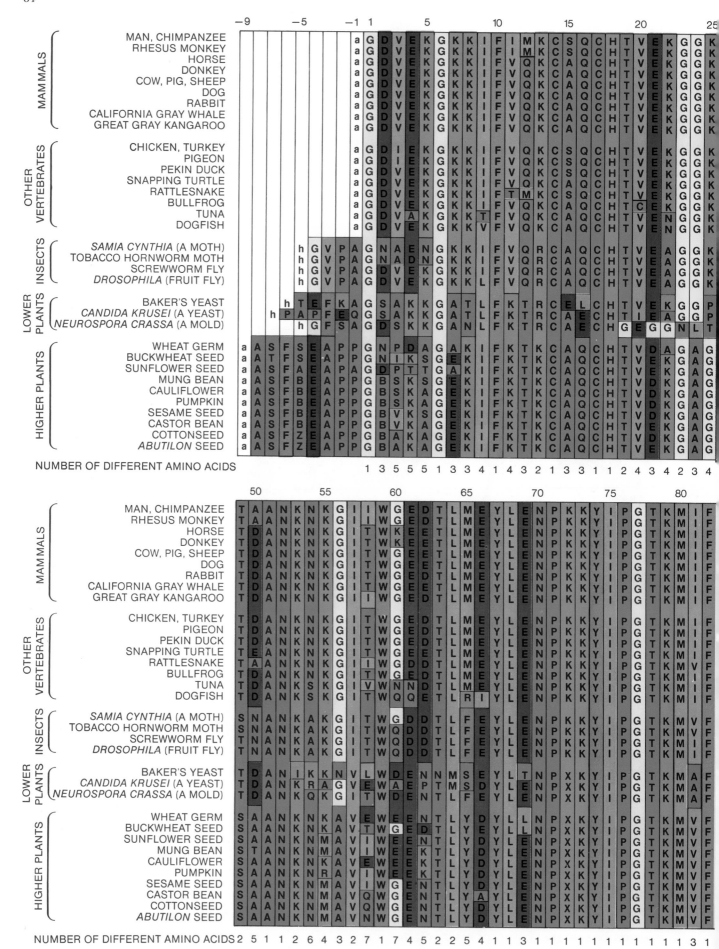

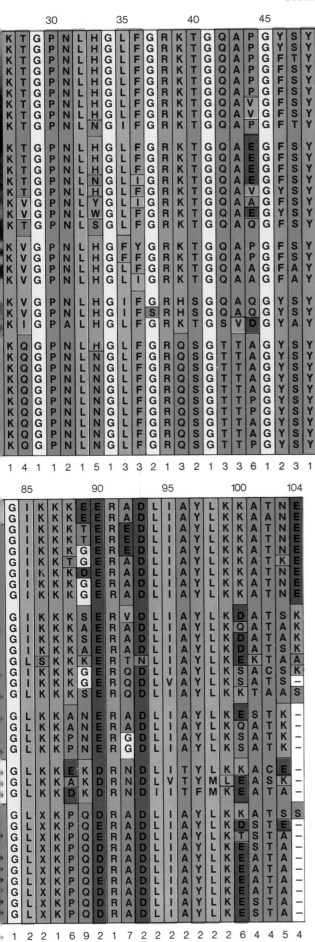

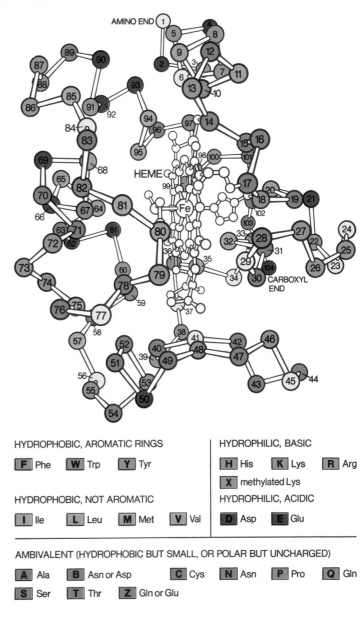

HYDROPHOBIC, AROMATIC RINGS

| F | Phe | W | Trp | Y | Tyr |

HYDROPHOBIC, NOT AROMATIC

| I | Ile | L | Leu | M | Met | V | Val |

HYDROPHILIC, BASIC

| H | His | K | Lys | R | Arg |
| X | methylated Lys |

HYDROPHILIC, ACIDIC

| D | Asp | E | Glu |

AMBIVALENT (HYDROPHOBIC BUT SMALL, OR POLAR BUT UNCHARGED)

| A | Ala | B | Asn or Asp | C | Cys | N | Asn | P | Pro | Q | Gln |
| S | Ser | T | Thr | Z | Gln or Glu |

NO SIDE CHAIN (HYDROGEN ATOM) | G | Gly |

COMPOSITION OF CYTOCHROME c IN 38 SPECIES is presented in the table at left. No other protein has been so fully analyzed for so many different organisms. The color code used here differs from the one used in the molecular skeleton on page 83. On these two pages color is employed to classify amino acids according to their chemical properties (*see key directly above*). Thus the three "oily" (hydrophobic) amino acids with aromatic benzene rings in their side chains (phenylalanine, tryptophan and tyrosine) are shown in red. Four other amino acids that are hydrophobic but nonaromatic are shown in orange. At the other extreme, amino acids that are hydrophilic, or water-loving, are shown in blue or violet. Amino acids that can be found in either aqueous or nonaqueous environments (and hence are ambivalent) are green or yellow. Polar amino acids can have asymmetric distributions of positive and negative charge. The detailed structure of the side chains of the amino acids can be found on page 87. It is easy to pick out from the table at left the amino acid sites where evolution has allowed no change or has allowed substitution only by chemically similar amino acids; these sites are identified by vertical bands of a single color. A letter *a* at the beginning of the chain indicates that a methyl group (CH_3) is attached to the amino end of the molecular chain. A letter *h* indicates that the methyl group is absent.

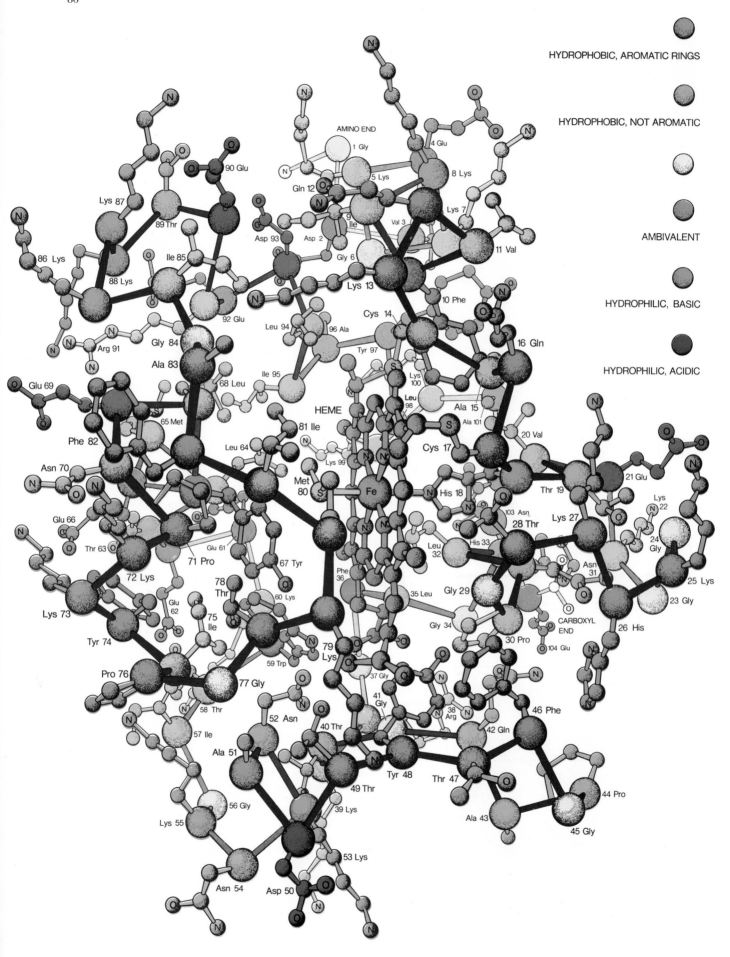

HYDROPHOBIC, AROMATIC RINGS

HYDROPHOBIC, NOT AROMATIC

AMBIVALENT

HYDROPHILIC, BASIC

HYDROPHILIC, ACIDIC

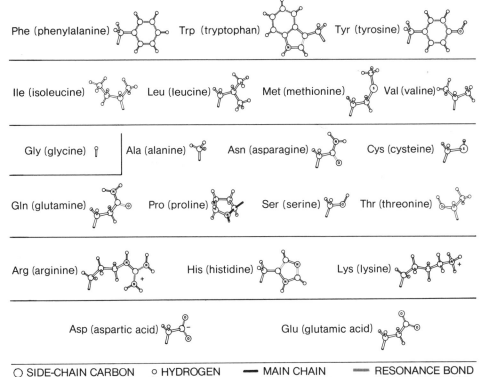

Phe (phenylalanine)   Trp (tryptophan)   Tyr (tyrosine)

Ile (isoleucine)   Leu (leucine)   Met (methionine)   Val (valine)

Gly (glycine)   Ala (alanine)   Asn (asparagine)   Cys (cysteine)

Gln (glutamine)   Pro (proline)   Ser (serine)   Thr (threonine)

Arg (arginine)   His (histidine)   Lys (lysine)

Asp (aspartic acid)   Glu (glutamic acid)

○ SIDE-CHAIN CARBON   ○ HYDROGEN   ━ MAIN CHAIN   ▬ RESONANCE BOND
Ⓝ NITROGEN   Ⓢ SULFUR   ═ SINGLE BOND   OF INTERMEDIATE
Ⓞ OXYGEN   ═ DOUBLE BOND   CHARACTER

CYTOCHROME c MOLECULE WITH SIDE CHAINS appears in the illustration on the opposite page. The picture shows the structure of horse-heart cytochrome c in the oxidized state as determined through X-ray crystallography by the author and his colleagues at the California Institute of Technology. Alpha-carbon atoms are numbered and the amide groups (—CO—NH—) connecting the alpha carbons are represented only by a solid bond, as in the preceding drawings. For clarity three side chains at the "back" of the molecule have been left out: leucine 35, phenylalanine 36 and leucine 98. The color coding follows the coding in the illustration on pages 84 and 85. One can see from the three-dimensional structure that side chains in the interior of the molecule, around the heme group (*crimson*), tend to be hydrophobic (*red and orange*), whereas amino acids with hydrophilic side chains (*blue and violet*) are found on the outside, where they are ordinarily in contact with water. A major exception to this rule is the hydrophobic side chain of phenylalanine 82, which sits on the surface of the molecule at the left of the heme. The region between the hydrophilic chains, and above isoleucine 81, is a cavity that is apparently open to solvent molecules. Lysine 13, above this cavity, is known to interact with a large oxidase complex when cytochrome c is oxidized. The structures of the side chains of 20 amino acids appear above.

within an organized nucleus, and their respiratory and photosynthetic machinery (if it is present) is similarly dispersed. Green algae and all the higher plants and animals are eukaryotes (cells with "good" nuclei); their DNA is organized within a nucleus, and their respiration is carried out in the organelles called mitochondria. In eukaryote plants photosynthesis is conducted in still other organelles called chloroplasts. Mitochondria are the powerhouse of all eukaryote cells. Their role is to break down the energy-rich molecules obtained from foods, combine them with oxygen and store the energy produced by harnessing it to synthesize molecules of adenosine triphosphate (ATP). The mitochondria of all eukaryotes are alike in their chemistry, as if once the optimum chemical mechanism had been arrived at it was never changed.

Biological oxidation involves at least a score of special enzymes that act first as acceptors and then as donors of the electrons or hydrogen atoms removed from food molecules. In the last part of the process one finds a series of cytochrome molecules (identified by various subscript letters), all of which incorporate a heme group containing iron, the same heme group found in hemoglobin. Electrons are passed down a chain of cytochrome molecules: from cytochrome b to cytochrome c_1, from cytochrome c_1 to cytochrome c, to cytochromes a and a_3 and finally to oxygen atoms, where they are combined with hydrogen ions to produce water. This is a stepwise process designed to release energy in small parcels rather than all at once. In the transfer of electrons from cytochrome b to cytochrome c_1 and again in the transfer between the cytochromes a, a_3 and oxygen, energy is channeled off to synthesize ATP, which acts as a general-purpose energy source for cell metabolism.

Most of the cytochromes are bound tightly to the mitochondrial membrane, but one of them, cytochrome c, can easily be solubilized in aqueous mediums and can be isolated in pure form. The other components can be isolated as multienzyme complexes: b and c_1 as a cytochrome reductase complex, and a and a_3 as a cytochrome oxidase. The reductase donates electrons to cytochrome c; the oxidase accepts them again. To illustrate how similar all eukaryotes really are to one another, it has been found that cytochrome c from any species of plant, animal or eukaryotic microorganism can react in the test tube with the cytochrome oxidase from any other species. Worm or primate, whale

teria and the blue-green algae [see "The Oldest Fossils," by Elso S. Barghoorn; SCIENTIFIC AMERICAN Offprint 895].

The more advanced forms of photosynthesis released a corrosive and poisonous gas into the atmosphere: oxygen. Some bacteria responded by retreating to oxygen-free corners of the planet, where their descendants are found today. Other bacteria and blue-green algae developed ways to neutralize gaseous oxygen by combining it with their own waste products. The next step was to harness the energy released by oxidation of these waste compounds. (If you are going to burn your garbage, you might as well keep warm by the fire.) This was the beginning of oxidation, or respiration, the second big breakthrough in increasing the supply of energy available to life on the earth.

When a yeast cell oxidizes sugars all the way to carbon dioxide and water instead of stopping short at ethyl alcohol, it gets 19 times as much energy per gram of fuel. When oxygen combines with lactic acid in the athlete's muscles and the cramps are dissipated, he receives a correspondingly greater energy return from his glucose. Any improvement in metabolism that multiplies the supply of energy available by such a large factor would be expected to have a revolutionary effect on the development of life. We now believe the specialization of cells and the appearance of multicelled plants and animals could only have come about in the presence of such a large new supply of energy.

Bacteria and blue-green algae are prokaryotes (prenuclear cells); their genetic material, DNA, is not confined or wheat are all alike under the mitochondrial membrane.

The Evolution of Cytochrome c

Since cytochrome c is so ancient and at the same time so small and easily purified, it has received much attention from protein chemists interested in the evolutionary process. The complete amino acid sequence of cytochrome c has been determined for more than 40 species of eukaryotic life. Thirty-eight of these sequences are compared in the illustration on pages 84 and 85. We have more information on the evolution of this molecule than on the evolution of any other protein.

Emanuel Margoliash of Northwestern University and Emil Smith of the University of California at Los Angeles were among the first to notice that the amino acid sequences from various species are different and that the degree of difference corresponds quite well with the distance that separates the two species on the evolutionary tree. Detailed computer analyses of these differences by Margoliash, by Walter Fitch of the University of Wisconsin and by others have led to the construction of elaborate family trees of living organisms entirely without recourse to the traditional anatomical data. The family trees agree remarkably well with those obtained from classical morphology; it is obvious that

comparison of amino acid sequence is a powerful tool for studying the process of evolution.

Another result may at first be surprising. Cytochrome c is still evolving slowly and is doing so at a rate that is approximately constant for all species, when the rate is averaged over geological time periods. This kind of analysis of molecular evolution was first carried out on hemoglobin a decade ago by Linus Pauling and Emile Zuckerkandl at the California Institute of Technology. If we compare hemoglobin and cytochrome c, we find that cytochrome c is changing much more slowly. Why should this be? The protein chains are synthesized from instructions that are embodied in DNA, and it is in the DNA that mutations take place. Do mutations occur more often in the DNA that makes hemoglobin than in the DNA that makes cytochrome c? There is no reason to think so. The explanation therefore must lie in the natural-selection, or screening, process that tests whether or not mutant molecules can do their job.

Before discussing the various "formulas" that have passed the test of making a successful cytochrome c, I shall describe briefly the structure of proteins. All protein molecules are built up by linking amino acids end to end. Each of the 20 different amino acids has a carboxyl group (–COOH) at one end and an amino group (–NH₂) at the other. To link the carboxyl group of one amino acid with the amino group of another amino acid a molecule of water must be removed, producing an amide linkage (–CO–NH–). Because only a part (although, to be sure, the distinctive part) of an amino acid enters a protein chain, the chemist refers to it as a "residue." Thus he speaks of a glycine residue or a phenylalanine residue at such-and-such a position in a protein chain.

The carbon adjacent to the amide linkage is called the alpha carbon. It is important because each amino acid has a distinctive side chain at this position. The side chain may be nothing more than a single atom of hydrogen (as it is in the case of the amino acid glycine) or it may consist of a number of atoms, including a six-carbon "aromatic" ring (as it does in the case of phenylalanine, tryptophan and tyrosine).

The 20 amino acids can be grouped into three broad classes, depending on the character of their side chains [see illustration on preceding page]. Five are hydrophilic, or water-loving, and tend to acquire either a positive or a negative charge when placed in aqueous solution; three of the five are basic in character

PLOT OF DISTRIBUTION OF ELECTRIC CHARGES on the back of horse-heart cytochrome c reveals that most of the 19 hydrophilic lysines (*color*), which carry positive charges (and hence are basic), are distributed on the two flanks of the molecule. Nine of 12 negatively charged (acidic) side chains (*gray*) are clustered in one zone in the upper center of the molecule. The electrically negative character of this zone has been maintained throughout evolution, although the specific locations of the acidic side groups vary. No organism, from wheat germ to man, has a cytochrome c with fewer than six acidic amino acids in this zone and no organism has more than five acidic amino acids everywhere e₄se on the molecule. Furthermore, these extreme values are not found in the same species. It is highly likely that these charged zones participate in binding cytochrome c to other large molecules.

(arginine, histidine and lysine) and the other two are acidic (aspartic acid and glutamic acid). Seven are not readily soluble in water and hence are termed hydrophobic; they include the three amino acids mentioned above that have rings in their side chains plus leucine, isoleucine, methionine and valine. The remaining eight amino acids react ambivalently to water: alanine, asparagine, cysteine, glutamine, glycine, proline, serine and threonine.

Now let us see how much the successful formulas for cytochrome c differ from species to species. The cytochrome c molecules of men and horses differ by 12 out of 104 amino acids. The cytochrome c's of the higher vertebrates—mammals, birds and reptiles—differ from the cytochrome c's of fishes by an average of 19 amino acids. The cytochrome c's of vertebrates and insects differ by an average of 27 amino acids; moreover, the cytochrome c molecules of insects and plants have a few more amino acid residues at the beginning of the chain than the equivalent molecules of vertebrates. The greatest disparity between two cytochrome c's is the one between man and the bread mold *Neurospora;* they differ at more than 40 percent of their amino acid positions. How can two molecules with such large differences in amino acid composition perform identical chemical functions?

We begin to see an answer when we look at where these changes are. Some parts of the amino acid sequence, as indicated on pages 84 and 85, never vary. Thirty-five of the 104 amino acid positions in cytochrome c are completely invariant in all known species, including a long sequence from residue 70 through residue 80. The 35 invariant sites are occupied by 15 different amino acids; they are shown in red in the structural drawing on page 3. Another 23 sites are occupied by only one of two different but closely similar amino acids. There are 18 different sets of interchangeable pairs at these 23 sites; they are shown in orange in the illustration. At 17 sites natural selection has evidently accepted only sets of three different amino acids; these 17 interchangeable triplets are colored yellow-green.

It was already known from sequence studies, before the X-ray structural analysis, that where such substitutions are allowed the interchangeable amino acids almost always have the same chemical character. In general all must be either hydrophilic or hydrophobic or else neutral with respect to water. Such interchanges are called conservative substitutions because they conserve the overall chemical nature of that part of the protein molecule.

In only a few places along the chain can radical changes be tolerated. Residue 89, for example, can be acidic (aspartic acid or glutamic acid), basic (lysine), polar but uncharged (serine, threonine, asparagine and glutamine), weakly hydrophobic (alanine) or devoid of a side chain (glycine). Almost the only type of side chain that appears to be forbidden at this point in the molecule is a large hydrophobic one. Such "indifferent" regions are rare, however, and cytochrome c overall is an evolutionarily conservative molecule.

We have no reason to think the gene for cytochrome c mutates more slowly than the gene for hemoglobin, or that the invariant, conservative and radical regions of the sequence reflect any difference in mutational rate within the cytochrome c gene. The mutations are presumably random, and what we see in these species comparisons are the molecules that are left after the rigid test of survivability has been applied. Invariant regions evidently are invariant because any mutational changes there are lethal and are weeded out. Conservative changes can be tolerated elsewhere as long as they preserve the essential chemical properties of the molecule at that point. Radical changes presumably indicate portions of the molecule that do not matter for the operation of the protein.

This is as far as we can go from sequence comparisons alone. The explanation of variability in terms of the essential or nonessential character of different parts of the molecule is plausible, yet science has always been plagued by plausible but incorrect hypotheses. To progress any further we need to know how the amino acid sequence is folded to make an operating molecule. In short, we need the three-dimensional structure of the protein.

The Molecule in Three Dimensions

With the active collaboration of Margoliash, who was then working at the Abbott Laboratories in North Chicago, I began the X-ray-crystallographic analysis of horse-heart cytochrome c at Cal Tech in 1963, with the sponsorship of the National Science Foundation and the National Institutes of Health. As cytochrome c transfers electrons in the mitochondrion, it oscillates between an oxidized form (ferricytochrome) and a reduced form (ferrocytochrome); the iron atom in the heme group is alternately in the +3 and +2 oxidation state.

We decided to begin our analysis with the oxidized form, a decision that was largely tactical since both oxidation states would ultimately be needed if we were to try to decipher the electron-transfer process.

In X-ray crystallography one directs a beam of X rays at a purified crystal of the substance under study and records the diffraction pattern produced as the beam strikes the sample from different angles. X rays entering the sample are themselves deflected at various angles by the distribution of electron charges within the crystal. Highly sophisticated computer programs have been devised for deducing from tens of thousands of items of X-ray-diffraction data the three-dimensional distribution of electronic charge. From this distribution one can infer, in turn, the distribution of the amino acid side chains in the protein molecule.

We obtained our first low-resolution map of the oxidized form of horse-heart cytochrome c five years ago and the first high-resolution map three years ago. These maps have been used to construct detailed three-dimensional models of the protein. One can also feed the three-dimensional coordinates into a computer and obtain simple ball-and-stick drawings that can be viewed stereoptically, enabling one to visualize the folded chain of the protein in three dimensions [see illustrations on next two pages]. Just a year ago we calculated the first high-resolution map for the reduced form of cytochrome c. We are now improving this model and comparing the two oxidation states.

Several striking features of the amino acid sequences of cytochrome c were in the back of our minds as we worked out the first high-resolution structure. We knew that the most strongly conserved sites throughout evolution were those occupied by three distinctive types of residue: the positively charged (basic) residues of lysine; the three hydrophobic and aromatic residues of phenylalanine, tryptophan and tyrosine, and the four hydrophobic but nonaromatic residues of leucine, isoleucine, methionine and valine. These sites can now be located with the help of the illustration on page 86, whose color coding differs from the coding of the illustration on page 83. Here hydrophobic residues are shown in warm colors (red and orange) whereas neutral residues and hydrophilic residues, both basic and acidic, are shown in cool colors (green, yellow, blue and violet).

It had been known from the chemical analysis of the molecule's amino acid se-

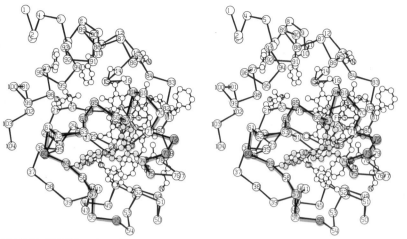

STEREOSCOPIC PAIR OF LEFT SIDE of oxidized cytochrome *c* molecule, drawn by computer, shows only a few key side chains for clarity. The main chain (*color*) from sites 55 to 75 defines a loop, the "left channel," which is filled with strongly hydrophobic side chains; their alpha carbons are in light color. Three of these side chains include aromatic rings: tryptophan 59, tyrosine 67 and tyrosine 74, also shown in light color. Alpha carbons with hydrophilic, positively charged side chains around the left channel are shown in dark color. This pair and one on opposite page can be viewed with standard stereoscopic viewer.

quence that the basic and hydrophobic groups tend to appear in clusters along the chain. For example, basic residues are found in the regions of sites 22 through 27, sites 38 and 39, sites 53 through 55 and sites 86 through 91. Hydrophobic residues are found in regions 9 through 11, 32 through 37, 80 through 85 and 94 through 98. The residues at sites 14 and 17 (cysteine) and site 18 (histidine) are invariant, which is understandable since they form bonds to the heme group. Less understandably, the long stretch from site 70 to site 80 is equally invariant. Before the structural evidence was available it had been suspected that methionine, at site 80, might be bonded to the iron atom on the other side of the heme from the histidine at site 18, but it was impossible to be sure from chemical evidence alone.

It was also known from chemical analysis that horse cytochrome *c* incorporates 12 glycines (the residues with only hydrogen as a side chain) and that these glycines were either invariant or else conserved in the great majority of species. It was known too that of the eight phenylalanines or tyrosines (with aromatic rings in their side chains) seven are either invariant in all species or replaceable only by one another. In the case of residue 36, phenylalanine or tyrosine is replaced in three species by isoleucine, whose side chain, although it is nonaromatic, is at least as large and hydrophobic as the side chains it replaces.

All these similarities and conservatisms were known before the X-ray analysis, but none could be explained in

terms of structure. It was assumed that every residue had been placed where it was by natural selection and that it contained potentially important information about the working parts of the cytochrome molecule. Natural selection, however, does not act on an amino acid sequence but rather on the folded and operating molecule in its association with other biological molecules. Having a sequence without the folding instructions is like having a list of parts without a blueprint of the entire machine.

Cytochrome *c* and Evolution

Now that the blueprint for cytochrome *c* is revealed, let us look more closely at its representation on page 83. To keep the illustration simple no side chains have been included except for those that are bonded to the heme group. Moreover, along the main chain the illustration depicts only the alpha-carbon atoms from which side chains would, if they were shown, branch off. The amide groups (–CO–NH–) that connect alpha carbons are represented simply by straight lines. The picture is therefore a simplified folding diagram of the cytochrome molecule.

We see that the flat heme group, a symmetrical rosette of carbon and nitrogen atoms with an atom of iron at its center, sits in a crevice with only one edge exposed to the outside world. If the heme participates directly in shuttling electrons in and out of the molecule, the transfer probably takes place along this edge. Cysteines 14 and 17 and histidine 18 hold the heme in place

from the right as depicted, and the other heme-binding group on the left is indeed methionine 80, as had been suspected.

It was known from earlier X-ray studies of proteins that sequences of amino acids frequently fold themselves into the helical configuration known as the alpha helix; in other cases the amino acids tend to assume a rippled or corrugated configuration called a beta sheet. Cytochrome *c* has no beta sheets and only two stretches of alpha helix, formed by residues 1 through 11 and 89 through 101. For the most part the protein chain is wrapped tightly around the heme group, leaving little room for the alpha and beta configurations that are prominent in other proteins.

Just as one can use cytochrome *c* to learn about evolution, one can also use evolution to learn about cytochrome *c*. As I have noted, the illustration on page 3 is color-coded to indicate the amount of variability in the kind of amino acid tolerated at each site. The structure is "hot" (red and orange) in the functionally important places in the molecule when differences among species are absent or rare, and it is "cool" (green, blue and violet) in regions that vary widely from one species to another and thus are presumably less important to a viable molecule of cytochrome *c*.

The heme crevice is hot, indicating that strong selection pressures tend to keep the environment of the heme group constant throughout evolution. The invariant residues 70 through 80 are also hot, and we now see that they are folded to make the left side of the molecule and the pocket in which the heme sits. The right side of the molecule is warm, consisting of sites where only one, two or three different amino acids are tolerated. The back of the molecule is its cool side; residues 58 and 60 and four more residues on the back of the alpha helix are each occupied by six or more different amino acids in various species. These are powerful clues to the important parts of the molecule, whether for electron transfer or for interaction with two large molecular complexes, the reductase and the oxidase.

How the Molecule Folds Itself

If we now turn to the illustration on page 86, which shows all the side chains of horse-heart cytochrome *c*, many of the evolutionary conservatisms become understandable. (As before, amide groups are still shown only as straight lines; their atomic positions are known but are not particularly relevant to this article.) In this illustration the colors are selected to classify the various sites according to the character of the amino acid tolerated (hydrophilic, hydrophobic or ambivalent); the same color coding applies in the illustration on pages 84 and 85 showing the amino acid sequences in the cytochrome *c*'s of 38 different species.

Nonpolar, hydrophobic groups are found predominantly on the inside of the molecule, away from the external aqueous world, whereas charged groups, acidic or basic, are always on the outside. This arrangement is a good example of the "oil drop" model of a folded protein. According to this model, when an amino acid chain is synthesized inside a cell, it is helped to fold in the proper way by the natural tendency of hydrophobic, or "oily," side chains to retreat as far as possible from the aqueous environment and cluster in the center of the molecule. An even stronger statement can be made: If it is necessary for the successful operation of a protein molecule that certain portions of the polypeptide chain be folded into the interior, then natural selection will favor the retention of hydrophobic side chains at that point so that the proper folding is achieved. A charged, or hydrophilic, side chain can be pushed into the interior of a protein molecule, but a considerable price must be paid in terms of energy. Thus in most cases the presence of a charged group at a given site helps to ensure that the chain at that point will be on the outside of the folded molecule. (Charged groups inside a protein are known only in one or two cases where they play a role in the catalytic mechanism of the protein.)

We can now see the reason for the evolutionary conservatism of hydrophobic side chains, and one of the reasons for the conservatism of the hydrophilic residue lysine: they help to make the molecule fold properly. Radical changes of side chain that prevent proper folding are lethal. No folding, no cytochrome; no cytochrome, no respiration; no respiration, no life. It is seldom that cause and effect in evolution are quite so clear-cut.

There is still more to the lysine story. The lysines are not only on the outside; they are clustered in two positively charged regions of the molecular surface, separated by another zone of negative charge. This segregation of charge has not been found in any other protein structure and, as we shall see, probably occurs because cytochrome *c* interacts with two molecular complexes (the reductase and the oxidase) rather than with small substrate molecules as an en-

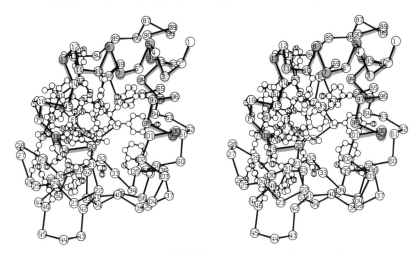

STEREOSCOPIC PAIR OF RIGHT SIDE of oxidized cytochrome *c* shows two sequences forming alpha helixes: the sequence from 1 through 11 and the sequence from 89 through 101. The two alpha helixes and the chain from 12 through 20 outline the right channel. Like the left channel it is lined with hydrophobic side groups, but it apparently contains a slot large enough to receive a hydrophobic side chain from another molecule. As in the stereoscopic drawing on the opposite page, alpha carbons with positively charged side chains around this channel are indicated in dark color; alpha carbons with strongly hydrophobic side chains are indicated in light color. The computer program for preparing the stereoscopic pictures was written by Carroll Johnson of the Oak Ridge National Laboratory.

zyme does. The charge arrangements are believed to be part of the process by which large molecules recognize each other.

Most of the 19 lysines are found on the left and right sides of the molecule, viewed from the back on page 88. The left and right sides of the molecule can be examined separately in the two stereoscopic pairs on these two pages. On the left side eight lysines surround a loop of chain from sites 55 through 75 that is tightly packed with hydrophobic groups, including the invariant tyrosine 74, tryptophan 59 and tyrosine 67 farther inside. Although we do not yet know the electron-transfer mechanism, it has been suggested that the aromatic rings of the three invariant residues could provide an inward path for the electron when cytochrome *c* is reduced. Another eight lysines are found on the right side, on the periphery of what appears to be a true channel large enough to hold a hydrophobic side chain of another large molecule. This right-side channel is bounded by the two alpha helixes and by the continuation from the first alpha helix through residues 12 to 20. Within this channel are found two large aromatic side chains: phenylalanine 10 (which cannot change) and tyrosine 97 (which can also be phenylalanine but nothing else). In summary, on the right side is a channel lined with hydrophobic groups (including two aromatic rings) and surrounded by an outer circle of positive charges. As someone in our laboratory remarked on looking at the

model, it resembles a docking ring for a spaceship.

This remark may not be entirely frivolous. It is known from chemical work that the attraction between cytochrome *c* and the cytochrome oxidase complex is largely electrostatic, involving negatively charged groups on the oxidase and positively charged basic groups on cytochrome *c*. Either the left or the right cluster of lysines must be involved in this binding. Moreover, Kazuo Okunuki of the University of Osaka has shown that if just one positive charge, lysine 13, is blocked with a bulky aromatic chemical group, the reactivity of cytochrome with its oxidase is cut in half. Chemically blocking lysine 13 means physically blocking the upper part of the heme crevice. Lysine 13 is closer to the right cluster of positive charges than to the left; thus it would appear more likely that the heme crevice and the right channel together are the portions of the molecular surface that "see" the oxidase complex.

What, then, are the roles of the positive zone on the left side and the negative patch at the rear? The positive zone, with its three aromatic rings, may be the binding site to the reductase; we know virtually nothing about the chemical nature of this binding. The negative patch may be a "trash dump," an unimportant part of the molecule's surface where there are enough negative charges to prevent an excessively positive overall charge. The fact that the six most variable amino acid sites are in this part

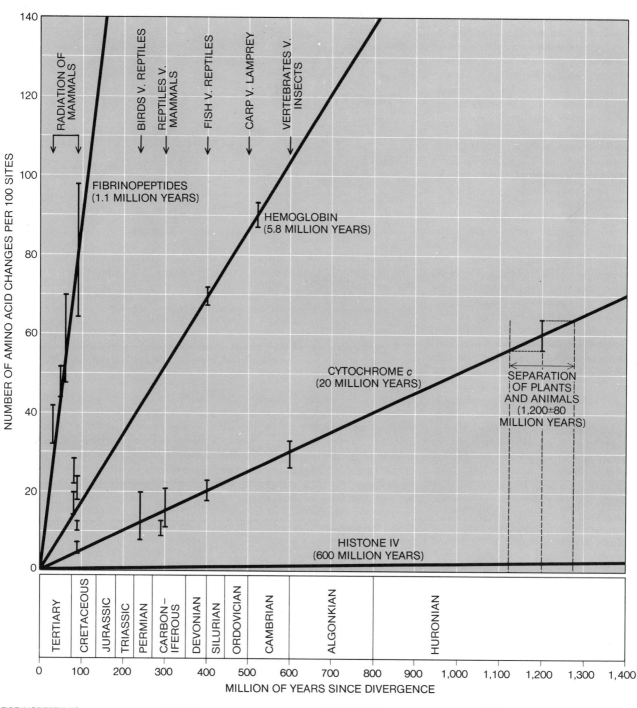

NUMBER OF AMINO ACID CHANGES PER 100 SITES

RADIATION OF MAMMALS

BIRDS V. REPTILES

REPTILES V. MAMMALS

FISH V. REPTILES

CARP V. LAMPREY

VERTEBRATES V. INSECTS

FIBRINOPEPTIDES
(1.1 MILLION YEARS)

HEMOGLOBIN
(5.8 MILLION YEARS)

CYTOCHROME c
(20 MILLION YEARS)

SEPARATION
OF PLANTS
AND ANIMALS
(1,200±80
MILLION YEARS)

HISTONE IV
(600 MILLION YEARS)

TERTIARY | CRETACEOUS | JURASSIC | TRIASSIC | PERMIAN | CARBON– IFEROUS | DEVONIAN | SILURIAN | ORDOVICIAN | CAMBRIAN | ALGONKIAN | HURONIAN

MILLION OF YEARS SINCE DIVERGENCE

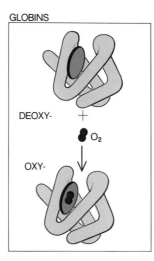

FIBRINOPEPTIDES

FIBRINOGEN

A   B

FIBRIN   +

A   B

PEPTIDES

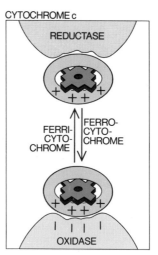

GLOBINS

DEOXY-   +

O₂

OXY-

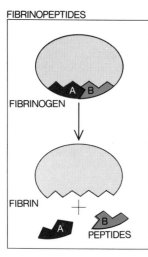

CYTOCHROME c

REDUCTASE

FERRI-
CYTO-
CHROME

FERRO-
CYTO-
CHROME

OXIDASE

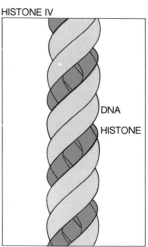

HISTONE IV

DNA

HISTONE

of the molecule would support such an idea.

On the other hand, it is equally possible that this collection of negative charges has a function. Acidic amino acids are actually conserved throughout the various species, although in a subtle way that was overlooked in the earlier sequence comparisons. Selection pressures have kept this zone of the molecular surface negative, even though the individual residues that carry the negative charges differ from one species to another. Because several sections of the protein chain bend into and out of this acidic region, the conservation of negative charge is not immediately obvious if one looks only at the stretched-out sequence. This is a good illustration of the principle of molecular evolution that natural selection acts on the folded, functioning protein and not on its amino acid sequence alone.

If we look carefully at where the glycine residues are, we can appreciate why such a large number are evolutionarily invariant. The heme group is so large that 104 amino acids are barely enough to wrap around it. There are many places where a chain comes too close to the heme or to another chain for a side chain to fit in. It is just at these points that we find the glycines with their single hydrogen atom as a side chain.

The last type of conservatism, the conservatism of the aromatic side chains, is more difficult to explain. Tyrosines and phenylalanines tend to occur in nearby pairs in the folded cytochrome c molecule: residues 10 and 97 in the right channel, 46 and 48 below the heme crevice, 67 and 74 along with tryptophan 59 in the hydrophobic left channel. Only residue 36, which can be tyrosine, phenylalanine or isoleucine, seems to have merely a space-filling role on the back of the molecule; it is an "oily brick."

The three aromatic rings in the left channel may be involved in electron transfer during reduction. The two rings in the right channel could also be employed in electron transfer, or might only help to define the hydrophobic slot in the middle of that channel. Tyrosine 48 at the bottom of the molecule helps to hold the heme in place by making a hydrogen bond to one of the heme's propionic acid side chains. In cytochrome c from the tuna and the bonito, where residue 46 is tyrosine, electron-density maps have shown that this residue also holds the heme by a hydrogen bond to its other propionic acid group. These two tyrosines, along with cysteines 14 and 17, help to lock the heme in place in a way not seen in hemoglobin or myoglobin.

Phenylalanine 82 is the enigma. Never tyrosine or anything else, it extends its oily side chain out into the aqueous world on the left side of the heme crevice, where it has no visible role. A price must be paid in energy for its being there. Why should such a large hydrophobic group be on the outside of the molecule, and why should it be absolutely unchanging through the entire course of evolution? Viewing the oxidized molecule alone, it is impossible to say, but when at the end of this article we look briefly at the recently revealed structure of reduced cytochrome, we shall see the answer fall into place at once.

The structural reasons for the evolutionary conservatism in cytochrome c throughout the history of eukaryotic life can now largely be explained. Cytochrome c is unique among the structurally analyzed proteins in that it has segregated regions of charge on its surface. The roles assigned to these regions in the foregoing discussions have been speculative and may be quite wrong. What we can be sure of is that these regions do have roles in the operation of the molecule. Chance alone, or even common ancestry, could not maintain these positive and negative regions, along with paired and exposed aromatic groups, in all species through more than a billion years of molecular evolution. The conservative sequences are shouting to us, "Look!" Now we have to be clever enough to know what to look for.

Rates of Protein Evolution

With this background we are equipped to return to a question raised earlier: What determines the rates of evolution of different proteins? We begin by making a graph where the vertical axis represents the average difference in amino acid sequence between two species of organism on two sides of an evolutionary branch point, for example the branch point between fish and reptiles or between reptiles and mammals. The horizontal axis represents the time elapsed since the divergence of the two lines as determined by the geological record. If such a graph is plotted for cytochrome c, one finds that all the branch points fall close to a straight line, indicating a constant average rate of evolutionary change [see illustration on opposite page].

How can this be? How can cytochrome c change at so nearly a constant rate during the long period in which the external morphology of the organism was diversifying toward the present-day cotton plant, bread mold, fruit fly, rattlesnake and chimpanzee? This is an illustration of a fundamental advantage of proteins as tools in studying evolution. Natural selection ultimately operates on populations of whole living organisms, the only criterion of success being the ability of the population to survive, reproduce and leave behind a new generation. The farther down toward the molecular level one goes in examining living organisms, the more similar they become and the less important are the morphological differences that separate a clam from a horse. One kind of chemical machinery can serve many diverse organisms. Conversely, one external change in an organism that can be acted on by natural selection is usually the effect not of a single enzyme molecule but of an entire set of metabolic pathways.

The observed uniform rate of change in cytochrome c simply means that the biochemistry of the respiratory package, the mitochondrion, is so well adjusted, and the mitochondrion is so well insulated from natural selection, that the selection pressures become smoothed out at the molecular level over time spans of millions of years. A factory can convert

RATES OF EVOLUTION OF PROTEINS (*opposite page*) can be inferred by plotting average differences in amino acid sequences between species on two sides of an evolutionary branch point that can be dated, for example the branch point between fish and reptiles or between reptiles and mammals. The average differences (*vertical axis*) have been corrected to allow for the occurrence of more than one mutation at a given amino acid site. The length of the vertical data bars indicates the experimental scatter. Times since the divergence of two lines of organisms from a common branch point (*horizontal axis*) have been obtained from the geological record. The drawings below the graph show schematically the function (described in the text) of the molecules whose evolutionary rate of change is plotted. The rate of change is proportional to the steepness of the curve. It can be represented by a number called the unit evolutionary period, which is the time required for the amino acid sequence of a protein to change by 1 percent after two evolutionary lines have diverged. For fibrinopeptides this period is about 1.1 million years, whereas for histone IV it is 600 million years. The probable reasons for these differences are discussed in the text.

from making military tanks to making sports cars and keep the same machine tools and power source. Similarly, a primitive eukaryote cell line can lead to such diverse organisms as sunflowers and mammals and still retain a common metabolic chemistry, including the respiratory package that comprises cytochrome c. One of the advantages of proteins in studying the process of evolution is just this relative insulation from the immediate effects of external selection. Protein structure is farther removed from selection pressures and closer to the sources of genetic variation in DNA than gross anatomical features or inherited behavior patterns are.

The only other proteins for which enough sequence information is available to allow this kind of analysis are hemoglobin and the fibrinopeptides: the short amino acid chains left over when fibrinogen is converted to fibrin in the process of blood clotting. One hemoglobin chain consists of approximately 140 amino acids. Fibrinopeptides A and B, on the other hand, consist of only about 20 amino acids, which are cut out of fibrinogen and discarded during the clotting process. The hemoglobins and the fibrinopeptides also appear to be evolving individually at a uniform average rate, but their rates are quite different. Whereas 20 million years are required to produce a change of 1 percent in the amino acid sequence of two diverging lines of cytochrome c, the same amount of change takes a little less than six million years in hemoglobins and just over one million years in the fibrinopeptides, as indicated on page 92. The approximate time required for a 1 percent change in sequence to appear between diverging lines of the same protein is defined as the unit evolutionary period. That period has been roughly estimated for a number of proteins for which only two or three sequences from different species are known. Most simple enzymes evolve approximately as fast as hemoglobin and much more rapidly than cytochrome c. Does all of this mean that the genes for these proteins are mutating at different rates? Are we looking at differences in variation or in selection?

Since there is no evidence to suggest variable rates of mutation, one asks what case can be made for differences in selection pressure among the different proteins. The case appears to be quite convincing [see inset figures in illustration on page 92]. The fibrinopeptides are "spacers" that prevent fibrinogen from adopting the fibrin configuration before the clotting mechanism is triggered. As long as they can be cut out by an enzyme when the time comes for the blood to clot, they would seem to have few other requirements. Thus one would expect a fibrinogen molecule to tolerate many random changes in the fibrinopeptide spacers. If the unit evolutionary period measures not the rate of appearance of mutations but the rate of appearance of harmless mutations, then it is not surprising that a 1 percent change can occur in the sequence of fibrinopeptides in just over a million years.

A successful hemoglobin molecule has more constraints. Each hemoglobin molecule embodies four heme groups that not only bind oxygen but also cooperate in such a way that the oxygen is released more rapidly into the cell when the local acidity, created by the presence of carbon dioxide, builds up. The structural basis for this "breathing" mechanism has only recently been explained with the help of X-ray crystallography by M. F. Perutz and his co-workers at the Medical Research Council Laboratory of Molecular Biology in England. If a random mutation is five times as likely to be harmful in hemoglobin as in the fibrinopeptides, one can account for hemoglobin's having a unit evolutionary period that is five times as long.

The chances of randomly damaging cytochrome c are evidently three to four times greater than they are for hemoglobin. Why should this be, and why should the unit evolutionary period for cytochrome c be greater than the period for enzymes of comparable size? The X-ray structure has given us a clue to the answer. Cytochrome c is a small protein that interacts over a large portion of its surface with molecular complexes that are larger than itself. It is virtually a "substrate" for the reductase and oxidase complexes. A large fraction of its surface is subject to strong conservative selection pressures because of the requirement that it mate properly with other large molecules, each with its own genetic blueprint. This evidently explains why the patches of positive and negative charge are preserved so faithfully throughout the history of eukaryotic life. Hemoglobin and most enzymes, in contrast, interact principally with smaller molecules: with oxygen in the case of hemoglobin or with small substrate molecules at the active sites of enzymes. As long as these restricted regions of the molecule are preserved the rest of the molecular surface is relatively free to change. Mutations are weeded out less rigorously and sequences diverge faster.

A satisfying confirmation of these ideas comes from the amino acid sequences of histone IV, one of the basic proteins that binds to DNA in the chromosomes and that may play a role in expressing or suppressing genetic information. When molecules of histone IV from pea seedlings and calf thymus are compared, one finds that they differ in only two of their 102 amino acids. If we adopt an approximate date of 1.2 billion years ago from the cytochrome study for the divergence of plants and animals, we find that histone IV has a unit evolutionary period of 600 million years. Clearly the conservative selection pressure on histone IV must be intense. Since histone IV participates in the control processes that are at the heart of the genetic mechanism, its sensitivity to random changes is hardly surprising.

The date of 1.2 billion years ago for the divergence of plants and animals is based on the cytochrome-sequence comparisons, assuming that the observed linear rate of evolution of cytochrome c in more recent times can be extrapolated back to that remote epoch. Is this a fair extrapolation? It probably is for cytochrome c because the biochemistry of the mitochondrion evolved still earlier; the great similarity in respiratory reactions among all eukaryotes argues that there has not been much innovation in cytochrome systems since. The respiratory chain had probably "settled down" by 1.2 billion years ago. It is reassuring that the cytochrome figure of 1.2 billion years is in harmony with the relatively scarce fossil record of Precambrian life.

If one accepts the provocative suggestion that eukaryotes developed from a symbiotic association of several prokaryotes, one of which was a respiring bacterium that became the ancestor of present-day mitochondria, one is obliged to conclude that the respiratory machinery had stabilized in essentially its present form before or during this symbiosis. The same thing cannot be said for hemoglobin and its probable ancestor, myoglobin. They can provide no clue to the date when animals and plants diverged, since the globins were evolving to play several different roles during and after this period, as multicelled organisms arose. In no sense had the globins settled down 1.2 million years ago. Nevertheless, if the right proteins are selected, and if the data are not overextended, it should be possible to use the rates of protein evolution to assign times to events in the evolution of life that have left only faint traces in the geological record.

So far we have mentioned the electron-transfer mechanism of cytochrome *c* only in passing, virtually ignoring the structure of the reduced molecule. The mechanism is another story in itself, and one that cannot yet be written. One hopes that the clues supplied by X-ray analysis will suggest the best chemical experiments to try next, in order to learn the mechanism of the oxidation-reduction process. The reduced cytochrome structure has been obtained so recently that it would be premature to base many deductions on it.

A Glimpse of Molecular Dynamics

One obvious structural feature, which undoubtedly has great physiological significance, is that in the reduced molecule the top of the heme crevice is closed. The chain from residues 80 to 83 swings to the right (as the molecule is depicted on page 86), the exposed phenylalanine 82 slips into the heme crevice to the left of the heme and nearly parallel to it, and the heme becomes less accessible to the outside world. The absolute preservation of this phenylalanine side chain throughout evolution, in an environment that is energetically unfavorable in the oxidized molecule, argues that closing of the heme crevice in the reduced molecule is important for its biological activity.

Several explanations might be offered. The aromatic ring of phenylalanine 82 may be part of the electron-transfer mechanism, or its removal from the heme crevice may be necessary to permit an electron-transferring group to enter beside the heme or just to approach the edge of the heme. At a minimum the refolding of the chain from residues 80 to 83 may be a "convulsive" motion that pushes the oxidase complex away from the protein after electron transfer is achieved by some other pathway.

This article has been speculative enough without making a choice between these or other alternatives. At this stage, as both oxidized and reduced cytochrome analyses are being extended to higher resolution, it is enough to say that we can see more refolding of the protein chain in passing between the two states than has been observed in any other protein. Phenylalanine 82 swings to an entirely new position and several other aromatic rings change orientation, including the three in the left channel. As the molecule is reduced, the right channel apparently is partly blocked by residues 20 and 21. We now have pictures of both strokes of a very ancient two-stroke molecular engine. We hope in time to be able to figure out how it operates.

9

THE STRUCTURE AND FUNCTION OF ANTIBODIES

GERALD M. EDELMAN

August 1970

The complete amino acid sequence of an immunoglobulin molecule has been determined, defining the structure of antibodies and providing information on their evolution and differentiation and how they work

"Immunity" is an everyday word, ordinarily applied to the elaborate set of responses by which the body defends itself against invading microorganisms or foreign tissues. There is much more to immunity than its clinical aspects, however. What we have come to know about the immune system and its key molecules, antibodies, makes it apparent that immunology bears directly on some very fundamental problems: the nature of the mechanisms whereby molecules recognize one another, the manner in which genes are expressed in higher organisms and the origin of a variety of disease states, including cancer. In one way or another the solution of these deep problems will require an understanding of the structure of antibody molecules.

Antibodies have been known since the classic studies of Emil von Behring in the late 19th century. Only recently, however, have we begun to understand how vertebrate organisms recognize the sometimes subtle chemical differences between their own molecules and foreign molecules, which are termed antigens. Our insights rest on three serial developments. The first was the demonstration by Karl Landsteiner in 1917 that animals could form antibodies against certain small organic chemicals of known structure. By manipulating and modifying these "haptens," Landsteiner and others showed that antibodies distinguish among different antigens by recognizing differences in their shape. Moreover, the early studies implied that the number of different antibodies any single animal can make must be very large indeed. If an animal could make antibodies that could specifically bind a synthetic hapten the animal or its ancestors had never encountered before in nature, it seemed likely that the animal could make antibodies to almost any foreign antigen. Further work has largely supported this inference: most vertebrates are indeed capable of making many thousands of different antibodies. The second fundamental development was the bold idea of selective immunity advanced in the late 1950's by Niels K. Jerne and Sir Macfarlane Burnet, who proposed that the body already has all the information for making any of its antibodies *before* it ever encounters an antigen. The third development was the analysis of the structure of antibodies, which has taken place largely in the past decade. Before discussing this last development in detail, it will be useful to outline the seminal idea of selective immunity and some of the advances made by cellular immunologists.

The main notion, as implied above, is that the cells of the antibody-forming system have among them all the information they need to make any antibody molecule before they ever encounter any antigen. The antigen molecule does not instruct antibody-producing cells to shape the antibody molecule to fit it. Instead it selects cells that are already making antibodies that happen to fit. Then it stimulates those cells to make large quantities of the antibodies.

Antibodies can be likened to ready-made suits. The antigen is a buyer who decides to pick a number of different suits that fit more or less well rather than instruct a tailor to make one suit to fit him to order. To be well satisfied, the buyer must patronize a store with a very large stock of suits in a great variety of sizes and styles. The immune system is like a store with an almost unlimited stock, one ready to please any possible customer. This analogy fails in one important respect: to be complete it should provide that after each somewhat different ready-made suit is picked the manufacturer would proceed to make thousands of exact copies of it.

In a simplified picture of the cellular mechanisms corresponding to this selective response, each cell makes only one kind of antibody, which has an antigen-binding site of a particular shape [*see top illustration on page 98*]. Presumably that antibody is located at the cell surface. An antigen injected into the body "tries on" different shapes. If a particular antibody "fits" more or less well, the cell making it divides and matures. Its progeny then make many more copies of the identical kinds of antibody, which may then be released into the blood to carry out their function of defense or of tissue rejection. Notice that this is a form of molecular recognition machine and that the specificity of recognition rests both on the presentation of a variety of antibodies and on the capacity of the cells to "amplify" the results of a recognition event. How are these requirements accomplished at a molecular level? In order to answer that question satisfactorily we must know a good deal about the structure of the antibody molecule itself.

A decade ago a number of experimental observations began to shed light on the details of antibody structure. The advance came when the results of sophisticated protein chemistry and genetic analysis fitted together with a classical observation about the products of a certain cancer. At the Rockefeller Institute in 1959 I found that antibody molecules consisted of polypeptide chains, or protein subunits, of more than one kind, and that the chains could be separated from one another by chemical means and studied in detail. These chains were called light chains and heavy chains because of the difference in their size, or molecular weight. At the same time R. R. Porter, now at the University of Oxford, showed that the antibody molecule could be cut into three different pieces by enzymes that cleave polypeptide chains. Two of them, termed "fragment antigen binding" (F*ab*), were identical and would still combine with antigen. The third, "fragment crystalline" (F*c*), was quite different: it would not combine with antigen but could be crystallized readily [see "The Structure of Antibodies," by R. R. Porter; SCIENTIFIC AMERICAN Offprint 1083]. The observations on the two polypeptide chains and on the two fragments were the starting point for a series of investigations in many laboratories into the details of antibody structure.

Antibodies, it was known, were a family of proteins with a number of properties in common, found in the gamma globulin fraction of the blood. Unlike all proteins whose structure had been determined, antibody molecules were known to be very "heterogeneous": no single sequence of amino acids—the building blocks of proteins—could represent the polypeptide chains of antibodies, as can be done, for example, in the case of a "homogeneous" protein such as insulin or hemoglobin. This fact, together with the large size of antibody molecules, made it infeasible to carry out detailed chemical analyses of such molecules or to determine exactly how antibodies that

bound to different antigens differed from one another. Fortunately the structural studies of the polypeptide chains of the immunoglobulin molecule led to a clue that made it possible to bypass the problem raised by the intrinsic heterogeneity of antibodies. The clue had to do with the nature of certain homogeneous proteins made by tumors of plasma cells, the cells that ordinarily produce the most antibodies.

Knowledge of these tumors goes back to 1847, when Henry Bence Jones, physician at St. George's Hospital in London, published a paper titled "On a new substance occurring in the Urine of a patient with Mollities Ossium." It

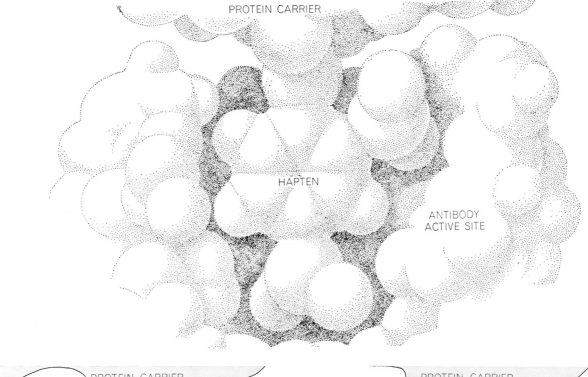

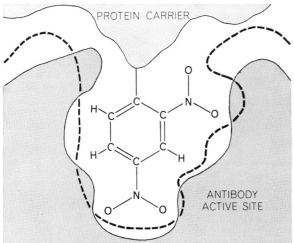

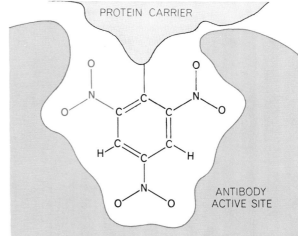

ANTIBODY recognizes a specific antigen by its fit. The top drawing shows the fit between the combining site on an antibody and a "hapten" antigen, a dinitrophenyl group, on a protein carrier. The bottom drawings show that two different antibody contours (*gray shape and black line*) can fit the dinitrophenyl group, one better than the other (*left*). If a third nitro group (*color*) were on the hapten (*right*), those two antibodies would not fit; a third antibody with a different antigen-combining site (*color*) would fit this picryl antigen. Because of picryl's similarity to dinitrophenyl, the third antibody would also fit the original hapten, but less precisely.

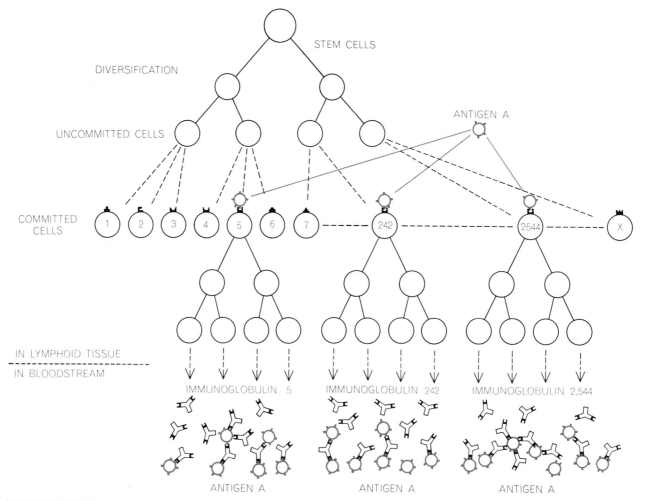

SELECTIVE-IMMUNITY THEORY holds that stem cells (precursors of antibody-producing cells) contain information for making all possible antibodies; at some point in embryonic development each is committed to producing a unique immunoglobulin (num-bers). These receptor antibodies can interact with various antigens. A single antigen (*color*) may be recognized by more than one antibody-producing cell. Interaction of an antigen with a cell stimulates proliferation of the cell and the synthesis of antibody.

began as follows: "On the 1st of November 1845 I received from Dr. Watson the following note, with a test tube containing a thick, yellow, semi-solid substance:—'The tube contains urine of very high specific gravity; when boiled it becomes highly opake; on the addition of nitric acid it effervesces, assumes a reddish hue, becomes quite clear, but, as it cools, assumes the consistence and appearance which you see: heat reliquifies it. What is it?' "

Jones verified the peculiar thermosolubility properties of the protein, subjected it to a careful elementary analysis and concluded that it was the "hydrated deutoxide of albumin."

In succeeding years many attempts were made to answer Dr. Watson's question in a definitive way, but although some 700 papers on the subject appeared

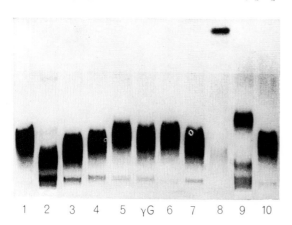

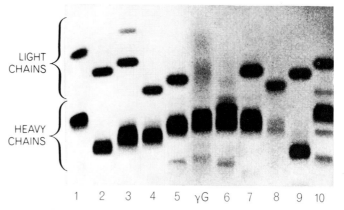

MYELOMA PROTEINS from different patients are compared with a normal human immunoglobulin fraction by electrophoresis on a starch gel. The intact immunoglobulins migrate in approximately the same way (*left*). After dissociation into light and heavy chains (*right*) the myeloma proteins show sharp bands in the light-chain position but the normal protein shows a diffuse zone there because it is a heterogeneous mixture of immunoglobulins. Protein No. 8 is a Bence Jones protein and it therefore shows no heavy chain.

in the ensuing century, Bence Jones proteins remained a kind of medical and biochemical curiosity except in the domain of practical diagnosis: the demonstration of the protein in the urine called for a diagnosis of multiple myeloma, the malignant disease of plasma cells that was formerly called mollities ossium. This disease is usually associated with malignant proliferation of plasma cells, excessive production of serum gamma globulins called myeloma proteins, bone lesions, disturbances of calcium metabolism, kidney disorders and often, of course, excretion of the characteristic Bence Jones proteins.

It seemed that these proteins had something to do with immunoglobulins, but the exact relation remained obscure. The 1959 finding that immunoglobulins contained multiple polypeptide chains [see bottom illustration on opposite page] suggested that Bence Jones proteins were homogeneous light chains of the myeloma protein made by the tumor but not incorporated into whole molecules. This hypothesis was confirmed by my student Joseph A. Gally and me in 1962. Because different Bence Jones proteins had different amino acid compositions, we compared their properties with those of light chains of antibodies. These comparisons were instrumental in suggesting that antibodies with different antigen specificities differ from one another in the sequence of amino acids of which they are composed. Moreover, because the light chains produced by the myeloma tumors were pure, available in large amounts and smaller than a whole immunoglobulin molecule, it became possible to study the details of their chemical structure.

Once it had become apparent that these proteins might provide a clue to the nature of antibody variability, a number of laboratories undertook the task of determining their exact amino acid sequence. Since no two individuals produce the same Bence Jones proteins, each laboratory reported a different sequence, but from the first report of partial sequences by Norbert G. D. Hilschmann and Lyman C. Craig at Rockefeller University in 1965 it became clear that these proteins had a singular structure. Each molecule contained about 214 amino acids, linked together in a polypeptide chain. (The amino acids are numbered starting from the end of the molecule that is made first, the amino terminus.) From position 109 on various Bence Jones proteins had essentially the same sequence, and accordingly this part of the molecule was called the constant region. In striking contrast, the sequence of the first 108 amino acids differed markedly from one Bence Jones protein to another, and this first part of the chain was designated the variable region. Concurrent studies in several laboratories suggested that the heavy chains of myeloma proteins also had variable and constant regions. The homogeneity of the constant region enabled Robert L. Hill and his associates at Duke University to determine the amino acid sequence of the Fc fragment of rabbit immunoglobulin.

It was against this background that my colleagues and I decided to attempt the determination of the complete amino acid sequence of a whole immunoglobulin molecule. The earlier structural studies had suggested an overall picture and we wanted to confirm and extend it in detail so that we could apply it to an analysis of the origin and function of antibodies. We obtained a large amount of plasma from a patient with multiple myeloma, because it was essential to have enough of at least one myeloma protein. As a matter of fact, we obtained two different proteins from different patients for purposes of comparison. The difficulties of our project were related to the enormous size of the molecule, which has 19,996 atoms and is larger in terms of the number of unique amino acid sequences than any protein that had been determined up to that time. Our approach was based on the pioneering methods first developed by Frederick Sanger to analyze the insulin molecule, and the challenge, successfully met last year, was whether these methods would suffice.

The immunoglobulin molecule is about 25 times as large as insulin. If we could break the immunoglobulin molecule, its chains and its fragments into small pieces about the size of the insulin molecule itself, then we could use standard methods for determining amino acid sequence. A useful tool for such protein surgery had already been devised by Erhard Gross and Bernhard

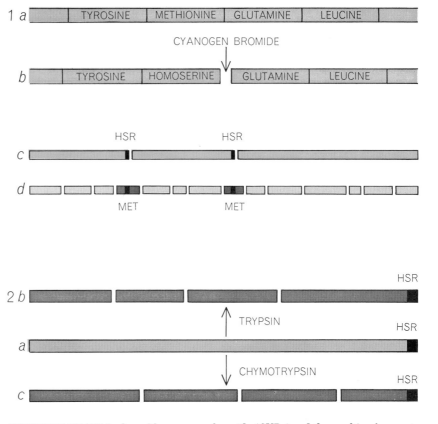

ANTIBODY CHAIN is cleaved by cyanogen bromide (CNBr) and the resulting fragments are ordered (1). The CNBr breaks a chain (a) at methionyl residues, which are converted into homoserine (b). To order CNBr fragments (c) the original chain is cleaved also with the enzyme trypsin, the tryptic fragments containing methionine are isolated (d) and their sequences are compared with those of the ends of the CNBr fragments. Then the amino acid sequence of each CNBr fragment must be determined (2). This is done by cleaving a CNBr fragment (a) with trypsin and determining the sequence of each tryptic peptide (b) by a chemical procedure. The tryptic peptides are ordered by comparison with the composition and partial sequences of different peptides (c) made by cleaving with chymotrypsin.

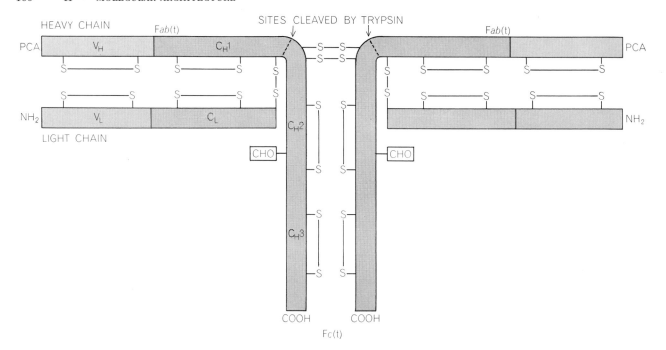

STRUCTURE of the immunoglobulin determined by the author and his colleagues shows two kinds of chain and regions in each. The protein can be cleaved into two antibody-binding fragments, F*ab*(t), and a "crystallizable" fragment, F*c*(t). Sulfur-sulfur bonds are designated —S—S—. Light chains have variable and constant regions (V_L and C_L). Heavy chains have a variable region (V_H) and a constant region divisible into three homology regions (C_H1, C_H2 and C_H3). CHO indicates carbohydrate. Chains have amino (NH_2), carboxyl ($COOH$) and pyrollidonecarboxylic acid (PCA) ends. V_H and V_L are homologous, as are C_L, C_H1, C_H2 and C_H3.

Witkop of the National Institutes of Health. It depended on cyanogen bromide, a reagent that selectively cleaves polypeptide chains at the positions occupied by the sulfur-containing amino acid methionine; because there were just a small number of methionines in the molecule, we could expect a decently small number of pieces. With this reagent we were able to cleave the heavy chain of the immunoglobulin into seven pieces and the light chain into three pieces [see *illustration on preceding page*]. Each piece was then separated from the others by chromatography. A key procedure in these separations was molecular sieving on Sephadex, a technique developed largely by Jerker O. Porath of the University of Uppsala, which speeded up the thousands of separations required to determine the structure of immunoglobulin.

After the fractionation of the pieces the next step was to establish their order.

AMINO ACID SEQUENCES (three-letter codes; positions as numbered)

V_L (1 TO 108):
ASP ILE GLN MET THR GLN SER PRO SER THR LEU SER ALA SER VAL GLY ASP ARG VAL THR ILE THR CYS ALA ALA SER GLN SER ILE ASN THR — — TRP LEU ALA TRP TYR GLN GLN LYS PRO GLY LYS ALA PRO LEU LEU MET THR LEU ALA SER SER — LEU GLU

V_H (1 TO 114):
PCA VAL GLN LEU VAL GLN SER GLY — ALA GLU VAL LYS LYS PRO GLY SER SER VAL LYS VAL SER CYS LYS ALA SER GLY GLY THR PHE SER ARG SER ALA ILE ILE THR VAL ARG GLY ALA PRO GLY GLN LEU GLY TRP MET GLY ILE VAL PRO MET PHE GLY PRO

C_L (109 TO 214):
THR VAL ALA ALA PRO SER VAL PRO ILE PRO PRO PRO SER ALA GLY GLN — — LEU LEU SER GLY THR ALA SER VAL VAL CYS LEU LEU ALA ALA PRO THR PRO ALA GLY ALA LEU VAL — — GLY THR LEU VAL ALA ALA ALA LEU GLY SER GLY ALA SER GLY GLY SER

C_H1 (119 TO 220):
SER THR LEU GLY PRO SER VAL PRO PRO LEU ALA PRO SER SER LEU SER — — THR SER GLY GLY THR ALA ALA LEU GLY CYS LEU VAL LEU ALA THR PRO PRO GLY PRO VAL THR VAL — — SER THR ALA SER — GLY ALA LEU THR SER GLY — VAL HIS THR PRO PHE

C_H2 (234 TO 341):
LEU LEU GLY GLY PRO SER VAL PRO LEU PRO PRO PRO LEU PRO LEU ALA THR LEU MET ILE SER ALA THR PRO GLY VAL THR CYS VAL VAL ALA VAL SER HIS GLY ALA PRO GLY VAL LEU PRO ALA THR THR VAL ALA GLY — VAL GLY VAL HIS ALA ALA LEU THR LYS SER

C_H3 (342 TO 446):
GLY PRO ALA GLY PRO GLY VAL THR THR LEU PRO PRO PRO SER ALA GLY GLY — — MET THR LEU ALA GLY VAL SER LEU THR CYS LEU VAL LEU GLY PRO THR PRO SER ALA ILE ALA VAL — — GLY THR GLY SER ALA ALA GLY GLY PRO GLY ALA THR LEU THR THR

AMINO ACID SEQUENCES of the variable regions (*top*) and of the constant homology regions (*bottom*) were fully determined. The extent of the homology between the two variable regions and among the four constant homology regions is indicated by the

This was done by cleaving the original chains not with cyanogen bromide but with enzymes that attack polypeptides at other specific sites. Those peptides that contained methionine and that would therefore overlap the cyanogen bromide fragments were then isolated. By comparing the two kinds of fragment we could see which ends of the cyanogen bromide fragments butted up against one another.

Each separate cyanogen bromide fragment could now be studied independently as if it were a separate small protein or polypeptide. Accordingly it was cleaved with enzymes into smaller peptides, which were separated. When small pure peptides were obtained, the sequence of their amino acids was determined directly by a chemical procedure. The order of the peptides was then established by breaking the whole cyanogen bromide fragment with a second enzyme that cleaved it at different sites and isolating a second set of peptides that overlapped the first set.

The sequence determination was thus a "two pass" procedure. In the first pass we obtained cyanogen bromide fragments and ordered them. In the second we treated each fragment as a separate protein, obtaining its peptides, ordering them and determining their amino acid sequence. When these tasks were finished, there remained the job of determining the location of the bonds between the sulfur atoms of the amino acid cysteine that helped to link the chains and parts of the chains together.

The completed structure showed that the antibody molecule differed from proteins that had been analyzed earlier not only in size but also in more unusual ways. Our molecule was what is classified as a γG, or gamma G, immunoglobulin molecule, an example of the most prevalent class of immunoglobulins. As earlier studies had suggested, such a molecule consists of two identical light and two identical heavy chains [see top illustration on opposite page]. The structure is symmetrical, each half consisting of one light and one heavy chain. Although the actual shape or three-dimensional structure of the chains is not known, it is established that they are held together by weak forces and by interchain sulfur-sulfur bonds between corresponding pairs of cysteines; similar intrachain bonds are formed within each chain at approximately equal intervals. The most striking feature of the structure is its division into two kinds of region, variable regions and constant regions, whose disposition is related to these intervals. The length of the variable regions was determined by comparing the amino acid sequences of the light and the heavy chains to the sequences of Bence Jones proteins and to the sequence of another heavy chain that was analyzed concurrently. As in the Bence Jones proteins, the regions are so named because in different antibodies the variable regions differ in the sequence of amino acids that make up the chain, whereas the constant regions have the same sequence in each of the major classes of antibodies (except for a single variable amino acid at position 191). It has now been firmly established that it is the different sequences in the variable regions that give different shapes to various antigen-binding sites. The variety of shapes provides for a range of specific interactions with a great variety of antigens, including small molecules, other proteins, carbohydrates and even DNA itself.

There is another feature of the structure that merits notice. Detailed examination of the constant regions showed evidence of internal periodicity, which had already been hinted at by the distribution of the sulfur-sulfur bonds. Por-

ALA	ALANINE		LEU	LEUCINE
ARG	ARGININE		LYS	LYSINE
ASN	ASPARAGINE		MET	METHIONINE
ASP	ASPARTIC ACID		PCA	PYROLLIDONECARBOXYLIC ACID
CYS	CYSTEINE		PHE	PHENYLALANINE
GLN	GLUTAMINE		PRO	PROLINE
GLU	GLUTAMIC ACID		SER	SERINE
GLY	GLYCINE		THR	THREONINE
HIS	HISTIDINE		TRP	TRYPTOPHAN
ILE	ISOLEUCINE		TYR	TYROSINE
			VAL	VALINE

coloring or shading of identical residues in each position; dark and light shading indicates identities that occur in pairs at one position. Gaps have been introduced to maximize the homology. The numbering across the top is that of positions of residues in light chains.

I

1	2	3	4	5	6	7	8	9	10	11	12	13	14	15	16	17	18	19	20	21	22	23
ASP	ILE	GLN	MET	THR	GLN	SER	PRO	SER	SER	LEU	SER	ALA	SER	VAL	GLY	ASP	ARG	VAL	THR	ILE	THR	CYS
ASP	ILE	GLN	MET	THR	GLN	SER	PRO	SER	SER	LEU	SER	ALA	SER	VAL	GLY	ASP	ARG	VAL	THR	ILE	THR	CYS
ASP	ILE	GLX	MET	THR	GLX	SER	PRO	SER	**THR**	LEU	SER	ALA	SER	VAL	GLY	ASP	ARG	VAL	THR	ILE	THR	CYS
ASP	ILE	GLX	MET	THR	GLN	SER	PRO	SER	SER	LEU	SER	ALA	SER	VAL	GLY	ASP	ARG	**ILE**	THR	ILE	THR	CYS
ASP	**VAL**	GLX	MET	THR	GLN	SER	PRO	SER	SER	LEU	SER	ALA	SER	VAL	GLY	ASP	ARG	VAL	THR	ILE	THR	CYS
ASP	ILE	GLN	MET	THR	GLN	SER	PRO	SER	SER	LEU	SER	ALA	SER	**LEU**	**ARG**	ASP	ARG	VAL	THR	ILE	THR	
ASP	ILE	GLN	MET	THR	GLN	SER	PRO	SER	SER	LEU	SER	**VAL**	SER	VAL	GLY	ASP	ARG	VAL	THR	ILE	**ALA**	
ASP	ILE	GLN	**LEU**	THR	GLN	SER	PRO	SER	**PHE**	LEU	SER	ALA	SER	VAL	GLY	ASP	ARG	VAL	THR	ILE	THR	
ASP	ILE	GLN	MET	THR	GLN	SER	PRO	SER	**THR**	LEU	SER	ALA	SER	VAL	GLY	ASP	ARG	VAL	THR	ILE	THR	
ASP	ILE	GLN	MET	THR	GLN	SER	PRO	SER	SER	LEU	SER	ALA	SER	VAL	GLY	ASP	ARG	VAL	THR	ILE	THR	
ASP	ILE	GLN	MET	THR	GLN	SER	PRO	SER	SER	LEU	SER	ALA	SER	VAL	GLY	ASP	ARG	VAL	THR	ILE	THR	

II

1	2	3	4	5	6	7	8	9	10	11	12	13	14	15	16	17	18	19	20	21	22	23
GLU	ILE	VAL	LEU	THR	GLN	SER	PRO	GLY	THR	LEU	SER	LEU	SER	PRO	GLY	**ASP**	ARG	ALA	THR	LEU	SER	CYS
GLU	ILE	VAL	LEU	THR	GLN	SER	PRO	GLY	THR	LEU	SER	LEU	SER	PRO	GLY	GLU	ARG	ALA	THR	LEU	SER	CYS
LYS	ILE	VAL	LEU	THR	GLN	SER	PRO	GLY	THR	LEU	SER	LEU	SER	PRO	GLY	GLU	ARG	ALA	THR	LEU	SER	
ASP	ILE	VAL	LEU	THR	GLN	SER	PRO	**ALA**	THR	LEU	SER	LEU	SER	PRO	GLY	GLU	ARG	ALA	THR	LEU	SER	
GLU	**MET**	VAL	**MET**	THR	GLN	SER	PRO	**ALA**	THR	LEU	SER	**MET**	SER	PRO	GLY	GLU	ARG	ALA	THR	LEU	SER	
GLU	ILE	VAL	LEU	THR	GLN	SER	PRO	GLY	THR	LEU	SER	LEU	SER	PRO	GLY	**ASP**	ARG	ALA	THR	LEU	SER	
GLU	ILE	VAL	LEU	THR	GLN	SER	PRO	**ALA**	THR	LEU	SER	LEU	SER	PRO	GLY	GLU	ARG	ALA	THR	LEU	SER	

III

1	2	3	4	5	6	7	8	9	10	11	12	13	14	15	16	17	18	19	20	21	22	23
ASP	ILE	VAL	LEU	THR	GLN	SER	PRO	LEU	SER	LEU	PRO	VAL	THR	PRO	GLY	GLU	PRO	ALA	SER	ILE	**THR**	CYS
(GLU) ASP	ILE	VAL	**MET**	THR	GLN	**THR**	PRO	LEU	SER	LEU	PRO	VAL	THR	PRO	GLY	GLU	PRO	ALA	SER	ILE	SER	CYS
ASP								SER	LEU	PRO	VAL	THR	PRO	GLY	GLU	PRO	ALA	SER	**THR**	SER	CYS

PATTERN OF VARIATION of three different subgroups of one class of light-chain variable regions yields clues to their genetic origin. Each line represents a partial sequence (the first 23 residues) determined in the laboratories of H. D. Niall and P. Edman, C. Milstein, Norbert G. D. Hilschmann, Frank W. Putnam or Lee Hood. Each subgroup (*roman numerals*) has a characteristic sequence, indicated in each case by a dominant color or shade of gray. Within each subgroup there are variations (*black*) that arose from mutations. (*GLX* refers to positions where it was not yet definitely established whether the residue was glutamic acid or glutamine.)

tions of the constant region of the heavy chain turn out to have homologous amino acid sequences, that is, sequences more similar than could occur by chance. These portions are designated C_H1, C_H2 and C_H3, and each is homologous also to the constant region of the light chain, C_L. It is these constant regions that carry out functions of the molecule other than the binding of antigens. For example, C_H2 is believed to be bound by members of a complex family of serum proteins known as complement, thus beginning the series of reactions that is capable of killing cells, one of the aspects of the immune response.

The homology of the constant regions and the somewhat weaker homology of the variable regions to one another is demonstrated by directly comparing their amino acid sequences [*see bottom*

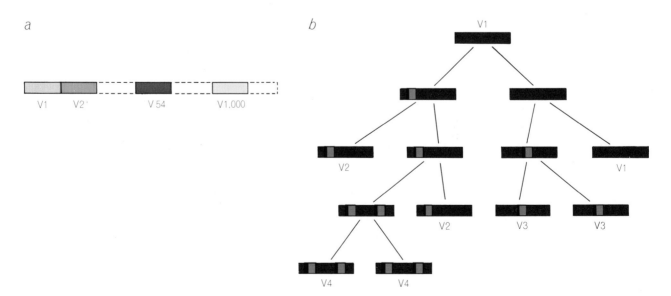

a

V1   V2   V54   V1,000

b

V1

V2   V1   V2   V3   V3   V4   V4

DIVERSITY of variable regions can be explained by three theories. A large number of genes, one for each variable region, could have arisen in the course of evolution (*a*). Alternatively, there could be one V gene, which mutates in an individual animal's body cells during development to produce the required variety (*b*). Finally, several V genes could evolve by mutation and be *illustration on pages 100 and 101*]. This is an unusual finding, and it means that the regions must be related in their evolutionary origins. It is likely that present-day antibodies have evolved by a process known as gene duplication. A primitive gene of a size sufficient to specify one homology region must have doubled and tripled, thereby forming a larger gene whose segments then became somewhat different from one another, as reflected in the sequences. By a similar process the genes for the two kinds of chain, heavy and light, appear to have had a common ancestor. This hypothesis, which was first suggested by Hill at Duke and S. Jonathan Singer of the University of California at San Diego, is strongly supported by the structure of the whole molecule.

A comparison of the variable regions with the four constant homology regions shows that although they have roughly the same length, they have few sequences in common. Did they also arise from the same original gene? Probably, far back in evolution, but if so, they must have diverged rapidly as they carried out different functions of the antibody. Indeed, studies by C. Milstein of the Medical Research Council in England and by Lee Hood of the National Institutes of Health indicate that in each individual there must be more than one gene for each variable (V) region. Earlier, pioneering genetic investigations by Jacques Oudin of the Pasteur Institute and by Rune Grubb of the University of Lund had laid the groundwork for the conclusion that there is only one gene for each constant (C) region. Since the polypeptide chains of antibodies appear to be made in one piece, as are other proteins, it seems that information from two genes is required to specify a single polypeptide chain. This is a unique situation, because in all proteins that have so far been investigated a single gene is enough to specify a single polypeptide chain.

The analysis of antibodies, then, poses two special problems: How can the V genes vary so that many different V regions are made in each individual? And how can such a V gene, which evolved to give the antibody system a range of different combining sites, be joined with a C gene that evolved to specify the constant portions of the chains and thus carry out effector functions?

Before attempting to suggest answers to these questions, let us look at the actual variation seen in V regions [*see top illustration on opposite page*]. The variation has several important characteristics. First, the genetic-code dictionary (in which each amino acid is coded for by a triplet of three DNA nucleotides) reveals that the variation arises from one-base changes in the code words for the amino acids in each variable position. This means that the variations were caused by mutations, just as in the case of other proteins. Second, not every position in the V region varies. For example, no one has ever observed that any of the cysteines that contribute to the sulfur bonds are missing or replaced by another amino acid. Third, certain positions seem to have more variations than others, although the number of examples is still too small for one to be completely sure of this. These last two observations mean that the variation is not random but is the result of some kind of selection. We can conclude that, as in other proteins, both mutation and selection are responsible for antibody variation.

The question about the origin of variability can be resolved into two more pointed questions. They are: Where and when do these processes occur? How many V genes are required? One theory suggests that the variation and selection occur during evolution, so that in an animal each different V region has a corresponding V gene. This would require a very large number of V genes in the germ cells. Another theory states that there is one V gene, which mutates not during evolution but somatically—in the body cells of the individual animal—and that the mutant cells are somehow selected. A third theory, which I favor, is that there are a few V genes, which have mutated and have been selected in evolution but which then recombine somatically in the cells of the animal to provide the broad variational pattern. This last theory has the advantage that the same processes that recombine the V genes could also accomplish the fusion of V and C genes that is required to make a single antibody chain; one can thus account with one mechanism for the two questions: How is antibody diversity created? How are V and C genes joined?

The mechanism may be one that is somewhat similar to mechanisms that have already been described for infection of the bacterium *Escherichia coli* by

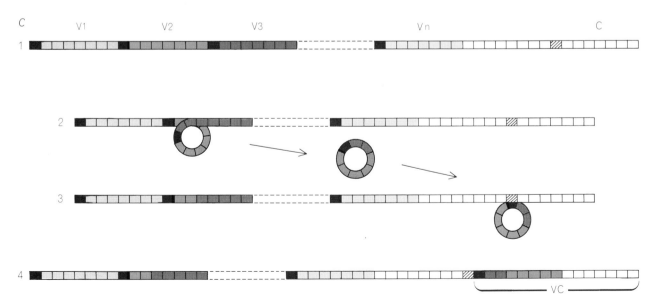

selected during evolution, and then be recombined in many different ways in the animal (*c*). In this process the evolved genes (*1*) might recombine to form a ring-shaped V-gene episome (*2*), a variant gene composed of sequences from adjacent V genes. The episome might be translocated and become integrated with the C gene (*3*) to form the complete VC gene that is expressed (*4*).

a

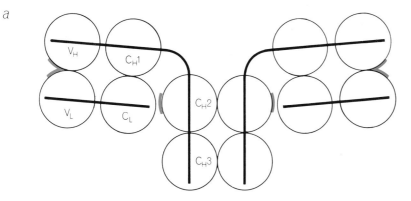

b

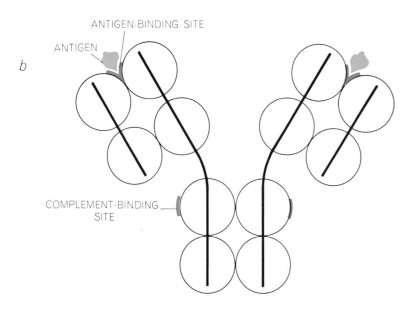

ANTIGEN-BINDING SITE

ANTIGEN

COMPLEMENT-BINDING
SITE

BINDING OF ANTIGEN may stabilize a change in the shape of the antibody that triggers a series of immune reactions. Here the antibody is drawn as a flexible grouping of compact domains, with the chain structure suggested by the heavy black lines (*a*). The antigen is bound by the variable regions, perhaps facilitating a pivoting movement of the molecule that exposes effector sites, such as the complement-binding site, in the C regions (*b*).

structure tell us about how these molecules actually carry out their functions? It is clear that antibodies have two kinds of task: first, recognizing the antigen, and second, doing something to the antigen or initiating a chain of cellular responses. The task of antigen-binding is delegated to the V regions. The more dynamic role of influencing cellular responses, the binding of complement and the initiation of processes that alter the antigen appears to be the function of the C regions. In some sense the antibody molecule must behave as a switch: binding the antigen must change the antibody's state in such a way as to "turn on" its effector functions. It is known that protein molecules can act as switches by changing their shape. There is now a hint that this may be the case for antibodies. When antibodies are viewed in the electron microscope after combination with antigens, their dimensions appear to be smaller than those of unbound immunoglobulins measured by X-ray scattering in solution. This raises the possibility that the binding of the antigen causes a rearrangement of the structure of the antibody molecule, which is known to be somewhat flexible. The rearrangement might consist, for example, of a pivoting movement involving part of the constant region of the heavy chains [*see illustration at left*]. Binding sites for complement might be exposed by this pivoting, as well as other sites for different effector functions. Similar mechanisms may be involved in triggering the antibody-producing cell to divide and mature.

As these hypotheses indicate, much remains to be done in the field of antibody structure. What has been learned indicates that the antibody system is special and that it may have evolved to solve its problem of molecular recognition in a unique way. There remains the intriguing possibility, however, that the special genetic mechanisms hinted at by the differentiation of antibody-producing cells will be found in other systems of cellular differentiation; certainly there are at least conceptual similarities to some other systems of pattern recognition, such as those of the central nervous system. In any event, whether the immune system turns out to be unique or representative of a more general type of evolutionary development, we can expect practical consequences of great significance for fields of study such as immune tolerance, organ transplantation and autoimmune disease to flow from a continuing analysis of the structure of antibodies.

the bacterial virus lambda. A piece of DNA (a V gene) could be removed from a row of V genes (each having evolved to be slightly different) and could then be inserted and fused with a C gene. If the DNA is removed as a ring, the process would effectively permute the sequence, leading to variation [*see bottom illustration on preceding two pages*]. As mentioned above, the alternative somatic theory is that a single V gene is mutated and then translocated, following which the cell making the VC product is selected for or against within the body. Which of these theories is correct remains to be defined, but both theories require a process of assembly of a VC gene from separate V and C genes. This requirement suggests that it may be by the translocation of genes that the cell achieves its goal of making just one kind of immunoglobulin. Molecular differentiation of this type is so far unique to

the immune system, but it may in fact turn out to be important in other systems of differentiation among higher organisms.

Although the molecular details of the mechanism of translocation and recombination remain hypothetical, there is some recent evidence that we are on the right track in concluding that antibody variation is somatic in origin. The evidence comes from studies of the genetics of mouse immunoglobulins done with my student Paul D. Gottlieb. These experiments are still in their early stages, however, and the actual mechanism of variation has so far not been demonstrated.

The origin of the required diversity and the restriction of that diversity to one kind of antibody for each cell seem to be mirrored in the structure of the antibody molecule. What does antibody

10

THE HEMOGLOBIN MOLECULE

M. F. PERUTZ
November 1964

*Its 10,000 atoms are assembled into four chains, each a
helix with several bends. The molecule has one shape
when ferrying oxygen molecules and a slightly different
shape when it is not*

In 1937, a year after I entered the University of Cambridge as a graduate student, I chose the X-ray analysis of hemoglobin, the oxygen-bearing protein of the blood, as the subject of my research. Fortunately the examiners of my doctoral thesis did not insist on a determination of the structure, otherwise I should have had to remain a graduate student for 23 years. In fact, the complete solution of the problem, down to the location of each atom in this giant molecule, is still outstanding, but the structure has now been mapped in enough detail to reveal the intricate three-dimensional folding of each of its four component chains of amino acid units, and the positions of the four pigment groups that carry the oxygen-combining sites.

The folding of the four chains in hemoglobin turns out to be closely similar to that of the single chain of myoglobin, an oxygen-bearing protein in muscle whose structure has been elucidated in atomic detail by my colleague John C. Kendrew and his collaborators. Correlation of the structure of the two proteins allows us to specify quite accurately, by purely physical methods, where each amino acid unit in hemoglobin lies with respect to the twists and turns of its chains.

Physical methods alone, however, do not yet permit us to decide which of the 20 different kinds of amino acid units occupies any particular site. This knowledge has been supplied by chemical analysis; workers in the U.S. and in Germany have determined the sequence of the 140-odd amino acid units along each of the hemoglobin chains. The combined results of the two different methods of approach now provide an accurate picture of many facets of the hemoglobin molecule.

In its behavior hemoglobin does not resemble an oxygen tank so much as a molecular lung. Two of its four chains shift back and forth, so that the gap between them becomes narrower when oxygen molecules are bound to the hemoglobin, and wider when the oxygen is released. Evidence that the chemical activities of hemoglobin and other proteins are accompanied by structural changes had been discovered before, but this is the first time that the nature of such a change has been directly demonstrated. Hemoglobin's change of shape makes me think of it as a breathing molecule, but paradoxically it expands, not when oxygen is taken up but when it is released.

When I began my postgraduate work in 1936 I was influenced by three inspiring teachers. Sir Frederick Gowland Hopkins, who had received a Nobel prize in 1929 for discovering the growth-stimulating effect of vitamins, drew our attention to the central role played by enzymes in catalyzing chemical reactions in the living cell. The few enzymes isolated at that time had all proved to be proteins. David Keilin, the discoverer of several of the enzymes that catalyze the processes of respiration, told us how the chemical affinities and catalytic properties of iron atoms were altered when the iron combined with different proteins. J. D. Bernal, the X-ray crystallographer, was my research supervisor. He and Dorothy Crowfoot Hodgkin had taken the first X-ray diffraction pictures of crystals of protein a year or two before I arrived, and they had discovered that protein molecules, in spite of their large size, have highly ordered structures. The wealth of sharp X-ray diffraction spots produced by a single crystal of an enzyme such as pepsin could be explained only if every one, or almost every one, of the 5,000 atoms in the pepsin molecule occupied a definite position that was repeated in every one of the myriad of pepsin molecules packed in the crystal. The notion is commonplace now, but it caused a sensation at a time when proteins were still widely regarded as "colloids" of indefinite structure.

In the late 1930's the importance of the nucleic acids had yet to be discovered; according to everything I had learned the "secret of life" appeared to be concealed in the structure of proteins. Of all the methods available in chemistry and physics, X-ray crystallography seemed to offer the only chance, albeit an extremely remote one, of determining that structure.

The number of crystalline proteins then available was probably not more than a dozen, and hemoglobin was an obvious candidate for study because of its supreme physiological importance, its ample supply and the ease with which it could be crystallized. All the same, when I chose the X-ray analysis of hemoglobin as the subject of my Ph.D. thesis, my fellow students regarded me with a pitying smile. The most complex organic substance whose structure had yet been determined by X-ray analysis was the molecule of the dye phthalocyanin, which contains 58 atoms. How could I hope to locate the thousands of atoms in the molecule of hemoglobin?

The Function of Hemoglobin

Hemoglobin is the main component of the red blood cells, which carry oxygen from the lungs through the arteries to the tissues and help to carry carbon dioxide through the veins back to the lungs. A single red blood cell contains about 280 million molecules of hemoglobin. Each molecule has 64,500 times the weight of a hydrogen atom and is

HEMOGLOBIN MOLECULE, as deduced from X-ray diffraction studies, is shown from above (*top*) and side (*bottom*). The drawings follow the representation scheme used in three-dimensional models built by the author and his co-workers. The irregular blocks represent electron-density patterns at various levels in the hemoglobin molecule. The molecule is built up from four subunits: two identical alpha chains (*light blocks*) and two identical beta chains (*dark blocks*). The letter "N" in the top view identifies the amino ends of the two alpha chains; the letter "C" identifies the carboxyl ends. Each chain enfolds a heme group (*colored disk*), the iron-containing structure that binds oxygen to the molecule.

made up of about 10,000 atoms of hydrogen, carbon, nitrogen, oxygen and sulfur, plus four atoms of iron, which are more important than all the rest. Each iron atom lies at the center of the group of atoms that form the pigment called heme, which gives blood its red color and its ability to combine with oxygen. Each heme group is enfolded in one of the four chains of amino acid units that collectively constitute the protein part of the molecule, which is called globin. The four chains of globin consist of two identical pairs. The members of one pair are known as alpha chains and those of the other as beta chains. Together the four chains contain a total of 574 amino acid units.

In the absence of an oxygen carrier a liter of arterial blood at body temperature could dissolve and transport no more than three milliliters of oxygen. The presence of hemoglobin increases this quantity 70 times. Without hemoglobin large animals could not get enough oxygen to exist. Similarly, hemoglobin is responsible for carrying more than 90 percent of the carbon dioxide transported by venous blood.

Each of the four atoms of iron in the hemoglobin molecule can take up one molecule (two atoms) of oxygen. The reaction is reversible in the sense that oxygen is taken up where it is plentiful, as in the lungs, and released where it is scarce, as in the tissues. The reaction is accompanied by a change in color: hemoglobin containing oxygen, known as oxyhemoglobin, makes arterial blood look scarlet; reduced, or oxygen-free, hemoglobin makes venous blood look purple. The term "reduced" for the oxygen-free form is really a misnomer because "reduced" means to the chemist that electrons have been added to an atom or a group of atoms. Actually, as James B. Conant of Harvard University demonstrated in 1923, the iron atoms in both reduced hemoglobin and oxyhemoglobin are in the same electronic condition: the divalent, or ferrous, state. They become oxidized to the trivalent, or ferric, state if hemoglobin is treated with a ferricyanide or removed from the red cells and exposed to the air for a considerable time; oxidation also occurs in certain blood diseases. Under these conditions hemoglobin turns brown and is known as methemoglobin, or ferrihemoglobin.

Ferrous iron acquires its capacity for binding molecular oxygen only through its combination with heme and globin. Heme alone will not bind oxygen, but the specific chemical environment of the globin makes the combination possible. In association with other proteins, such as those of the enzymes peroxidase and catalase, the same heme group can exhibit quite different chemical characteristics.

The function of the globin, however, goes further. It enables the four iron atoms within each molecule to interact in a physiologically advantageous manner. The combination of any three of the iron atoms with oxygen accelerates the combination with oxygen of the fourth; similarly, the release of oxygen by three of the iron atoms makes the fourth cast off its oxygen faster. By tending to make each hemoglobin molecule carry either four molecules of oxygen or none, this interaction ensures efficient oxygen transport.

I have mentioned that hemoglobin also plays an important part in bearing carbon dioxide from the tissues back to the lungs. This gas is not borne by the iron atoms, and only part of it is bound directly to the globin; most of it is taken up by the red cells and the noncellular fluid of the blood in the form of bicarbonate. The transport of bicarbonate is facilitated by the disappearance of an acid group from hemoglobin for each molecule of oxygen discharged. The reappearance of the acid group when oxygen is taken up again in the lungs sets in motion a series of chemical reactions that leads to the discharge of carbon dioxide. Conversely, the presence of bicarbonate and lactic acid in the tissues accelerates the liberation of oxygen.

Breathing seems so simple, yet it appears as if this elementary manifestation of life owes its existence to the interplay of many kinds of atoms in a giant molecule of vast complexity. Elucidating the structure of the molecule should tell us not only what the molecule looks like but also how it works.

The Principles of X-Ray Analysis

The X-ray study of proteins is sometimes regarded as an abstruse subject comprehensible only to specialists, but the basic ideas underlying our work are so simple that some physicists find them boring. Crystals of hemoglobin and other proteins contain much water and, like living tissues, they tend to lose their regularly ordered structure on drying. To preserve this order during X-ray analysis crystals are mounted wet in small glass capillaries. A single crystal is then illuminated by a narrow beam of X rays that are essentially all of one wavelength. If the crystal is kept stationary, a photographic film placed behind it will often exhibit a pattern of

X-RAY DIFFRACTION PATTERN was made from a single crystal of hemoglobin that was rotated during the photographic exposure. Electrons grouped around the centers of the atoms in the crystal scatter the incident X rays, producing a symmetrical array of spots. Spots that are equidistant from the center and opposite each other have the same density.

spots lying on ellipses, but if the crystal is rotated in certain ways, the spots can be made to appear at the corners of a regular lattice that is related to the arrangement of the molecules in the crystal [see illustration above]. Moreover, each spot has a characteristic intensity that is determined in part by the arrangement of atoms inside the molecules. The reason for the different intensities is best explained in the words of W. L. Bragg, who founded X-ray analysis in 1913—the year after Max von Laue had discovered that X rays are diffracted by crystals—and who later succeeded Lord Rutherford as Cavendish Professor of Physics at Cambridge:

"It is well known that the form of the lines ruled on a [diffraction] grating has an influence on the relative intensity of the spectra which it yields. Some spectra may be enhanced, or reduced, in intensity as compared with others. Indeed, gratings are sometimes ruled in such a way that most of the energy is thrown into those spectra which it is most desirable to examine. The form of the line on the grating does not influence the positions of the spec-

tra, which depend on the number of lines to the centimetre, but the individual lines scatter more light in some directions than others, and this enhances the spectra which lie in those directions.

"The structure of the group of atoms which composes the unit of the crystal grating influences the strength of the various reflexions in exactly the same way. The rays are diffracted by the electrons grouped around the centre of each atom. In some directions the atoms conspire to give a strong scattered beam, in others their effects almost annul each other by interference. The exact arrangement of the atoms is to be deduced by comparing the strength of the reflexions from different faces and in different orders."

Thus there should be a way of reversing the process of diffraction, of proceeding backward from the diffraction pattern to an image of the arrangement of atoms in the crystal. Such an image can actually be produced, somewhat laboriously, as follows. It will be noted that spots on opposite sides of the center of an X-ray picture have the same

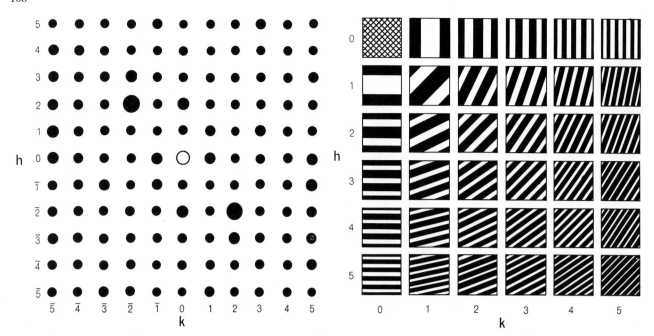

INTERPRETATION OF X-RAY IMAGE can be done with a special optical device to generate a set of diffraction fringes (*right*) from the spots in an X-ray image (*left*). Each pair of symmetrically related spots produces a unique set of fringes. Thus the spots in-

dexed $2,\bar{2}$ and $\bar{2},2$ yield the fringes indexed 2,2. A two-dimensional image of the atomic structure of a crystal can be generated by printing each set of fringes on the same sheet of photographic paper. But the phase problem (*below*) must be solved first.

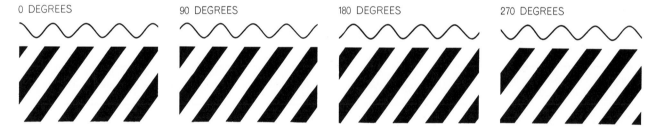

PHASE PROBLEM arises because the spots in an X-ray image do not indicate how the fringes are related in phase to an arbitrarily chosen common origin. Here four identical sets of fringes are

related by different phases to the point of origin at the top left corner. The phase marks the distance of the wave crest from the origin, measured in degrees. One wavelength is 360 degrees.

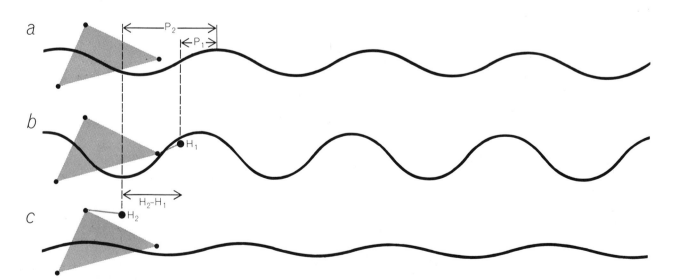

HEAVY-ATOM REPLACEMENT METHOD provides information about phases by changing the intensities of the X-ray diffraction pattern. In *a* a highly oversimplified protein (a triangle of three atoms) scatters a sinusoidal wave that represents the amplitude and phase of a single set of fringes. In *b* and *c*, after heavy atoms H_1

and H_2 are attached to the protein in different positions, the wave is changed in amplitude and phase. The heavy atoms can serve as points of common origin for measuring the magnitude of the phases (P_1 and P_2) of waves scattered by the unaltered protein. The distance between H_1 and H_2 must be accurately known.

degree of intensity. With the aid of a simple optical device each symmetrically related pair of spots can be made to generate a set of diffraction fringes, with an amplitude proportional to the square root of the intensity of the spots. The device, which was invented by Bragg and later developed by H. Lipson and C. A. Taylor at the Manchester College of Science and Technology, consists of a point source of monochromatic light, a pair of plane-convex lenses and a microscope. The pair of spots in the diffraction pattern is represented by a pair of holes in a black mask that is placed between the two lenses. If the point source is placed at the focus of one of the lenses, the waves of parallel light emerging from the two holes will interfere with one another at the focus of the second lens, and their interference pattern, or diffraction pattern, can be observed or photographed through the microscope.

Imagine that each pair of symmetrically related spots in the X-ray picture is in turn represented by a pair of holes in a mask, and that its diffraction fringes are photographed. Each set of fringes will then be at right angles to the line joining the two holes, and the distance between the fringes will be inversely proportional to the distance between the holes. If the spots are numbered from the center along two mutually perpendicular lines by the indices h and k, the relation between any pair of spots and its corresponding set of fringes would be as shown in the top illustration on the opposite page.

The Phase Problem

An image of the atomic structure of the crystal can be generated by printing each set of fringes in turn on the same sheet of photographic paper, or by superposing all the fringes and making a print of the light transmitted through them. At this point, however, a fatal complication arises. In order to obtain the right image one would have to place each set of fringes correctly with respect to some arbitrarily chosen common origin [*see middle illustration on opposite page*]. At this origin the amplitude of any particular set of fringes may show a crest or trough or some intermediate value. The distance of the wave crest from the origin is called the phase. It is almost true to say that by superposing sets of fringes of given amplitude one can generate an infinite number of different images, depending on the choice of phase for each set of fringes. By itself the X-ray picture tells us only about the amplitudes and nothing about the phases of the fringes to be generated by each pair of spots, which means that half the information needed for the production of the image is missing.

The missing information makes the diffraction pattern of a crystal like a hieroglyphic without a key. Having spent years hopefully measuring the intensities of several thousand spots in the diffraction pattern of hemoglobin, I found myself in the tantalizing position of an explorer with a collection of tablets engraved in an unknown script. For some time Bragg and I tried to develop methods for deciphering the phases, but with only limited success. The solution finally came in 1953, when I discovered that a method that had been developed by crystallographers for solving the phase problem in simpler structures could also be applied to proteins.

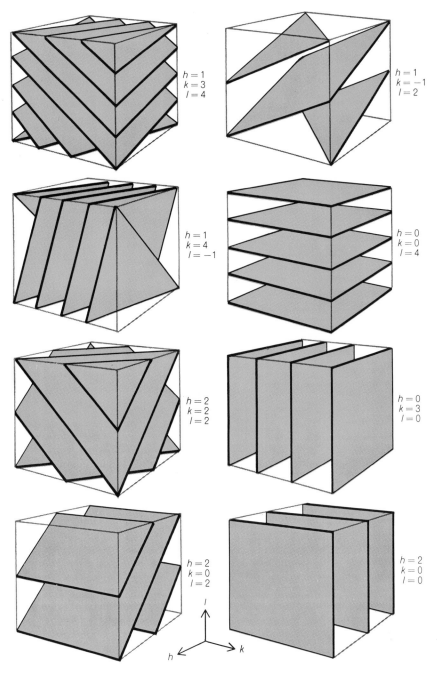

THREE-DIMENSIONAL FRINGES are needed to build up an image of protein molecules. For this purpose many different X-ray diffraction images are prepared and symmetrically related pairs of spots are indexed in three dimensions: h, k and l and $\bar{h}$, $\bar{k}$ and $\bar{l}$. Each pair of spots yields a three-dimensional fringe like those shown here. Fringes from thousands of spots must be superposed in proper phase to build up an image of the molecule.

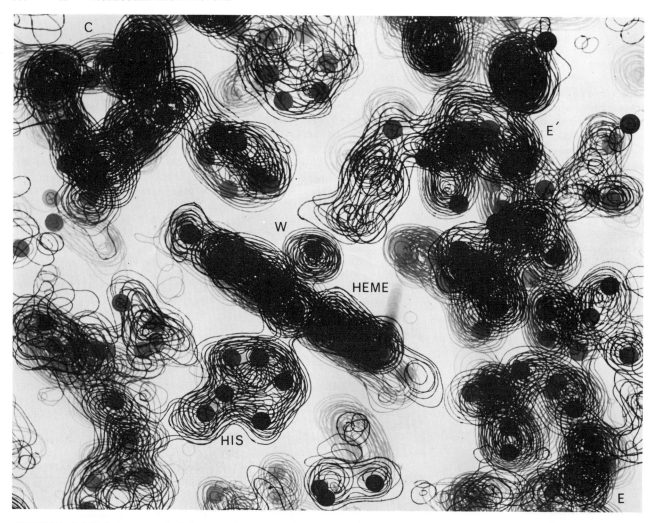

CONTOUR MAPS, drawn on stacked sheets of clear plastic, show a portion of the myoglobin molecule as revealed by superposition of three-dimensional fringe patterns. The maps were made by John C. Kendrew and his associates at the University of Cambridge. Myoglobin is very similar to the beta chain of hemoglobin. The heme group is seen edge on. *His* is an amino acid subunit of histidine that is attached to the iron atom of the heme group. *W* is a water molecule linked to the iron atom. The region between *E* and *E'* represents amino acid subunits arranged in an alpha helix. *C* is an alpha helix seen end on. The black dots mark atomic positions.

In this method the molecule of the compound under study is modified slightly by attaching heavy atoms such as those of mercury to definite positions in its structure. The presence of a heavy atom produces marked changes in the intensities of the diffraction pattern, and this makes it possible to gather information about the phases. From the difference in amplitude in the absence or presence of a heavy atom, the distance of the wave crest from the heavy atom can be determined for each set of fringes. Thus with the heavy atom serving as a common origin the magnitude of the phase can be measured. The bottom illustration on page 108 shows how the phase of a single set of fringes, represented by a sinusoidal wave that is supposedly scattered by the oversimplified protein molecule, can be measured from the increase in amplitude produced by the heavy atom H_1.

Unfortunately this still leaves an am-biguity of sign; the experiment does not tell us whether the phase is to be measured from the heavy atom in the forward or the backward direction. If n is the number of diffracted spots, an ambiguity of sign in each set of fringes would lead to 2^n alternative images of the structure. The Dutch crystallographer J. M. Bijvoet had pointed out some years earlier in another context that the ambiguity could be resolved by examining the diffraction pattern from a second heavy-atom compound.

The bottom illustration on page 108 shows that the heavy atom H_2, which is attached to the protein in a position different from that of H_1, diminishes the amplitude of the wave scattered by the protein. The degree of attenuation allows us to measure the distance of the wave crest from H_2. It can now be seen that the wave crest must be in front of H_1; otherwise its distance from H_1 could not be reconciled with its distance from H_2. The final answer depends on knowing the length and direction of the line joining H_2 to H_1. These quantities are best calculated by a method that does not easily lend itself to exposition in nonmathematical language. It was devised by my colleague Michael G. Rossmann.

The heavy-atom method can be applied to hemoglobin by attaching mercury atoms to the sulfur atoms of the amino acid cysteine. The method works, however, only if this attachment leaves the structure of the hemoglobin molecules and their arrangement in the crystal unaltered. When I first tried it, I was not at all sure that these stringent demands would be fulfilled, and as I developed my first X-ray photograph of mercury hemoglobin my mood alternated between sanguine hopes of immediate success and desperate forebodings of all the possible causes of failure. When the diffraction spots appeared in exactly the same position as in the mercury-free protein but with slightly altered intensities, just as I had hoped, I rushed off to Bragg's room in jubilant excitement, expecting that the structure of hemoglobin and of many other proteins would soon be determined. Bragg shared my excitement, and luckily neither of us anticipated the formidable technical difficulties that were to hold us up for another five years.

Resolution of the Image

Having solved the phase problem, at least in principle, we were confronted with the task of building up a structural image from our X-ray data. In simpler structures atomic positions can often be found from representations of the structure projected on two mutually perpendicular planes, but in proteins a three-dimensional image is essential. This can be attained by making use of the three-dimensional nature of the diffraction pattern. The X-ray diffraction pattern on page 107 can be regarded as a section through a sphere that is filled with layer after layer of diffraction spots. Each pair of spots can be made to generate a set of three-dimensional fringes like the ones shown on page 6. When their phases have been measured, they can be superposed by calculation to build up a three-dimensional image of the protein. The final image is represented by a series of sections through the molecule, rather like a set of microtome sections through a piece of tissue, only on a scale 1,000 times smaller [see illustration on opposite page].

The resolution of the image is roughly equal to the shortest wavelength of the fringes used in building it up. This means that the resolution increases with the number of diffracted spots included in the calculation. If the image is built up from part of the diffraction pattern only, the resolution is impaired.

In the X-ray diffraction patterns of protein crystals the number of spots runs into tens of thousands. In order to determine the phase of each spot accurately, its intensity (or blackness) must be measured accurately several times over: in the diffraction pattern from a crystal of the pure protein and in the patterns from crystals of several compounds of the protein, each with heavy atoms attached to different positions in the molecule. Then the results have to be corrected by various geometric factors before they are finally used to build up an image through the super-

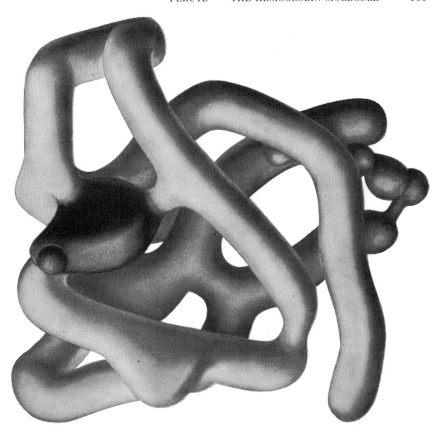

MYOGLOBIN MOLECULE, as first reconstructed at low resolution by Kendrew and his co-workers in 1957, had this rather repulsive visceral appearance. The sausage-like knot marks the path of the amino acid chain of the molecule. The dark disklike shape (here placed at an incorrect angle) is the heme group. A more detailed and more correct view of myoglobin, as seen from the other side, appears at bottom right on the next page.

position of tens of thousands of fringes. In the final calculation tens of millions of numbers may have to be added or subtracted. Such a task would have been quite impossible before the advent of high-speed computers, and we have been fortunate in that the development of computers has kept pace with the expanding needs of our X-ray analyses.

While I battled with technical difficulties of various sorts, my colleague John Kendrew successfully applied the heavy-atom method to myoglobin, a protein closely related to hemoglobin [see "The Three-dimensional Structure of a Protein Molecule," by John C. Kendrew; SCIENTIFIC AMERICAN Offprint 121]. Myoglobin is simpler than hemoglobin because it consists of only one chain of amino acid units and one heme group, which binds a single molecule of oxygen. The complex interaction phenomena involved in hemoglobin's dual function as a carrier of oxygen and of carbon dioxide do not occur in myoglobin, which acts simply as an oxygen store.

Together with Howard M. Dintzis and G. Bodo, Kendrew was brilliantly successful in managing to prepare as

many as five different crystalline heavy-atom compounds of myoglobin, which meant that the phases of the diffraction spots could be established very accurately. He also pioneered the use of high-speed computers in X-ray analysis. In 1957 he and his colleagues obtained the first three-dimensional representation of myoglobin [see illustration on this page].

It was a triumph, and yet it brought a tinge of disappointment. Could the search for ultimate truth really have revealed so hideous and visceral-looking an object? Was the nugget of gold a lump of lead? Fortunately, like many other things in nature, myoglobin gains in beauty the closer you look at it. As Kendrew and his colleagues increased the resolution of their X-ray analysis in the years that followed, some of the intrinsic reasons for the molecule's strange shape began to reveal themselves. This shape was found to be not a freak but a fundamental pattern of nature, probably common to myoglobins and hemoglobins throughout the vertebrate kingdom.

In the summer of 1959, nearly 22 years after I had taken the first X-ray

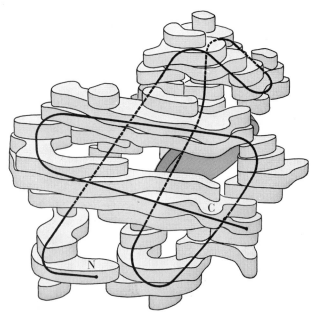

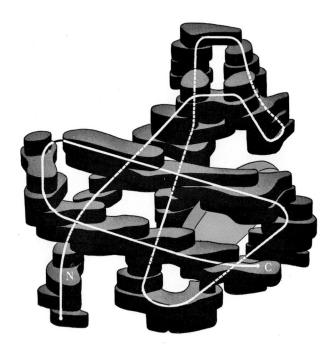

HEMOGLOBIN CHAINS, alpha at left and beta at right, are redrawn from models built by the author and his colleagues. The superposed lines show the course of the central chain. A heme group (*color*) is partly visible, tucked in the back of each model.

pictures of hemoglobin, its structure emerged at last. Michael Rossmann, Ann F. Cullis, Hilary Muirhead, Tony C. T. North and I were able to prepare a three-dimensional electron-density map of hemoglobin at a resolution of 5.5 angstrom units, about the same as that obtained for the first structure of myoglobin two years earlier. This resolution is sufficient to reveal the shape of the chain forming the backbone of a protein molecule but not to show the position of individual amino acids.

As soon as the numbers printed by the computer had been plotted on contour maps we realized that each of the four chains of hemoglobin had a shape closely resembling that of the single chain of myoglobin. The beta chain and myoglobin look like identical twins, and the alpha chains differ from them merely by a shortcut across one small loop [*see illustration below*].

Kendrew's myoglobin had been extracted from the muscle of the sperm whale; the hemoglobin we used came from the blood of horses. More recent observations indicate that the myoglobins of the seal and the horse, and the hemoglobins of man and cattle, all have the same structure. It seems as though the apparently haphazard and irregular folding of the chain is a pattern specifically devised for holding a heme group in place and for enabling it to carry oxygen.

What is it that makes the chain take up this strange configuration? The extension of Kendrew's analysis to a high-

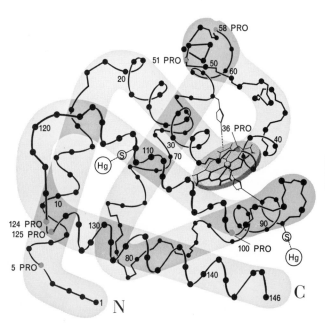

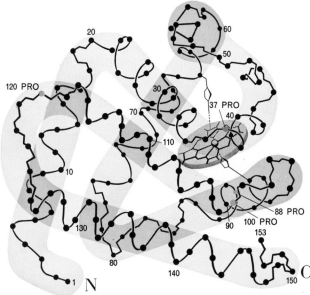

BETA CHAIN AND MYOGLOBIN appear at left and right. Every 10th amino acid subunit is marked, as are proline subunits (*color*), which often coincide with turns in the chain. Balls marked "Hg" show where mercury atoms can be attached to sulfur atoms (*S*).

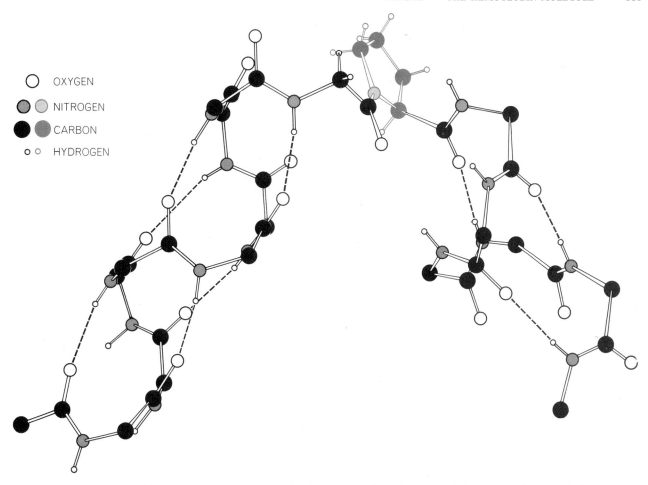

CORNER IN HEMOGLOBIN MOLECULE occurs where a subunit of the amino acid proline (*color*) falls between two helical regions in the beta chain. The chain is shown bare; all hydrogen atoms and amino acid side branches, except for proline, are removed.

er resolution shows that the chain of myoglobin consists of a succession of helical segments interrupted by corners and irregular regions. The helical segments have the geometry of the alpha helix predicted in 1951 by Linus Pauling and Robert B. Corey of the California Institute of Technology. The heme group lies embedded in a fold of the chain, so that only its two acid groups protrude at the surface and are in contact with the surrounding water. Its iron atom is linked to a nitrogen atom of the amino acid histidine.

I have recently built models of the alpha and beta chains of hemoglobin and found that they follow an atomic pattern very similar to that of myoglobin. If two protein chains look the same, one would expect them to have much the same composition. In the language of protein chemistry this implies that in the myoglobins and hemoglobins of all vertebrates the 20 different kinds of amino acid should be present in about the same proportion and arranged in similar sequence.

Enough chemical analyses have been done by now to test whether or not this is true. Starting at the Rockefeller Institute and continuing in our laboratory, Allen B. Edmundson has determined the sequence of amino acid units in the molecule of sperm-whale myoglobin. The sequences of the alpha and beta chains of adult human hemoglobin have been analyzed independently by Gerhardt Braunitzer and his colleagues at the Max Planck Institute for Biochemistry in Munich, and by William H. Konigsberg, Robert J. Hill and their associates at the Rockefeller Institute. Fetal hemoglobin, a variant of the human adult form, contains a chain known as gamma, which is closely related to the beta chain. Its complete sequence has been analyzed by Walter A. Schroeder and his colleagues at the California Institute of Technology. The sequences of several other species of hemoglobin and that of human myoglobin have been partially elucidated.

The sequence of amino acid units in proteins is genetically determined, and changes arise as a result of mutation. Sickle-cell anemia, for instance, is an inherited disease due to a mutation in one of the hemoglobin genes. The mutation causes the replacement of a single amino acid unit in each of the beta chains. (The glutamic acid unit normally present at position No. 6 is replaced by a valine unit.) On the molecular scale evolution is thought to involve a succession of such mutations, altering the structure of protein molecules one amino acid unit at a time. Consequently when the hemoglobins of different species are compared, we should expect the sequences in man and apes, which are close together on the evolutionary scale, to be very similar, and those of mammals and fishes, say, to differ more widely. Broadly speaking, this is what is found. What was quite unexpected was the degree of chemical diversity ·among the amino acid sequences of proteins of similar three-dimensional structure and closely related function. Comparison of the known hemoglobin and myoglobin sequences shows only 15 positions—no more than one in 10—where the same amino acid unit is present in all species. In all the other positions one or more replacements have occurred in the course of evolution.

What mechanism makes these diverse

chains fold up in exactly the same way? Does a template force them to take up this configuration, like a mold that forces a car body into shape? Apart from the topological improbability of such a template, all the genetic and physico-chemical evidence speaks against it, suggesting instead that the chain folds up spontaneously to assume one specific structure as the most stable of all possible alternatives.

Possible Folding Mechanisms

What is it, then, that makes one particular configuration more stable than all others? The only generalization to emerge so far, mainly from the work of Kendrew, Herman C. Watson and myself, concerns the distribution of the so-called polar and nonpolar amino acid units between the surface and the interior of the molecule.

Some of the amino acids, such as glutamic acid and lysine, have side groups of atoms with positive or negative electric charge, which strongly attract the surrounding water. Amino acid side groups such as glutamine or tyrosine, although electrically neutral as a whole, contain atoms of nitrogen or oxygen in which positive and negative charges are sufficiently separated to form dipoles; these also attract water, but not so strongly as the charged groups do. The attraction is due to a separation of charges in the water molecule itself, making it dipolar. By attaching themselves to electrically charged groups, or to other dipolar groups, the water molecules minimize the strength of the electric fields surrounding these groups and stabilize the entire structure by lowering the quantity known as free energy.

The side groups of amino acids such as leucine and phenylalanine, on the other hand, consist only of carbon and hydrogen atoms. Being electrically neutral and only very weakly dipolar, these groups repel water as wax does. The reason for the repulsion is strange and intriguing. Such hydrocarbon groups, as they are called, tend to disturb the haphazard arrangement of the liquid water molecules around them, making it ordered as it is in ice. The increase in order makes the system less stable; in physical terms it leads to a reduction of the quantity known as entropy, which is the measure of the disorder in a system. Thus it is the water molecules' anarchic distaste for the orderly regimentation imposed on them by the hydrocarbon side groups that forces these side groups to turn away from water and to stick to one another.

Our models have taught us that most electrically charged or dipolar side groups lie at the surface of the protein molecule, in contact with water. Nonpolar side groups, in general, are either confined to the interior of the molecule or so wedged into crevices on its surface as to have the least contact with water. In the language of physics, the distribution of side groups is of the kind leading to the lowest free energy and the highest entropy of the protein molecules and the water around them. (There is a reduction of entropy due to the orderly folding of the protein chain itself, which makes the system less stable, but this is balanced, at moderate temperatures, by the stabilizing contributions of the other effects just described.) It is too early to say whether these are the only generalizations to be made about the forces that stabilize one particular configuration of the protein chain in preference to all others.

At least one amino acid is known to be a misfit in an alpha helix, forcing the chain to turn a corner wherever the unit occurs. This is proline [see illustration on preceding page]. There is, however, only one corner in all the hemoglobins and myoglobins where a proline is always found in the same position: position No. 36 in the beta chain and No. 37 in the myoglobin chain [see bottom illustration on page 112]. At other corners the appearance of prolines is haphazard and changes from species to species. Elkan R. Blout of the Harvard Medical School finds that certain amino acids such as valine or threonine, if present in large numbers, inhibit the formation of alpha helices, but these do not seem to have a decisive influence in myoglobin and hemoglobin.

Since it is easier to determine the sequence of amino acid units in proteins than to unravel their three-dimensional structure by X rays, it would be useful to be able to predict the structure from the sequence. In principle enough is probably known about the forces between atoms and about the way they tend to arrange themselves to make such predictions feasible. In practice the enormous number of different ways in which a long chain can be twisted still makes the problem one of baffling complexity.

Assembling the Four Chains

If hemoglobin consisted of four identical chains, a crystallographer would expect them to lie at the corners of a regular tetrahedron. In such an arrangement each chain can be brought into congruence with any of its three neighbors by a rotation of 180 degrees about one of three mutually perpendicular

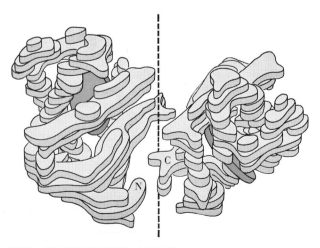

FOUR CHAINS OF HEMOGLOBIN are arranged in symmetrical fashion. Two alpha chains (*left*) and two beta chains (*right*) face each other across an axis of symmetry (*broken vertical lines*). In the assembled molecule the two alpha chains are inverted over the two beta chains and nested down between them. When arranged in this manner, the four chains lie at the corners of a tetrahedron.

axes of symmetry. Since the alpha and beta chains are chemically different, such perfect symmetry is unattainable, but the actual arrangement comes very close to it. As a first step in the assembly of the molecule two alpha chains are placed near a twofold symmetry axis, so that a rotation of 180 degrees brings one chain into congruence with its partner [see illustration on opposite page].

Next the same is done with the two beta chains. One pair, say the alpha chains, is then inverted and placed over the top of the other pair so that the four chains lie at the corners of a tetrahedron. A true twofold symmetry axis now passes vertically through the molecule, and "pseudo-axes" in two directions perpendicular to the first relate the alpha to the beta chains. Thus the arrangement is tetrahedral, but because of the chemical differences between the alpha and beta chains the tetrahedron is not quite regular.

The result is an almost spherical molecule whose exact dimensions are $64 \times 55 \times 50$ angstrom units. It is astonishing to find that four objects as irregular as the alpha and beta chains can fit together so neatly. On formal grounds one would expect a hole to pass through the center of the molecule because chains of amino acid units, being asymmetrical, cannot cross any symmetry axis. Such a hole is in fact found [see top illustration on page 106].

The most unexpected feature of the oxyhemoglobin molecule is the way the four heme groups are arranged. On the basis of their chemical interaction one would have expected them to lie close together. Instead each heme group lies in a separate pocket on the surface of the molecule, apparently unaware of the existence of its partners. Seen at the present resolution, therefore, the structure fails to explain one of the most important physiological properties of hemoglobin.

In 1937 Felix Haurowitz, then at the German University of Prague, discovered an important clue to the molecular explanation of hemoglobin's physiological action. He put a suspension of needle-shaped oxyhemoglobin crystals away in the refrigerator. When he took the suspension out some weeks later, the oxygen had been used up by bacterial infection and the scarlet needles had been replaced by hexagonal plates of purple reduced hemoglobin. While Haurowitz observed the crystals under the microscope, oxygen penetrated between the slide and the cover slip, causing the purple plates to dissolve

and the scarlet needles of hemoglobin to re-form. This transformation convinced Haurowitz that the reaction of hemoglobin with oxygen must be accompanied by a change in the structure of the hemoglobin molecule. In myoglobin, on the other hand, no evidence for such a change has been detected.

Haurowitz' observation and the enigma posed by the structure of oxyhemoglobin caused me to persuade a graduate student, Hilary Muirhead, to attempt an X-ray analysis at low resolution of the reduced form. For technical reasons human rather than horse hemoglobin was used at first, but we have now found that the reduced hemoglobins of man and the horse have very similar structures, so that the species does not matter here.

Unlike me, Miss Muirhead succeeded in solving the structure of her protein in time for her Ph.D. thesis. When we examined her first electron-density maps, we looked for two kinds of structural change: alterations in the folding

of the individual chains and displacements of the chains with respect to each other. We could detect no changes in folding large enough to be sure that they were not due to experimental error. We did discover, however, that a striking displacement of the beta chains had taken place. The gap between them had widened and they had been shifted sideways, increasing the distance between their respective iron atoms from 33.4 to 40.3 angstrom units [see illustration on page 116]. The arrangement of the two alpha chains had remained unaltered, as far as we could judge, and the distance between the iron atoms in the beta chains and their nearest neighbors in the alpha chains had also remained the same. It looked as though the two beta chains had slid apart, losing contact with each other and somewhat changing their points of contact with the alpha chains.

F. J. W. Roughton and others at the University of Cambridge suggest that the change to the oxygenated form of

RESIDUE NUMBER	HEMOGLOBIN			MYOGLOBIN
	ALPHA	BETA	GAMMA	
81	MET	LEU	LEU	HIS
82	PRO	LYS	LYS	GLU
83	ASN	GLY	GLY	ALA
84	ALA	THR	THR	GLU
85	LEU	PHE	PHE	LEU
86	SER	ALA	ALA	LYS
87	ALA	THR	GLN	PRO
88	LEU	LEU	LEU	LEU
89	SER	SER	SER	ALA
90	ASP	GLU	GLU	GLN
91	LEU	LEU	LEU	SER
92	HIS	HIS	HIS	HIS
93	ALA	CYS	CYS	ALA
94	HIS	ASP	ASN	THR
95	LYS	LYS	LYS	LYS
96	LEU	LEU	LEU	HIS
97	ARG	HIS	HIS	LYS
98	VAL	VAL	VAL	ILEU
99	ASP	ASP	ASP	PRO
100	PRO	PRO	PRO	ILEU
101	VAL	GLU	GLU	LYS
102	ASP	ASN	ASN	TYR

ALA ALANINE	GLY GLYCINE	PRO PROLINE
ARG ARGININE	HIS HISTIDINE	SER SERINE
ASN ASPARAGINE	ILEU ISOLEUCINE	THR THREONINE
ASP ASPARTIC ACID	LEU LEUCINE	TYR TYROSINE
CYS CYSTEINE	LYS LYSINE	VAL VALINE
GLN GLUTAMINE	MET METHIONINE	
GLU GLUTAMIC ACID	PHE PHENYLALANINE	

AMINO ACID SEQUENCES are shown for corresponding stretches of the alpha and beta chains of hemoglobin from human adults, the gamma chain that replaces the beta chain in fetal human hemoglobin and sperm-whale myoglobin. Colored bars show where the same amino acid units are found either in all four chains or in the first three. Site numbers for the alpha chain and myoglobin are adjusted slightly because they contain a different number of amino acid subunits overall than do the beta and gamma chains. Over their full length of more than 140 subunits the four chains have only 20 amino acid subunits in common.

hemoglobin takes place after three of the four iron atoms have combined with oxygen. When the change has occurred, the rate of combination of the fourth iron atom with oxygen is speeded up several hundred times. Nothing is known as yet about the atomic mechanism that sets off the displacement of the beta chains, but there is one interesting observation that allows us at least to be sure that the interaction of the iron atoms and the change of structure do not take place unless alpha and beta chains are both present.

Certain anemia patients suffer from a shortage of alpha chains; the beta chains, robbed of their usual partners, group themselves into independent assemblages of four chains. These are known as hemoglobin *H* and resemble normal hemoglobin in many of their properties. Reinhold Benesch and Ruth E. Benesch of the Columbia University College of Physicians and Surgeons have discovered, however, that the four iron atoms in hemoglobin *H* do not interact, which led them to predict that the combination of hemoglobin *H* with oxygen should not be accompanied by a change of structure. Using crystals grown by Helen M. Ranney of the Albert Einstein College of Medicine, Lelio Mazzarella and I verified this prediction. Oxygenated and reduced hemoglobin *H* both resemble normal human reduced hemoglobin in the arrangement of the four chains.

The rearrangement of the beta chains must be set in motion by a series of atomic displacements starting at or near the iron atoms when they combine with oxygen. Our X-ray analysis has not yet reached the resolution needed to discern these, and it seems that a deeper understanding of this intriguing phenomenon may have to wait until we succeed in working out the structures of reduced hemoglobin and oxyhemoglobin at atomic resolution.

Allosteric Enzymes

There are many analogies between the chemical activities of hemoglobin and those of enzymes catalyzing chemical reactions in living cells. These analogies lead one to expect that some enzymes may undergo changes of structure on coming into contact with the substances whose reactions they catalyze. One can imagine that the active sites of these enzymes are moving mechanisms rather than static surfaces magically endowed with catalytic properties.

Indirect and tentative evidence suggests that changes of structure involv-

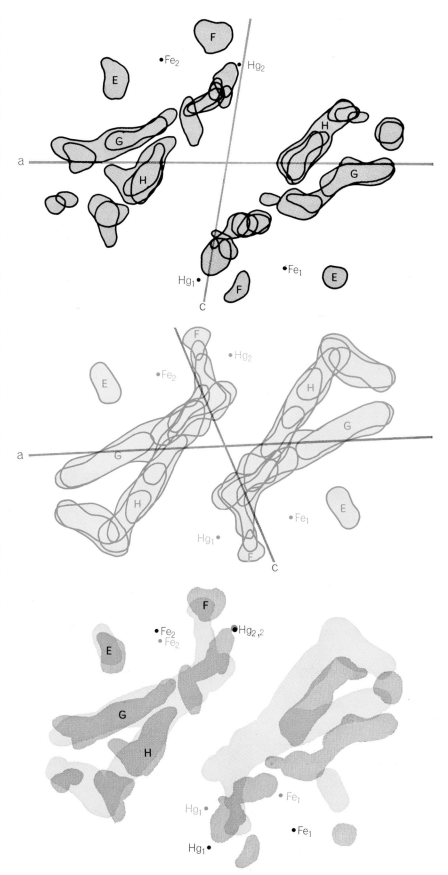

MOVEMENT OF HEMOGLOBIN CHAINS was discovered by comparing portions of the two beta chains in "reduced" (oxygen-free) human hemoglobin (*top*) with the same portions of horse hemoglobin containing oxygen (*middle*). The bottom illustration shows the outlines of the top and middle pictures superposed so that the mercury atoms (Hg_2) and helical regions (*E, F, G, H*) of the two chains at left coincide. The iron atoms (Fe_2) do not quite match. The chains at right are now seen to be shifted with respect to each other.

ing a rearrangement of subunits like that of the alpha and beta chains of hemoglobin do indeed occur and that they may form the basis of a control mechanism known as feedback inhibition. This is a piece of jargon that biochemistry has borrowed from electrical engineering, meaning nothing more complicated than that you stop being hungry when you have had enough to eat.

Constituents of living matter such as amino acids are built up from simpler substances in a series of small steps, each step being catalyzed by an enzyme that exists specifically for that purpose. Thus a whole series of different enzymes may be needed to make one amino acid. Such a series of enzymes appears to have built-in devices for ensuring the right balance of supply and demand. For example, in the colon bacillus the amino acid isoleucine is made from the amino acid threonine in several steps. The first enzyme in the series has an affinity for threonine: it catalyzes the removal of an amino group from it. H. Edwin Umbarger of the Long Island Biological Association in Cold Spring Harbor, N.Y., discovered that the action of the enzyme is inhibited by isoleucine, the end product of the last enzyme in the series. Jean-Pierre Changeux of the Pasteur Institute later showed that isoleucine acts not, as one might have expected, by blocking the site on the enzyme molecule that would otherwise combine with threonine but probably by combining with a different site on the molecule.

The two sites on the molecule must therefore interact, and Jacques Monod, Changeux and François Jacob have suggested that this is brought about by a rearrangement of subunits similar to that which accompanies the reaction of hemoglobin with oxygen. The enzyme is thought to exist in two alternative structural states: a reactive one when the supply of isoleucine has run out and an unreactive one when the supply exceeds demand. The discoverers have coined the name "allosteric" for enzymes of this kind.

The molecules of the enzymes suspected of having allosteric properties are all large ones, as one would expect them to be if they are made up of several subunits. This makes their X-ray analysis difficult. It may not be too hard to find out, however, whether or not a change of structure occurs, even if it takes a long time to unravel it in detail. In the meantime hemoglobin will serve as a useful model for the behavior of more complex enzyme systems.

11

THE THREE-DIMENSIONAL STRUCTURE
OF AN ENZYME MOLECULE

DAVID C. PHILLIPS

November 1966

*The arrangement of atoms in an enzyme molecule has
been worked out for the first time. The enzyme is
lysozyme, which breaks open cells of bacteria. The
study has also shown how lysozyme performs its task*

One day in 1922 Alexander Fleming was suffering from a cold.
This is not unusual in London,
but Fleming was a most unusual man
and he took advantage of the cold in a
characteristic way. He allowed a few
drops of his nasal mucus to fall on a
culture of bacteria he was working with
and then put the plate to one side to
see what would happen. Imagine his
excitement when he discovered some
time later that the bacteria near the
mucus had dissolved away. For a while
he thought his ambition of finding a
universal antibiotic had been realized.
In a burst of activity he quickly established that the antibacterial action of
the mucus was due to the presence in
it of an enzyme; he called this substance
lysozyme because of its capacity to lyse,
or dissolve, the bacterial cells. Lysozyme was soon discovered in many tissues and secretions of the human body,
in plants and most plentifully of all in
the white of egg. Unfortunately Fleming found that it is not effective against
the most harmful bacteria. He had to
wait seven years before a strangely
similar experiment revealed the existence of a genuinely effective antibiotic: penicillin.

Nevertheless, Fleming's lysozyme has
proved a more valuable discovery than
he can have expected when its properties were first established. With it,
for example, bacterial anatomists have
been able to study many details of bacterial structure [see "Fleming's Lysozyme," by Robert F. Acker and S. E.
Hartsell; SCIENTIFIC AMERICAN, June,
1960]. It has now turned out that
lysozyme is the first enzyme whose
three-dimensional structure has been

determined and whose properties are
understood in atomic detail. Among
these properties is the way in which the
enzyme combines with the substance on
which it acts—a complex sugar in the
wall of the bacterial cell.

Like all enzymes, lysozyme is a protein. Its chemical makeup has been
established by Pierre Jollès and his
colleagues at the University of Paris
and by Robert E. Canfield of the Columbia University College of Physicians
and Surgeons. They have found that
each molecule of lysozyme obtained
from egg white consists of a single
polypeptide chain of 129 amino acid
subunits of 20 different kinds. A peptide bond is formed when two amino
acids are joined following the removal of
a molecule of water. It is customary to
call the portion of the amino acid in-

corporated into a polypeptide chain a
residue, and each residue has its own
characteristic side chain. The 129-resi-
due lysozyme molecule is cross-linked
in four places by disulfide bridges
formed by the combination of sulfur-
containing side chains in different parts
of the molecule [*see illustration on opposite page*].

The properties of the molecule cannot
be understood from its chemical constitution alone; they depend most critically on what parts of the molecule are
brought close together in the folded
three-dimensional structure. Some form
of microscope is needed to examine the
structure of the molecule. Fortunately one is effectively provided by the
techniques of X-ray crystal-structure
analysis pioneered by Sir Lawrence
Bragg and his father Sir William Bragg.

ALA	ALANINE	GLY	GLYCINE	PRO	PROLINE
ARG	ARGININE	HIS	HISTIDINE	SER	SERINE
ASN	ASPARAGINE	ILEU	ISOLEUCINE	THR	THREONINE
ASP	ASPARTIC ACID	LEU	LEUCINE	TRY	TRYPTOPHAN
CYS	CYSTEINE	LYS	LYSINE	TYR	TYROSINE
GLN	GLUTAMINE	MET	METHIONINE	VAL	VALINE
GLU	GLUTAMIC ACID	PHE	PHENYLALANINE		

TWO-DIMENSIONAL MODEL of the lysozyme molecule is shown on the opposite page.
Lysozyme is a protein containing 129 amino acid subunits, commonly called residues (*see
key to abbreviations above*). These residues form a polypeptide chain that is cross-linked at
four places by disulfide (–S–S–) bonds. The amino acid sequence of lysozyme was determined independently by Pierre Jolles and his co-workers at the University of Paris and by
Robert E. Canfield of the Columbia University College of Physicians and Surgeons. The
three-dimensional structure of the lysozyme molecule has now been established with the
help of X-ray crystallography by the author and his colleagues at the Royal Institution in
London. A painting of the molecule's three-dimensional structure appears on pages 4 and
5. The function of lysozyme is to split a particular long-chain molecule, a complex sugar,
found in the outer membrane of many living cells. Molecules that are acted on by enzymes
are known as substrates. The substrate of lysozyme fits into a cleft, or pocket, formed by the
three-dimensional structure of the lysozyme molecule. In the two-dimensional model on
the opposite page the amino acid residues that line the pocket are shown in dark green.

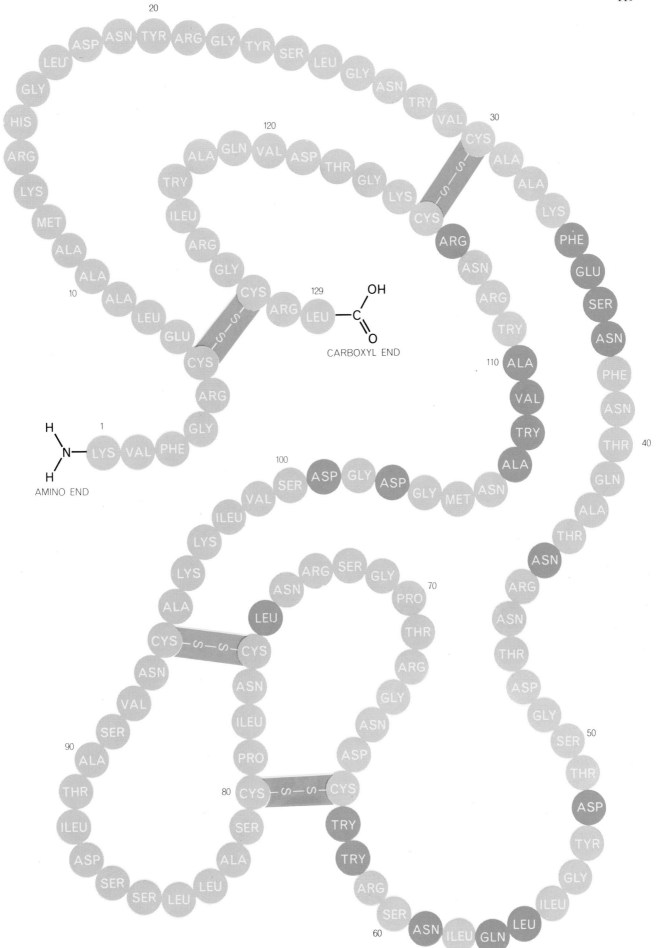

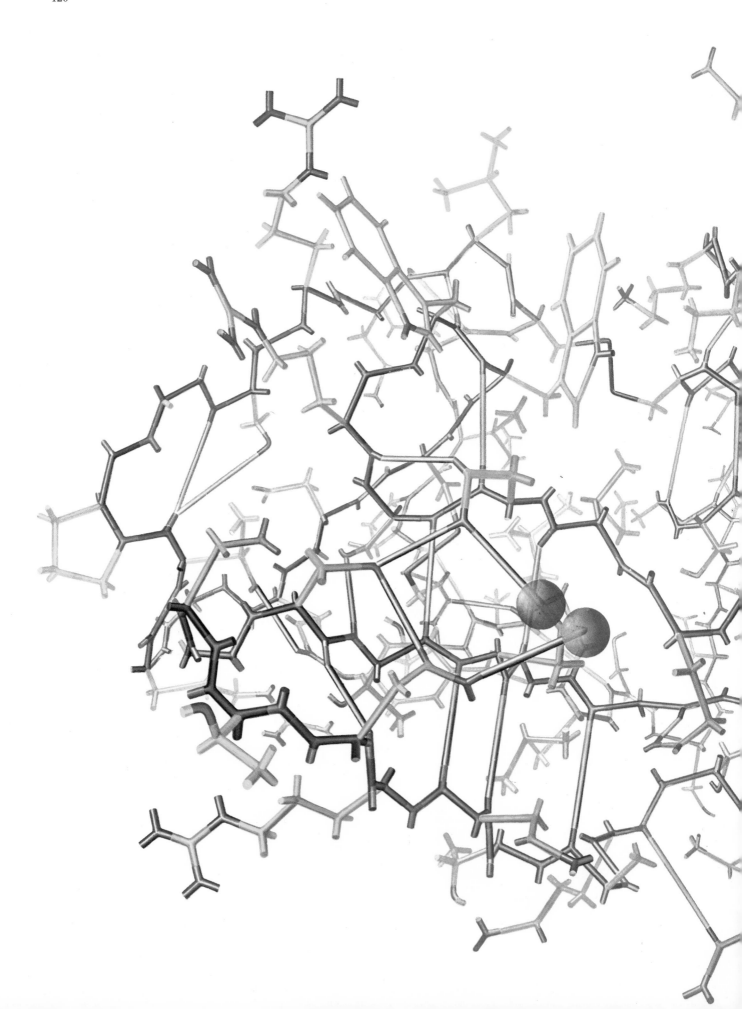

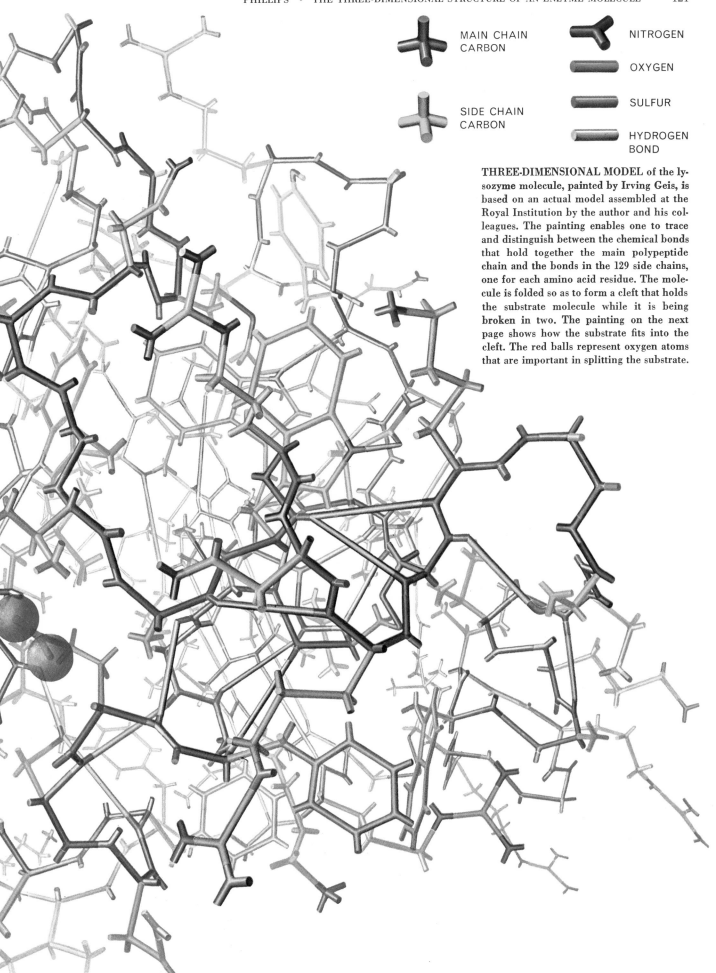

MAIN CHAIN
CARBON

SIDE CHAIN
CARBON

NITROGEN

OXYGEN

SULFUR

HYDROGEN
BOND

THREE-DIMENSIONAL MODEL of the lysozyme molecule, painted by Irving Geis, is based on an actual model assembled at the Royal Institution by the author and his colleagues. The painting enables one to trace and distinguish between the chemical bonds that hold together the main polypeptide chain and the bonds in the 129 side chains, one for each amino acid residue. The molecule is folded so as to form a cleft that holds the substrate molecule while it is being broken in two. The painting on the next page shows how the substrate fits into the cleft. The red balls represent oxygen atoms that are important in splitting the substrate.

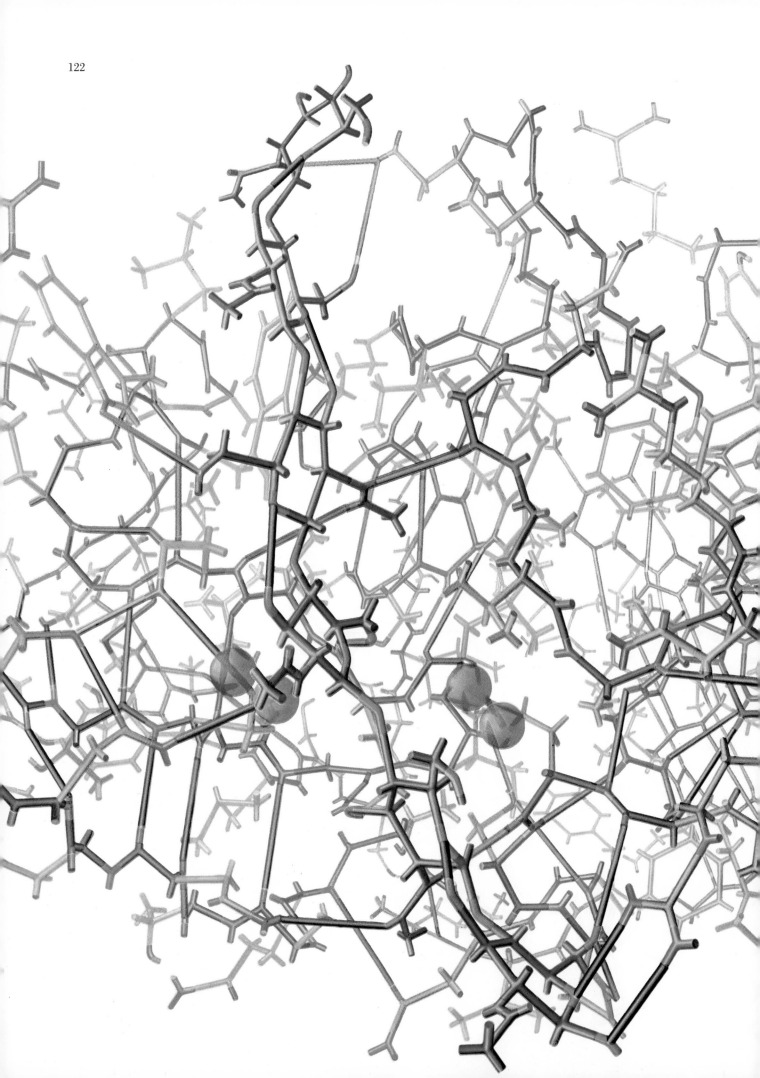

The difficulties of examining molecules in atomic detail arise, of course, from the fact that molecules are very small. Within a molecule each atom is usually separated from its neighbor by about 1.5 angstrom units (1.5×10^{-8} centimeter). The lysozyme molecule, which contains some 1,950 atoms, is about 40 angstroms in its largest dimension. The first problem is to find a microscope in which the atoms can be resolved from one another, or seen separately.

The resolving power of a microscope depends fundamentally on the wavelength of the radiation it employs. In general no two objects can be seen separately if they are closer together than about half this wavelength. The shortest wavelength transmitted by optical microscopes (those working in the ultraviolet end of the spectrum) is about 2,000 times longer than the distance between atoms. In order to "see" atoms one must use radiation with a much shorter wavelength: X rays, which have a wavelength closely comparable to interatomic distances. The employment of X rays, however, creates other difficulties: no satisfactory way has yet been found to make lenses or mirrors that will focus them into an image. The problem, then, is the apparently impossible one of designing an X-ray microscope without lenses or mirrors.

Consideration of the diffraction theory of microscope optics, as developed by Ernst Abbe in the latter part of the 19th century, shows that the problem can be solved. Abbe taught us that the formation of an image in the microscope can be regarded as a two-stage process. First, the object under examination scatters the light or other radiation falling on it in all directions, forming a diffraction pattern. This pattern arises because the light waves scattered from different parts of the object combine so as to produce a wave of large or small amplitude in any direction

according to whether the waves are in or out of phase—in or out of step—with one another. (This effect is seen most easily in light waves scattered by a regularly repeating structure, such as a diffraction grating made of lines scribed at regular intervals on a glass plate.) In the second stage of image formation, according to Abbe, the objective lens of the microscope collects the diffracted waves and recombines them to form an image of the object. Most important, the nature of the image depends critically on how much of the diffraction pattern is used in its formation.

X-Ray Structure Analysis

In essence X-ray structure analysis makes use of a microscope in which the two stages of image formation have been separated. Since the X rays cannot be focused to form an image directly, the diffraction pattern is recorded and the image is obtained from it by calculation. Historically the method was not developed on the basis of this reasoning, but this way of regarding it (which was first suggested by Lawrence Bragg) brings out its essential features and also introduces the main difficulty of applying it. In recording the intensities of the diffracted waves, instead of focusing them to form an image, some crucial information is lost, namely the phase relations among the various diffracted waves. Without this information the image cannot be formed, and some means of recovering it has to be found. This is the well-known phase problem of X-ray crystallography. It is on the solution of the problem that the utility of the method depends.

The term "X-ray crystallography" reminds us that in practice the method was developed (and is still applied) in the study of single crystals. Crystals suitable for study may contain some

10^{15} identical molecules in a regular array; in effect the molecules in such a crystal diffract the X radiation as though they were a single giant molecule. The crystal acts as a three-dimensional diffraction grating, so that the waves scattered by them are confined to a number of discrete directions. In order to obtain a three-dimensional image of the structure the intensity of the X rays scattered in these different directions must be measured, the phase problem must be solved somehow and the measurements must be combined by a computer.

The recent successes of this method in the study of protein structures have depended a great deal on the development of electronic computers capable of performing the calculations. They are due most of all, however, to the discovery in 1953, by M. F. Perutz of the Medical Research Council Laboratory of Molecular Biology in Cambridge, that the method of "isomorphous replacement" can be used to solve the phase problem in the study of protein crystals. The method depends on the preparation and study of a series of protein crystals into which additional heavy atoms, such as atoms of uranium, have been introduced without otherwise affecting the crystal structure. The first successes of this method were in the study of sperm-whale myoglobin by John C. Kendrew of the Medical Research Council Laboratory and in Perutz' own study of horse hemoglobin. For their work the two men received the Nobel prize for chemistry in 1962 [see "The Three-dimensional Structure of a Protein Molecule," by John C. Kendrew, SCIENTIFIC AMERICAN Offprint 121, and the article "The Hemoglobin Molecule," by M. F. Perutz, beginning on page 105].

Because the X rays are scattered by the electrons within the molecules, the image calculated from the diffraction pattern reveals the distribution of electrons within the crystal. The electron density is usually calculated at a regular array of points, and the image is made visible by drawing contour lines through points of equal electron density. If these contour maps are drawn on clear plastic sheets, one can obtain a three-dimensional image by assembling the maps one above the other in a stack. The amount of detail that can be seen in such an image depends on the resolving power of the effective microscope, that is, on its "aperture," or the extent of the diffraction pattern that has been included in the formation of the image. If the waves diffracted through sufficiently high angles are included

MODEL OF SUBSTRATE shows how it fits into the cleft in the lysozyme molecule. All the carbon atoms in the substrate are shown in purple. The portion of the substrate in intimate contact with the underlying enzyme is a polysaccharide chain consisting of six ringlike structures, each a residue of an amino-sugar molecule. The substrate in the model is made up of six identical residues of the amino sugar called N-acetylglucosamine (NAG). In the actual substrate every other residue is an amino sugar known as N-acetylmuramic acid (NAM). The illustration is based on X-ray studies of the way the enzyme is bound to a trisaccharide made of three NAG units, which fills the top of the cleft; the arrangement of NAG units in the bottom of the cleft was worked out with the aid of three-dimensional models. The substrate is held to the enzyme by a complex network of hydrogen bonds. In this style of model-making each straight section of chain represents a bond between atoms. The atoms themselves lie at the intersections and elbows of the structure. Except for the four red balls representing oxygen atoms that are active in splitting the polysaccharide substrate, no attempt is made to represent the electron shells of atoms because they would merge into a solid mass.

(corresponding to a large aperture), the atoms appear as individual peaks in the image map. At lower resolution groups of unresolved atoms appear with characteristic shapes by which they can be recognized.

The three-dimensional structure of lysozyme crystallized from the white of hen's egg has been determined in atomic detail with the X-ray method by our group at the Royal Institution in Lon-

don. This is the laboratory in which Humphry Davy and Michael Faraday made their fundamental discoveries during the 19th century, and in which the X-ray method of structure analysis was developed between the two world wars by the brilliant group of workers led by William Bragg, including J. D. Bernal, Kathleen Lonsdale, W. T. Astbury, J. M. Robertson and many others. Our work on lysozyme was begun in 1960

when Roberto J. Poljak, a visiting worker from Argentina, demonstrated that suitable crystals containing heavy atoms could be prepared. Since then C. C. F. Blake, A. C. T. North, V. R. Sarma, Ruth Fenn, D. F. Koenig, Louise N. Johnson and G. A. Mair have played important roles in the work.

In 1962 a low-resolution image of the structure was obtained that revealed the general shape of the molecule and

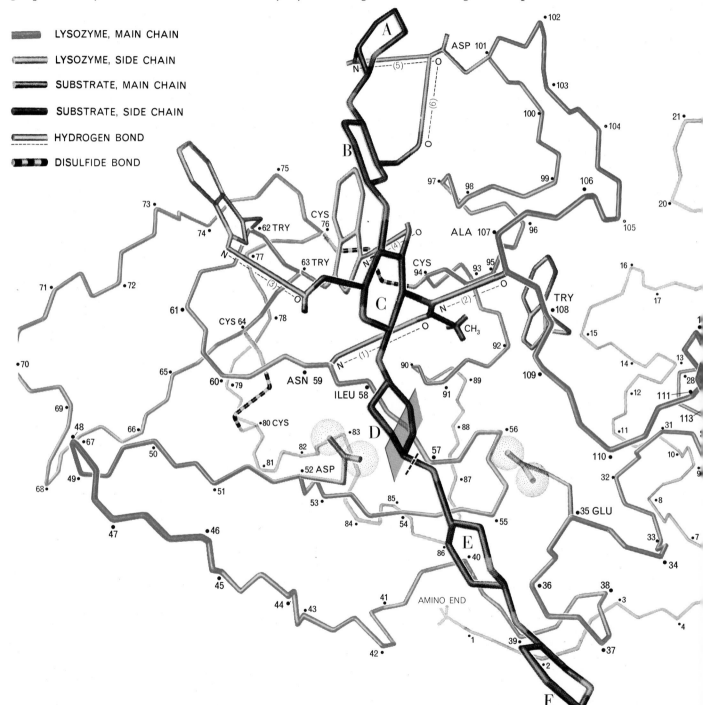

MAP OF LYSOZYME AND SUBSTRATE depicts in color the central chain of each molecule. Side chains have been omitted except for those that produce the four disulfide bonds clipping the lysozyme molecule together and those that supply the terminal connections for hydrogen bonds holding the substrate to the lysozyme. The top three rings of the substrate (A, B, C) are held to the underlying enzyme by six principal hydrogen bonds, which are identified by number to key with the description in the text. The lyso-

showed that the arrangement of the polypeptide chain is even more complex than it is in myoglobin. This low-resolution image was calculated from the amplitudes of about 400 diffraction maxima measured from native protein crystals and from crystals containing each of three different heavy atoms. In 1965, after the development of more efficient methods of measurement and computation, an image was calculated on the basis of nearly 10,000 diffraction maxima, which resolved features separated by two angstroms. Apart from showing a few well-separated chloride ions, which are present because the lysozyme is crystallized from a solution containing sodium chloride, the two-angstrom image still does not show individual atoms as separate maxima in the electron-density map. The level of resolution is high enough, however, for many of the groups of atoms to be clearly recognizable.

The Lysozyme Molecule

The main polypeptide chain appears as a continuous ribbon of electron density running through the image with regularly spaced promontories on it that are characteristic of the carbonyl groups (CO) that mark each peptide bond. In some regions the chain is folded in ways that are familiar from theoretical studies of polypeptide configurations and from the structure analyses of myoglobin and fibrous proteins such as the keratin of hair. The amino acid residues in lysozyme have now been designated by number; the residues numbered 5 through 15, 24 through 34 and 88 through 96 form three lengths of "alpha helix," the conformation that was proposed by Linus Pauling and Robert B. Corey in 1951 and that was found by Kendrew and his colleagues to be the most common arrangement of the chain in myoglobin. The helixes in lysozyme, however, appear to be somewhat distorted from the "classical" form, in which four atoms (carbon, oxygen, nitrogen and hydrogen) of each peptide group lie in a plane that is parallel to the axis of the alpha helix. In the lysozyme molecule the peptide groups in the helical sections tend to be rotated slightly in such a way that their CO groups point outward from the helix axes and their imino groups (NH) inward.

The amount of rotation varies, being slight in the helix formed by residues 5 through 15 and considerable in the one formed by residues 24 through 34. The effect of the rotation is that each NH group does not point directly at the CO group four residues back along the chain but points instead between the CO groups of the residues three and four back. When the NH group points directly at the CO group four residues back, as it does in the classical alpha helix, it forms with the CO group a hydrogen bond (the weak chemical bond in which a hydrogen atom acts as a bridge). In the lysozyme helixes the hydrogen bond is formed somewhere between two CO groups, giving rise to a structure intermediate between that of an alpha helix and that of a more symmetrical helix with a three-fold symmetry axis that was discussed by Lawrence Bragg, Kendrew and Perutz in 1950. There is a further short length of helix (residues 80 through 85) in which the hydrogen-bonding arrangement is quite close to that in the three-fold helix, and also an isolated turn (residues 119 through 122) of three-fold helix. Furthermore, the peptide at the far end of helix 5 through 15 is in the conformation of the three-fold helix, and the hydrogen bond from its NH group is made to the CO three residues back rather than four.

Partly because of these irregularities in the structure of lysozyme, the proportion of its polypeptide chain in the alpha-helix conformation is difficult to calculate in a meaningful way for comparison with the estimates obtained by other methods, but it is clearly less than half the proportion observed in myoglobin, in which helical regions make up about 75 percent of the chain. The lysozyme molecule does include, however, an example of another regular conformation predicted by Pauling and Corey. This is the "antiparallel pleated sheet," which is believed to be the basic structure of the fibrous protein silk and in which, as the name suggests, two lengths of polypeptide chain run parallel to each other in opposite directions. This structure again is stabilized by hydrogen bonds between the NH and CO groups of the main chain. Residues 41 through 45 and 50 through 54 in the lysozyme molecule form such a structure, with the connecting residues 46 through 49 folded into a hairpin bend between the two lengths of comparatively extended chain. The remainder of the polypeptide chain is folded in irregular ways that have no simple short description.

Even though the level of resolution achieved in our present image was not enough to resolve individual atoms, many of the side chains characteristic of the amino acid residues were readily identifiable from their general shape. The four disulfide bridges, for example, are marked by short rods of high electron density corresponding to the two relatively dense sulfur atoms within them. The six tryptophan residues also were easily recognized by the extended electron density produced by the large double-ring structures in their

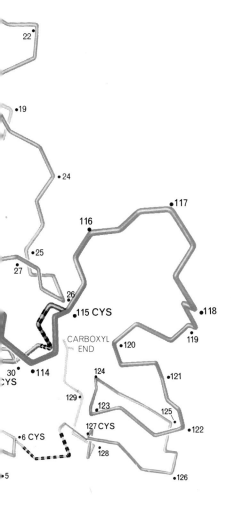

zyme molecule fulfills its function when it cleaves the substrate between the *D* and the *E* ring. Note the distortion of the *D* ring, which pushes four of its atoms into a plane.

FIRST 56 RESIDUES in lysozyme molecule contain a higher proportion of symmetrically organized regions than does all the rest of the molecule. Residues 5 through 15 and 24 through 34 (*right*) form two regions in which hydrogen bonds (*gray*) hold the residues in a helical configuration close to that of the "classical" alpha helix. Residues 41 through 45 and 50 through 54 (*left*) fold back against each other to form a "pleated sheet," also held together by hydrogen bonds. In addition the hydrogen bond between residues 1 and 40 ties the first 40 residues into a compact structure that may have been folded in this way before the molecule was fully synthesized (*see illustration at the bottom of these two pages*).

side chains. Many of the other residues also were easily identifiable, but it was nevertheless most important for the rapid and reliable interpretation of the image that the results of the chemical analysis were already available. With their help more than 95 percent of the atoms in the molecule were readily identified and located within about .25 angstrom.

Further efforts at improving the accuracy with which the atoms have been located is in progress, but an almost complete description of the lysozyme molecule now exists [*see illustration on pages 120 and 121*]. By studying it and the

results of some further experiments we can begin to suggest answers to two important questions: How does a molecule such as this one attain its observed conformation? How does it function as an enzyme, or biological catalyst?

Inspection of the lysozyme molecule immediately suggests two generalizations about its conformation that agree well with those arrived at earlier in the study of myoglobin. It is obvious that certain residues with acidic and basic side chains that ionize, or dissociate, on contact with water are all on the surface of the molecule more or less readily accessible to the surrounding

liquid. Such "polar" side chains are hydrophilic—attracted to water; they are found in aspartic acid and glutamic acid residues and in lysine, arginine and histidine residues, which have basic side groups. On the other hand, most of the markedly nonpolar and hydrophobic side chains (for example those found in leucine and isoleucine residues) are shielded from the surrounding liquid by more polar parts of the molecule. In fact, as was predicted by Sir Eric Rideal (who was at one time director of the Royal Institution) and Irving Langmuir, lysozyme, like myoglobin, is quite well described as an oil drop with a polar coat. Here it is important to note that the environment of each molecule in the crystalline state is not significantly different from its natural environment in the living cell. The crystals themselves include a large proportion (some 35 percent by weight) of mostly watery liquid of crystallization. The effect of the surrounding liquid on the protein conformation thus is likely to be much the same in the crystals as it is in solution.

It appears, then, that the observed conformation is preferred because in it the hydrophobic side chains are kept out of contact with the surrounding liquid whereas the polar side chains are generally exposed to it. In this way the system consisting of the protein and the solvent attains a minimum free energy, partly because of the large number of favorable interactions of like groups within the protein molecule and between it and the surrounding liquid, and partly because of the relatively high disorder of the water molecules that are in contact only with other polar groups of atoms.

Guided by these generalizations, many workers are now interested in the possibility of predicting the conforma-

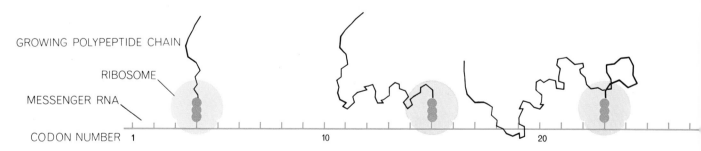

FOLDING OF PROTEIN MOLECULE may take place as the growing polypeptide chain is being synthesized by the intracellular particles called ribosomes. The genetic message specifying the amino acid sequence of each protein is coded in "messenger" ribonucleic acid (RNA). It is believed several ribosomes travel simultaneously along this long-chain molecule, reading the message as they go.

tion of a protein molecule from its chemical formula alone [see "Molecular Model-building by Computer," by Cyrus Levinthal; SCIENTIFIC AMERICAN Offprint 1043]. The task of exploring all possible conformations in the search for the one of lowest free energy seems likely, however, to remain beyond the power of any imaginable computer. On a conservative estimate it would be necessary to consider some 10^{129} different conformations for the lysozyme molecule in any general search for the one with minimum free energy. Since this number is far greater than the number of particles in the observable universe, it is clear that simplifying assumptions will have to be made if calculations of this kind are to succeed.

The Folding of Lysozyme

For some time Peter Dunnill and I have been trying to develop a model of protein-folding that promises to make practicable calculations of the minimum energy conformation and that is, at the same time, qualitatively consistent with the observed structure of myoglobin and lysozyme. This model makes use of our present knowledge of the way in which proteins are synthesized in the living cell. For example, it is well known, from experiments by Howard M. Dintzis and by Christian B. Anfinsen and Robert Canfield, that protein molecules are synthesized from the terminal amino end of their polypeptide chain. The nature of the synthetic mechanism, which involves the intracellular particles called ribosomes working in collaboration with two forms of ribonucleic acid ("messenger" RNA and "transfer" RNA), is increasingly well understood in principle, although the detailed environment of the growing protein chain remains unknown. Nevertheless,

it seems a reasonable assumption that, as the synthesis proceeds, the amino end of the chain becomes separated by an increasing distance from the point of attachment to the ribosome, and that the folding of the protein chain to its native conformation begins at this end even before the synthesis is complete. According to our present ideas, parts of the polypeptide chain, particularly those near the terminal amino end, may fold into stable conformations that can still be recognized in the finished molecule and that act as "internal templates," or centers, around which the rest of the chain is folded [see illustration at bottom of these two pages]. It may therefore be useful to look for the stable conformations of parts of the polypeptide chain and to avoid studying all the possible conformations of the whole molecule.

Inspection of the lysozyme molecule provides qualitative support for these ideas [see top illustration on opposite page]. The first 40 residues from the terminal amino end form a compact structure (residues 1 and 40 are linked by a hydrogen bond) with a hydrophobic interior and a relatively hydrophilic surface that seems likely to have been folded in this way, or in a simply related way, before the molecule was fully synthesized. It may also be important to observe that this part of the molecule includes more alpha helix than the remainder does.

These first 40 residues include a mixture of hydrophobic and hydrophilic side chains, but the next 14 residues in the sequence are all hydrophilic; it is interesting, and possibly significant, that these are the residues in the antiparallel pleated sheet, which lies out of contact with the globular submolecule formed by the earlier residues. In the light of our model of protein fold-

ing the obvious speculation is that there is no incentive to fold these hydrophilic residues in contact with the first part of the chain until the hydrophobic residues 55 (isoleucine) and 56 (leucine) have to be shielded from contact with the surrounding liquid. It seems reasonable to suppose that at this stage residues 41 through 54 fold back on themselves, forming the pleated-sheet structure and burying the hydrophobic side chains in the initial hydrophobic pocket.

Similar considerations appear to govern the folding of the rest of the molecule. In brief, residues 57 through 86 are folded in contact with the pleated-sheet structure so that at this stage of the process—if indeed it follows this course—the folded chain forms a structure with two wings lying at an angle to each other. Residues 86 through 96 form a length of alpha helix, one side of which is predominantly hydrophobic, because of an appropriate alternation of polar and nonpolar residues in that part of the sequence. This helix lies in the gap between the two wings formed by the earlier residues, with its hydrophobic side buried within the molecule. The gap between the two wings is not completely filled by the helix, however; it is transformed into a deep cleft running up one side of the molecule. As we shall see, this cleft forms the active site of the enzyme. The remaining residues are folded around the globular unit formed by the terminal amino end of the polypeptide chain.

This model of protein-folding can be tested in a number of ways, for example by studying the conformation of the first 40 residues in isolation both di-

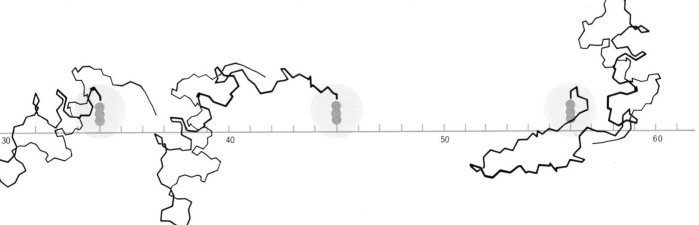

Presumably the messenger RNA for lysozyme contains 129 "codons," one for each amino acid. Amino acids are delivered to the site of synthesis by molecules of "transfer" RNA (dark color). The illustration shows how the lysozyme chain would lengthen as a ribosome travels along the messenger RNA molecule. Here, hypothetically, the polypeptide is shown folding directly into its final shape.

rectly (after removal of the rest of the molecule) and by computation. Ultimately, of course, the model will be regarded as satisfactory only if it helps us to predict how other protein molecules are folded from a knowledge of their chemical structure alone.

The Activity of Lysozyme

In order to understand how lysozyme brings about the dissolution of bacteria we must consider the structure of the bacterial cell wall in some detail. Through the pioneer and independent studies of Karl Meyer and E. B. Chain, followed up by M. R. J. Salton of the University of Manchester and many others, the structures of bacterial cell walls and the effect of lysozyme on them are now quite well known. The important part of the cell wall, as far as lysozyme is concerned, is made up of glucose-like amino-sugar molecules linked together into long polysaccharide chains, which are themselves cross-connected by short lengths of polypeptide chain. This part of each cell wall probably forms one enormous molecule—a "bag-shaped macromolecule," as W. Weidel and H. Pelzer have called it.

The amino-sugar molecules concerned in these polysaccharide structures are of two kinds; each contains an acetamido (–NH · CO · CH$_3$) side group, but one of them contains an additional major group, a lactyl side chain [see illustration below]. One of these amino sugars is known as N-acetylglucosamine (NAG) and the other as N-acetylmuramic acid (NAM). They occur alternately in the polysaccharide chains, being connected by bridges that include an oxygen atom (glycosidic linkages) between carbon atoms 1 and 4 of consecutive sugar rings; this is the same linkage that joins glucose residues in cellulose. The polypeptide chains that cross-connect these polysaccharides are attached to the NAM residues through the lactyl side chain attached to carbon atom 3 in each NAM ring.

Lysozyme has been shown to break the linkages in which carbon 1 in NAM is linked to carbon 4 in NAG but not the other linkages. It has also been shown to break down chitin, another common natural polysaccharide that is found in lobster shell and that contains only NAG.

Ever since the work of Svante Arrhenius of Sweden in the late 19th century enzymes have been thought to work by forming intermediate compounds with their substrates: the substances whose chemical reactions they catalyze. A proper theory of the enzyme-substrate complex, which underlies all present thinking about enzyme activity, was clearly propounded by Leonor Michaelis and Maude Menton in a remarkable paper published in 1913. The idea, in its simplest form, is that an enzyme molecule provides a site on its surface to which its substrate molecule can bind in a quite precise way. Reactive groups of atoms in the enzyme then promote the required chemical reaction in the substrate. Our immediate objective, therefore, was to find the structure of a reactive complex between lysozyme and its polysaccha-ride substrate, in the hope that we would then be able to recognize the active groups of atoms in the enzyme and understand how they function.

Our studies began with the observation by Martin Wenzel and his colleagues at the Free University of Berlin that the enzyme is prevented from functioning by the presence of NAG itself. This small molecule acts as a competitive inhibitor of the enzyme's activity and, since it is a part of the large substrate molecule normally acted on by the enzyme, it seems likely to do this by binding to the enzyme in the way that part of the substrate does. It prevents the enzyme from working by preventing the substrate from binding to the enzyme. Other simple amino-sugar molecules, including the trisa charide made of three NAG units, behave in the same way. We therefore decided to study the binding of these sugar molecules to the lysozyme molecules in our crystals in the hope of learning something about the structure of the enzyme-substrate complex itself.

My colleague Louise Johnson soon found that crystals containing the sugar molecules bound to lysozyme can be prepared very simply by adding the sugar to the solution from which the lysozyme crystals have been grown and in which they are kept suspended. The small molecules diffuse into the protein crystals along the channels filled with water that run through the crystals. Fortunately the resulting change in the crystal structure can be studied quite simply. A useful image of the electron-density changes can be calculated from

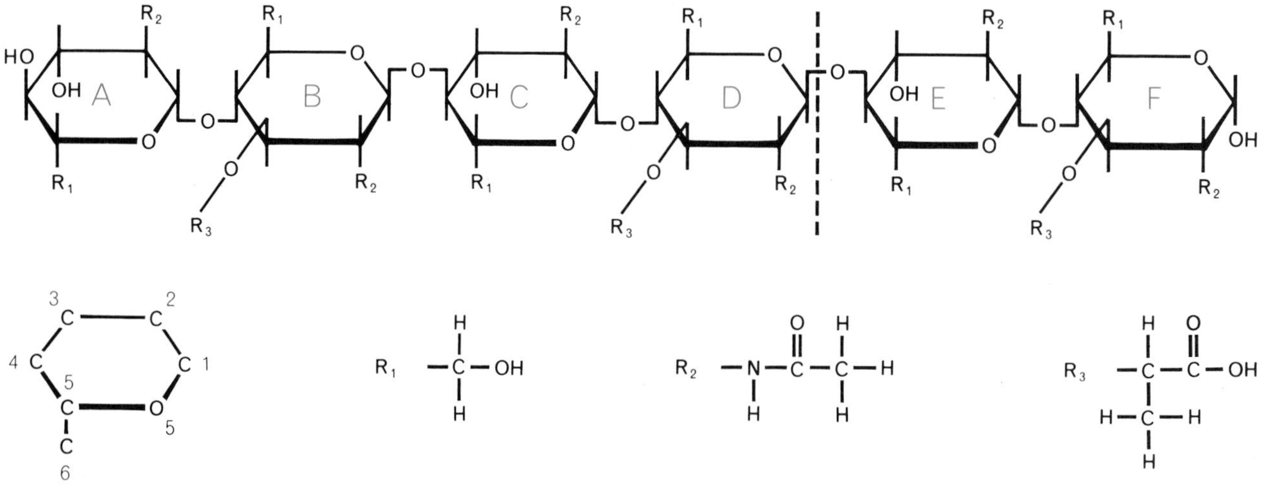

POLYSACCHARIDE MOLECULE found in the walls of certain bacterial cells is the substrate broken by the lysozyme molecule. The polysaccharide consists of alternating residues of two kinds of amino sugar: N-acetylglucosamine (NAG) and N-acetylmuramic acid (NAM). In the length of polysaccharide chain shown here A, C and E are NAG residues; B, D and F are NAM residues. The inset at left shows the numbering scheme for identifying the principal atoms in each sugar ring. Six rings of the polysaccharide fit into the cleft of the lysozyme molecule, which effects a cleavage between rings D and E (see illustration on pages 124 and 125).

measurements of the changes in amplitude of the diffracted waves, on the assumption that their phase relations have not changed from those determined for the pure protein crystals. The image shows the difference in electron density between crystals that contain the added sugar molecules and those that do not.

In this way the binding to lysozyme of eight different amino sugars was studied at low resolution (that is, through the measurement of changes in the amplitude of 400 diffracted waves). The results showed that the sugars bind to lysozyme at a number of different places in the cleft of the enzyme. The investigation was hurried on to higher resolution in an attempt to discover the exact nature of the binding. Happily these studies at two-angstrom resolution (which required the measurement of 10,000 diffracted waves) have now shown in detail how the trisaccharide made of three NAG units is bound to the enzyme.

The trisaccharide fills the top half of the cleft and is bound to the enzyme by a number of interactions, which can be followed with the help of the illustration on pages 124 and 125. In this illustration six important hydrogen bonds, to be described presently, are identified by number. The most critical of these interactions appear to involve the acetamido group of sugar residue C [third from top], whose carbon atom 1 is not linked to another sugar residue. There are hydrogen bonds from the CO group of this side chain to the main-chain NH group of amino acid residue 59 in the enzyme molecule [bond No. 1] and from its NH group to the main-chain CO group of residue 107 (alanine) in the enzyme molecule [bond No. 2]. Its terminal CH_3 group makes contact with the side chain of residue 108 (tryptophan). Hydrogen bonds [No. 3 and No. 4] are also formed between two oxygen atoms adjacent to carbon atoms 6 and 3 of sugar residue C and the side chains of residues 62 and 63 (both tryptophan) respectively. Another hydrogen bond [No. 5] is formed between the acetamido side chain of sugar residue A and residue 101 (aspartic acid) in the enzyme molecule. From residue 101 there is a hydrogen bond [No. 6] to the oxygen adjacent to carbon atom 6 of sugar residue B. These polar interactions are supplemented by a large number of nonpolar interactions that are more difficult to summarize briefly. Among the more important nonpolar interactions, however, are those between sugar residue B and the ring system of residue

62; these deserve special mention because they are affected by a small change in the conformation of the enzyme molecule that occurs when the trisaccharide is bound to it. The electron-density map showing the change in electron density when tri-NAG is bound in the protein crystal reveals clearly that parts of the enzyme molecule have moved with respect to one another. These changes in conformation are largely restricted to the part of the enzyme structure to the left of the cleft, which appears to tilt more or less as a whole in such a way as to close the cleft slightly. As a result the side chain of residue 62 moves about .75 angstrom toward the position of sugar residue B. Such changes in enzyme conformation have been discussed for some time, notably by Daniel E. Koshland, Jr., of the University of California at Berkeley, whose "induced fit" theory of the enzyme-substrate interaction is supported in some degree by this observation in lysozyme.

The Enzyme-Substrate Complex

At this stage in the investigation excitement grew high. Could we tell how the enzyme works? I believe we can. Unfortunately, however, we cannot see this dynamic process in our X-ray images. We have to work out what must happen from our static pictures. First of all it is clear that the complex formed by tri-NAG and the enzyme is not the enzyme-substrate complex involved in catalysis because it is stable. At low concentrations tri-NAG is known to behave as an inhibitor rather than as a substrate that is broken down; clearly we have been looking at the way in which it binds as an inhibitor. It is noticeable, however, that tri-NAG fills only half of the cleft. The possibility emerges that more sugar residues, filling the remainder of the cleft, are required for the formation of a reactive enzyme-substrate complex. The assumption here is that the observed binding of tri-NAG as an inhibitor involves interactions with the enzyme molecule that also play a part in the formation of the functioning enzyme-substrate complex.

Accordingly we have built a model that shows that another three sugar residues can be added to the tri-NAG in such a way that there are satisfactory interactions of the atoms in the proposed substrate and the enzyme. There is only one difficulty: carbon atom 6 and its adjacent oxygen atom in sugar residue D make uncomfortably close contacts

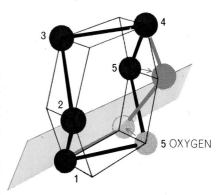

"CHAIR" CONFIGURATION (gray) is that normally assumed by the rings of amino sugar in the polysaccharide substrate. When bound against the lysozyme, however, the D ring is distorted (color) so that carbon atoms 1, 2 and 5 and oxygen atom 5 lie in a plane. The distortion evidently assists in breaking the substrate below the D ring.

with atoms in the enzyme molecule, unless this sugar residue is distorted a little out of its most stable "chair" conformation into a conformation in which carbon atoms 1, 2 and 5 and oxygen atom 5 all lie in a plane [see illustration above]. Otherwise satisfactory interactions immediately suggest themselves, and the model falls into place.

At this point it seemed reasonable to assume that the model shows the structure of the functioning complex between the enzyme and a hexasaccharide. The next problem was to decide which of the five glycosidic linkages would be broken under the influence of the enzyme. Fortunately evidence was at hand to suggest the answer. As we have seen, the cell-wall polysaccharide includes alternate sugar residues of two kinds, NAG and NAM, and the bond broken is between NAM and NAG. It was therefore important to decide which of the six sugar residues in our model could be NAM, which is the same as NAG except for the lactyl side chain appended to carbon atom 3. The answer was clear-cut. Sugar residue C cannot be NAM because there is no room for this additional group of atoms. Therefore the bond broken must be between sugar residues B and C or D and E. We already knew that the glycosidic linkage between residues B and C is stable when tri-NAG is bound. The conclusion was inescapable: the linkage that must be broken is the one between sugar residues D and E.

Now it was possible to search for the origin of the catalytic activity in the neighborhood of this linkage. Our task was made easier by the fact that John A.

Rupley of the University of Arizona had shown that the chemical bond broken under the influence of lysozyme is the one between carbon atom 1 and oxygen in the glycosidic link rather than the link between oxygen and carbon atom 4. The most reactive-looking group of atoms in the vicinity of this bond are the side chains of residue 52 (aspartic acid) and residue 35 (glutamic acid).

One of the oxygen atoms of residue 52 is about three angstroms from carbon atom 1 of sugar residue D as well as from the ring oxygen atom 5 of that residue. Residue 35, on the other hand, is about three angstroms from the oxygen in the glycosidic linkage. Furthermore, these two amino acid residues have markedly different environments. Residue 52 has a number of polar neighbors and appears to be involved in a network of hydrogen bonds linking it with residues 46 and 59 (both asparagine) and, through them, with residue 50 (serine). In this environment residue 52 seems likely to give up a terminal hydrogen atom and thus be negatively charged under most conditions, even when it is in a markedly acid solution, whereas residue 35, situated in a nonpolar environment, is likely to retain its terminal hydrogen atom.

A little reflection suggests that the concerted influence of these two amino

acid residues, together with a contribution from the distortion to sugar residue D that has already been mentioned, is enough to explain the catalytic activity of lysozyme. The events leading to the rupture of a bacterial cell wall probably take the following course [see illustration on this page].

First, a lysozyme molecule attaches itself to the bacterial cell wall by interacting with six exposed amino-sugar residues. In the process sugar residue D is somewhat distorted from its usual conformation.

Second, residue 35 transfers its terminal hydrogen atom in the form of a hydrogen ion to the glycosidic oxygen, thus bringing about cleavage of the bond between that oxygen and carbon atom 1 of sugar residue D. This creates a positively charged carbonium ion (C^+) where the oxygen has been severed from carbon atom 1.

Third, this carbonium ion is stabilized by its interaction with the negatively charged aspartic acid side chain of residue 52 until it can combine with a hydroxyl ion (OH^-) that happens to diffuse into position from the surrounding water, thereby completing the reaction. The lysozyme molecule then falls away, leaving behind a punctured bacterial cell wall.

It is not clear from this description that the distortion of sugar residue D plays any part in the reaction, but in fact it probably does so for a very interesting reason. R. H. Lemieux and G. Huber of the National Research Council of Canada showed in 1955 that when a sugar molecule such as NAG incorporates a carbonium ion at the carbon-1 position, it tends to take up the same conformation that is forced on ring D by its interaction with the enzyme molecule. This seems to be an example, therefore, of activation of the substrate by distortion, which has long been a favorite idea of enzymologists. The binding of the substrate to the enzyme itself favors the formation of the carbonium ion in ring D that seems to play an important part in the reaction.

It will be clear from this account that although lysozyme has not been seen in action, we have succeeded in building up a detailed picture of how it may work. There is already a great deal of chemical evidence in agreement with this picture, and as the result of all the work now in progress we can be sure that the activity of Fleming's lysozyme will soon be fully understood. Best of all, it is clear that methods now exist for uncovering the secrets of enzyme action.

SPLITTING OF SUBSTRATE BY LYSOZYME is believed to involve the proximity and activity of two side chains, residue 35 (glutamic acid) and residue 52 (aspartic acid). It is proposed that a hydrogen ion (H^+) becomes detached from the OH group of residue 35 and attaches itself to the oxygen atom that joins rings D and E, thus breaking the bond between the two rings. This leaves carbon atom 1 of the D ring with a positive charge, in which form it is known as a carbonium ion. It is stabilized in this condition by the negatively charged side chain of residue 52. The surrounding water supplies an OH^- ion to combine with the carbonium ion and an H^+ ion to replace the one lost by residue 35. The two parts of the substrate then fall away, leaving the enzyme free to cleave another polysaccharide chain.

Storage, Replication, and Expression
of Genetic Information

12

THE DISCOVERY OF DNA

ALFRED E. MIRSKY
June 1968

*In 1869 Friedrich Miescher found a substance in white
blood cells that he called "nuclein." Cell biologists
saw that it was a constituent of chromosomes and
hence must play a major role in heredity*

Deoxyribonucleic acid was discovered in 1869 by Friedrich Miescher. The reader may find it surprising that DNA, which has been the focus of so much recent work in biology, was isolated almost a century ago. If so, he will find it even more surprising that the function of DNA as the substance in chromosomes that transmits hereditary characteristics was recognized only a few years later by a number of biologists (not, however, by Miescher). Here I shall recall how it happened that Miescher discovered "nuclein" (as he called it) and what it meant to him and his contemporaries, and tell something of the investigative history of DNA until, some 25 years ago, it was conclusively shown to be the genetic material.

Miescher was the son of a Swiss physician who practiced in Basel and taught at the university there, and he followed his father into medicine. As a medical student he came under the influence of his uncle Wilhelm His, professor of anatomy and one of the outstanding investigators and teachers of his time. That influence was profound and lifelong. Miescher's later views on biology are often best understood in the light of His's attitudes and ideas; Miescher's writings are available today largely because His gathered and published the younger man's letters and papers after Miescher's death in 1895 at the age of 51.

His urged Miescher to go into histochemistry, the study of the chemical composition of tissues, because in "my own histological investigations I was constantly reminded that the ultimate problems of tissue development would be solved on the basis of chemistry." Miescher took the advice, and after receiving his degree in 1868 he went to the University of Tübingen first to learn organic chemistry and then to work in the laboratory of the biochemist F. Hoppe-Seyler. (The first laboratory devoted entirely to biochemistry, it was located incongruously in an ancient castle overlooking the Neckar River. In later years Miescher liked to tell his students how the narrow, deep-set windows and the dark vault of his room reminded him of a medieval alchemist's laboratory.) Within a few months, in the course of experiments on cells in pus, he discovered DNA.

Pus may seem an unlikely material for a study of cell composition, but Miescher considered the white blood cells present in pus to be among the simplest of animal cells, and there was a fresh supply of pus every day in the Tübingen surgical clinic. He was given the bandages removed from postoperative wounds, and he washed the white cells from the bandages for experiments. If the bandages were washed with ordinary saline solution, the cells swelled to form a gelatinous mass. In a dilute sodium sulfate solution, on the other hand, the cells were preserved and sedimented rapidly, making it easy to separate them from the blood serum and other material in pus.

Miescher undertook a general study of the chemical composition of the white cells. First he extracted them in various ways—with salt solutions, acid, alkali and alcohol. Earlier workers, including Hoppe-Seyler, had extracted pus cells with concentrated salt solutions and had obtained a gelatinous material that reminded them of myosin, the protein of muscle. Miescher got the same result. What is of interest to us is that the gelatinous substance consists largely of DNA! This did not become known until 1942, when it was demonstrated that concentrated neutral salt solutions are an exceedingly useful medium for the extraction of polymerized (and therefore gelatinous) DNA. To extract DNA under these conditions, however, one needs a centrifuge. If Miescher had had a centrifuge, it is quite possible that he would have obtained DNA in its natural form instead of in the depolymerized form he eventually discovered.

Miescher's route to DNA was necessarily more indirect, as is frequently the case for the pathfinder. When he extracted pus cells with dilute alkali, he obtained a substance that precipitated on the addition of acid and redissolved when a trace of alkali was added. At this point Miescher noted: "According to recognized histochemical data I had to ascribe such material to the nuclei... and I therefore tried to isolate the nuclei." The isolation of the nucleus—or of any other cell organelle—had not been attempted before, as far as we know, although the nucleus had been identified in 1831. Miescher's primary observation, on which the isolation of nuclei depended, was that dilute hydrochloric acid dissolves most of the materials of a cell, leaving the nuclei behind. (This observation is the basis of what is still a valuable procedure for isolating nuclei.)

When Miescher examined the isolated nuclei under his microscope, he could still see contamination, which he suspected was cell protein. To obtain clean nuclei he therefore added the protein-digesting enzyme pepsin to the dilute hydrochloric acid he was using. More precisely (since this was in 1868), he made a hydrochloric acid extract of pig's stomach and applied it to pus cells. The isolated nuclei so prepared were somewhat shrunken but were clean enough for chemical study. The next step was to extract the isolated nuclei (rather than the whole cells) with dilute alkali. The extracted material precipitated on the addition of acid and redissolved readily in alkali. Miescher analyzed this material into its elements (finding, for example, 14 percent nitrogen and 2.5 percent phosphorus) and studied some of its other properties. He came to the conclusion that it did not fit into any known group of substances: it was a substance "*sui generis*" and he called it nuclein. The analytical data indicate that somewhat less than 30 percent of Miescher's first nuclein preparation consisted of DNA.

The year with Hoppe-Seyler was now ended. Hoppe-Seyler, who agreed that a new substance had been discovered, was sufficiently interested to repeat the preparation of nuclein after Miescher's departure, and sufficiently cautious to delay the publication of Miescher's paper until he had satisfied himself that the work was sound. The paper was published in 1871.

Miescher left Tübingen in the fall of 1869 to spend his vacation at home in Basel. He decided that during the two-month vacation he would broaden his study of nuclein by looking for it in various cells. The first material he chose was the hen's egg, because His had recently claimed that the microscopic particles called yolk platelets were genuine cells of connective-tissue origin. This was a controversial claim, and Miescher believed a biochemical approach—the search for nuclein in the platelets—would help to establish its validity. Following the procedure he had worked out for pus cells, he soon isolated from yolk platelets what he took to be nuclein. The phosphorus content and some other properties of platelet nuclein did differ somewhat from those of pus nuclein, but Miescher (like many another investigator!) was determined to find what he was looking for, and he was convinced that this was nuclein. In spite of the curious appearance of the platelets, he wrote, "nobody would any longer deny that they have genuine nuclei, because it is not in the optical properties but in the chemical nature of a structure that its role in the molecular events of a cell's life is rooted."

Miescher wrote up his new work and sent it off to Hoppe-Seyler; it was published along with the Tübingen research. The paper on pus cells remains a classic but the one on yolk platelets is forgotten. It was wrong. The microscopists, considering what Miescher called the "optical properties" of platelets, never did accept the idea that they were cells, and in time detailed chemical analysis showed (and Miescher had to agree) that what he had taken for platelet nuclein in fact had a very different composition. It is only recently that careful microscopic observation has demonstrated that yolk platelets are derived from the subcellular particles called mitochondria; they contain only a trace of DNA, whereas Miescher thought they contained a large amount of nuclein.

Having learned biochemistry at Tübingen, Miescher next spent a year at the University of Leipzig in Carl Lud-

CASTLE ON THE NECKAR RIVER in Germany was where nuclein, or deoxyribonucleic acid (DNA), was discovered. The castle housed F. Hoppe-Seyler's biochemical laboratory at the University of Tübingen, where Friedrich Miescher was a postdoctoral student.

wig's laboratory, a world center of physiology. Here he became convinced of the central role of the physical sciences in biology and in particular learned the importance of developing new instrumentation for research.

Miescher returned to Basel in 1870 and soon began the investigation on which he did his finest work: an analysis of the nuclein and other components of salmon spermatozoa. Wilhelm His introduced him to the salmon fishery then flourishing along the banks of the Rhine at Basel. The salmon, having swum all the way up from the North Sea to spawn, were sexually mature, so that huge quantities of ripe sperm were available. The nucleus is extremely large in any sperm cell, and among spermatozoa the salmon's is remarkable: its nucleus accounts for more than 90 percent of its mass. For a young investigator who had recently discovered nuclein in pus cells washed out of bandages, the sperm of the Rhine salmon must have seemed to present a God-given opportunity, and Miescher

seized it. He was an intense and rapid worker. Within a year and a half he had completed most of his investigation—in spite of the fact that his laboratory was so crowded with medical students that he could only do his chemical analyses, an essential part of the job, at night and on Sundays. His classic paper on the sperm cell of the salmon was published in 1874.

By acidifying a suspension of sperm cells, Miescher first caused the cells to aggregate and settle. Then he treated them with hydrochloric acid (omitting the pepsin he had used with pus cells) and extracted from the sperm an organic base with a high content of nitrogen. It accounted for 27 percent of the mass of the sperm, and he called it "protamine." Then he treated the residue of the sperm with dilute alkali. This extracted the nuclein, which accounted for 49 percent of the sperm's mass.

Miescher saw that nuclein was an acid containing a number of acid groups (a "polybasic" acid) and that it combined with protamine, a base, to form an insoluble salt in the nucleus. He experimented with modifying the chemical equilibrium of the nuclein-protamine combination, a problem that still interests biochemists today. He found, for example, that if sperm washed with acetic acid and alcohol were treated with a sodium chloride solution, much of the protamine was released from combination and passed into solution. If *fresh* sperm were treated with the same salt solution, however, the material became a lumpy gel that could almost be cut with a pair of scissors, as in the case of the pus cells treated with salt solution. The reason was that the DNA in the preparation was present as an extremely long linear polymer. Miescher, as I have mentioned, never did isolate DNA in its natural state and learn the importance of its fibrous structure. He did, however, get some idea that nuclein consisted of large molecules because he found that it would not pass through a parchment filter, whereas protamine would do so readily.

Although most of Miescher's work was done on salmon sperm, he also investigated the sperm of other species, notably the carp and the bull. He was disappointed not to find protamine in either, or even in unripe salmon sperm, and in a letter to His he referred to its presence in ripe salmon sperm as a "miserable special case." Ten years later, however, the biochemist Albrecht Kossel discovered histone, a base analo-

FRIEDRICH MIESCHER, the discoverer of DNA, was born in 1844 and died in 1895. This portrait is the frontispiece of a collection of his letters and papers published posthumously.

gous to protamine, in the nucleus of red blood cells, and soon it was found also in the nucleus of lymphocytes, white cells from the thymus gland. Today it is known that either histones or protamines are present in saltlike combination with nuclein in the nuclei of all plant and animal cells.

Miescher's analyses of protamine and nuclein into their constituent elements were done with great care, and his results for nuclein are very close to those of later workers for what came to be called nucleic acid. (That term was introduced by the biochemist Richard Altmann in 1889. Altmann's method of preparation was different, but Miescher recognized that the substance was the same, and he did not object to the new name because he had been aware that nuclein was an acid.) To obtain good analytical results Miescher considered it essential to isolate his nuclein at low temperatures. He analyzed nuclein in the fall and winter, sometimes working from five in the morning until late at night in unheated rooms. (Later in his life, when he again took up his study of salmon sperm, he went back to spending long hours in the cold. His health, which had always been delicate, gave way and he died of a chest ailment.)

The process of elementary analysis was particularly important for Miescher because it was his main guide to the composition and identification of a substance. He could not know, for example, that protamine is a protein, and that it is basic and has a nitrogen content because it contains large amounts of the basic amino acid arginine. He knew that phosphoric acid was responsible for the acidity of sperm nuclein, and that nuclein was a substance *sui generis* quite distinct from proteins, but he did not know that nucleic acids had two kinds of subunit: purines and pyrimidines. The detailed chemistry of nucleic acids and the proteins associated with them in the cell nucleus was worked out by later biochemists, beginning with Kossel in 1884. It took longer still, as we shall see, to understand the biological function of these substances.

Miescher's work on nuclein led him to a deep interest in the life of the Rhine salmon. In conducting what he called his "sperm campaign" almost every year he was witness to one of the most impressive events of animal life: the migration of the salmon 500 miles from the sea on their way to spawn in the headwaters of the Rhine. Between 1874 and 1880 Miescher devoted himself to the physiology of the salmon, examining

more than 2,000 fish. He learned that the salmon spent eight to 10 months in fresh water, and that during this period, although they were extremely active, they did not feed; they did not, indeed, secrete any digestive juices. And yet during this time an extraordinary metabolic rearrangement occurred: a massive increase in the size of the sex organs. Miescher found that when female salmon left the sea, their ovaries accounted for .5 percent of their body weight; by the time the fish reached Basel the ovaries represented 26 percent of their weight. The growth, he discovered, was at the expense of the large "lateral trunk" muscle; the loss of protein and phosphate by this muscle was sufficient to account for the growth of the ovaries. From physiology Miescher moved on to ecology, making a number of suggestions for the conservation of the Rhine salmon.

In 1872, at the age of 28, Miescher had been appointed professor of physiology at Basel, succeeding His, who had moved to Leipzig. Miescher became a leading figure at the university. In time he built a new physiological institute (which he named the Vesalianum in honor of Andreas Vesalius, the 16th-century Belgian anatomist who had spent six months in Basel while his *De humani corporis fabrica* was being printed there), stocked it with new precision instruments as they became available and guided it in researches on such classical problems of physiology as respiration, the circulation and the effects of high altitude. From time to time he undertook public service projects. The canton of Basel asked his advice on the nutrition of inmates of prisons and other institutions, and he also gave a series of public lectures on nutrition and home economics. Living in a country that produced milk but did not consume much of it, he was aware of milk's nutritional value, and in one lecture he heaped scorn on the "sordid avarice" of peasants who withheld milk from their children in order to make "the last drop" into salable cheese. "If you ask these pale and feeble people what they eat," he said, "they reply: potatoes, coffee, more potatoes and schnapps to keep down hunger." Miescher's social concern led a colleague to comment that he would surely have been active in politics had he not been hard of hearing.

All his life Miescher's primary and absorbing interest was nuclein. From time to time he went back to investigating its chemistry; he was also preoccupied with the question of its biological

function. During his work on the white cells of pus he had made no mention, either in his letters or in the paper published in 1871, of the possible function of the nucleus. This is hardly surprising, since at that time and for some years to come the role of the nucleus in the life of the cell was simply not understood. As late as 1882 the leading cell biologist Walther Flemming, describing the latest work on the nucleus, conceded that "concerning the biological significance of the nucleus we remain completely in the dark."

When Miescher came to the investigation of sperm, however, he began to ask questions about the role of nuclein in fertilization and about the nature of fertilization. Does the sperm contain certain special substances that are effective in fertilization? Willy Kühne, the Heidelberg biochemist who had just introduced the word "enzyme," suggested that there might be enzymes in sperm. Since salmon sperm seemed to be a clean and uncontaminated material, Miescher looked in it for enzymes, but he failed to find anything that appeared to be promising. Then he went on to say: "If one wants to assume that a single substance ... is the specific cause of fertilization, then one should undoubtedly first of all think of nuclein." Coming on this passage written in 1874, the reader holds his breath for a moment—but only for a moment, because Miescher turned away from the idea. He supposed, one must remember, that the egg contained a rich supply of nuclein in its yolk granules. And in his opinion there was no special characteristic that distinguished sperm nuclein from the great mass of egg nuclein. Indeed, he believed "the riddle of fertilization is not hidden in a particular substance"; the sperm is acting as a whole through the cooperation of all its parts, and if one considers the magnitude of the sperm's contribution to heredity, it must work in a most complex way. This line of thought led Miescher to think of fertilization as a physical procedure in which a certain movement (*Bewegung*) of the sperm is transmitted to the egg. In casting about for the nature of the movement, Miescher pointed to the "molecular process" that occurs when a nerve stimulates a muscle as perhaps being analogous to the effect of the sperm on the egg.

When Miescher invoked physical motion to explain fertilization, it is clear that he was thinking of kinetic theory, which at that time was in its formative period. By 1892 another physicochemical approach to fertilization appealed to

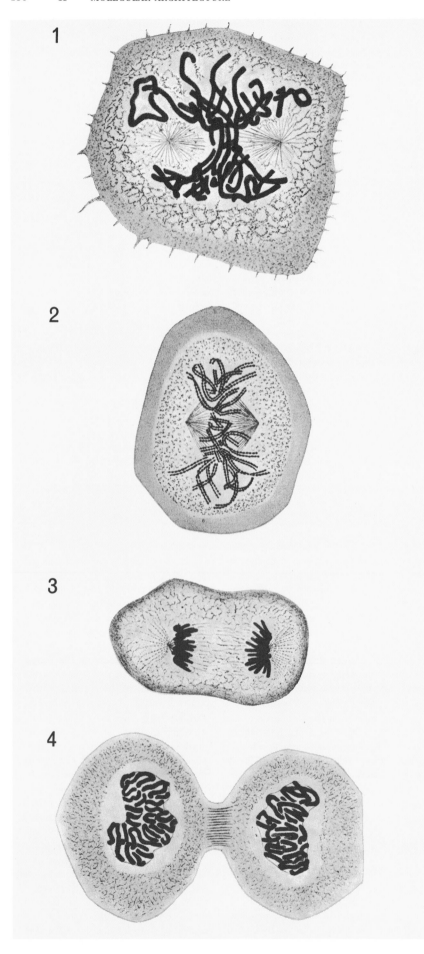

him and he wrote: "The key to sexuality lies for me in stereochemistry," that is, in the varying positions of the asymmetric carbon atoms in molecules. Miescher's desire to find explanations in physical science did not lead him to an understanding of the biological role of nuclein. At about the same time, however, another group of biologists—who were not interested in kinetic theory or stereochemistry—succeeded in laying some of the foundations of biology and at the same time recognizing the fundamental role of nuclein.

Miescher was nurtured in the great 19th-century school of "molecular biology." Both His and Carl Ludwig (in whose laboratory Miescher had spent a year) were leaders of the movement to analyze vital phenomena on the basis of physical science. Other outstanding figures in the movement were Claude Bernard and the neurophysiologists Emil du Bois-Reymond and Hermann von Helmholtz. The contributions of these men taken together constitute the foundation of modern physiology. Ludwig's work on kidney function, to take an example, is the basis of modern kidney physiology. In 1885 Michael Foster, professor of physiology at the University of Cambridge, spoke of "a new molecular physiology," maintaining that "the more these molecular problems of physiology ... are studied, the stronger becomes the conviction that the consideration of what we call 'structure' and 'composition' must, in harmony with the modern teachings of physics, be approached under the dominant conception of modes of motion.... The phenomena in question are the result not of properties of kinds of matter, in the vulgar sense of these words, but of kinds of motion." It was from this point of view that His considered fertilization to be a process in which the sperm communicates a mode

IN CELL DIVISION it is the chromosomes that provide continuity, as is shown in these drawings of salamander epithelial cells published in 1882 by Walther Flemming. In an early stage of cell division the chromosomes (which were known to contain nuclein) are ribbon-like and a "spindle" (two rayed structures) is beginning to form (1). The chromosomes seem to split (2); actually each has replicated lengthwise and the members of each resulting pair are separating. Each member goes to a different pole of the cell, so that two equal complements of chromosomes are established (3). Nuclear membranes surround each complement and the cell divides into two identical daughter cells (4).

of motion to the egg—a view that, as we have seen, Miescher accepted.

The biologists who at this time succeeded in discovering what actually happens in fertilization, and thereupon recognized the role of Miescher's nuclein, were (with several notable exceptions) not interested in physical science and in modes of motion. They were the founders of cell biology. The essential step taken by these biologists was the minute observation of what actually happens when a sperm fertilizes an egg. Although salmon sperm was an ideal material for Miescher's experiments on the chemistry of the sperm nucleus, it was of little value for observing the act of fertilization. For this purpose Oskar Hertwig of the University of Berlin and the Swiss biologist Hermann Fol chose the sea urchin and the starfish. In their classic experiments in the late 1870's they observed that the sperm cell penetrates the egg and that the sperm nucleus then fuses with the egg nucleus. At the same time these experiments were being done Flemming, in another series of observations, described the changes that occur in the nucleus during cell division [see illustration, page 136]. He was able to show that chromosomes provide the continuing elements from one generation of cells to the next and that they do so by replicating during cell division.

The observations on fertilization by Hertwig and Fol and those on cell division by Flemming were brought together by Edouard van Beneden of the University of Liège in a wonderful series of observations on fertilization in the threadworm *Ascaris*, a parasite of horses. *Ascaris*, unlike most other animals, has nuclei that break down before the egg and sperm nuclei fuse, so that its chromosomes can be seen with unusual clarity. Chromosome behavior can be followed readily because there are only two chromosomes in each nucleus. The male and female chromosomes, indistinguishable from each other, are brought together but do not fuse. Each chromosome replicates and soon there are two cells, formed in the way Flemming described for cell division. Van Beneden saw that in fertilization, as in cell division, continuity depends on chromosomes: the sperm's contribution to fertilization is ·a set of chromosomes homologous with those present in the egg [see illustration on next two pages]. (In the course of this work, which was published in 1883, van Beneden discovered meiosis, the halving of chromosome number that precedes the bringing together of egg and sperm chro-

mosomes in fertilization.)

Two years earlier Miescher's nuclein had been brought into the picture, not by Miescher but by an obscure young botanist named E. Zacharias. He showed that the characteristic material of chromosomes either was nuclein or was intimately associated with it. In his work on cell division Flemming had relied extensively on stains to make the nucleus and chromosomes readily visible. In the nucleus of cells that were not in the process of division there was a somewhat formless structure that took up certain stains; the same stains were taken up by the rod-shaped chromosomes as they emerged from the nucleus during mitosis. The material that took the stain was called chromatin, and Zacharias identified it as nuclein by following the same procedure that had led Miescher to the discovery of nuclein. He found that when a cell was digested with pepsin–hydrochloric acid its nucleus remained and retained its ability to be stained. If, however, the digested cell was extracted with dilute alkali (which, as Miescher had shown, removes nuclein), then no stainable material remained. Zacharias carried out these tests on an exceedingly wide range of cells, both plant and animal, with essentially uniform results. He tested cells in the process of division and found that chromatin could be stained after pepsin-digestion but not after extraction in alkali. Moreover, the spindle that forms during mitosis did not stain, and it was removed by pepsin–hydrochloric acid digestion. All of this pointed to the coupling of nuclein and chromatin. Zacharias' conclusions were quickly accepted by Flemming and many others.

In 1884 and 1885 four biologists published papers that summarized and interpreted the work of the preceding decade, which had been so crowded with discoveries that those who participated felt the swift movement of events in much the same way that biologists do nowadays. Of the four summarizing accounts, three were by zoologists (Hertwig, Albrecht Kölliker and August Weismann) and one by a botanist (Eduard Strasburger). It was clear that an understanding of fertilization was at the same time an understanding of heredity. Continuity from one generation of an organism to the next was accomplished by the chromosomes in the nuclei of egg and sperm; continuity from one cell generation to the next was also accomplished by chromosomes in mitosis. At this point (in 1884) we suddenly come very close to what is one of the corner-

stones of current biology: "I believe that I have at least made it highly probable," Hertwig said, "that nuclein is the substance that is responsible not only for fertilization but also for the transmission of hereditary characteristics.... Furthermore, nuclein is in an organized state before, during and after fertilization, so that fertilization is at the same time both a morphological and a physicochemical event."

Even before Hertwig recognized nuclein as the genetic material, the botanist Julius von Sachs had (in 1882) not only suggested this role for nuclein but also gone a step further in pointing out that the nucleins of egg and sperm could hardly be identical—that the nuclein brought into the egg by the sperm must be different from the nuclein already there. And so by 1885, only a decade after Miescher's paper on salmon sperm and 14 years after his first publication on nuclein, a number of biologists had reached a point of view that is at the heart of our present conception of DNA.

What was the attitude of the discoverer of nuclein? As far as we can tell from Miescher's letters (and those to His freely express his views on many problems of biology), he did not accept the new ideas. He doubted the association of nuclein with chromatin, which was an essential element in the ideas expressed by Hertwig and the others. Miescher had a rather low estimate of the value of staining. He wrote in a letter of 1890, "Here once again I must defend my skin against the guild of dyers who suppose there is nothing else [in the sperm head] but chromatin," and in a paper published posthumously he referred disparagingly to Zacharias' work on chromatin. Miescher's attitude was also conditioned by his belief (expressed in several letters in 1892 and 1893) that he had discovered a new substance in the sperm head, which he proposed calling "karyogen." It was this phosphorus-free substance (the nature of which is a mystery today), and not nuclein, that in his opinion was responsible for the special chromatin stain.

In a letter of 1893 Miescher wrote: "The speculations of Weismann and others are afflicted with half-chemical concepts, which are partly unclear and partly derived from an outmoded kind of chemistry. When, as is quite possible, a protein molecule has 40 asymmetric carbon atoms so that there can be a billion isomers,...my [stereochemical] theory is better suited than any other to account for the unimaginable diversity required by our knowledge of heredity." Weismann's speculations were based on the most advanced biology of the time; Miescher preferred to base his speculations on recent advances in chemistry and paid relatively little attention to advances in biology. Time—and the rediscovery of Mendel's "units" of heredity and their identification with genes by Thomas Hunt Morgan—vindicated the idea, forcefully expressed by Weismann, that chromosomes transmit heredity.

Time did not, however, deal so consistently with the idea that nuclein is the material in chromosomes that transmits heredity. That idea appeared in the 1880's, as we have seen, and was widely held in the 1890's. As late as 1895 the American cytologist E. B. Wilson wrote: "Now, chromatin is known to be closely similar to, if not identical with, a substance known as nuclein. . . . And thus we reach the remarkable conclusion that in-

heritance may, perhaps, be effected by the physical transmission of a particular chemical compound from parent to offspring." In the next few years doubts arose concerning this conclusion because the amount of chromatin in the nucleus seemed too unstable to provide continuity; the amount varied considerably with the cycle of cell division and with changes in the physiological state of the cell. In the second edition of his book *The Cell in Development and Heredity,* published in 1900, Wilson described this fluctuation in staining but was able to explain it: "We may infer that the original chromosomes contain a high percentage of nucleinic acid; that their growth and loss of staining power is due to a combination with a large amount of albuminous substance . . . ; that their final diminution in size and resumption of staining power is caused by a giving up of the albumi-

nous constituent."

This analysis corresponds exactly with our present understanding, but it was soon to succumb to the apparent evidence of staining. When the large "lampbrush" chromosomes present in certain egg-cell precursors came under intensive study in the 1890's, there seemed to be no chromatin in them at all, and surely if chromatin (or nuclein) was the hereditary material, it must be present in an unbroken line from one cell generation to the next. In 1909 Strasburger wrote: "Chromatin cannot itself be the hereditary substance [because] the amount of it is subject to considerable variation in the nucleus, according to its stage of development." Strasburger was an eminent authority. So was Wilson, and the third edition of his book, which was published in 1925 and influenced a generation of biologists, took the Strasburger view:

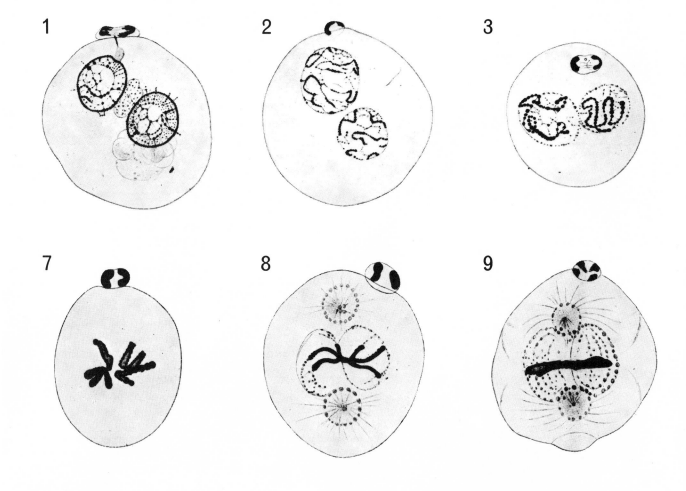

IN FERTILIZATION TOO chromosomes provide the continuity, as shown in drawings of the fertilized egg of *Ascaris,* a parasitic worm, published in 1883 by Edouard van Beneden. Sperm and egg nuclei approach each other (1). The nuclear membrane breaks down and the chromosomes become clearly visible; there are two chromosomes in each nucleus, half the normal "diploid" number in that the loss of staining in the enlarged chromosomes indicates "a progressive accumulation of protein components and a giving up, or even a complete loss, of nuclein." Wilson emphasized, using italics, "These facts afford conclusive proof that *the individuality and genetic continuity of chromosomes does not depend upon a persistence of chromatin.*" Biologists maintained this position for a generation.

Then, in 1948 and 1949, groups at the Rockefeller Institute and in France independently measured the quantity of DNA in cell nuclei. They found that the amount of DNA per set of chromosomes is in general constant in the different cell types of any organism, even when there are striking differences in the intensity of staining, and that the amount of protein associated with a fixed quantity of DNA may vary considerably and so account for variations in staining capacity. Moreover, the DNA content per chromosome set is a characteristic of each particular species. The doubts and questions concerning staining that had been raised by Miescher and many others were essentially resolved by these measurements. Now the biological role of DNA, proposed in 1884 and 1885 when the chromosome theory of heredity was formulated, received solid support: It was demonstrated that the chromosome complements of egg and sperm carry identical amounts of DNA, which are combined at fertilization and then carried by successive replications to all cells of the organism. The continuity of chromosomes at fertilization and cell division has always been an essential element in the chromosome theory of heredity; the associated continuity of DNA (Miescher's nuclein) points to it as providing the molecular basis for heredity in the chromosomes.

Several years before this point was finally established investigators at the Rockefeller Institute—following a line of investigation with a different historical background—found that hereditary traits could be transmitted from one strain of bacteria to another by the transfer of DNA. Nucleic acid was thus shown to be the genetic material. It should be pointed out that if this substance had been discovered in bacteria, it would never have been called nucleic acid, because a bacterial cell does not have a formed nucleus. The use of such words as "nucleic acid" and "chromosome" in work on bacteria is a constant reminder that the contributions of Friedrich Miescher and his contemporaries form the background for the study of heredity in all living cells.

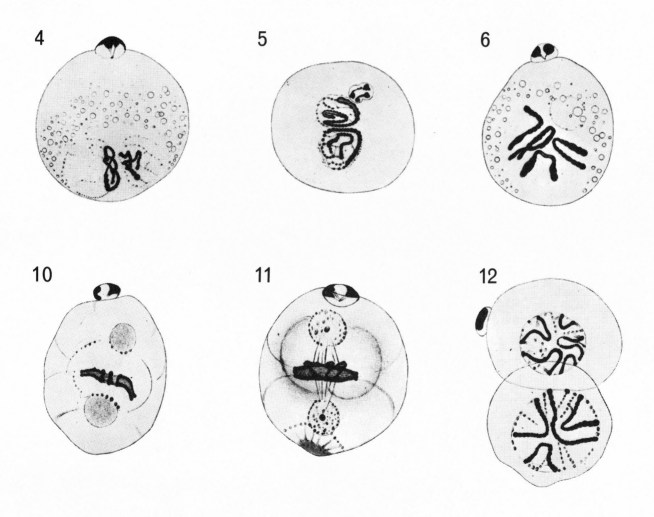

an *Ascaris* cell (*2, 3, 4*). The four chromosomes come together (*5, 6*). Each of the chromosomes has previously doubled; the replicated pairs separate slightly (*7*). Two "centrioles" appear and a spindle forms between them, and cell division begins (*8, 9, 10, 11*). Two cells are finally formed, each with a nucleus containing four chromosomes, two derived from the sperm and two from egg (*12*).

13

THE SYNTHESIS OF DNA

ARTHUR KORNBERG
October 1968

Test-tube synthesis of the double helix that controls heredity climaxes a half-century of effort by biochemists to re-create biologically active giant molecules outside the living cell

My colleagues and I first undertook to synthesize nucleic acids outside the living cell, with the help of cellular enzymes, in 1954. A year earlier James Watson and Francis Crick had proposed their double-helix model of DNA, the nucleic acid that conveys genetic information from generation to generation in all organisms except certain viruses. We attained our goal within a year, but not until some months ago—14 years later—were we able to report a completely synthetic DNA, made with natural DNA as a template, that has the full biological activity of the native material.

Our starting point was an unusual single-strand form of DNA found in the bacterial virus designated ϕX174. The single strand is in the form of a closed loop. When ϕX174 infects cells of the bacterium *Escherichia coli,* the single-strand loop of DNA serves as the template that directs enzymes in the synthesis of a second loop of DNA. The two loops form a ring-shaped double helix similar to the DNA helixes found in bacterial cells and higher organisms. In our laboratory at the Stanford University School of Medicine we succeeded in reconstructing the synthesis of the single-strand DNA copies of viral DNA and finally in making a completely synthetic double helix. The way now seems open for the synthesis of DNA from other sources: viruses associated with human disease, bacteria, multicellular organisms and ultimately the DNA of vertebrates such as mammals.

An Earlier Beginning

The story of the cell-free synthesis of DNA does not start with the revelation of the structure of DNA in 1953. It begins around 1900 with the biochemical understanding of how the fermentation of fruit juices yields alcohol. Some 40 years earlier Louis Pasteur had convinced his contemporaries that the living yeast cell played an essential role in the fermentation process. Then Eduard Buchner observed in 1897 that a cell-free juice obtained from yeast was just as effective as intact cells for converting sugar to alcohol. This observation opened the era of modern biochemistry.

During the first half of this century biochemists resolved the overall conversion of sucrose to alcohol into a sequence of 14 reactions, each catalyzed by a specific enzyme. When this fermentation proceeds in the absence of air, each molecule of sucrose consumed gives rise to four molecules of adenosine triphosphate (ATP), the universal currency of energy exchange in living cells. The energy represented by the fourfold output of ATP per molecule of sucrose is sufficient to maintain the growth and multiplication of yeast cells. When the fermentation takes place in air, the oxidation of sucrose goes to completion, yielding carbon dioxide and water along with 18 times as much energy as the anaerobic process does. This understanding of how the combustion of sugar provides energy for cell metabolism was succeeded by similar explanations of how enzymes catalyze the oxidation of fatty acids, amino acids and the subunits of nucleic acids for the energy needs of the cell.

By 1950 the enzymatic dismantling of large molecules was well understood. Little thought or effort had yet been invested, however, in exploring how the cell makes large molecules out of small ones. In fact, many biochemists doubted that biosynthetic pathways could be suc-

DOUBLE HELIX, the celebrated model of deoxyribonucleic acid (DNA) proposed in 1953 by James D. Watson and F. H. C. Crick, consists of two strands held together by crossties (*color*) that spell out a genetic message, unique for each organism. The Watson-Crick model explained for the first time how each crosstie consists of two subunits, called bases, that form obligatory pairs (*see illustrations on page 142*). Thus each strand of the double helix and its associated sequence of bases is complementary to the other strand and its bases. Consequently each strand can serve as a template for the reconstruction of the other strand.

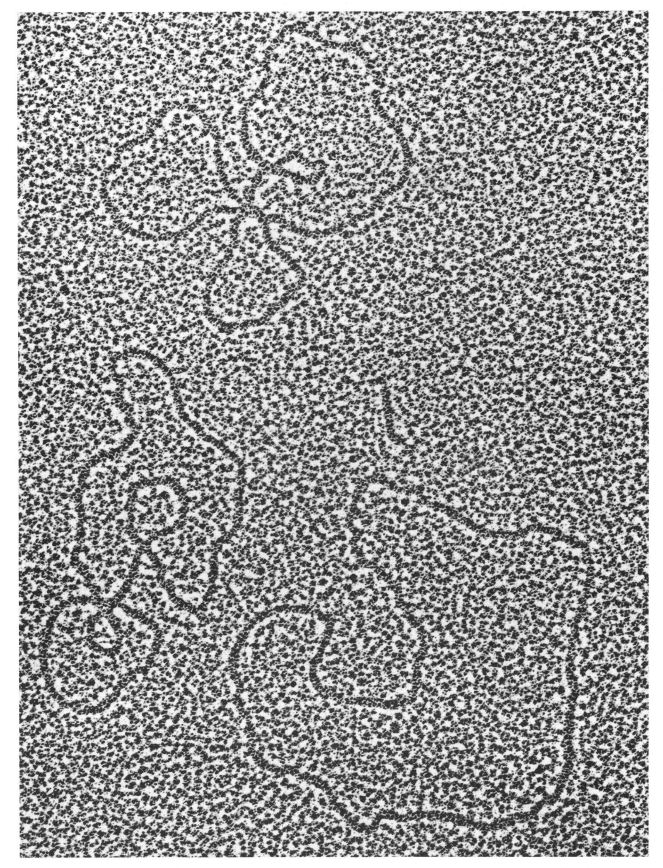

THREE CLOSED LOOPS OF DNA, each a complete double helix, are shown in this electron micrograph made in the author's laboratory at the Stanford University School of Medicine. One strand of each loop is the natural single-strand DNA of the bacterial virus ϕX174, which served as a template for the test-tube synthesis, carried out by enzymes, of a synthetic complementary strand. The hybrid molecules are biologically active. The enlargement is about 200,000 diameters. Each loop contains some 5,500 pairs of bases. If enlarged to the scale of the model on the opposite page, each loop of DNA would form a circle roughly 150 feet in circumference.

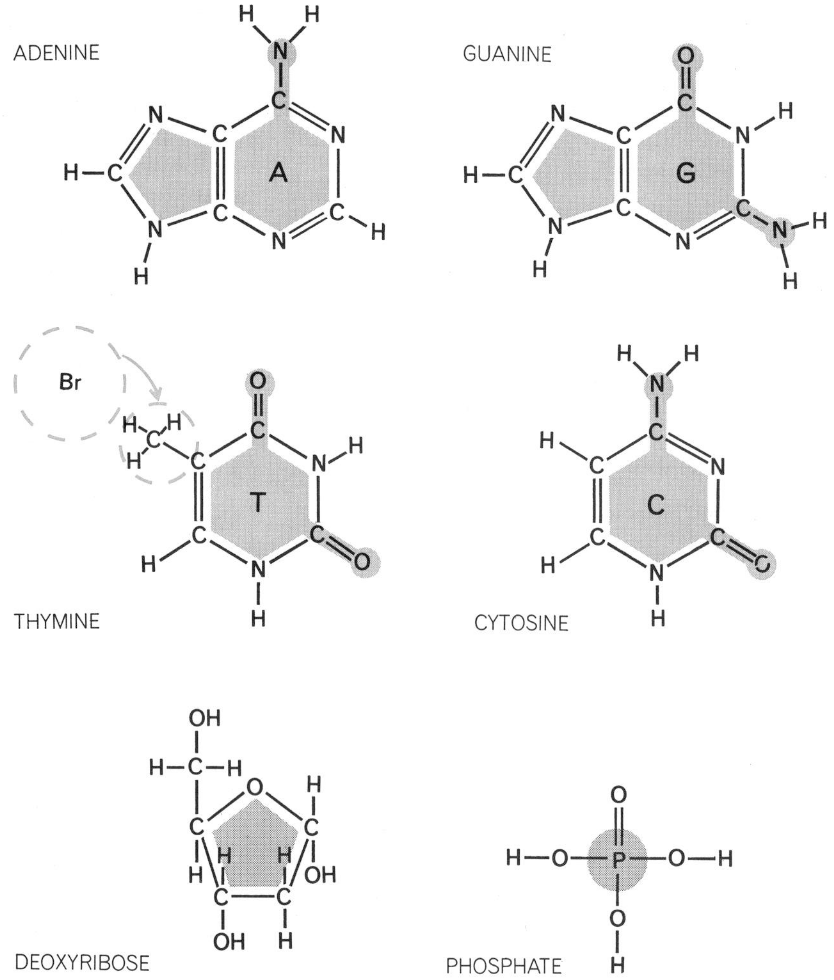

ADENINE

GUANINE

THYMINE

CYTOSINE

DEOXYRIBOSE

PHOSPHATE

DNA CONSTITUENTS are bases of four kinds, deoxyribose (a sugar) and a simple phosphate. The bases are adenine (*A*) and thymine (*T*), which form one obligatory pair, and guanine (*G*) and cytosine (*C*), which form another. Deoxyribose and phosphate form the backbone of each strand of the DNA molecule. The bases provide the code letters of the genetic message. For purposes of tagging synthetic DNA, thymine can be replaced by 5′-bromouracil, which contains a bromine atom where thymine contains a lighter CH₃ group.

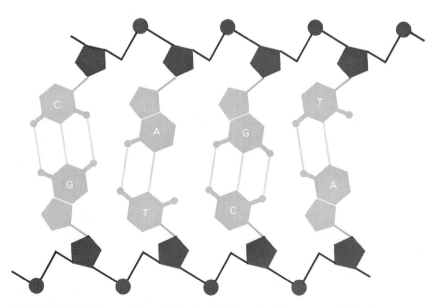

DNA STRUCTURE resembles a ladder in which the side pieces consist of alternating units of deoxyribose and phosphate. The rungs are formed by the bases paired in a special way, A with T and G with C, and held together respectively by two and three hydrogen bonds.

cessfully reconstructed in cell-free systems. Since then nearly two decades of intensive study have been devoted to the cell-free biosynthesis of large molecules. Two things above all have been made clear.

The first is that large molecules can be assembled in cell-free systems with the aid of purified enzymes and coenzymes. The second is that the routes of biosynthesis are different from those of degradation. Some biochemists had speculated that the routes of breakdown were really two-way streets whose flow might somehow be reversed. Now we know that the molecular traffic in cells flows on distinctive and divided highways. All cells have the enzymatic machinery to manufacture most of the subunits of large molecules from simple nutrients such as glucose, ammonia and carbon dioxide. Cells also have the capacity to salvage preformed subunits when they are available. On the basis of what has been learned the prospects are that in this century biochemists will assemble in the test tube complex viruses and major components of the cell. Perhaps the next century will bring the synthesis of a complete cell.

The Nucleotides

My co-workers and I were at Washington University in St. Louis when we made our first attempts to synthesize a nucleic acid in the test tube. By that time the constituents of nucleic acid were well known [*see illustrations at left*]. If one regards DNA as a chain made up of repeating links, the basic link is a structure known as a nucleotide [*see illustration on opposite page*]. It consists of a phosphate group attached to the five-carbon sugar deoxyribose, which is linked in turn to one of four different nitrogen-containing bases. The four bases are adenine (A), thymine (T), guanine (G) and cytosine (C). In the double helix of DNA the phosphate and deoxyribose units alternate to form the two sides of a twisted ladder. The rungs joining the sides consist of two bases: A is invariably linked to T and G is invariably linked to C. This particular pairing arrangement was the key insight of the Watson-Crick model. It means that if the two strands of the helix are separated, uncoupling the paired bases, each half can serve as a template for re-creating the missing half. Thus if the bases projecting from a single strand follow the sequence A, G, G, C, A, T..., one immediately knows that the complementary bases on the missing strand are T, C, C, G, T, A.... This base-pairing mechanism enables the cell to make accurate copies of the DNA molecule however many times the cell may divide.

When a strand of DNA is taken apart link by link (by treatment with acid or certain enzymes), the phosphate group of the nucleotide may be found attached to carbon No. 3 of the five-carbon deoxyribose sugar. Such a structure is called a 3′-nucleoside monophosphate. We judged, however, that better subunits for purposes of synthesis would be the 5′-nucleoside monophosphates, in which the phosphate linkage is to carbon No. 5 of deoxyribose.

This judgment was based on two lines of evidence. The first had just emerged from an understanding of how the cell itself made nucleotides from glucose, ammonia, carbon dioxide and amino acids. John M. Buchanan of the Massachusetts Institute of Technology had shown that nucleotides containing the bases A and G were naturally synthesized with a 5′ linkage. Our own work had shown the same thing for nucleotides containing T and C. The second line of evidence came from earlier studies my group had conducted at the National Institutes of Health. We had found that certain coenzymes, the simplest molecules formed from two nucleotides, were elaborated from 5′ nucleotide units. For the enzymatic linkage to take place the phosphate of the nucleotide had to be activated by an additional phosphate group [see illustration on next page]. Thus it seemed reasonable that activated 5′ nucleotides (nucleoside 5′ triphosphates) might combine with each other, under the proper enzymatic guidance, to form long chains of nucleic-acid.

Our initial attempts at nucleic acid synthesis relied principally on two techniques. The first involved the use of radioactive atoms to label the nucleotide so that we could detect the incorporation of even minute amounts of it into nucleic acid. We sought the enzymatic machinery for synthesizing nucleic acids in the juices of the thymus gland, bone marrow and bacterial cells. Unfortunately such extracts also have a potent capacity for degrading nucleic acids. We added our labeled nucleotides to a pool of nucleic acids and hoped that a few synthesized molecules containing a labeled nucleotide would survive by being mixed into the pool. Even if there were net destruction of the pool of nucleic acids, the synthesis of a few molecules trapped in this pool might still be detected. The second technique exploited the fact that the nucleic acid could be precipitated by making the medium strongly acidic, whereas the nucleotide

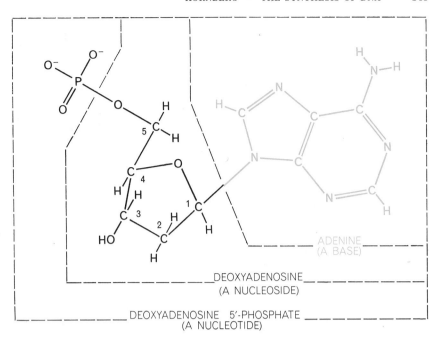

DNA BUILDING BLOCK, the monomer from which DNA polymers are constructed, is termed a nucleotide. There are four nucleotides, one for each of the four bases A, T, G and C. Deoxyadenosine 5′-phosphate, the nucleotide incorporating adenine, is shown here. If the phosphate group is replaced by a hydrogen atom, the structure is called a nucleoside.

precursors remained behind in solution.

Our first experiments with animal-cell extracts were uniformly negative. Therefore we turned to E. coli, which has the virtue of reproducing once every 20 minutes. Here we saw a glimmer. In samples to which we had added a quantity of labeled nucleotides whose radioactive atoms disintegrated at the rate of a million per minute we detected about 50 radioactive disintegrations per minute in the nucleic acid fraction that was precipitated by acid. Although the amount of nucleotide incorporated into nucleic acid was minuscule, it was nonetheless significantly above the level of background "noise." Through this tiny crack we tried to drive a wedge. The hammer was enzyme purification, a technique that had matured during the elucidation of alcoholic fermentation.

DNA Polymerase

In these experiments Uriel Littauer, a Fellow of the Weizmann Institute in Israel, and I observed the incorporation of adenylate (a nucleotide) from ATP into ribonucleic acid (RNA), in which the five-carbon sugar in the backbone of the chain is ribose rather than deoxyribose. Actually the first definitive demonstration of synthesis of an RNA-like molecule in a cell-free system had been achieved in the laboratory of Severo Ochoa in 1955. Working at the New York University School of Medicine, he

and Marianne Grunberg-Manago were investigating an aspect of energy metabolism and made the unexpected observation that one of the reactants, adenosine diphosphate (ADP), had been polymerized by cell juices into a chain of adenylates resembling RNA.

In our first attempts to achieve DNA synthesis in a cell-free system we used the deoxyribonucleoside called deoxythymidine. To Morris E. Friedkin, who was then at Washington University, we are grateful not only for supplying the radioactively labeled compound but also for the knowledge that the compound was readily incorporated into DNA by bone marrow cells and other animal cells. We were hopeful that extracts of E. coli would be able to incoporate deoxythymidine into nucleic acid by converting it first into the 5′ deoxynucleotide and then activating the deoxynucleotide to the triphosphate form. I found this to be the case. In subsequent months Ernest Simms and I were able to prepare separately deoxythymidine 5′-triphosphate and the other deoxynucleoside triphosphates, using enzymes or chemical synthetic routes. (In what follows the various deoxynucleosides in their 5′ triphosphate form will be designated simply by the initial of the base followed by an asterisk. Thus deoxythymidine 5′-triphosphate will be T*.)

In November, 1955, I. Robert Lehman, who is now at Stanford, started on the purification of the enzyme system in *E. coli* extracts that is responsible for converting T* into DNA. We were joined by Maurice J. Bessman some weeks later. Those were eventful days in which the enzyme, now given the name DNA polymerase, was progressively separated from other large molecules. With each step in purification the character of this DNA synthetic reaction became clearer. By June, 1956, when we participated at a conference on the chemical basis of heredity held at Johns Hopkins University, we could report two important facts about DNA synthesis in vitro, although we still lacked the answers to many important questions.

We reported first that preformed DNA had to be present along with DNA polymerase, and that all four of the de-

oxynucleotides that occur in DNA (A, G, T and C) had to be furnished in the activated triphosphate form. We also reported that DNA from virtually any source—virus, bacterium or animal—could serve with the *E. coli* enzyme. What we still did not know was whether the synthetic DNA was a new molecule or an extension of a preexisting one. There were other questions. Did the synthetic DNA have the same chemical backbone and physical structure as natural DNA? Did it have a chemical composition typical of DNA, in which A equals T and G equals C, and in which, therefore, A plus G equals T plus C? Finally, and crucially: Did the chemical composition of the synthetic DNA reflect the composition of the particular

natural DNA used to direct the reaction?

During the next three years these questions and related ones were resolved by the efforts of Julius Adler, Sylvy Kornberg and Steven B. Zimmerman. The synthetic DNA was shown to be a molecule with the chemical structure typical of DNA and the same ratio of A-T pairs to G-C pairs as the particular DNA used to prime, or direct, the reaction [*see illustration, page 146*]. The relative starting amounts of the four deoxynucleoside triphosphates had no influence whatever on the composition of the new DNA. The composition of the synthetic DNA was determined solely by the composition of the DNA that served as a template. An interesting illustration of this last fact justifies a slight digression.

Howard K. Schachman of the University of California at Berkeley spent his sabbatical year of 1957–1958 with us at Washington University examining the physical properties of the synthetic DNA. It had the high viscosity, the comparatively slow rate of sedimentation and other physical properties typical of natural DNA. The new DNA, like the natural one, was therefore a long, fibrous polymer molecule. Moreover, the longer the mixture of active ingredients was allowed to incubate, the greater the viscosity of the product was; this was direct evidence that the synthetic DNA was continuing to grow in length and in amount. However, we were startled to find one day that viscosity developed in a control test tube that lacked one of the essential triphosphates, G*. To be sure, no reaction was observed during the standard incubation period of one or two hours. On prolonging the incubation for several more hours, however, a viscous substance materialized!

Analysis proved this substance to be a DNA that contained only A and T nucleotides. They were arranged in a perfect alternating sequence. The isolated polymer, named dAT, behaved like any other DNA in directing DNA synthesis: it led to the immediate synthesis of more dAT polymer. Would any G* and C* be polymerized if these nucleotides were present in equal or even far greater amounts than A* and T* in a synthesis directed by dAT polymer? We found no detectable incorporation of G or C under conditions that would have measured the inclusion of even one G for every 100,-000 A or T nucleotides polymerized. Thus DNA polymerase rarely, if ever, made the mistake of matching G or C with A or T.

The DNA of a chromosome is a linear array of many genes. Each gene, in turn,

DEOXYADENOSINE   5'-PHOSPHATE

ATP

ATP

DEOXYADENOSINE   5'-TRIPHOSPHATE
(A*)

ACTIVATED BUILDING BLOCK is required when synthesizing DNA on a template of natural DNA with the aid of enzymes. The activated form of the nucleotide containing adenine is deoxyadenosine 5'-triphosphate, symbolized in this article by "A*." It is made from deoxyadenosine 5'-monophosphate by two different enzymes in two steps. Each step involves the donation of a terminal phosphate group from adenosine triphosphate (ATP).

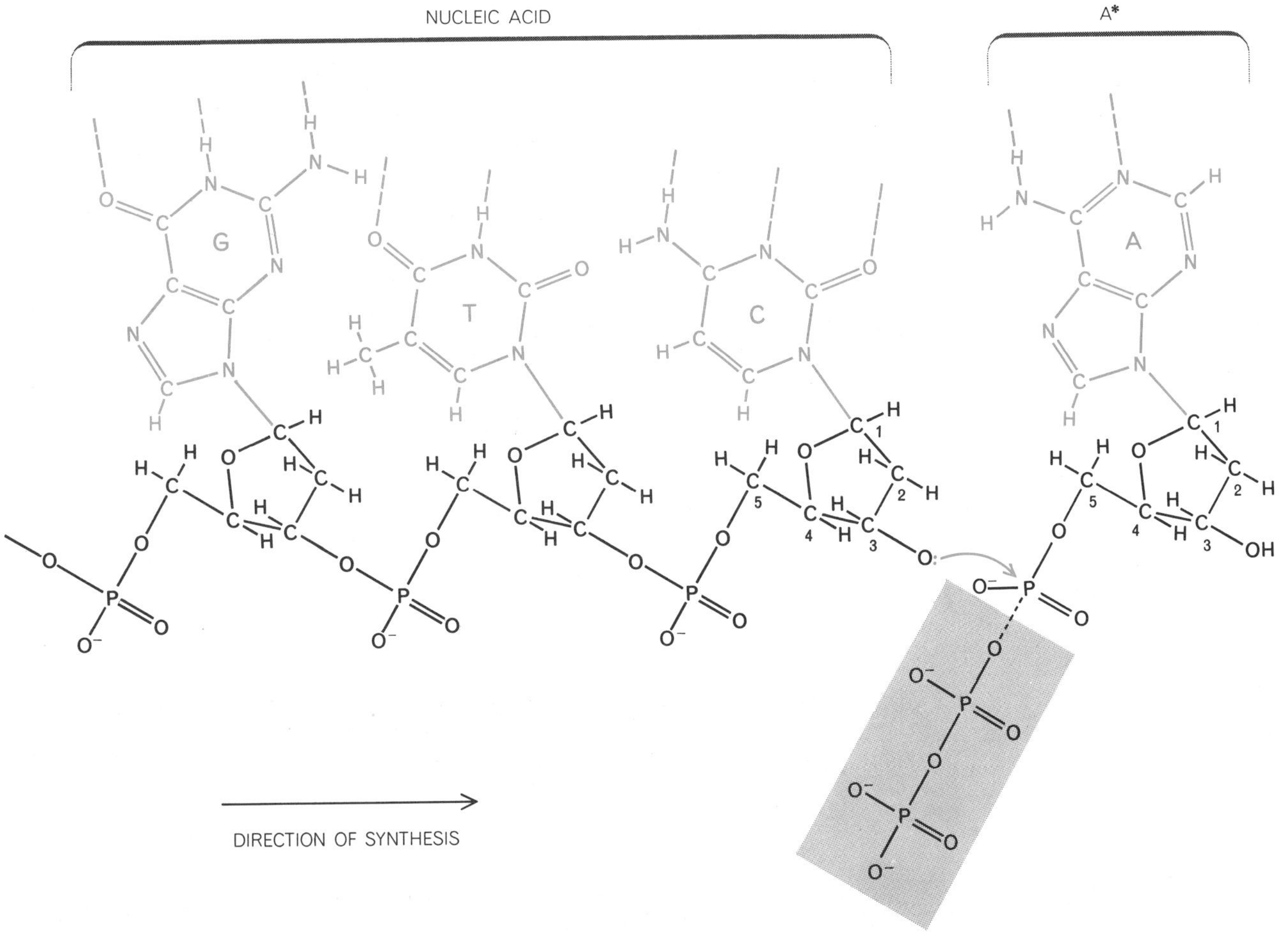

NUCLEIC ACID

A*

DIRECTION OF SYNTHESIS

SYNTHESIS OF DNA involves the stepwise addition of activated nucleotides to the growing polymer chain. In this illustration deoxyadenosine 5'-triphosphate (A*) is being coupled through a phosphodiester bond that links the 3' carbon in the deoxyribose portion of the last nucleotide in the growing chain to the 5' carbon in the deoxyribose portion of the newest member of the chain.

is a chain of about 1,000 nucleotides in a precisely defined sequence, which when translated into amino acids spells out a particular protein or enzyme. Does DNA polymerase in its test-tube synthesis of DNA accurately copy the sequential arrangement of nucleotides by base-pairing (A = T, G = C) without errors of mismatching, omission, commission or transposition? Unfortunately techniques are not available for determining the precise sequence of nucleotides of even short DNA chains. Because it is impossible to spell out the base sequence of natural DNA or any copy of it, we have resorted to two other techniques to test the fidelity with which DNA polymerase copies the template DNA. One is "nearest neighbor" analysis. The other is the duplication of genes with demonstrable biological activity.

Nearest-Neighbor Analysis

The nearest-neighbor analysis devised by John Josse, A. Dale Kaiser and myself in 1959 determines the relative frequency with which two nucleotides can end up side by side in a molecule of synthetic DNA. There are 16 possible combinations in all. There are four possible nearest-neighbor sequences of A (AA, AG, AT and AC), four for G (GA, GG, GT and GC) and similarly four for T and four for C. How can the frequency of these dinucleotide sequences be determined in a synthetic DNA chain? The procedure is to use a triphosphate labeled with a radioactive phosphorus atom in conducting the synthesis and to treat the synthesized DNA with a specific enzyme that cleaves the DNA and leaves the radioactive phosphorus atom attached to its nearest neighbor. For example, DNA synthesis is carried out with A* labeled in the innermost phosphate group, the group that will be included in the DNA product. This labeled phosphate group now forms a normal linkage (10^{16} times in a typical experiment!) with the nucleotide next to it in the chain—its nearest neighbor [*see*

illustration on page 147]. After the synthetic DNA is isolated it is subjected to degradation by an enzyme that cleaves every bond between the 5' carbon of deoxyribose and the phosphate, leaving the radioactive phosphorus atom attached to the neighboring nucleotide rather than to the one (A) to which it had originally been attached. The nucleotides of the degraded DNA are readily separated by electrophoresis or paper chromatography into the four types of which DNA is composed: A, G, T and C. Radioactive assay establishes the radioactive phosphorus content in each of these nucleotides and at once indicates the frequency with which A is next to A, to G, to T and to C.

The entire experiment is repeated, this time with the radioactive label in G* instead of A*. The second experiment yields the frequency of GA, GG, GT and GC dinucleotides. Two more experiments with radioactive T* and C* complete the analysis and establish the 16 possible nearest-neighbor frequencies.

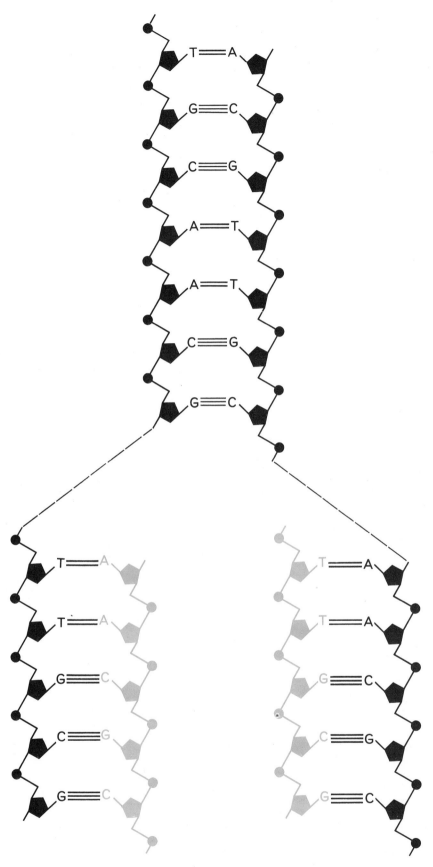

SYNTHESIS OF DUPLEX CHAIN OF DNA yields two hybrid molecules, consisting of a parental strand and a daughter strand, that are identical with each other and with the original duplex molecule. During the replicating process the parental duplex (*black*) separates into two strands, each of which then serves as the template for assembly of a daughter strand (*color*). The pairing of A with T and G with C guarantees faithful reproduction.

Many such experiments were performed with DNA templates obtained from viruses, bacteria, plants and animals. The DNA of each species guided the synthesis of DNA with what proved to be a distinctive assortment of nearest-neighbor frequencies. What is more, when a synthetic DNA was used as a template for a new round of replication, it gave rise to DNA with a nearest-neighbor frequency distribution identical with itself. Among the other insights obtained from these analyses was the recognition of a basic fact about the structure of the double helix. In replication the direction of the DNA chain being synthesized was found to run opposite to that of its template. By inference we can conclude that the chains of the double helix in natural DNA, as surmised by Watson and Crick, must also run in opposite directions.

Even with considerable care the accuracy of nearest-neighbor frequency analysis cannot be better than about 98 percent. Consequently we were still left with major uncertainties as to the precision of copying chains that contain 1,000 nucleotides or more, corresponding to the length of genes. An important question thus remained unanswered: Does DNA that is synthesized on a genetically or biologically active template duplicate the activity of that template?

One way to recognize the biological activity of bacterial DNA is to see if it can carry out "transformation," a process in which DNA from one species of bacteria alters the genetic endowment of a second species. For example, DNA from a strain of *Bacillus subtilis* resistant to streptomycin can be assimilated by a strain susceptible to the antibiotic, whereupon the recipient bacterium and all its descendants carry the trait of resistance to streptomycin. In other words, DNA molecules carrying the genes for a particular characteristic can be identified by their capacity for assimilation into the chromosome of a cell that previously lacked that trait. Yet when DNA was synthesized on a template of DNA that had transforming ability, the synthetic product invariably lacked that ability.

Part of the difficulty in synthesizing biologically active DNA lay in the persistence of trace quantities of nuclease enzymes in our DNA polymerase preparations. Nucleases are enzymes that degrade DNA. The introduction by a nuclease of one break in a long chain of DNA is enough to destroy its genetic activity. Further purification of DNA polymerase was indicated. Efforts over several years by Charles C. Richardson, Thomas Jovin, Paul T. Englund and LeRoy L. Bertsch resulted in a new procedure that was both simple and efficient. Finally, in April, 1967, with the assistance of the personnel and large-scale equipment of the New England Enzyme Center (sponsored by the National Institutes of Health at the Tufts University School of Medicine), we processed 100 kilograms of *E. coli* bacterial paste and obtained about half a gram of pure enzyme, free of the nuclease that puts random breaks in a DNA chain.

Unfortunately even this highly purified DNA polymerase has proved incapable of producing a biologically active DNA from a template of bacterial DNA. The difficulty, we believe, is that the DNA we extract from a bacterium such as *B. subtilis* provides the enzyme with a poor template. A proper template would be the natural chromosome, which is a double-strand loop about one millimeter in circumference. During its isolation from the bacterium the chromosome is broken, probably at random, into 100 or more fragments. The manner in which DNA polymerase and its related enzymes go about the replication of a DNA molecule as large and complex as the *B. subtilis* chromosome is the subject of current study in many laboratories.

The Virus φX174

It occurred to us in 1964 that the problem of synthesizing biologically active DNA might be solved by dealing with a simpler form of DNA that also has genetic activity. This is represented in viruses, such as φX174, whose DNA core is a single-strand loop. This "chromosome" not only is simpler in structure but also it is so small (about two microns in circumference) that it is fairly easy to extract without breakage. We also knew from the work of Robert L. Sinsheimer at the California Institute of Technology that when the DNA of φX174 invades *E. coli*, the first stage of infection involves the "subversion" of one of the host's enzymes to convert the single-strand loop into a double-strand helical loop. Sinsheimer called this first-stage product a "replicative form." Could the host enzyme that copies the viral DNA be the same DNA polymerase we had isolated from *E. coli?*

In undertaking the problem of copying a closed-loop DNA we could foresee some serious obstacles. Would it be possible for DNA polymerase to orient itself and start replication on a DNA

template if the template had no ends? Shashanka Mitra and later Peter Reichard succeeded in finding conditions under which the enzyme, as judged by electron microscope pictures, appeared to copy the single-strand loop. We then wondered if in spite of appearances in the electron micrographs, the DNA of φX174 was really just a simple loop. Perhaps, as had been suggested by other workers, it was really more like a necklace with a clasp, the clasp consisting of substances unrelated to the nucleotides we were supplying. Finally, we were aware from Sinsheimer's work that the DNA of φX174 had to be a completely closed loop in order to be in-

fectious. We knew that our polymerase could only catalyze the synthesis of linear DNA molecules. How could we synthesize a genuinely closed loop? We were still missing either the clasplike component to insert into our product or, if the clasp was a mistaken hypothesis, a new kind of enzyme to close the loop.

Fortunately the missing factor was provided for us by work carried on independently in five different laboratories. The discovery in 1966 of a polynucleotide-joining enzyme was made almost simultaneously by Martin F. Gellert and his co-workers at the National Institutes of Health, by Richardson and Bernard Weiss at the Harvard Medical

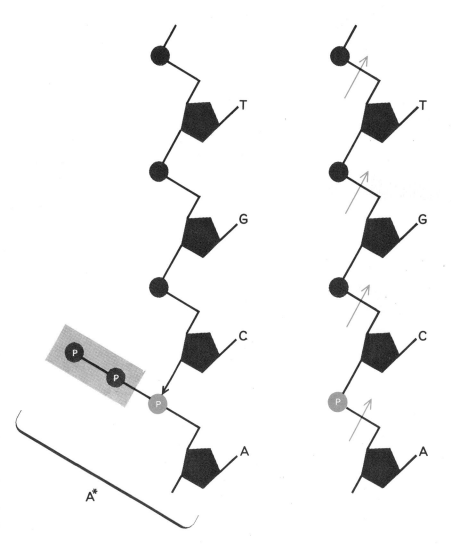

"NEAREST NEIGHBOR" ANALYSIS can reveal how often any of the four bases is located next to any other base in a single strand of synthetic DNA. Thus one can learn how often A is next to A, T, G or C, and so on. A radioactive phosphorus atom (*color*) is placed in the innermost position of one of the activated nucleotides, for example A*. The finished DNA molecule is then treated with an enzyme (*right*) that cleaves the chain between every phosphate and the 5′ carbon of the adjacent deoxyribose. Thus the phosphate is separated from the nucleotide on which it entered the chain and ends up attached to the nearest neighbor instead, C in the above example. The four kinds of nucleotide are separated by paper chromatography and the radioactivity associated with each is measured. The experiment is repeated with radioactive phosphorus linked to the other activated nucleotides.

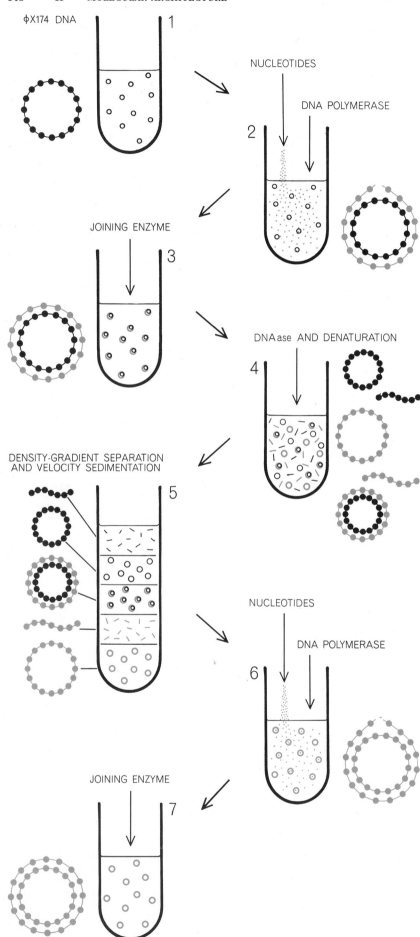

φX174 DNA

1

NUCLEOTIDES

DNA POLYMERASE

2

JOINING ENZYME

3

DNAase AND DENATURATION

4

DENSITY-GRADIENT SEPARATION
AND VELOCITY SEDIMENTATION

5

NUCLEOTIDES

DNA POLYMERASE

6

JOINING ENZYME

7

School, by Jerard Hurwitz and his colleagues at the Albert Einstein College of Medicine in New York, by Lehman and Baldomero M. Olivera at Stanford and by Nicholas R. Cozzarelli in my own group. It was the Lehman-Olivera preparation that we now employed in our experiments.

The polynucleotide-joining enzyme has the ability to repair "nicks" in the DNA strand. The nicks occur where there is a break in the sugar-phosphate backbone of one strand of the DNA molecule. The enzyme can repair a break only if all the nucleotides are intact and if what is missing is the covalent bond in the DNA backbone between a sugar and the neighboring phosphate. Provided with the joining enzyme, we were now in a position to find out whether it could work in conjunction with DNA polymerase to synthesize a completely circular and biologically active virus DNA.

By using the DNA of φX174 as a template we gained an important advantage over experiments based on transforming ability. Even if we were successful in synthesizing a DNA with transforming activity, this would still be of relatively limited significance. We could then say only that a restricted section of the DNA—a section as small as a part of a gene—had been assimilated by the recipient cell to replace a comparable section of its chromosome, substituting a proper sequence for a defective or incorrect one. However, Sinsheimer had demonstrated with the DNA of φX174 that a change in even one of its 5,500 nucleotides is sufficient to make the virus noninfective. Therefore the demonstration of infectivity in a completely synthetic

SYNTHESIS OF φX174 DNA was accomplished by the following steps. Circular single-strand φX174 DNA, tagged with tritium, served as a template (1). Activated nucleotides containing A, G, C and 5'-bromouracil instead of T were added to the template, together with DNA polymerase. One of the activated nucleotides was tagged with radioactive phosphorus. The DNA synthesized on the template was complete but not yet joined in a loop (2). The loop was closed by the joining enzyme (3). Enough nuclease was now added to cut one strand in about half of all the duplex loops (4). This left a mixture of complete duplex loops, template loops, synthetic loops, linear template strands and linear synthetic strands. Since the synthetic strands contained 5'-bromouracil, they were heavier than the template strands and could be separated by centrifugation (5). The synthetic loops were then isolated and used as templates for making wholly synthetic duplex loops (6 and 7).

virus DNA would conclusively prove that we had carried out virtually error-free synthesis of this large number of nucleotides, comprising the five or six genes that carry out the virus's biological function.

In less than a year the test-tube synthesis of φX174 DNA was achieved. The steps can be summarized as follows. Template DNA was obtained from φX174 and labeled with tritium, the radioactive isotope of hydrogen. Tritium would thereafter provide a continuing label identifying the template. To the template were added DNA polymerase, purified joining enzyme and a cofactor (diphosphopyridine nucleotide), together with A*, T*, G* and C*. One of the nucleoside triphosphates was labeled with radioactive phosphorus. The radioactive phosphorus would thus provide a label for synthetic material analogous to the tritium label for the template. The interaction of the reagents then proceeded until the number of nucleotide units polymerized was exactly equal to the number of nucleotides in the template DNA. This equality was readily determined by comparing the radioactivity from the tritium in the template with the radioactivity from the phosphorus in the nucleotides provided for synthesis.

Such comparison showed that the experiments had progressed to an extent adequate for the formation of complementary loops of synthetic DNA. Complementary loops were designated (−) to distinguish them from the template loop (+). We had to demonstrate that the synthetic (−) loops were really loops. Had the polymerase made a full turn around the template and had the two ends of the chain been united by the joining enzyme? Several physical measurements, including electron microscopy, assured us that our product was a closed loop coiled tightly around the virus-DNA template and that it was identical in size and other details with the replicative form of DNA that appears in the infected cells. We could now exclude the possibility that some clasp material different from the nucleotide-containing compounds we had employed was involved in closing the virus-DNA loop.

The critical questions remaining were whether the synthetic (−) loops had biological activity—that is, infectivity—and whether the synthetic loops could in turn act as templates for the formation of a completely synthetic "duplex" DNA analogous to the replicative forms that were produced naturally inside infected cells. In order to answer the first of these questions we had to isolate the synthetic DNA strands from the partially synthetic duplexes. For reasons that will be apparent below, we substituted bromouracil, a synthetic but biologically active analogue of thymine, for thymine [see top illustration, page 142]. We then introduced just enough nuclease to produce a single nick in one strand of about half the population of molecules. The duplex loops that had been nicked would release a single linear strand of DNA; these single strands could be separated from their circular companions and from unnicked duplex loops by heating. Thus we were left with a mixture that contained (+) template loops, (−) synthetic loops, (+) template linear forms, (−) synthetic linear forms—all in about equal quantities—and full duplex loops.

It was at this point that the substitution of bromouracil for thymine became useful. Because bromouracil contains a bromine atom in place of the methyl group of thymine, it is heavier than thymine. Therefore a molecule containing bromouracil can be separated from one containing thymine by high-speed centrifugation in a heavy salt solution (the density-gradient technique perfected by Jerome R. Vinograd of Cal Tech). In this system the denser a substance is, the lower in the centrifuge tube it will settle. Thus from top to bottom of the centrifuge tube we obtained fractions containing the light single strands of thymine-containing (+) template DNA, the duplex hybrids of intermediate weight and finally the single-strand synthetic (−) DNA "weighted down" with bromouracil. The reliability of this fractionation was confirmed by three separate peaks of radioactivity corresponding to each of the fractions. We were further reassured by observations that the mean density of each fraction corresponded almost exactly to the mean density of standard samples of virus DNA containing bromouracil or thymine.

Still another physical technique involving density-gradient sedimentation was employed to separate the synthetic linear forms from the synthetic circular forms. The circular forms could then be used in tests of infectivity, by methods previously developed by Sinsheimer to demonstrate the infectivity of circular φX174 DNA. We tested our (−) loops by incubating them with E. coli cells whose walls had been removed by the action of the enzyme lysozyme. Infectivity is assayed by the ability of the virus to lyse, or dissolve, these cells when they are "plated" on a nutrient medium. Our synthetic loops showed almost exactly the same patterns of infectivity as their natural counterparts had. Their biological activity was now demonstrated.

One further set of experiments remained in which the (−) synthetic loops were employed as the template to determine if we could produce completely synthetic duplex circular forms analogous to the replicative forms found in cells infected with natural φX174 virus. Because the synthetic (−) loops were labeled with radioactive phosphorus, this time we added tritium to one of the nucleotide-containing subunits (C*). The remaining procedures were essentially the same as the ones described above, and we did produce fully synthetic duplex loops of φX174. The (+) loops were then separated and were found to be identical in all respects with the (+) loops of natural φX174 virus. Their infectivity could also be demonstrated. Sinsheimer had previously shown that, under these assay conditions, a change in a single nucleotide of the virus gave rise to a mutant of markedly decreased infectivity. Therefore the correspondence between the infectivity of our synthetic forms and their natural counterparts attested to the precision of the enzymatic operation.

Future Directions

The total synthesis of infective virus DNA by DNA polymerase with the four deoxynucleoside triphosphates not only demonstrates the capacity of this enzyme to copy a small chromosome (of five or six genes) without error but also shows that this chromosome, at least, is as simple and straightforward as a linear sequence of the standard four deoxynucleotide units. It is a long step to the human chromosome, some 10,000 times larger, yet we are encouraged to extrapolate our current conceptions of nucleotide composition and nucleotide linkage from the tiny φX174 chromosome to larger ones.

What are the major directions this research will take? I see at least three immediate and productive paths. One is the exploration of the physical and chemical nature of DNA polymerase in order to understand exactly how it performs its error-free replication of DNA. Without this knowledge of the structure of the enzyme and how it operates under defined conditions in the test tube, our understanding of the intracellular behavior of the enzyme will be incomplete.

A second direction is to clarify the

control of DNA replication in the cell and in the animal. Why is DNA synthesis arrested in a mature liver cell and what sets it in motion 24 hours after part of the liver is removed surgically? What determines the slow rate of DNA replication in adult cells compared with the rate in embryonic or cancer cells? The time is ripe for exploration of the factors that govern the initiation and rate of DNA synthesis in the intact cell and animal. Finally, there are now prospects of applying our knowledge of DNA structure and synthesis directly to human welfare. This is the realm of genetic engineering, and it is our collective responsibility to see that we exploit our great opportunities to improve the quality of human life.

An obvious area for investigation would be the synthesis of the polyoma virus, a virus known to induce a variety of malignant tumors in several species of rodents. Polyoma virus in its infective form is made up of duplex circular DNA and presumably replicates in this form on entering the cell. On the basis of our experience it would appear quite feasible to synthesize polyoma virus DNA. If this synthesis is accomplished, there would seem to be many opportunities for modifying the virus DNA and thus determining where in the chromosome its tumor-producing capacity lies. With this knowledge it might prove possible to modify the virus in order to control its tumor-producing potential.

Our speculations can extend even to large DNA molecules. For example, if a failure in the production of insulin were to be traced to a genetic deficit, then administration of the appropriate synthetic DNA might conceivably provide a cure for diabetes. Of course, a system for delivering the corrective DNA to the cells must be devised. Even this does not seem inconceivable. The extremely interesting work of Stanfield Rogers at the Oak Ridge National Laboratory suggests a possibility. Rogers has shown that the Shope papilloma virus, which is not pathogenic in man, is capable of inducing production of the enzyme arginase in rabbits at the same time that it induces tumors. Rogers found that in the blood of laboratory investigators working with the virus there is a significant reduction of the amino acid arginine, which is destroyed by arginase. This is apparently an expression of enhanced arginase activity. Might it not be possible, then, to use similar nonpathogenic viruses to carry into man pieces of DNA capable of replacing or repairing defective genes?

14

THE BACTERIAL CHROMOSOME

JOHN CAIRNS
January 1966

*When bacterial DNA is labeled with radioactive atoms,
it takes its own picture. Autoradiographs reveal that
the bacterial chromosome is a single very long DNA
molecule and show how it is duplicated*

The information inherited by living things from their forebears is inscribed in their deoxyribonucleic acid (DNA). It is written there in a decipherable code in which the "letters" are the four subunits of DNA, the nucleotide bases. It is ordered in functional units—the genes—and thence translated by way of ribonucleic acid (RNA) into sequences of amino acids that determine the properties of proteins. The proteins are, in the final analysis, the executors of each organism's inheritance.

The central event in the passage of genetic information from one generation to the next is the duplication of DNA. This cannot be a casual process. The complement of DNA in a single bacterium, for example, amounts to some six million nucleotide bases; this is the bacterium's "inheritance." Clearly life's security of tenure derives in large measure from the precision with which DNA can be duplicated, and the manner of this duplication is therefore a matter of surpassing interest. This article deals with a single set of experiments on the duplication of DNA, the antecedents to them and some of the speculations they have provoked.

When James D. Watson and Francis H. C. Crick developed their two-strand model for the structure of DNA, they saw that it contained within it the seeds of a system for self-duplication. The two strands, or polynucleotide chains, were apparently related physically to each other by a strict system of *complementary* base pairing. Wherever the nucleotide base adenine occurred in one chain, thymine was present in the other; similarly, guanine was always paired with cytosine. These rules meant that the sequence of bases in each chain inexorably stipulated the sequence in the other; each chain, on its own, could generate the entire sequence of base pairs. Watson and Crick therefore suggested that accurate duplication of DNA could occur if the chains separated and each then acted as a template on which a new complementary chain was laid down. This form of duplication was later called "semiconservative" because it supposed that although the individual parental chains were conserved during duplication (in that they were not thrown away), their association ended as part of the act of duplication.

The prediction of semiconservative replication soon received precise experimental support. Matthew S. Meselson and Franklin W. Stahl, working at the California Institute of Technology, were able to show that each molecule of DNA in the bacterium *Escherichia coli* is composed of equal parts of newly synthesized DNA and of old DNA that was present in the previous generation [*see top illustration on page 153*]. They realized they had not proved that the two parts of each molecule were in fact two chains of the DNA duplex, because they had not established that the molecules they were working with consisted of only two chains. Later experiments, including some to be described in this article, showed that what they were observing was indeed the separation of the two chains during duplication.

The Meselson-Stahl experiment dealt with the end result of DNA duplication. It gave no hint about the mechanism that separates the chains and then supervises the synthesis of the new chains. Soon, however, Arthur Kornberg and his colleagues at Washington University isolated an enzyme from *E. coli* that, if all the necessary precursors were provided, could synthesize in the test tube chains that were complementary in base sequence to any DNA offered as a template. It was clear, then, that polynucleotide chains could indeed act as templates for the production of complementary chains and that this kind of reaction could be the normal process of duplication, since the enzymes for carrying it out were present in the living cell.

Such, then, was the general background of the experiments I undertook beginning in 1962 at the Australian National University. My object was simply (and literally) to look at molecules of DNA that had been caught in the act of duplication, in order to find out which of the possible forms of semiconservative replication takes place in the living cell: how the chains of parent DNA are arranged and how the new chains are laid down [*see bottom illustration on next page*].

Various factors dictated that the experiments should be conducted with *E. coli*. For one thing, this bacterium was known from genetic studies to have only one chromosome; that is, its DNA is contained in a single functional unit in which all the genetic markers are arrayed in sequence. For another thing, the duplication of its chromosome was known to occupy virtually the entire cycle of cell division, so that one could be sure that every cell in a rapidly multiplying culture would contain replicating DNA.

Although nothing was known about the number of DNA molecules in the *E. coli* chromosome (or in any other complex chromosome, for that matter), the dispersal of the bacterium's DNA among its descendants had been shown to be semiconservative. For this and other reasons it seemed likely that the

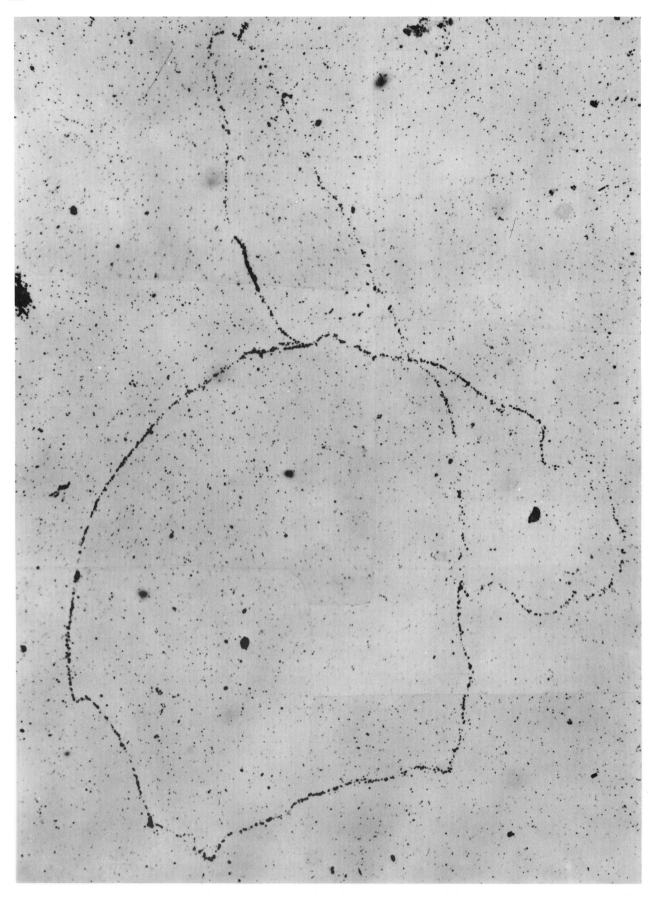

AUTORADIOGRAPH shows a duplicating chromosome from the bacterium *Escherichia coli* enlarged about 480 diameters. The DNA of the chromosome is visible because for two generations it incorporated a radioactive precursor, tritiated thymine. The thymine reveals its presence as a line of dark grains in the photographic emulsion. (Scattered grains are from background radiation.) The diagram on the opposite page shows how the picture is interpreted as demonstrating the manner of DNA duplication.

bacterial chromosome would turn out to be a single very large molecule. All the DNA previously isolated from bacteria had, to be sure, proved to be in molecules much smaller than the total chromosome, but a reason for this was suggested by studies by A. D. Hershey of the Carnegie Institution Department of Genetics at Cold Spring Harbor, N.Y. He had pointed out that the giant molecules of DNA that make up the genetic complement of certain bacterial viruses had been missed by earlier workers simply because they are so large that they are exceedingly fragile. Perhaps the same thing was true of the bacterial chromosome.

If so, the procedure for inspecting the replicating DNA of bacteria would have to be designed to cater for an exceptionally fragile molecule, since the bacterial chromosome contains some 20 times more DNA than the largest bacterial virus. It would have to be a case of looking but not touching. This was not as onerous a restriction as it may sound. The problem was, after all, a topographical one, involving delineation of strands of parent DNA and newly synthesized DNA. There was no need for manipulation, only for visualization.

Although electron microscopy is the

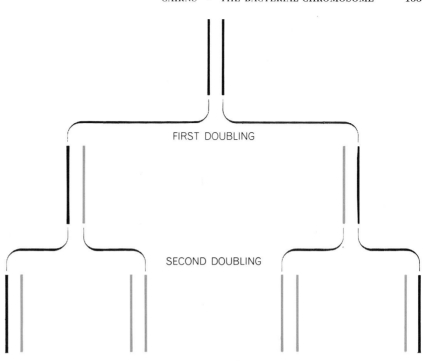

SEMICONSERVATIVE DUPLICATION was confirmed by the Meselson-Stahl experiment, which showed that each DNA molecule is composed of two parts: one that is present in the parent molecule, the other comprising new material synthesized when the parent molecule is duplicated. If radioactive labeling begins with the first doubling, the unlabeled (*black*) and labeled (*colored*) nucleotide chains of DNA form two-chain duplexes as shown here.

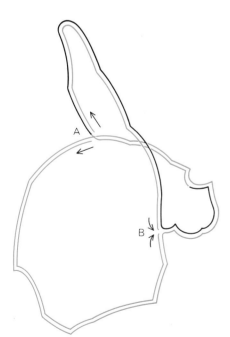

INTERPRETATION of autoradiograph on opposite page is based on the varying density of the line of grains. Excluding artifacts, dense segments represent doubly labeled DNA duplexes (*two colored lines*), faint segments singly labeled DNA (*color and black*). The parent chromosome, labeled in one strand and part of another, began to duplicate at *A*; new labeled strands have been laid down in two loops as far as *B*.

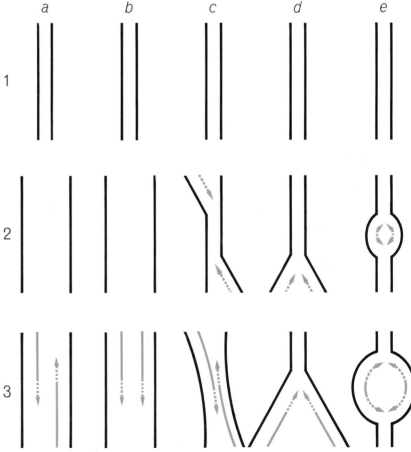

DUPLICATION could proceed in various ways (*a–e*). In these examples parental chains are shown as black lines and new chains as colored lines. The arrows show the direction of growth of the new chains, the newest parts of which are denoted by broken-line segments.

obvious way to get a look at a large molecule, I chose autoradiography in this instance because it offered certain peculiar advantages (which will become apparent) and because it had already proved to be the easier, albeit less accurate, technique for displaying large DNA molecules. Autoradiography capitalizes on the fact that electrons emitted by the decay of a radioactive isotope produce images on certain kinds of photographic emulsion. It is possible, for example, to locate the destination within a cell of a particular species of molecule by labeling such molecules with a radioactive atom, feeding them to the cell and then placing the cell in contact with an emulsion; a developed grain in the emulsion reveals the presence of a labeled molecule [see "Autobiographies of Cells," by Renato Baserga and Walter E. Kisieleski; SCIENTIFIC AMERICAN Offprint 165].

It happens that the base thymine, which is solely a precursor of DNA, is susceptible to very heavy labeling with tritium, the radioactive isotope of hydrogen. Replicating DNA incorporates the labeled thymine and thus becomes visible in autoradiographs. I had been able to extend the technique to demonstrating the form of individual DNA molecules extracted from bacterial viruses. This was possible because, in spite of the poor resolving power of autoradiography (compared with electron microscopy), molecules of DNA are so extremely long in relation to the resolving power that they appear as a linear array of grains. The method grossly exaggerates the apparent width of the DNA, but this is not a serious fault in the kind of study I was undertaking.

The general design of the experiments called for extracting labeled DNA from bacteria as gently as possible and then mounting it—without breaking the DNA molecules—for autoradiography. What I did was kill bacteria that had been fed tritiated thymine for various periods and put them, along with the enzyme lysozyme and an excess of unlabeled DNA, into a small capsule closed on one side by a semipermeable membrane. The enzyme, together with a detergent diffused into the chamber, induced the bacteria to break open and discharge their DNA. After the detergent, the enzyme and low-molecular-weight cellular debris had been diffused out of the chamber, the chamber was drained, leaving some of the DNA deposited on the membrane [see illustration below]. Once dry, the membrane was coated with a photographic emulsion sensitive to electrons emitted by the tritium and was left for two months. I hoped by this procedure to avoid subjecting the DNA to appreciable turbulence and so to find

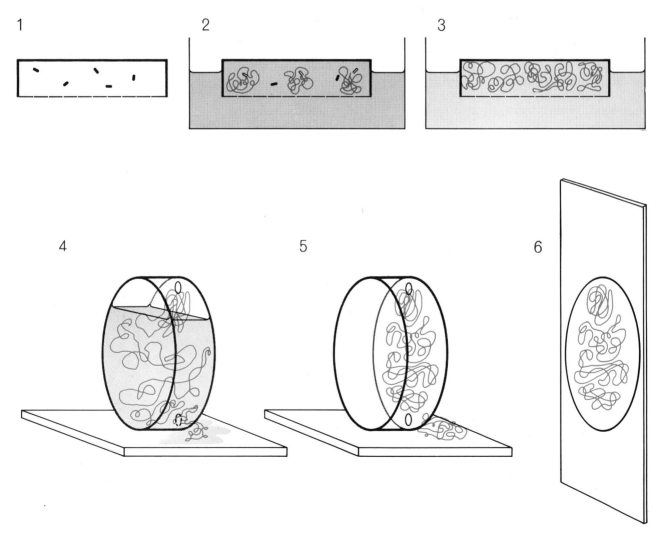

AUTORADIOGRAPHY EXPERIMENT begins with bacteria whose DNA has been labeled with radioactive thymine. The bacteria and an enzyme are placed in a small chamber closed by a semipermeable membrane (1). Detergent diffused into the chamber causes the bacteria to discharge their contents (2). The detergent and cellular debris are washed away by saline solution diffused through the chamber (3). The membrane is then punctured. The saline drains out slowly (4), leaving some unbroken DNA molecules (color) clinging to the membrane (5). The membrane, with DNA, is placed on a microscope slide and coated with emulsion (6).

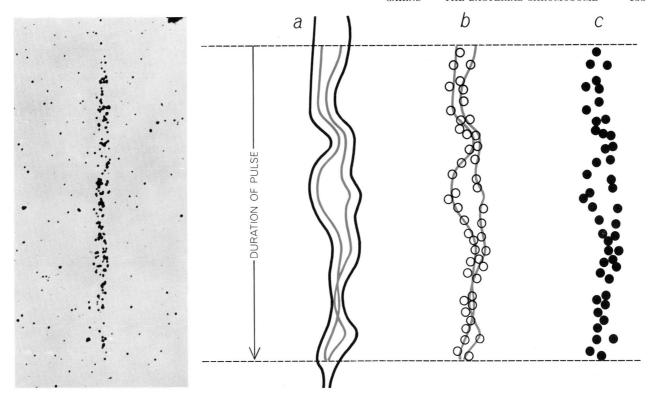

DURATION OF PULSE

a   *b*   *c*

DNA synthesized in *E. coli* fed radioactive thymine for three minutes is visible in an autoradiograph, enlarged 1,200 diameters, as an array of heavy black grains (*left*). The events leading to the autoradiograph are shown at right. The region of the DNA chains synthesized during the "pulse-labeling" is radioactive and is shown in color (*a*). The radioactivity affects silver grains in the photographic emulsion (*b*). The developed grains appear in the autoradiograph (*c*), approximately delineating the new chains of DNA.

some molecules that—however big—had not been broken and see their form. Inasmuch as *E. coli* synthesizes DNA during its entire division cycle, some of the extracted DNA should be caught in the act of replication. (Since there was an excess of unlabeled DNA present, any tendency for DNA to produce artificial aggregates would not produce a spurious increase in the size of the labeled molecules or an alteration in their form.)

It is the peculiar virtue of autoradiography that one sees only what has been labeled; for this reason the technique can yield information on the history as well as the form of a labeled structure. The easiest way to determine which of the schemes of replication was correct was to look at bacterial DNA that had been allowed to duplicate for only a short time in the presence of labeled thymine. Only the most recently made DNA would be visible (corresponding to the broken-line segments in the bottom illustration on page 153), and so it should be possible to determine if the two daughter molecules were being made at the same point or in different regions of the parent molecule. A picture obtained after labeling bacteria for

three minutes, or a tenth of a generation-time [*at left in illustration above*], makes it clear that two labeled structures are being made in the same place. This place is presumably a particular region of a larger (unseen) parent molecule [*see diagrams at right in illustration above*].

The autoradiograph also shows that at least 80 microns (80 thousandths of a millimeter) of the DNA has been duplicated in three minutes. Since duplication occupies the entire generation-time (which was about 30 minutes in these experiments), it follows that the process seen in the autoradiograph could traverse at least 10 × 80 microns, or about a millimeter, of DNA between one cell division and the next. This is roughly the total length of the DNA in the bacterial chromosome. The autoradiograph therefore suggests that the entire chromosome may be duplicated at a single locus that can move fast enough to traverse the total length of the DNA in each generation.

Finally, the autoradiograph gives evidence on the semiconservative aspect of duplication. Two structures are being synthesized. It is possible to estimate how heavily each structure is labeled (in

terms of grains produced per micron of length) by counting the number of exposed grains and dividing by the length. Then the density of labeling can be compared with that of virus DNA labeled similarly but uniformly, that is, in both of its polynucleotide chains. It turns out that each of the two new structures seen in the picture must be a single polynucleotide chain. If, therefore, the picture is showing the synthesis of two daughter molecules from one parent molecule, it follows that each daughter molecule must be made up of one new (labeled) chain and one old (unlabeled) chain—just as Watson and Crick predicted.

The "pulse-labeling" experiment just described yielded information on the isolated regions of bacterial DNA actually engaged in duplication. To learn if the entire chromosome is a single molecule and how the process of duplication proceeds it was necessary to look at DNA that had been labeled with tritiated thymine for several generations. Moreover, it was necessary to find, in the jumble of chromosomes extracted from *E. coli*, autoradiographs of unbroken chromosomes that were disen-

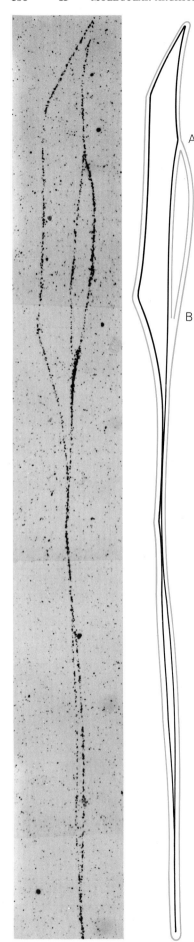

tangled enough to be seen as a whole. Rather than retrace all the steps that led, after many months, to satisfactory pictures of the entire bacterial chromosome in one piece, it is simpler to present two sample autoradiographs and explain how they can be interpreted and what they reveal.

The autoradiographs on page 152 and at the left show bacterial chromosomes in the process of duplication. All that is visible is labeled, or "hot," DNA; any unlabeled, or "cold," chain is unseen. A stretch of DNA duplex labeled in only one chain ("hot-cold") makes a faint trace of black grains. A duplex that is doubly labeled ("hot-hot") shows as a heavier trace. The autoradiographs therefore indicate, as shown in the diagrams that accompany them, the extent to which new, labeled polynucleotide chains have been laid down along labeled or unlabeled parent chains. Such data make it possible to construct a bacterial family history showing the process of duplication over several generations [see illustration on opposite page].

The significant conclusions are these:

1. The chromosome of *E. coli* apparently contains a single molecule of DNA roughly a millimeter in length and with a calculated molecular weight of about two billion. This is by far the largest molecule known to occur in a biological system.

2. The molecule contains two polynucleotide chains, which separate at the time of duplication.

3. The molecule is duplicated at a single locus that traverses the entire length of the molecule. At this point both new chains are being made: two chains are becoming four. This locus has come to be called the replicating "fork" because that is what it looks like.

4. Replicating chromosomes are not Y-shaped, as would be the case for a linear structure [see "d" in bottom illustration on page 153]. Instead the three ends of the Y are joined: the ends of the daughter molecules are joined to each other and to the far end of the parent molecule. In other words, the chromo-

COMPLETE CHROMOSOME is seen in this autoradiograph, enlarged about 370 diameters. Like the chromosome represented on pages 152 and 153, this one is circular, although it happens to have landed on the membrane in a more compressed shape and some segments are tangled. Whereas the first chromosome was more than halfway through the duplication process, this one is only about one-sixth duplicated (from A to B).

some is circular while it is being duplicated.

It is hard to conceive of the behavior of a molecule that is about 1,000 times larger than the largest protein and that exists, moreover, coiled inside a cell several hundred times shorter than itself. Apart from this general problem of comprehension, there are two special difficulties inherent in the process of DNA duplication outlined here. Both have their origin in details of the structure of DNA that I have not yet discussed.

The first difficulty arises from the opposite polarities of the two polynucleotide chains [see illustration on page 158]. The deoxyribose-phosphate backbone of one chain of the DNA duplex has the sequence $-O-C_3-C_4-C_5-O-P-O-C_3-C_4-C_5-O-P-\ldots$ (The C_3, C_4 and C_5 are the three carbon atoms of the deoxyribose that contribute to the backbone.) The other chain has the sequence $-P-O-C_5-C_4-C_3-O-P-O-C_5-C_4-C_3-O-\ldots$

If both chains are having their complements laid down at a single locus moving in one particular direction, it follows that one of these new chains must grow by repeated addition to the C_3 of the preceding nucleotide's deoxyribose and the other must grow by addition to a C_5. One would expect that two different enzymes should be needed for these two quite different kinds of polymerization. As yet, however, only the reaction that adds to chains ending in C_3 has been demonstrated in such experiments as Kornberg's. This fact had seemed to support a mode of replication in which the two strands grew in opposite directions [see "a" and "c" in bottom illustration on page 153]. If the single-locus scheme is correct, the problem of opposite polarities remains to be explained.

The second difficulty, like the first, is related to the structure of DNA. For the sake of simplicity I have been representing the DNA duplex as a pair of chains lying parallel to each other. In actuality the two chains are wound helically around a common axis, with one complete turn for every 10 base pairs, or 34 angstrom units of length (34 ten-millionths of a millimeter). It would seem, therefore, that separation of the chains at the time of duplication, like separation of the strands of an ordinary rope, must involve rotation of the parent molecule with respect to the two daughter molecules. Moreover, this rotation must be very rapid. A fast-multiplying bacterium can divide every 20 minutes;

during this time it has to duplicate—and consequently to unwind—about a millimeter of DNA, or some 300,000 turns. This implies an average unwinding rate of 15,000 revolutions per minute.

At first sight it merely adds to the difficulty to find that the chromosome is circular while all of this is going on. Obviously a firmly closed circle— whether a molecule or a rope—cannot be unwound. This complication is worth worrying about because there is increasing evidence that the chromosome of *E. coli* is not exceptional in its circularity. The DNA of numerous viruses has been shown either to be circular or to become circular just before replication begins. For all we know, circularity may therefore be the rule rather than the exception.

There are several possible explanations for this apparent impasse, only one of which strikes me as plausible.

First, one should consider the possibility that there is no impasse—that in the living cell the DNA is two-stranded but not helical, perhaps being kept that way precisely by being in the form of a circle. (If a double helix cannot be unwound when it is firmly linked into a circle, neither can relational coils ever be introduced into a pair of uncoiled circles.) This hypothesis, however, requires a most improbable structure for two-strand DNA, one that has not been observed. And it does not really avoid the unwinding problem because there would still have to be some mechanism for making nonhelical circles out of the helical rods of DNA found in certain virus particles.

Second, one could avoid the unwinding problem by postulating that at least one of the parental chains is repeatedly broken and reunited during replication, so that the two chains can be separated over short sections without rotation of the entire molecule. One rather sentimental objection to this hypothesis (which was proposed some time ago) is that it is hard to imagine such cavalier and hazardous treatment being meted out to such an important molecule, and one so conspicuous for its stability. A second objection is that it does not explain circularity.

The most satisfactory solution to the unwinding problem would be to find some reason why the ends of the chromosome actually *must* be joined together. This is the case if one postulates that there is an active mechanism for unwinding the DNA, distinct from the mechanism that copies the unwound

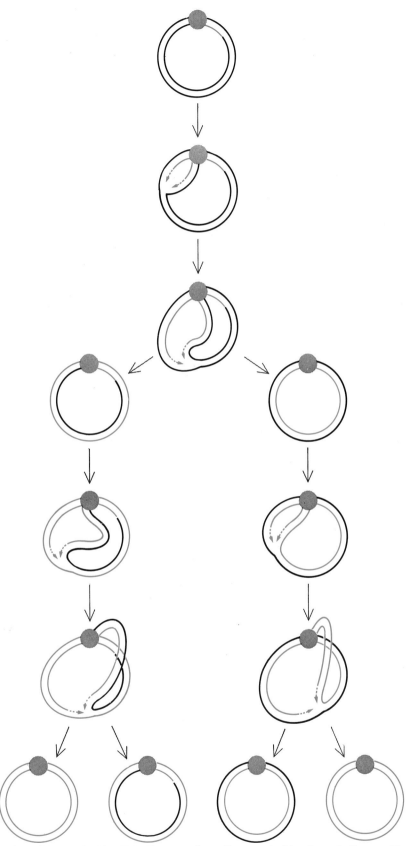

BACTERIAL DNA MOLECULE apparently replicates as in this schematic diagram. The two chains of the circular molecule are represented as concentric circles, joined at a "swivel" (*gray spot*). Labeled DNA is shown in color; part of one chain of the parent molecule is labeled, as are two generations of newly synthesized DNA. Duplication starts at the swivel and, in these drawings, proceeds counterclockwise. The arrowheads mark the replicating "fork": the point at which DNA is being synthesized in each chromosome. The drawing marked *A* is a schematic rendering of the chromosome in the autoradiograph on page 152.

158

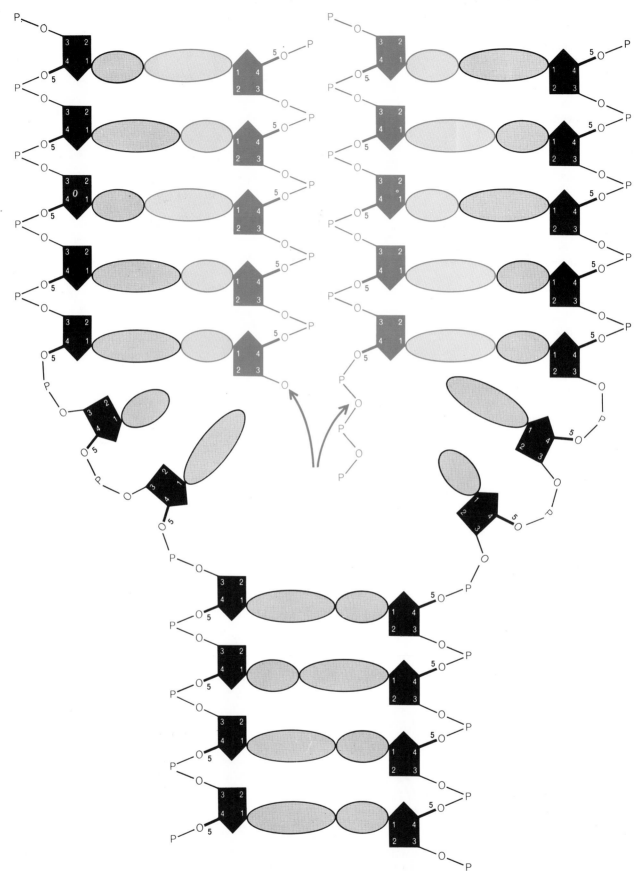

OPPOSITE POLARITIES of the two parental chains of the DNA duplex result in opposite polarities and different directions of growth in the two new chains (*color*) being laid down as complements of the old ones during duplication. Note that the numbered carbon atoms (*1 to 5*) in the deoxyribose rings (*solid black*) are in different positions in the two parental chains and therefore in the two new chains. As the replicating fork moves downward, the new chain that is complementary to the left parental chain must grow by addition to a C_3, the other new chain by addition to a C_5, as shown by the arrows. The elliptical shapes are the four bases.

chains. Now, any active unwinding mechanism must rotate the parent molecule with respect to the two new molecules—must hold the latter fast, in other words, just as the far end of a rope must be held if it is to be unwound. A little thought will show that this can be most surely accomplished by a machine attached, directly or through some common "ground," to the parent molecule and to the two daughters [*see illustration below*]. Every turn taken by such a machine would inevitably unwind the parent molecule one turn.

Although other kinds of unwinding machine can be imagined (one could be situated, for example, at the replicating fork), a practical advantage of this particular hypothesis is that it accounts for circularity. It also makes the surprising —and testable—prediction that any irreparable break in the parent molecule will instantly stop DNA synthesis, no matter how far the break is from the replicating fork. If this prediction is fulfilled, and the unwinding machine acquires the respectability that at present it lacks, we may find ourselves dealing with the first example in nature of something equivalent to a wheel.

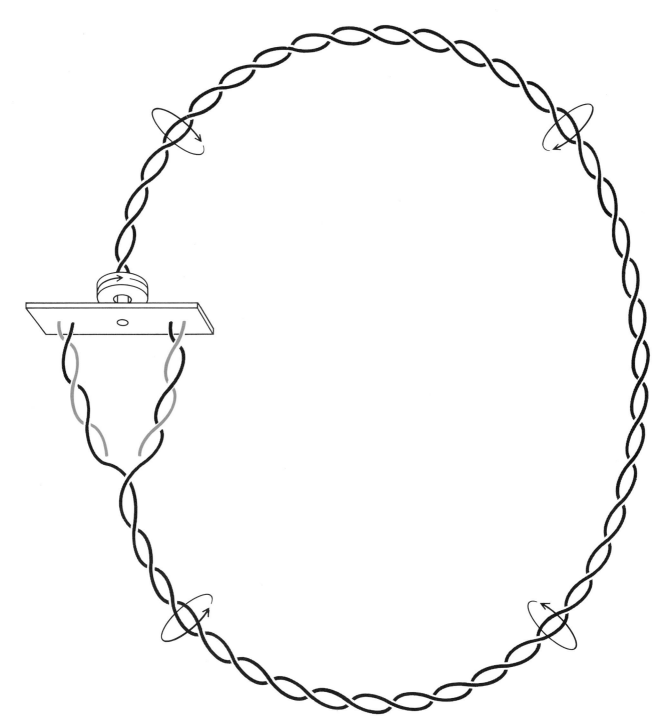

POSSIBLE MECHANISM for unwinding the DNA double helix is a swivel-like machine to which the end of the parent molecule and also the ends of the two daughter molecules are joined. The torque imparted by this machine is considered to be transmitted along the parent molecule, producing unwinding at the replicating fork. If this is correct, chromosome breakage should halt duplication.

15

THE FINE STRUCTURE OF THE GENE

SEYMOUR BENZER
January 1962

The question "What is a gene?" has bothered geneticists for fifty years. Recent work with a small bacterial virus has shown how to split the gene and make detailed maps of its internal structure

Much of the work of science takes the form of map making. As the tools of exploration become sharper, they reveal finer and finer details of the region under observation. In the December, 1961 issue of *Scientific American* John C. Kendrew of the University of Cambridge described the mapping of the molecule of the protein myoglobin, revealing a fantastically detailed architecture. A living organism manufactures thousands of different proteins, each to precise specifications. The "blueprints" for all this detail are stored in coded form within the genes. In this article we shall see how it is possible to map the internal structure of a single gene, with the revelation of detail comparable to that in a protein.

It has been known since about 1913 that the individual active units of heredity—the genes—are strung together in one-dimensional array along the chromosomes, the threadlike bodies in the nucleus of the cell. By crossing such organisms as the fruit fly *Drosophila*, geneticists were able to draw maps showing the linear order of various genes that had been marked by the occurrence of mutations in the organism. Most geneticists regarded the gene as a more or less indivisible unit. There seemed to be no way to attack the questions "Exactly what is a gene? Does it have an internal structure?"

In recent years it has become apparent that the information-containing part of the chromosomal chain is in most cases a giant molecule of deoxyribonucleic acid, or DNA. (In some viruses the hereditary material is ribonucleic acid, or RNA.) Indeed, the threadlike molecule of DNA can be seen in the electron microscope [*see bottom illustration on opposite page*]. For obtaining information about the fine structure of DNA, however, modern methods of genetic analysis are a more powerful tool than even the electron microscope.

It is important to understand why this fine structure is not revealed by conventional genetic mapping, as is done with fruit flies. Genetic mapping is possible because the chromosomes sometimes undergo a recombination of parts called crossing over. By this process, for example, two mutations that are on different chromosomes in a parent will sometimes emerge on the same chromosome in the progeny. In other cases the progeny will inherit a "standard" chromosome lacking the mutations seen in the parent. It is as if two chromosomes lying side by side could break apart at any point and recombine to form two new chromosomes, each made up of parts derived from the original two. As a matter of chance two points far apart will recombine frequently; two points close together will recombine rarely. By carrying out many crosses in a large population of fruit flies one can measure the frequency—meaning the ease—with which different genes will recombine, and from this one can draw a map showing the parts in correct linear sequence. This technique has been used to map the chromosomes of many organisms. Why not, then, use the technique to map mutations inside the gene? The answer is that points within the same gene are so close together that the chance of detecting recombination between them would be exceedingly small.

In the study of genetics, however, everything hinges on the choice of a suitable organism. When one works with fruit flies, one deals with at most a few thousand individuals, and each generation takes roughly 20 days. If one works with a microorganism, such as a bacterium or, better still, a bacterial virus (bacteriophage), one can deal with billions of individuals, and a generation takes only minutes. One can therefore perform in a test tube in 20 minutes an experiment yielding a quantity of genetic data that would require, if humans were used, the entire population of the earth. Moreover, with microorganisms special tricks enable one to select just those individuals of interest from a population of a billion. By exploiting these advantages it becomes possible not only to split the gene but also to map it in the utmost detail, down to the molecular limits of its structure.

Replication of a Virus

An extremely useful organism for this fine-structure mapping is the T4 bacteriophage, which infects the colon bacillus. T4 is one of a family of viruses that has been most fruitfully exploited by an entire school of molecular biologists founded by Max Delbrück of the California Institute of Technology. The T4 virus and its relatives each consist of a head, which looks hexagonal in electron micrographs, and a complex tail by which the virus attaches itself to the bacillus wall [*see top illustration on opposite page*]. Crammed within the head of the virus is a single long-chain molecule of DNA having a weight about 100 million times that of the hydrogen atom. After a T4 virus has attached itself to a bacillus, the DNA molecule enters the cell and dictates a reorganization of the cell machinery to manufacture 100 or so copies of the complete virus. Each copy consists of the DNA and at least six distinct protein components. To make these components the invading DNA specifies the formation of a series of special enzymes, which themselves are proteins. The entire process is controlled by the battery of genes that constitutes the DNA molecule.

According to the model for DNA de-

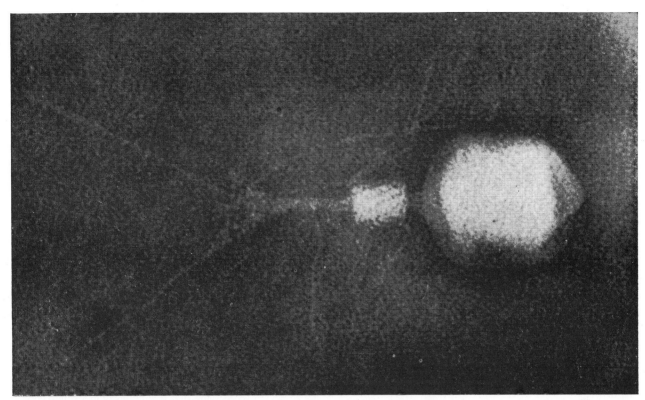

T2 BACTERIOPHAGE, magnified 500,000 diameters, is a virus that contains in its head complete instructions for it own replication. To replicate, however, it must find a cell of the colon bacillus into which it can inject a giant molecule of deoxyribonucleic acid (DNA). This molecule, comprising the genes of the phage, sub- verts the machinery of the cell to make about 100 copies of the complete phage. The mutations that occasionally arise in the DNA molecule during replication enable the geneticist to map the detailed structure of individual genes. The electron micrograph was made by S. Brenner and R. W. Horne at the University of Cambridge.

MOLECULE OF DNA is the fundamental carrier of genetic information. This electron micrograph shows a short section of DNA from calf thymus; its length is roughly that of the rII region in the DNA of T4 phage studied by the author. The DNA molecule in the phage would be about 30 feet long at this magnification of 150,000 diameters. The white sphere, a polystyrene "measuring stick," is 880 angstrom units in diameter. The electron micrograph was made by Cecil E. Hall of the Massachusetts Institute of Technology.

vised by James D. Watson and F. H. C. Crick, the DNA molecule resembles a ladder that has been twisted into a helix. The sides of the ladder are formed by alternating units of deoxyribose sugar groups and phosphate groups. The rungs, which join two sugar units, are composed of pairs of nitrogenous bases: either adenine paired with thymine or guanine paired with cytosine. The particular sequence of bases provides the genetic code of the DNA in a given organism.

The DNA in the T4 virus contains some 200,000 base pairs, which, in amount of information, corresponds to much more than that contained in this article. Each base pair can be regarded as a letter in a word. One word (of the DNA code) may specify which of 20-odd amino acids is to be linked into a polypeptide chain. An entire paragraph might be needed to specify the sequence of amino acids for a polypeptide chain that has functional activity. Several polypeptide

units may be needed to form a complex protein.

One can imagine that "typographical" errors may occur when DNA molecules are being replicated. Letters, words or sentences may be transposed, deleted or even inverted. When this occurs in a daily newspaper, the result is often humorous. In the DNA of living organisms typographical errors are never funny and are often fatal. We shall see how these errors, or mutations, can be used to analyze a small portion of the genetic information carried by the T4 bacteriophage.

Genetic Mapping with Phage

Before examining the interior of a gene let us see how genetic experiments are performed with bacteriophage. One starts with a single phage particle. This provides an important advantage over higher organisms, where two different individuals are required and the male and female may differ in any number of respects besides their sex. Another simplification is that phage is haploid, meaning that it contains only a single copy of its hereditary information, so that none of its genes are hidden by dominance effects. When a population is grown from a single phage particle, using a culture of sensitive bacteria as fodder, almost all the descendants are identical, but an occasional mutant form arises through some error in copying the genetic information. It is precisely these errors in reproduction that provide the key to the genetic analysis of the structure [see upper illustration on pages 164 and 165].

Suppose that two recognizably different kinds of mutant have been picked up; the next step is to grow a large population of each. This can be done in two test tubes in a couple of hours. It is now easy to perform a recombination experiment. A liquid sample of each phage population is added to a culture of bacterial cells in a test tube. It is arranged that the phage particles outnumber the bacterial cells at least three to one, so that each cell stands a good chance of being infected by both mutant forms of phage DNA. Within 20 minutes about 100 new phage particles are formed within each cell and are released when the cell bursts. Most of the progeny will resemble one or the other parent. In a few of them, however, the genetic information from the two parents may have been recombined to form a DNA molecule that is not an exact copy of the molecule possessed by either parent but a combination of the two. This new recombinant phage particle can carry

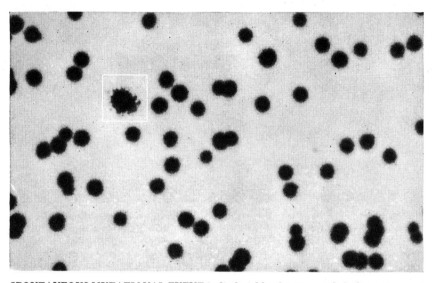

SPONTANEOUS MUTATIONAL EVENT is disclosed by the one mottled plaque (square) among dozens of normal plaques produced when standard T4 phage is "plated" on a layer of colon bacilli of strain B. Each plaque contains some 10 million progeny descended from a single phage particle. The plaque itself represents a region in which cells have been destroyed. Mutants found in abnormal plaques provide the raw material for genetic mapping.

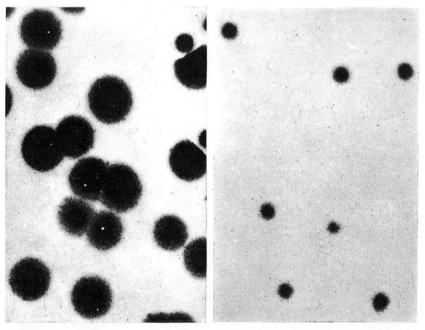

DUPLICATE REPLATINGS of mixed phage population obtained from a mottled plaque, like that shown at top of page, give contrasting results, depending on the host. Replated on colon bacilli of strain B (left), rII mutants produce large plaques. If the same mixed population is plated on strain K (right), only standard type of phage produce plaques.

both mutations or neither of them [*see lower illustration on next two pages*].

When this experiment is done with various kinds of mutant, some of the mutant genes tend to recombine almost independently, whereas others tend to be tightly linked to each other. From such experiments Alfred D. Hershey and Raquel Rotman, working at Washington University in St. Louis, were able to construct a genetic map for phage showing an ordered relationship among the various kinds of mutation, as had been done earlier with the fruit fly *Drosophila* and other higher organisms. It thus appears that the phage has a kind of chromosome —a string of genes that controls its hereditary characteristics.

One would like to do more, however, than just "drosophilize" phage. One would like to study the internal structure of a single gene in the phage chromosome. This too can be done by recombination experiments, but instead of choosing mutants of different kinds one chooses mutants that look alike (that is, have modifications of what is apparently the same characteristic), so that they are likely to contain errors in one or another part of the same gene.

Again the problem is to find an experimental method. When looking for mutations in fruit flies, say a white eye or a bent wing, one has to examine visually every fruit fly produced in the experiment. When working with phage, which reproduce by the billions and are invisible except by electron microscopy, the trick is to find a macroscopic method for identifying just those individuals in which recombination has occurred.

Fortunately in the T4 phage there is a class of mutants called *r*II mutants that can be identified rather easily by the appearance of the plaques they form on a given bacterial culture. A plaque is a clear region produced on the surface of a culture in a glass dish where phage particles have multiplied and destroyed the bacterial cells. This makes it possible to count individual phage particles without ever seeing them. Moreover, the shape and size of the plaques are hereditary characteristics of the phage that can be easily scored. A plaque produced in several hours will contain about 10 million phage particles representing the progeny of a single particle. T4 phage of the standard type can produce plaques on either of two bacterial host strains, B or K. The standard form of T4 occasionally gives rise to *r*II mutants that are easily noticed because they produce a distinctive plaque on B cultures. The key to the whole mapping technique is that

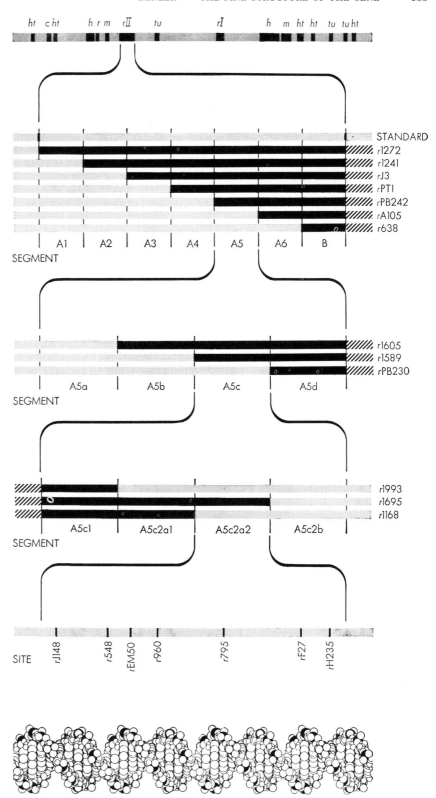

MAPPING TECHNIQUE localizes the position of a given mutation in progressively smaller segments of the DNA molecule contained in the T4 phage. The *r*II region represents to start with only a few per cent of the entire molecule. The mapping is done by crossing an unknown mutant with reference mutants having deletions (*dark gray tone*) of known extent in the *r*II region (*see illustration of method on page 166*). The order and spacing of the seven mutational sites in the bottom row are still tentative. Each site probably represents the smallest mutable unit in the DNA molecule, a single base pair. The molecular segment (*extreme bottom*), estimated to be roughly in proper scale, contains a total of about 40 base pairs.

these mutants do not produce plaques on K cultures.

Nevertheless, an *r*II mutant can grow normally on bacterial strain K if the cell is simultaneously infected with a particle of standard type. Evidently the standard DNA molecule can perform some function required in K that the mutants cannot. This functional structure has been traced to a small portion of the DNA molecule, which in genetic maps of the T4 phage is designated the *r*II region.

To map this region one isolates a number of independently arising *r*II mutants (by removing them from mutant plaques visible on B) and crosses them against one another. To perform a cross, the two mutants are added to a liquid culture of B cells, thereby providing an opportunity for the progeny to recombine portions of genetic information from either parent. If the two mutant versions are due to typographical errors in different parts of the DNA molecule, some individuals of standard type may be regenerated. The standards will produce plaques on the K culture, whereas the mutants cannot. In this way one can easily detect a single recombinant among a billion progeny. As a consequence one can "resolve" two *r*II mutations that are extremely close together. This resolving power is enough to distinguish two mutations that are only one base pair apart in the DNA molecular chain.

What actually happens in the recombination of phage DNA is still a matter of conjecture. Two defective DNA molecules may actually break apart and rejoin to form one nondefective molecule, which is then replicated. Some recent evidence strongly favors this hypothesis. Another possibility is that in the course of replication a new DNA molecule arises from a process that happens to copy only the good portions of the two mutant molecules. The second process is called copy choice. An analogy for the two different processes can be found in the methods available for making a good tape recording of a musical performance from two tapes having defects in different places. One method is to cut the defects out of the two tapes and splice the good sections together. The second method (copy choice) is to play the two tapes and record the good sections on a third tape.

Mapping the *r*II Mutants

A further analogy with tape recording will help to explain how it has been established that the *r*II region is a simple linear structure. Given three tapes, each with a blemish or deletion in a different place, labeled *A*, *B* and *C*, one can imagine the deletions so located that deletion *B* overlaps deletion *A* and deletion *C*, but that *A* and *C* do not overlap each other. In such a case a good performance can be re-created only by recombining *A* and *C*. In mutant forms of phage DNA containing comparable deletions the existence of overlapping can be established by recombination experiments of just the same sort.

To obtain such deletions in phage one looks for mutants that show no tendency to revert to the standard type when they reproduce. The class of nonreverting mutants automatically includes those in which large alterations or deletions have occurred. (By contrast, *r*II mutants that revert spontaneously behave as if their alterations were localized at single points). The result of an exhaustive study covering hundreds of nonreverting *r*II mutants shows that all can be represented as containing deletions of one size or another in a single linear structure. If the structure were more complex, containing, for example, loops or branches, some mutations would have been expected to overlap in such a way as to make it impossible to represent them in a linear map. Although greater complexity cannot be absolutely excluded, all observations to date are satisfied by the postulate of simple linearity.

Now let us consider the *r*II mutants that do, on occasion, revert spontaneously when they reproduce. Conceivably they arise when the DNA molecule of the phage undergoes an alteration of a single base pair. Such "point" mutants are those that must be mapped if one is to probe the fine details of genetic structure. However, to test thousands of point mutants against one another for recombination in all possible pairs would

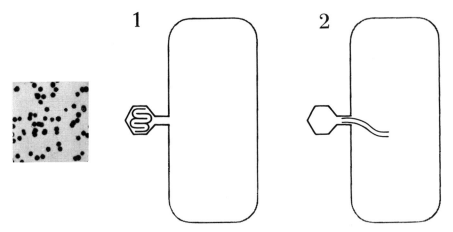

REPLICATION AND MUTATION occur when a phage particle infects a bacillus cell. The experiment begins by isolating a few standard particles from a normal plaque (*photograph at far left*) and growing billions of progeny in a broth culture of strain B colon bacilli. A sample of the broth is then spread on a Petri dish containing the same strain, on which the

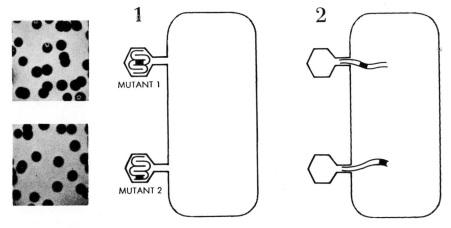

PROCESS OF RECOMBINATION permits parts of the DNA of two different phage mutants to be reassembled in a new DNA molecule that may contain both mutations or neither of them. Mutants obtained from two different cultures (*photographs at far left*) are introduced into a broth of strain B colon bacilli. Crossing occurs (*1*) when DNA from each mutant type require millions of crosses. Mapping of point mutations by such a procedure would be totally impractical.

The way out of this difficulty is to make use of mutants of the nonreverting type, whose deletions divide up the *r*II region into segments. Each point mutant is tested against these reference deletions. The recombination test gives a negative result if the deletion overlaps the point mutation and a positive result (over and above the "noise" level due to spontaneous reversion of the point mutant) if it does not overlap. In this way a mutation is quickly located within a particular segment of the map. The point mutation is then tested against a second group of reference mutants that divide this segment into smaller segments, and so on [*see illustration on pages 168 and 169*]. A point mutation can be assigned by this method to any of 80-odd ordered segments.

The final step in mapping is to test against one another only the group of mutants having mutations within each segment. Those that show recombination are concluded to be at different sites, and each site is then named after the mutant indicating it. (The mutants themselves have been assigned numbers according to their origin and order of discovery.) Finally, the order of the sites within a segment can be established by making quantitative measurements of recombination frequencies with respect to one another and neighbors outside the segment.

The Functional Unit

Thus we have found that the hereditary structure needed by the phage to multiply in colon bacilli of strain K consists of many parts distinguishable by mutation and recombination. Is this region to be thought of as one gene (because it controls one characteristic) or as hundreds of genes? Although mutation at any one of the sites leads to the same observed physiological defect, it does not necessarily follow that the entire structure is a single functional unit. For instance, growth in strain K could require a series of biochemical reactions, each controlled by a different portion of the region, and the absence of any one of the steps would suffice to block the final result. It is therefore of interest to see whether or not the *r*II region can be subdivided into parts that function independently.

This can be done by an experiment known as the *cis-trans* comparison. It will be recalled that the needed function can be supplied to a mutant by simultaneous infection of the cell with standard phage; the standard type supplies an intact copy of the genetic structure, so that

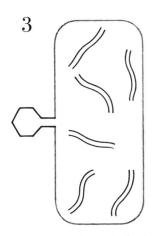 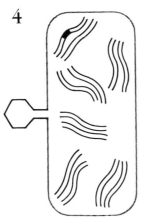 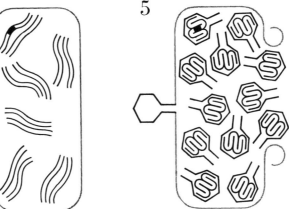 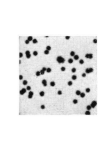

mutants and standard phage produce different plaque types. The diagrams show a bacillus infected by a single standard phage. The DNA molecule from the phage enters the cell (*2*) and is replicated (*3 and 4*). Among scores of perfect replicas, one may contain a mutation (*dark patch*). Encased in protein jackets, the phage particles finally burst out of the cell (*5*). When a mutant arises during development of a plaque, the mixture of its mutant progeny and standard types makes plaque look mottled (*photograph at right*).

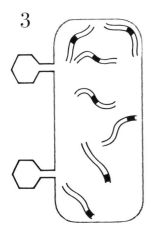 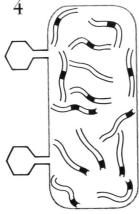 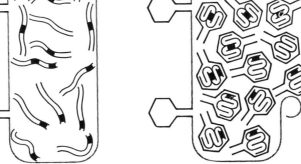 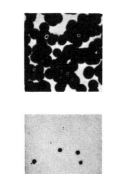

infects a single bacillus. Most of the DNA replicas are of one type or the other, but occasionally recombination will produce either a double mutant or a standard recombinant containing neither mutation. When the progeny of the cross are plated on strain B (*top photograph at far right*), all grow successfully, producing many plaques. Plated on strain K, only the standard recombinants are able to grow (*bottom photograph at right*). A single standard recombinant can be detected among as many as 100 million progeny.

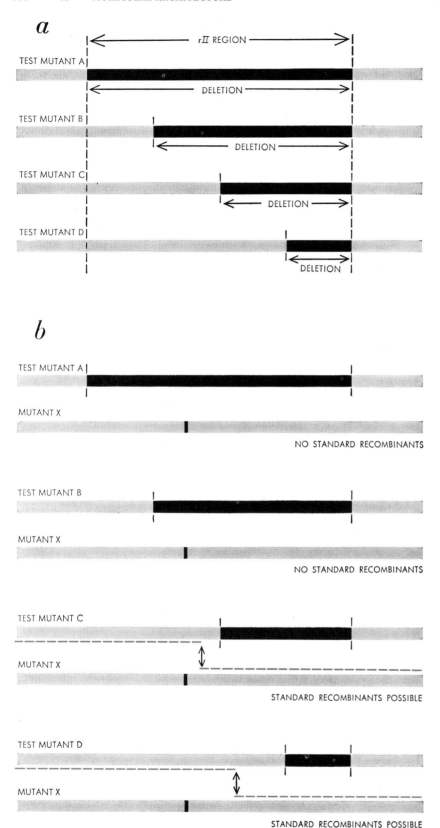

DELETION MAPPING is done by crossing an unknown mutant with a selected group of reference mutants (*four at top*) whose DNA molecules contain deletions—or what appear to be deletions—of known length in the rII region. Thus when mutant *X* is crossed with test mutants *A* and *B*, no standard recombinants are observed because both copies of the DNA molecule are defective at the same place. When *X* is crossed with *C* and *D*, however, standard recombinants can be formed, as indicated by broken lines and arrows. By using other reference mutants with appropriate deletions the location of *X* can be further narrowed.

it does not matter what defect the rII mutant has and both types are enabled to reproduce. Now suppose the intact structure of the standard type could be split into two parts. If this were to destroy the activity, the two parts could be regarded as belonging to a single functional unit. Although the experiment as such is not feasible, one can do the next best thing. That is to supply piece *A* intact by means of a mutant having a defect in piece *B*, and to use a mutant with a defect in piece *A* to supply an intact piece *B*. If the two pieces *A* and *B* can function independently, the system should be active, since each mutant supplies the function lacking in the other. If, however, both pieces must be together to be functional, the split combination should be inactive.

The actual experimental procedure is as follows. Let us imagine that one has identified two mutational sites in the rII region, *X* and *Y*, and that one wishes to know if they lie within the same functional unit. The first step is to infect cells of strain K with the two different mutants, *X* and *Y;* this is called the *trans* test because the mutations are borne by different DNA molecules. Now in K the decision as to whether or not the phage will function occurs very soon after infection and *before* there is any opportunity for recombination to take place. To carry out a control experiment one needs a double mutant (obtainable by recombination) that contains both *X* and *Y* within a single phage particle. When cells of strain K are infected with the double mutant and the standard phage, the experiment is called the *cis* test since one of the infecting particles contains both mutations in a single DNA molecule. In this case, because of the presence of the standard phage, normal replication is expected and provides the control against which to measure the activity observed in the *trans* test. If, in the *trans* test, the phage fails to function or shows only slight activity, one can conclude that *X* and *Y* fall within the same functional unit. If, on the other hand, the phage develops actively, it is probable (but not certain) that the sites lie in different functional units. (Certainty in this experiment is elusive because the products of two defective versions of the same functional unit, tested in a *trans* experiment, will sometimes produce a partial activity, which may be indistinguishable from that produced by a *cis* experiment.)

As applied to rII mutants, the test divides the structure into two clear-cut parts, each of which can function independently of the other. The functional units have been called cistrons, and we say that the rII region is composed of an A cistron and a B cistron.

We have, then, genetic units of various sizes: the small units of mutation and recombination, much larger cistrons and finally the rII region, which includes both cistrons. Which one of these shall we call the gene? It is not surprising to find geneticists in disagreement, since in classical genetics the term "gene" could apply to any one of these. The term "gene" is perfectly acceptable so long as one is working at a higher level of integration, at which it makes no difference which unit is being referred to. In describing data on the fine level, however, it becomes essential to state unambiguously which operationally defined unit one is talking about. Thus in describing experiments with rII mutants one can speak of the rII "region," two rII "cistrons" and many rII "sites."

Some workers have proposed using the word "gene" to refer to the genetic unit that provides the information for one enzyme. But this would imply that one should not use the word "gene" at all, short of demonstrating that a specific enzyme is involved. One would be quite hard pressed to provide this evidence in the great majority of cases in which the term has been used, as, for example, in almost all the mutations in *Drosophila*. A genetic unit should be defined by a genetic experiment. The absurdity of doing otherwise can be seen by imagining a biochemist describing an enzyme as that which is made by a gene.

We have seen that the topology of the rII region is simple and linear. What can be said about its topography? Are there local differences in the properties of the various parts? Specifically, are all the subelements equally mutable? If so, mutations should occur at random throughout the structure and the topography would then be trivial. On the other hand, sites or regions of unusually high or low mutability would be interesting topographic features. To answer this question one isolates many independently arising rII mutants and maps each one to see if mutations tend to occur more frequently at certain points than at others. Each mutation is first localized into a main segment, then into a smaller segment, and finally mutants of the same small segment are tested against each other. Any that show recombination are said to define different sites. If two or more reverting mutants are found to show no detectable recombination with each other, they are considered to be

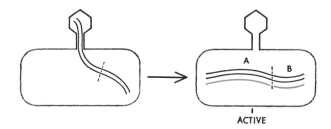

PHAGE ACTIVITY requires that the coded information inside functional units of the DNA molecule be available intact. The rII region consists of two functional units called A cistron and B cistron. When both are present intact (*right*), the phage actively replicates inside colon bacillus of strain K. Colored lines indicate effective removal of coded information.

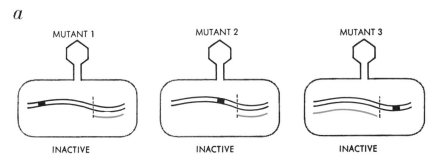

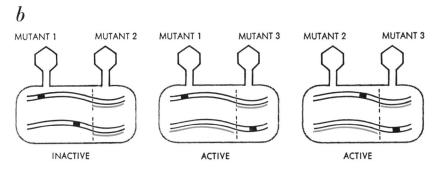

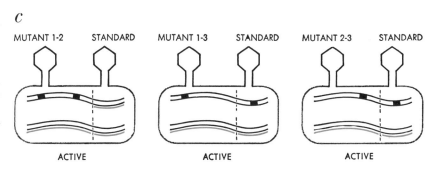

CIS-TRANS TEST determines the size of functional units. In bacillus of strain K, T4 phage is active only if both A and B cistrons are provided intact; hence mutants *1*, *2* and *3* are inactive. (The sites of mutation have been previously established.) Tests with the three mutants taken two at a time (*b*) show that sites *1* and *2* must be in the same cistron. A test of each mutant with standard phage (*c*) provides a control; in this case all are active.

repeats, and one of them is chosen to represent the site in further tests. A set of distinct sites is thereby obtained, each with its own group of repeats. The designation of a mutant as a repeat is, of course, tentative, since in principle it remains possible that a more sensitive test could show some recombination.

The illustration on the next two pages shows a map of the rII region with each occurrence of a spontaneous mutation indicated by a square. These mutations, as well as other data from induced mutations, subdivide the map into more than 300 distinct sites, and the distribution of repeats is indeed far from random. The topography for spontaneous mutation is evidently quite complex, the structure consisting of elements with widely different mutation rates.

Spontaneous mutation is a chronic disease; a spontaneous mutant is simply one for which the cause is unknown. By using chemical mutagens such as nitrous acid or hydroxylamine, or physical agents such as ultraviolet light, one can alter the DNA in a more controlled manner and induce mutations specifically. A method of inducing specific mutations has long been the philosophers' stone of genetics. What the genetic alchemist desired, however, was an effect that could be directed at the gene controlling a particular characteristic. Chemical mutagenesis is highly specific but not in this way. When Rose Litman and Arthur B. Pardee at the University of California discovered the mutagenic effect of 5-bromouracil on phage, they regarded it as a nonspecific mutagen because mutations were induced that affected a wide assortment of different phage characteristics. This nonspecificity resulted because each functional gene is a structure with many parts and is bound to contain a number of sites that are responsive to any particular mutagen. Therefore the rate at which mutation is

DELETION MAP shows the reference mutants that divided the rII region into 80 segments. These mutants behave as if various sections of the DNA molecule had been deleted or inactivated, and as a class they do not revert, or back-mutate, spontaneously to produce standard phage. Mutants that do revert usually act as if the mutation is localized at a single point on the DNA molecule. Where this point falls in the rII region is determined by systematically crossing the revertible mutant with these reference deletion mutants, as illustrated on page 166. The net result is to assign the point mutation to smaller and smaller segments of the map.

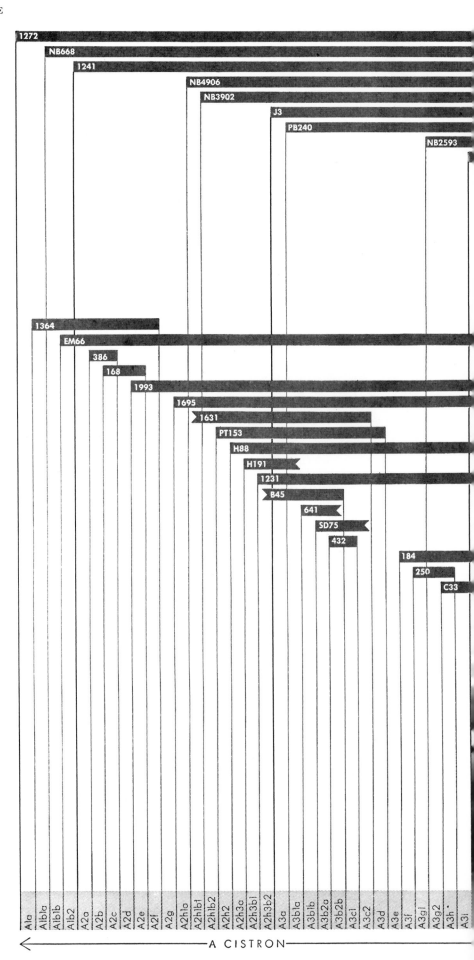

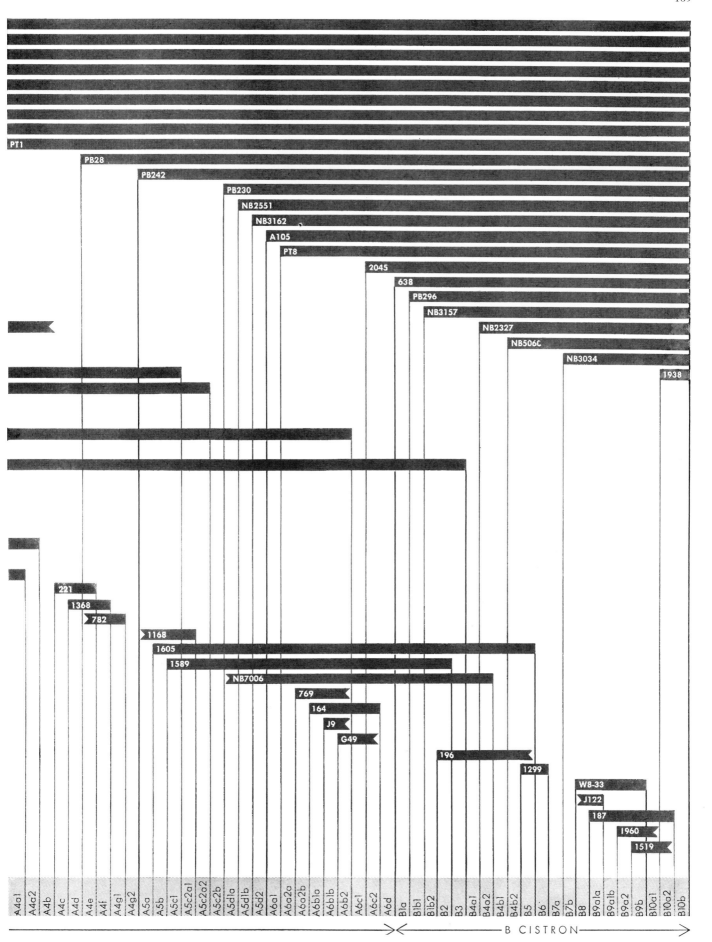

induced in various genes is more or less the same. By fine-structure genetic analysis, however, Ernst Freese and I, working in our laboratory at Purdue University, have found that 5-bromouracil increases the mutation rate at certain sites by a factor of 10,000 or more over the spontaneous rate, while producing no noticeable change at some other sites. This indicates a high degree of specificity indeed, but at the level within the cis-tron. Furthermore, other mutagens specifically alter other sites. The response of part of the B cistron to a variety of mutagens is shown in the illustration on the following two pages.

Each site in the genetic map can, then, be characterized by its spontaneous mutability and by its response to various mutagens. By this means many different kinds of site have been found. Some response patterns are represented at only a single site in the entire structure; for example, the prominent spontaneous hot spot in segment B4. This is at first surprising, because according to the Watson-Crick model for DNA the structure should consist of only two types of element, adenine-thymine (AT) pairs and guanine-hydroxymethylcytosine (GC) pairs. One possible explanation for the uneven reactivity among various sites is that the response may depend not

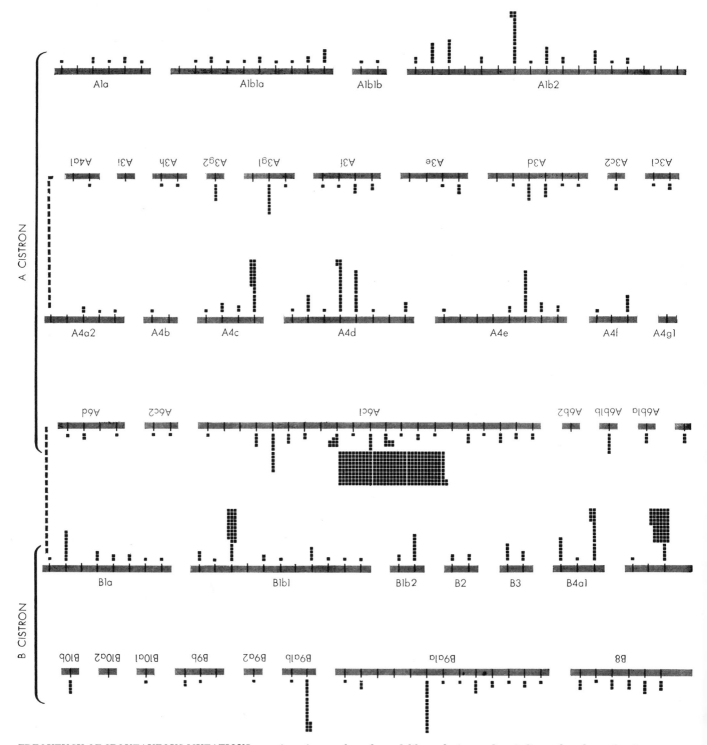

FREQUENCY OF SPONTANEOUS MUTATIONS at various sites is shown in this complete map of the *r*II region. Alternate rows have been deliberately inverted to indicate that the region is a continuous molecular thread. Each spontaneous mutation at a site

only on the particular base pair at a site but also very much on the type and arrangement of neighboring base pairs.

Once a site is identified it can be further characterized by the ease with which a particular mutagen makes reverse mutations produce phage of standard type. Combining such studies with studies of the chemical mechanism of mutagenesis, it may be possible eventually to translate the genetic map, bit by bit, into the actual base sequence.

Saturation of the Map

How far is the map from being run into the ground? Since many of the sites are represented by only one occurrence of a mutation, it is clear that there must still exist some sites with zero occurrences, so that the map as it stands is not saturated. From the statistics of the distribution it can be estimated that there must exist, in addition to some 350 sites now known, at least 100 sites not yet discovered. The calculation provides only a minimum estimate; the true number is probably larger. Therefore the map at the present time cannot be more than 78 per cent saturated.

Everything that we have learned about the genetic fine structure of T4 phage is compatible with the Watson-

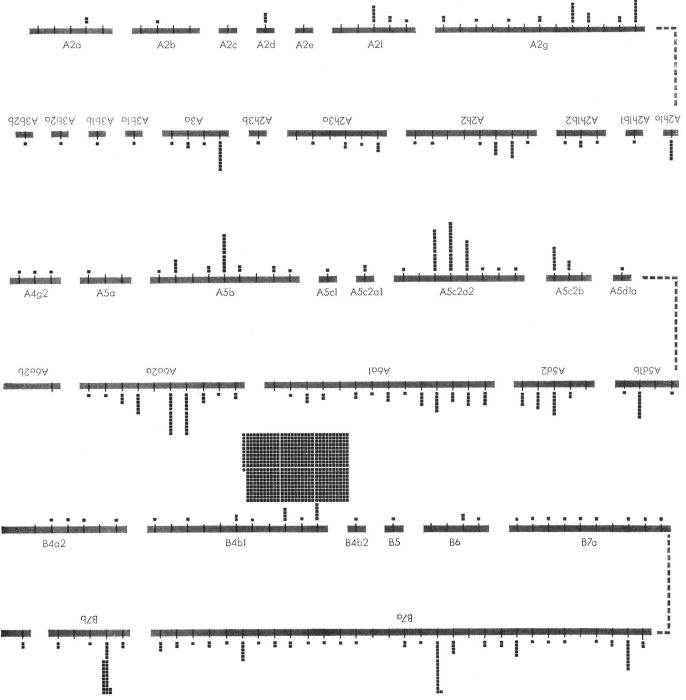

is represented by a small black square. Sites without squares are known to exist because they can be induced to mutate by use of chemical mutagens or ultraviolet light (*see illustration on next two pages*), but they have not been observed to mutate spontaneously.

172

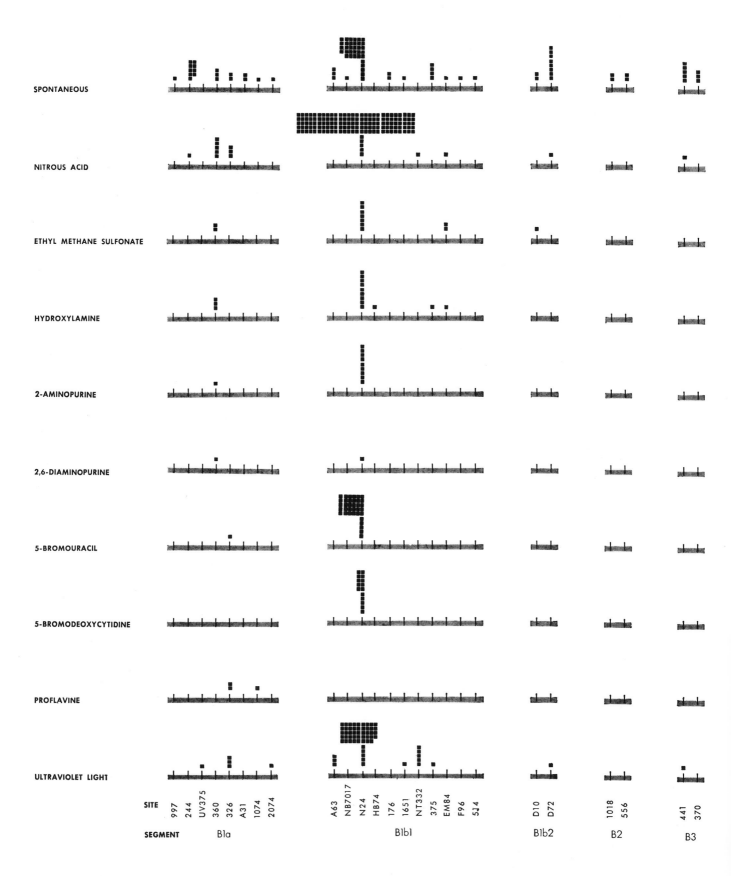

RESPONSE OF PHAGE TO MUTAGENS is shown for a portion of the B cistron. The total number of mutations studied is not the same for each mutagen. It is clear, nevertheless, that mutagenic action is highly specific at certain sites. For example, site EM26,

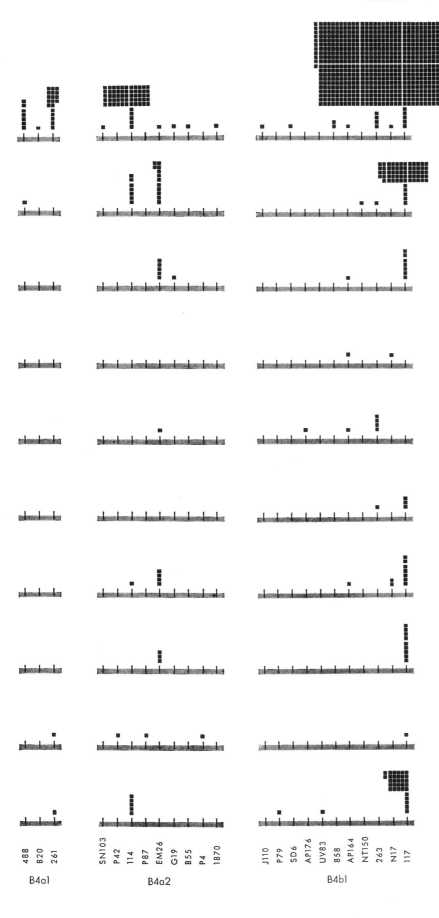

488 B20 261

B4a1

SN103 P42 114 P87 EM26 G19 B55 P4 1870

B4a2

J110 P79 SD6 AP176 UV83 858 AP164 NT150 263 N17 117

B4b1

Crick model of the DNA molecule. In this model the genetic information is contained in the specific order of bases arranged in a linear sequence. The four-letter language of the bases must somehow be translated into the 20-letter language of the amino acids, so that at least several base pairs must be required to specify one amino acid, and an entire polypeptide chain should be defined by a longer segment of DNA. Since the activity of the resulting enzyme, or other protein, depends on its precise structure, this activity should be impaired by any of a large number of changes in the DNA base sequence leading to amino acid substitutions.

One can also imagine that certain changes in base sequence can lead to a "nonsense" sequence that does not specify any amino acid, and that as a result the polypeptide chain cannot be completed. Thus the genetic unit of function should be vulnerable at many different points within a segment of the DNA structure. Considering the monotonous structure of the molecule, there is no obvious reason why recombination should not be possible at every link in the molecular chain, although not necessarily with the same probability. In short, the Watson-Crick model leads one to expect that the functional units— the genes of traditional genetics—should consist of linear segments that can be finely dissected by mutation and recombination.

Mapping Other Genes

The genetic results fully confirm these expectations. All mutations can in fact be represented in a strictly linear map, the functional units correspond to sharply defined segments, and each functional unit is divisible by mutation and recombination into hundreds of sites. Mutations are induced specifically at certain sites by agents that interact with the DNA bases. Although the data on mutation rates are complex, it is quite probable that they can be explained by interactions between groups of base pairs.

In confining this investigation to rII mutants of T4, attention has been focused on a tiny bit of hereditary material constituting only a few per cent of the genetic structure of a virus, enabling the exploration to be carried almost to the limits of the molecular structure. Similar results are being obtained in many other microorganisms and even in higher organisms such as corn. Given techniques for handling cells in culture in the appropriate way, man too may soon be a subject for genetic fine-structure analysis.

which resists spontaneous mutation, responds readily to certain mutagens. However, site 117 in segment B4b1 is more apt to mutate spontaneously than in response to a mutagen.

16

THE NUCLEOTIDE SEQUENCE OF A NUCLEIC ACID

ROBERT W. HOLLEY

February 1966

For the first time the specific order of subunits in one of the giant molecules that participate in the synthesis of protein has been determined. The task took seven years

Two major classes of chainlike molecules underlie the functioning of living organisms: the nucleic acids and the proteins. The former include deoxyribonucleic acid (DNA), which embodies the hereditary message of each organism, and ribonucleic acid (RNA), which helps to translate that message into the thousands of different proteins that activate the living cell. In the past dozen years biochemists have established the complete sequence of amino acid subunits in a number of different proteins. Much less is known about the nucleic acids.

Part of the reason for the slow progress with nucleic acids was the unavailability of pure material for analysis. Another factor was the large size of most nucleic acid molecules, which often contain thousands or even millions of nucleotide subunits. Several years ago, however, a family of small molecules was discovered among the ribonucleic

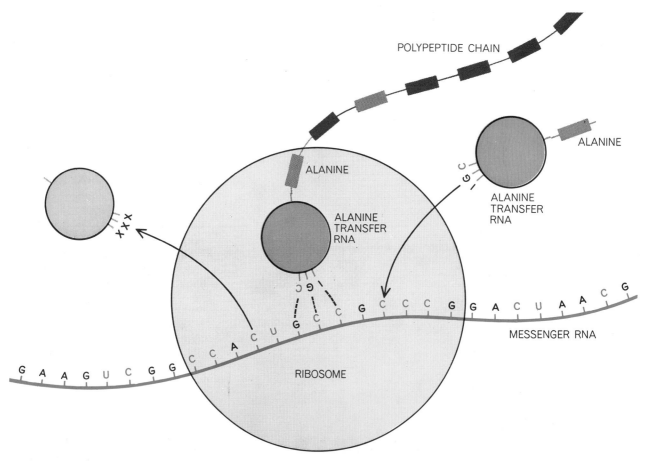

ROLE OF TRANSFER RNA is to deliver a specific amino acid to the site where "messenger" RNA and a ribosome (which also contains RNA) collaborate in the synthesis of a protein. As it is being synthesized a protein chain is usually described as a polypeptide. Each amino acid in the polypeptide chain is specified by a triplet code, or codon, in the molecular chain of messenger RNA. The diagram shows how an "anticodon" (presumably I—G—C) in alanine transfer RNA may form a temporary bond with the codon for alanine (G—C—C) in the messenger RNA. While so bonded the transfer RNA also holds the polypeptide chain. Each transfer RNA is succeeded by another one, carrying its own amino acid, until the complete message in the messenger RNA has been "read."

HYPOTHETICAL MODELS of alanine transfer ribonucleic acid (RNA) show three of the many ways in which the molecule's linear chain might be folded. The various letters represent nucleotide subunits; their chemical structure is given at the top of the next two pages. In these models it is assumed that certain nucleotides, such as C—G and A—U, will pair off and tend to form short double-strand regions. Such "base-pairing" is a characteristic feature of nucleic acids. The arrangement at the lower left shows how two of the large "leaves" of the "clover leaf" model may be folded together. The triplet I—G—C is the presumed anticodon shown in the illustration on the opposite page. The region containing the sequence G—T—Ψ—C—G may be common to all transfer RNA's.

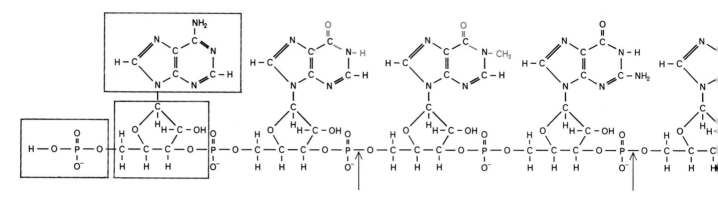

ADENYLIC ACID   INOSINIC ACID   1-METHYLINOSINIC ACID   GUANYLIC ACID   1-METHYLGUANYL

NUCLEOTIDE SUBUNITS found in alanine transfer RNA include the four commonly present in RNA (A, G, C, U), plus seven others that are variations of the standard structures. Ten of these 11 different nucleotide subunits are assembled above as if they were linked together in a single RNA chain. The chain begins at the left with a phosphate group (*outlined by a small rectangle*) and is followed by a ribose sugar group (*large rectangle*); the two groups alternate to form the backbone of the chain. The chain ends at the right with

acids. My associates and I at the U.S. Plant, Soil and Nutrition Laboratory and Cornell University set ourselves the task of establishing the nucleotide sequence of one of these smaller RNA molecules—a molecule containing fewer than 100 nucleotide subunits. This work culminated recently in the first determination of the complete nucleotide sequence of a nucleic acid.

The object of our study belongs to a family of 20-odd molecules known as transfer RNA's. Each is capable of recognizing one of the 20 different amino acids and of transferring it to the site where it can be incorporated into a growing polypeptide chain. When such a chain assumes its final configuration, sometimes joining with other chains, it is called a protein.

At each step in the process of protein

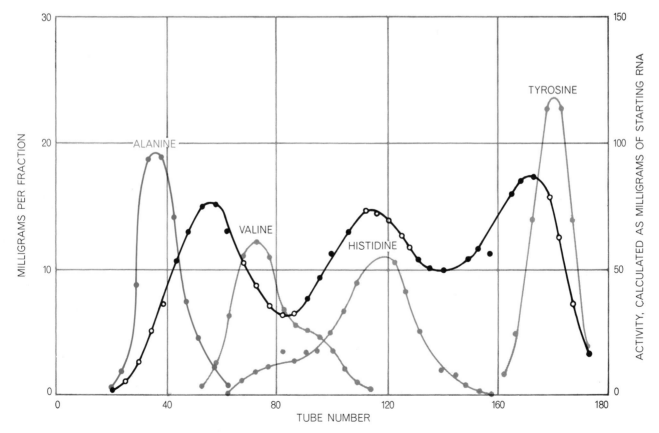

COUNTERCURRENT DISTRIBUTION PATTERN shows two steps in the separation of alanine transfer RNA, as carried out in the author's laboratory. After the first step the RNA content in various collection tubes, measured by ultraviolet absorption, follows the black curve. Biological activity, indicated by the amount of a given amino acid incorporated into polypeptide chains, follows the colored curves. Pure transfer RNA's of four types can be obtained by reprocessing the tubes designated by open circles.

ID N²-DIMETHYLGUANYLIC ACID — CYTIDYLIC ACID — URIDYLIC ACID — ▲RIBOTHYMIDYLIC ACID — DIHYDROURIDYLIC ACID — MIXTURE OF URIDYLIC AND DIHYDROURIDYLIC ACIDS — PSEUDOURIDYLIC ACID

a hydroxyl (OH) group. Each nucleotide subunit consists of a phosphate group, a ribose sugar group and a base. The base portion in the nucleotide at the far left, adenylic acid, is outlined by a large rectangle. In the succeeding bases the atomic variations are shown in color. The base structures without color are those commonly found in RNA. Black arrows show where RNA chains can be cleaved by the enzyme takadiastase ribonuclease T1. Colored arrows show where RNA chains can be cleaved by pancreatic ribonuclease.

synthesis a crucial role is played by the structure of the various RNA's. "Messenger" RNA transcribes the genetic message for each protein from its original storage site in DNA. Another kind of RNA—ribosomal RNA—forms part of the structure of the ribosome, which acts as a jig for holding the messenger RNA while the message is transcribed into a polypeptide chain [see illustration on page 174]. In view of the various roles played by RNA in protein synthesis, the structure of RNA molecules is of considerable interest and significance.

The particular nucleic acid we chose for study is known as alanine transfer RNA—the RNA that transports the amino acid alanine. It was isolated from commercial baker's yeast by methods I shall describe later. Preliminary analyses indicated that the alanine transfer RNA molecule consisted of a single chain of approximately 80 nucleotide subunits. Each nucleotide, in turn, consists of a ribose sugar, a phosphate group and a distinctive appendage termed a nitrogen base. The ribose sugars and phosphate groups link together to form the backbone of the molecule, from which the various bases protrude [see illustration at top of these two pages].

The problem of structural analysis is fundamentally one of identifying each base and determining its place in the sequence. In practice each base is usually isolated in combination with a unit of ribose sugar and a unit of phosphate, which together form a nucleotide. Formally the problem is analogous to de-

termining the sequence of letters in a sentence.

It would be convenient if there were a way to snip off the nucleotides one by one, starting at a known end of the chain and identifying each nucleotide as it appeared. Unfortunately procedures of this kind have such a small yield at each step that their use is limited. The alternative is to break the chain at particular chemical sites with the help of enzymes. This gives rise to small fragments whose nucleotide composition is amenable to analysis. If the chain can be broken up in various ways with different enzymes, one can determine how the fragments overlap and ultimately piece together the entire sequence.

One can visualize how this might work by imagining that the preceding sentence has been written out several times, in a continuous line, on different strips of paper. Imagine that each strip has been cut in a different way. In one case, for example, the first three words "If the chain" and the next three words "can be broken" might appear on separate strips of paper. In another case one might find that "chain" and "can" were together on a single strip. One would immediately conclude that the group of three words ending with "chain" and the group beginning with "can" form a continuous sequence of six words. The concept is simple; putting it into execution takes a little time.

For cleaving the RNA chain we used two principal enzymes: pancreatic ribonuclease and an enzyme called takadiastase ribonuclease T1, which was discovered by the Japanese workers K. Sato-Asano and F. Egami. The first

enzyme cleaves the RNA chain immediately to the right of pyrimidine nucleotides, as the molecular structure is conventionally written. Pyrimidine nucleotides are those nucleotides whose bases contain the six-member pyrimidine ring, consisting of four atoms of carbon and two atoms of nitrogen. The two pyrimidines commonly found in RNA are cytosine and uracil. Pancreatic ribonuclease therefore produces fragments that terminate in pyrimidine nucleotides such as cytidylic acid (C) or uridylic acid (U).

The second enzyme, ribonuclease T1, was employed separately to cleave the RNA chain specifically to the right of nucleotides containing a structure of the purine type, such as guanylic acid (G). This provided a set of short fragments distinctively different from those produced by the pancreatic enzyme.

The individual short fragments were isolated by passing them through a thin glass column packed with diethylaminoethyl cellulose—an adaptation of a chromatographic method devised by R. V. Tomlinson and G. M. Tener of the University of British Columbia. In general the short fragments migrate through the column more rapidly than the long fragments, but there are exceptions [see illustration on next page]. The conditions most favorable for this separation were developed in our laboratories by Mark Marquisee and Jean Apgar.

The nucleotides in each fragment were released by hydrolyzing the fragment with an alkali. The individual nucleotides could then be identified by paper chromatography, paper electrophoresis and spectrophotometric analysis. This procedure was sufficient to establish the sequence of each of the dinucleotides, because the right-hand member of the pair was determined by the particular enzyme that had been used to produce the fragment. To establish the sequence of nucleotides in larger fragments, however, required special techniques.

Methods particularly helpful in the separation and identification of the fragments had been previously described by Vernon M. Ingram of the Massachusetts Institute of Technology, M. Laskowski, Sr., of the Marquette University School of Medicine, K. K. Reddi of Rockefeller University, G. W. Rushizky and Herbert A. Sober of the National Institutes of Health, the Swiss worker M. Staehelin and Tener.

For certain of the largest fragments, methods described in the scientific literature were inadequate and we had to develop new stratagems. One of these involved the use of an enzyme (a phosphodiesterase) obtained from snake venom. This enzyme removes nucleotides one by one from a fragment, leaving a mixture of smaller fragments of all possible intermediate lengths. The mixture can then be separated into fractions of homogeneous length by passing it through a column of diethylaminoethyl cellulose [*see illustration on opposite page*]. A simple method is available for determining the terminal nucleotide at the right end of each fraction of homogeneous length. With this knowledge, and knowing the length of each fragment, one can establish the sequence of nucleotides in the original large fragment.

A summary of all the nucleotide sequences found in the fragments of transfer RNA produced by pancreatic ribonuclease is shown in Table 1 on page 180. Determination of the structure of the fragments was primarily the work of James T. Madison and Ada Zamir, who were postdoctoral fellows in my laboratory. George A. Everett of the Plant, Soil and Nutrition Laboratory helped us in the identification of the nucleotides.

Much effort was spent in determining the structure of the largest fragments and in identifying unusual nucleotides not heretofore observed in RNA molecules. Two of the most difficult to identify were 1-methylinosinic acid and 5,6-dihydrouridylic acid. (In the illustrations these are symbolized respectively by I^m and U^h.)

Because a free 5'-phosphate group (p) is found at one end of the RNA molecule (the left end as the structure is conventionally written) and a free 3'-hydroxyl group (OH) is found at the other end, it is easy to pick out from Table 1 and Table 2 the two sequences that form the left and right ends of the alanine transfer RNA molecule. The left end has the structure pG—G—G—C— and the right end the structure U—C—C—A—C—COH. (It is known, however, that the active molecule ends in C—C—AOH.)

The presence of unusual nucleotides

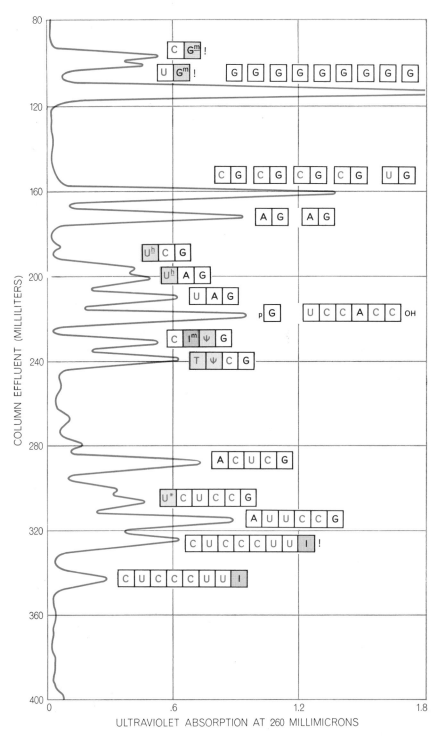

SEPARATION OF RNA FRAGMENTS is accomplished by chromatography carried out in a long glass column packed with diethylaminoethyl cellulose. The curve shows the separation achieved when the column input is a digest of alanine transfer RNA produced by takadiastase ribonuclease T1, an enzyme that cleaves the RNA into 29 fragments. The exclamation point indicates fragments whose terminal phosphate has a cyclical configuration. Such fragments travel faster than similar fragments that end in a noncyclical phosphate.

and unique short sequences made it clear that certain of the fragments found in Table 1 overlapped fragments found in Table 2. For example, there is only one inosinic acid nucleotide (I) in the molecule, and this appears in the sequence I–G–C– in Table 1 and in the sequence C–U–C–C–C–U–U–I– in Table 2. These two sequences must therefore overlap to produce the overall sequence C–U–C–C–C–U–U–I–G–C–. The information in Table 1 and Table 2 was combined in this way to draw up Table 3, which accounts for all 77 nucleotides in 16 sequences [*see illustration on page 181*].

With the knowledge that two of the 16 sequences were at the two ends, the structural problem became one of determining the positions of the intermediate 14 sequences. This was accomplished by isolating still larger fragments of the RNA.

In a crucial experiment John Robert Penswick, a graduate student at Cornell, found that a very brief treatment of the RNA with ribonuclease T1 at 0 degrees centigrade in the presence of magnesium ions splits the molecule at one position. The two halves of the molecule could be separated by chromatography. Analyses of the halves established that the sequences listed in the first column of Table 3 are in the left half of the molecule and that those in the second column are in the right half.

Using a somewhat more vigorous but still limited treatment of the RNA with ribonuclease T1, we then obtained and analyzed a number of additional large fragments. This work was done in collaboration with Jean Apgar and Everett. To determine the structure of a large fragment, the fragment was degraded completely with ribonuclease T1, which yielded two or more of the fragments previously identified in Table 2. These known sequences could be put together, with the help of various clues, to obtain the complete sequence of the large fragment. The process is similar to putting together a jigsaw puzzle [*see illustrations on pages 182 and 183*].

As an example of the approach that was used, the logical argument is given in detail for Fragment *A*. When Fragment *A* was completely degraded by ribonuclease T1, we obtained seven small fragments: three G–'s, C–G–, U–G–, U–G^m– and pG–. (G^m is used in the illustrations to represent 1-methylguanylic acid, another of the unusual nucleotides in alanine transfer RNA.) The presence of pG– shows that Frag-

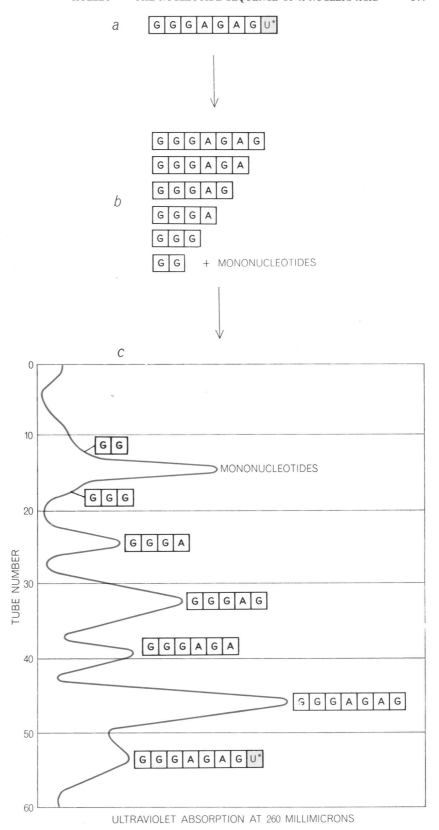

NEW DEGRADATION METHOD was developed in the author's laboratory to determine the sequence of nucleotides in fragments five to eight subunits in length. The example above begins with a fragment of eight subunits from which the terminal phosphate has been removed (*a*). When the fragment is treated with phosphodiesterase found in snake venom, the result is a mixture containing fragments from one to eight subunits in length (*b*). These are separated by chromatography (*c*). When the material from each peak is hydrolyzed, the last nucleoside (a nucleotide minus its phosphate) at the right end of the fragment is released and can be identified. Thus each nucleotide in the original fragment can be determined.

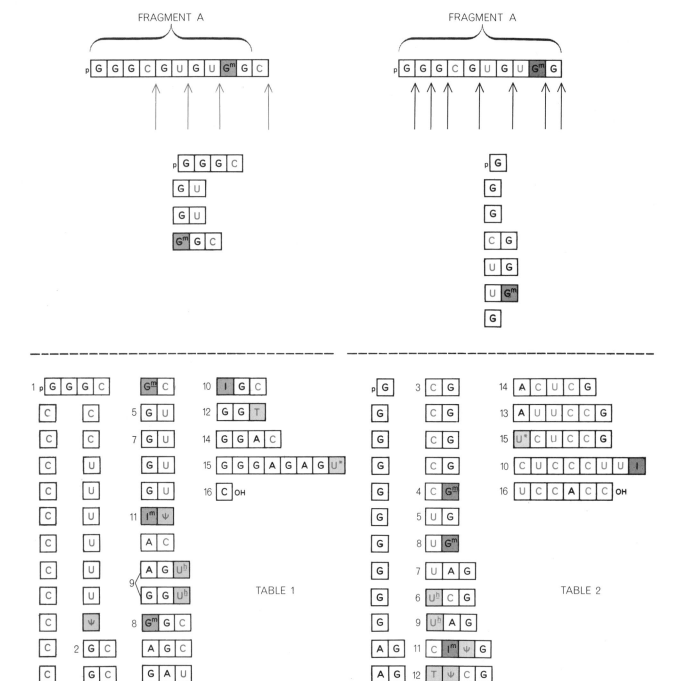

ACTION OF TWO DIFFERENT ENZYMES is reflected in these two tables. Table 1 shows the fragments produced when alanine transfer RNA is completely digested by pancreatic ribonuclease, which cleaves the molecule to the right of nucleotides containing bases with pyrimidine structures (C, U, Uʰ, ψ and T). The diagram at top left shows how pancreatic ribonuclease would cleave the first

11 nucleotides of alanine transfer RNA. The diagram at top right shows how the same region would be digested by takadiastase ribonuclease T1. Table 2 contains the fragments produced by this enzyme; they all end in nucleotides whose bases contain purine structures (G, Gᵐ, G*m* and I). The numbers indicate which ones appear in the consolidated list in Table 3 on the opposite page.

ment *A* is from the left end of the molecule. Since it is already known from Table 3 that the left terminal sequence is pG–G–G–C–, the positions of two of the three G–'s and C–G– are known; the terminal five nucleotides must be pG–G–G–C–G–.

The positions of the remaining G–, U–G– and U–Gᵐ– are established by the following information. Table 3

shows that the U–Gᵐ– is present in the sequence U–Gᵐ–G–C–. Since there is only one C in Fragment *A*, and its position is already known, Fragment *A* must terminate before the C of the U–Gᵐ–G–C– sequence. Therefore the U–G– must be to the left of the U–Gᵐ–, and the structure of Fragment *A* can be represented as pG–G–G–C–G–...U–G–...U–Gᵐ–, with one G–

remaining to be placed. If the G– is placed to the left or the right of the U–G– in this structure, it would create a G–G–U– sequence. If such a sequence existed in the molecule, it would have appeared as a fragment when the molecule was treated with pancreatic ribonuclease; Table 1 shows that it did not do so. Therefore the remaining G– must be to the right of the Gᵐ–, and the sequence of Fragment A is pG—G—G—C—G—U—G—U—G^m—G—.

Using the same procedure, the entire structure of alanine transfer RNA was worked out. The complete nucleotide sequence of alanine transfer RNA is shown at the top of the next two pages.

The work on the structure of this molecule took us seven years from start to finish. Most of the time was consumed in developing procedures for the isolation of a single species of transfer RNA from the 20 or so different transfer RNA's present in the living cell. We finally selected a fractionation technique known as countercurrent distribution, developed in the 1940's by Lyman C. Craig of the Rockefeller Institute.

This method exploits the fact that similar molecules of different structure will exhibit slightly different solubilities if they are allowed to partition, or distribute themselves, between two nonmiscible liquids. The countercurrent technique can be mechanized so that the mixture of molecules is partitioned hundreds or thousands of times, while the nonmiscible solvents flow past each other in a countercurrent pattern. The solvent system we adopted was composed of formamide, isopropyl alcohol and a phosphate buffer, a modification of a system first described by Robert C. Warner and Paya Vaimberg of New York University. To make the method applicable for fractionating transfer RNA's required four years of work in collaboration with Jean Apgar, B. P. Doctor and Susan H. Merrill of the Plant, Soil and Nutrition Laboratory. Repeated countercurrent extractions of the transfer RNA mixture gave three of the RNA's in a reasonably homogeneous state: the RNA's that transfer the amino acids alanine, tyrosine and valine [see bottom illustration on page 176].

The starting material for the countercurrent distributions was crude transfer RNA extracted from yeast cells using phenol as a solvent. In the course of the structural work we used about 200 grams (slightly less than half a pound) of mixed transfer RNA's isolated from 300 pounds of yeast. The total amount of purified alanine transfer RNA we had to work with over a three-year period was one gram. This represented a practical compromise between the difficulty of scaling up the fractionation procedures and scaling down the techniques for structural analysis.

Once we knew the complete sequence, we could turn to general questions about the structure of transfer RNA's. Each transfer RNA presumably embodies a sequence of three subunits (an "anticodon") that forms a temporary bond with a complementary sequence of three subunits (the "codon") in messenger RNA. Each codon triplet identifies a specific amino acid [see "The Genetic Code: II," by Marshall W. Nirenberg; SCIENTIFIC AMERICAN Offprint 153].

An important question, therefore, is which of the triplets in alanine transfer RNA might serve as the anticodon for the alanine codon in messenger RNA. There is reason to believe the anticodon is the sequence I—G—C, which is found in the middle of the RNA molecule. The codon corresponding to I—G—C could be the triplet G—C—C or perhaps G—C—U, both of which act as code words for alanine in messenger RNA. As shown in the illustration on page 174, the I—G—C in the alanine transfer RNA is upside down when it makes contact with the corresponding codon in messenger RNA. Therefore when alanine transfer RNA is delivering its amino acid cargo and is temporarily held by hydrogen bonds to messenger RNA, the I would pair with C (or U) in the messenger, G would pair with C, and C would pair with G.

We do not know the three-dimensional structure of the RNA. Presumably there is a specific form that interacts with the messenger RNA and ribosomes. The illustration on page 175 shows three hypothetical structures for alanine transfer RNA that take account of the propensity of certain bases to pair with other bases. Thus adenine pairs with uracil and cytosine with guanine. In the three hypothetical structures the I—G—C sequence is at an exposed position and could pair with messenger RNA.

The small diagram on page 175 indicates a possible three-dimensional folding of the RNA. Studies with atomic models suggest that single-strand regions of the structure are highly flexible. Thus in the "three-leaf-clover" configuration it is possible to fold one side leaf on top of the other, or any of the leaves back over the stem of the molecule.

One would also like to know whether or not the unusual nucleotides are concentrated in some particular region of the molecule. A glance at the sequence shows that they are scattered throughout the structure; in the three-leaf-clo-

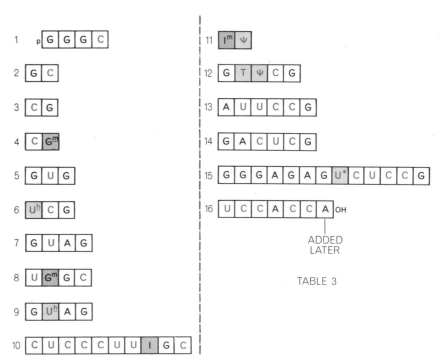

CONSOLIDATED LIST OF SEQUENCES accounts for all 77 nucleotides in alanine transfer RNA. The consolidated list is formed by selecting the largest fragments in Table 1 and Table 2 (opposite page) and by piecing together fragments that obviously overlap. Thus Fragment 15 has been formed by joining two smaller fragments, keyed by the number 15, in Table 1 and Table 2 on the opposite page. Since the entire molecule contains only one U*, the two fragments must overlap at that point. The origin of the other fragments in Table 3 can be traced in similar fashion. A separate experiment in which the molecule was cut into two parts helped to establish that the 10 fragments listed in the first column are in the left half of the molecule and that the six fragments in the second column are in the right half.

FRAGMENT A

FRAGMENT B

FRAGMENT E

COMPLETE MOLECULE of alanine transfer RNA contains 77 nucleotides in the order shown. The final sequence required a care-ful piecing together of many bits of information (*see illustration at bottom of these two pages*). The task was facilitated by degrada-

ver model, however, the unusual nucleotides are seen to be concentrated around the loops and bends.

Another question concerns the presence in the transfer RNA's of binding sites, that is, sites that may interact specifically with ribosomes and with the enzymes involved in protein synthesis. We now know from the work of Zamir and Marquisee that a particular sequence containing pseudouridylic acid (Ψ), the sequence G–T–Ψ–C–G, is found not only in the alanine transfer RNA but also in the transfer RNA's for tyrosine and valine. Other studies suggest that it may be present in all the transfer RNA's. One would expect such common sites to serve a common function; binding the transfer RNA's to the ribosome might be one of them.

Work that is being done in many

FRAGMENT A

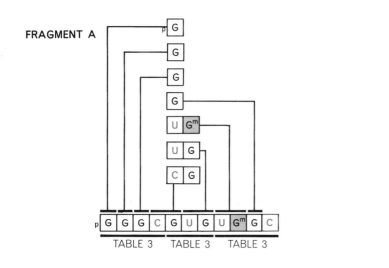

FRAGMENT B

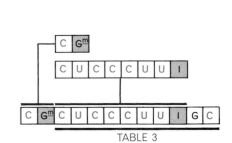

FRAGMENT E

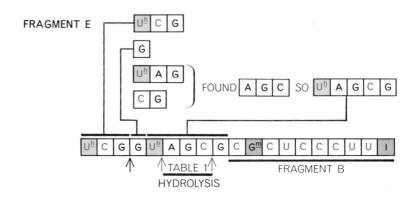

REMAINDER OF LEFT HALF OF MOLECULE

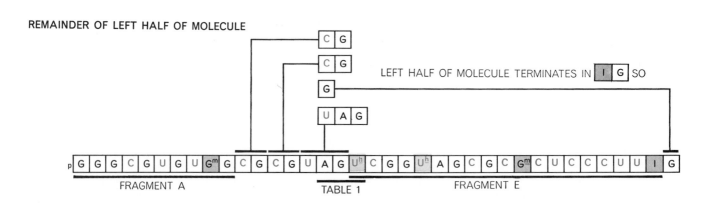

tion experiments that cleaved this molecule into several large fragments (*A, B, C, D, E, F, G*), and by the crucial discovery that

the molecule could be divided almost precisely into two halves. The division point is marked by the "gutter" between these two pages.

laboratories around the world indicates that alanine transfer RNA is only the first of many nucleic acids for which the nucleotide sequences will be known. In the near future it should be possible to identify those structural features that are common to various transfer RNA's,

and this should help greatly in defining the interactions of transfer RNA's with messenger RNA, ribosomes and enzymes involved in protein synthesis. Further in the future will be the description of the nucleotide sequences of the nucleic acids—both DNA and RNA—

that embody the genetic messages of the viruses that infect bacteria, plants and animals. Much further in the future lies the decoding of the genetic messages of higher organisms, including man. The work described in this article is a step toward that distant goal.

FRAGMENT C

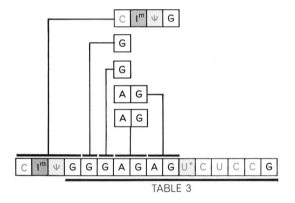

TABLE 3

FRAGMENT D

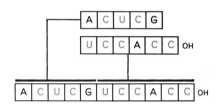

FRAGMENT F

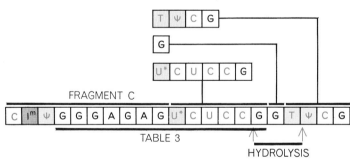

FRAGMENT G

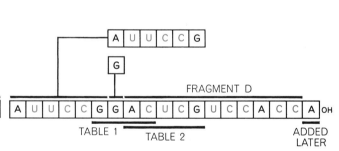

ASSEMBLY OF FRAGMENTS resembled the solving of a jigsaw puzzle. The arguments that established the sequence of nucleotides in Fragment *A* are described in the text. Fragment *B* contains two subfragments. The larger is evidently Fragment 10 in Table 3, which ends in G—C—. This means that the C—Gᵐ— fragment must go to the left. Fragment *E* contains Fragment *B* plus four smaller fragments. It can be shown that *E* ends with I—, therefore the four small pieces are again to the left. A pancreatic digest yielded A—G—C—, thus serving to connect Uʰ—A—G— and C—G—. A partial digestion with ribonuclease T1 removed Uʰ—C—G—, showing it to be at the far left. The remaining G— must follow immediately or a pancreatic digest would have yielded a G—G—C— sequence, which it did not. Analyses of Fragments *A* and *E* accounted for everything in the left half of the molecule except for four small pieces. The left half of the molecule was shown to terminate in I—G—, thus the remaining three pieces are between *A* and *E*. Table 1 shows that one Uʰ is preceded by A—G—, therefore U—

A—G— must be next to *E*. The two remaining C—G—'s must then fall to the left of U—A—G—. Fragment *C* contains five pieces. Table 3 (Fragment 15) shows that the two A—G—'s are next to U* and that the two G—'s are to the left of them. It is also clear that C—Iᵐ—Ψ—G— cannot follow U*, therefore it must be to the left. Fragment *D* contains two pieces; the OH group on one of them shows it to be to the right. Fragment *F* contains Fragment *C* plus three extra pieces. These must all lie to the right since hydrolysis with pancreatic ribonuclease gave G—G—T— and not G—T—, thus establishing that the single G— falls as shown. Fragment *G* gave *D* plus two pieces, which must both lie to the left (because of the terminal Cₒₕ). Table 1 shows a G—G—A—C— sequence, which must overlap the A—C— in A—C—U—C—G— and the G— at the right end of the A—U—U—C—C—G—. Fragments *F* and *G* can join in only one way to form the right half of the molecule. The molecule is completed by the addition of a final Aₒₕ, which is missing as the alanine transfer RNA is separated from baker's yeast.

17

THE GENETIC CODE

F. H. C. CRICK

October 1962

*How does the order of bases in a nucleic acid
determine the order of amino acids in a protein? It
seems that each amino acid is specified by a triplet of
bases, and that triplets are read in simple sequence*

Within the past year important progress has been made in solving the "coding problem." To the biologist this is the problem of how the information carried in the genes of an organism determines the structure of proteins.

Proteins are made from 20 different kinds of small molecule—the amino acids—strung together into long polypeptide chains. Proteins often contain several hundred amino acid units linked together, and in each protein the links are arranged in a specific order that is genetically determined. A protein is therefore like a long sentence in a written language that has 20 letters.

Genes are made of quite different long-chain molecules: the nucleic acids DNA (deoxyribonucleic acid) and, in some small viruses, the closely related RNA (ribonucleic acid). It has recently been found that a special form of RNA, called messenger RNA, carries the genetic message from the gene, which is located in the nucleus of the cell, to the surrounding cytoplasm, where many of the proteins are synthesized [see "Messenger RNA," by Jerard Hurwitz and J. J. Furth; SCIENTIFIC AMERICAN Offprint 119].

The nucleic acids are made by joining up four kinds of nucleotide to form a polynucleotide chain. The chain provides a backbone from which four kinds of side group, known as bases, jut at regular intervals. The order of the bases, however, is not regular, and it is their precise sequence that is believed to carry the genetic message. The coding problem can thus be stated more explicitly as the problem of how the sequence of the four bases in the nucleic acid determines the sequence of the 20 amino acids in the protein.

The problem has two major aspects, one general and one specific. Specifically one would like to know just what sequence of bases codes for each amino acid. Remarkable progress toward this goal was reported early this year by Marshall W. Nirenberg and J. Heinrich Matthaei of the National Institutes of Health and by Severo Ochoa and his colleagues at the New York University School of Medicine. [Editor's note: Brief accounts of this work appeared in "Science and the Citizen" for February and March. This article is a companion to one by Nirenberg, which deals with the biochemical aspects of the genetic code].

The more general aspect of the coding problem, which will be my subject, has to do with the length of the genetic coding units, the way they are arranged in the DNA molecule and the way in which the message is read out. The experiments I shall report were performed at the Medical Research Council Laboratory of Molecular Biology in Cambridge, England. My colleagues were Mrs. Leslie Barnett, Sydney Brenner, Richard J. Watts-Tobin and, more recently, Robert Shulman.

The organism used in our work is the bacteriophage T4, a virus that infects the colon bacillus and subverts the biochemical machinery of the bacillus to make multiple copies of itself. The infective process starts when T4 injects its genetic core, consisting of a long strand of DNA, into the bacillus. In less than 20 minutes the virus DNA causes the manufacture of 100 or so copies of the complete virus particle, consisting of a DNA core and a shell containing at least six distinct protein components. In the process the bacillus is killed and the virus particles spill out. The great value of the T4 virus for genetic experiments is that many generations and billions of individuals can be produced in a short time. Colonies containing mutant individuals can be detected by the appearance of the small circular "plaques" they form on culture plates. Moreover, by the use of suitable cultures it is possible to select a single individual of interest from a population of a billion.

Using the same general technique, Seymour Benzer of Purdue University was able to explore the fine structure of the A and B genes (or cistrons, as he prefers to call them) found at the "rII" locus of the DNA molecule of T4 [see the article "The Fine Structure of the Gene," by Seymour Benzer, beginning on page 160]. He showed that the A and B genes, which are next to each other on the virus chromosome, each consist of some hundreds of distinct sites arranged in linear order. This is exactly what one would expect if each gene is a segment, say 500 or 1,000 bases long, of the very long DNA molecule that forms the virus chromosome [see illustration on opposite page]. The entire DNA molecule in T4 contains about 200,000 base pairs.

The Usefulness of Mutations

From the work of Benzer and others we know that certain mutations in the A and B region made one or both genes inactive, whereas other mutations were only partially inactivating. It had also been observed that certain mutations were able to suppress the effect of harmful mutations, thereby restoring the function of one or both genes. We suspected that the various—and often puzzling—consequences of different kinds of mutation might provide a key to the nature of the genetic code.

We therefore set out to re-examine the effects of crossing T4 viruses bearing mutations at various sites. By growing two different viruses together in a common culture one can obtain "recombinants" that have some of the properties

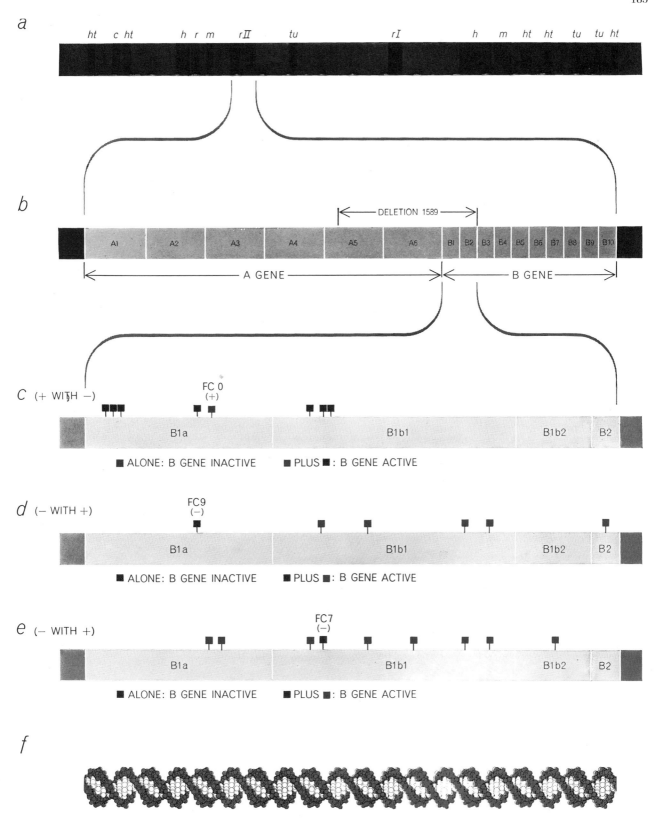

rII REGION OF THE T4 VIRUS represents only a few per cent of the DNA (deoxyribonucleic acid) molecule that carries full instructions for creating the virus. The region consists of two genes, here called A and B. The A gene has been mapped into six major segments, the B gene into 10 (*b*). The experiments reported in this article involve mutations in the first and second segments of the B genes. The B gene is inactivated by any mutation that adds a molecular subunit called a base (*colored square*) or removes one (*black square*). But activity is restored by simultaneous addition and removal of a base, as shown in *c*, *d* and *e*. An explanation for this recovery of activity is illustrated on page 188. The molecular representation of DNA (*f*) is estimated to be approximately in scale with the length of the B1 and B2 segments of the B gene. The two segments contain about 100 base pairs.

of one parent and some of the other. Thus one defect, such as the alteration of a base at a particular point, can be combined with a defect at another point to produce a phage with both defects [*see upper illustration below*]. Alternatively, if a phage has several defects, they can be separated by being crossed

with the "wild" type, which by definition has none. In short, by genetic methods one can either combine or separate different mutations, provided that they do not overlap.

Most of the defects we shall be considering are evidently the result of adding or deleting one base or a small group

of bases in the DNA molecule and not merely the result of altering one of the bases [*see lower illustration on this page*]. Such additions and deletions can be produced in a random manner with the compounds called acridines, by a process that is not clearly understood. We think they are very small additions or deletions, because the altered gene seems to have lost its function completely; mutations produced by reagents capable of changing one base into another are often partly functional. Moreover, the acridine mutations cannot be reversed by such reagents (and vice versa). But our strongest reason for believing they are additions or deletions is that they can be combined in a way that suggests they have this character.

To understand this we shall have to go back to the genetic code. The simplest sort of code would be one in which a small group of bases stands for one particular acid. This group can scarcely be a pair, since this would yield only 4×4, or 16, possibilities, and at least 20 are needed. More likely the shortest code group is a triplet, which would provide $4 \times 4 \times 4$, or 64, possibilities. A small group of bases that codes one amino acid has recently been named a codon.

The first definite coding scheme to be proposed was put forward eight years ago by the physicist George Gamow, now at the University of Colorado. In this code adjacent codons overlap as illustrated on the following page. One consequence of such a code is that only certain amino acids can follow others. Another consequence is that a change in a single base leads to a change in three adjacent amino acids. Evidence gathered since Gamow advanced his ideas makes an overlapping code appear unlikely. In the first place there seems to be no restriction of amino acid sequence in any of the proteins so far examined. It has also been shown that typical mutations change only a single amino acid in the polypeptide chain of a protein. Although it is theoretically possible that the genetic code may be partly overlapping, it is more likely that adjacent codons do not overlap at all.

Since the backbone of the DNA molecule is completely regular, there is nothing to mark the code off into groups of three bases, or into groups of any other size. To solve this difficulty various ingenious solutions have been proposed. It was thought, for example, that the code might be designed in such a way that if the wrong set of triplets were chosen, the message would always be complete nonsense and no protein would

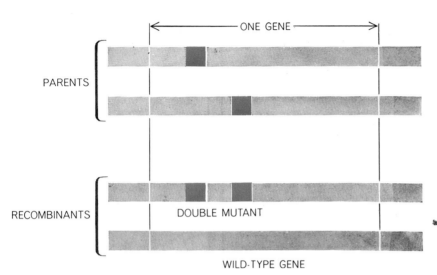

GENETIC RECOMBINATION provides the means for studying mutations. Colored squares represent mutations in the chromosome (DNA molecule) of the T4 virus. Through genetic recombination, the progeny can inherit the defects of both parents or of neither.

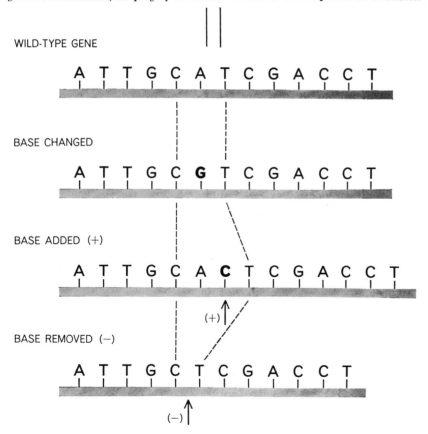

TWO CLASSES OF MUTATION result from introducing defects in the sequence of bases (A, T, G, C) that are attached to the backbone of the DNA molecule. In one class a base is simply changed from one into another, as A into G. In the second class a base is added or removed. Four bases are adenine (A), thymine (T), guanine (G) and cytosine (C).

be produced. But it now looks as if the most obvious solution is the correct one. That is, the message begins at a fixed starting point, probably one end of the gene, and is simply read three bases at a time. Notice that if the reading started at the wrong point, the message would fall into the wrong sets of three and would then be hopelessly incorrect. In fact, it is easy to see that while there is only one correct reading for a triplet code, there are two incorrect ones.

If this idea were right, it would immediately explain why the addition or the deletion of a base in most parts of the gene would make the gene completely nonfunctional, since the reading of the genetic message from that point onward would be totally wrong. Now, although our single mutations were always without function, we found that if we put certain pairs of them together, the gene would work. (In point of fact we picked up many of our functioning double mutations by starting with a nonfunctioning mutation and selecting for the rare second mutation that restored gene activity, but this does not affect our argument.) This enabled us to classify all our mutations as being either plus or minus. We found that by using the following rules we could always predict the behavior of any pair we put together in the same gene. First, if plus is combined with plus, the combination is nonfunctional. Second, if minus is combined with minus, the result is nonfunctional. Third, if plus is combined with minus, the combination is nonfunctional if the pair is too widely separated and functional if the pair is close together.

The interesting case is the last one. We could produce a gene that functioned, at least to some extent, if we combined a plus mutation with a minus mutation, provided that they were not too far apart.

To make it easier to follow, let us assume that the mutations we called plus really had an extra base at some point and that those we called minus had lost a base. (Proving this to be the case is rather difficult.) One can see that, starting from one end, the message would be read correctly until the extra base was reached; then the reading would get out of phase and the message would be wrong until the missing base was reached, after which the message would come back into phase again. Thus the genetic message would not be wrong over a long stretch but only over the short distance between the plus and the minus. By the same sort of argument one can see that for a triplet code the combination plus with plus or minus with

minus should never work [*see illustration on following page*].

We were fortunate to do most of our work with mutations at the left-hand end of the B gene of the *r*II region. It appears that the function of this part of the gene may not be too important, so that it may not matter if part of the genetic message in the region is incorrect. Even so, if the plus and minus are too far apart, the combination will not work.

Nonsense Triplets

To understand this we must go back once again to the code. There are 64 possible triplets but only 20 amino acids to be coded. Conceivably two or more triplets may stand for each amino acid. On the other hand, it is reasonable to expect that at least one or two triplets may not represent an amino acid at all but have some other meaning, such as "Begin here" or "End here." Although such hypothetical triplets may have a meaning of some sort, they have been named nonsense triplets. We surmised that sometimes the misreading produced in the region lying between a plus and a minus mutation might by chance give rise to a nonsense triplet, in which case the gene might not work.

We investigated a number of plus-with-minus combinations in which the distance between plus and minus was relatively short and found that certain combinations were indeed inactive when we might have expected them to function. Presumably an intervening nonsense triplet was to blame. We also found cases in which a plus followed by a minus worked but a minus followed by a plus did not, even though the two mutations appeared to be at the same sites, although in reverse sequence. As I have indicated, there are two wrong ways to read a message; one arises if the plus is to the left of the minus, the other if the plus is to the right of the minus. In cases where plus with minus gave rise to an active gene but minus with plus did not, even when the mutations evidently occupied the same pairs of sites, we concluded that the intervening misreading produced a nonsense triplet in one case but not in the other. In confirmation of this hypothesis we have been able to modify such nonsense triplets by mutagens that turn one base into another, and we have thereby restored the gene's activity. At the same time we have been able to locate the position of the nonsense triplet.

Recently we have undertaken one

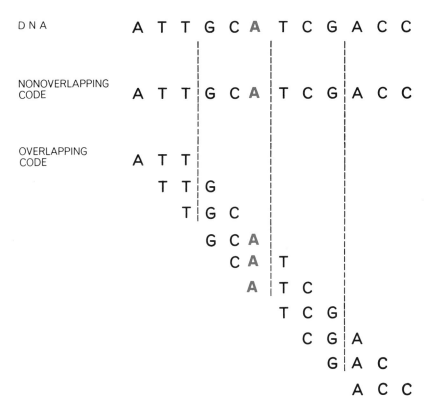

PROPOSED CODING SCHEMES show how the sequence of bases in DNA can be read. In a nonoverlapping code, which is favored by the author, code groups are read in simple sequence. In one type of overlapping code each base appears in three successive groups.

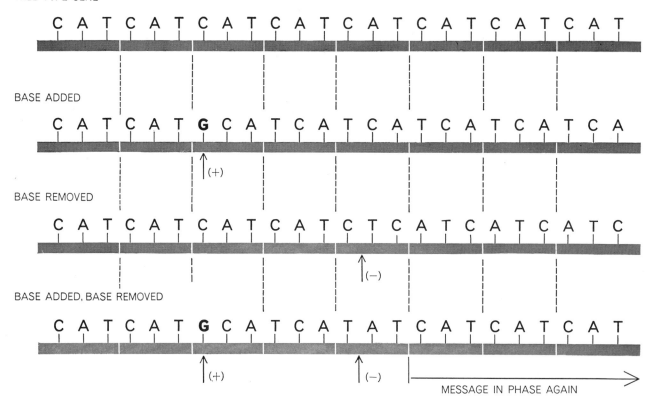

EFFECT OF MUTATIONS that add or remove a base is to shift the reading of the genetic message, assuming that the reading begins at the left-hand end of the gene. The hypothetical message in the wild-type gene is CAT, CAT ... Adding a base shifts the reading to TCA, TCA ... Removing a base makes it ATC, ATC ... Addition and removal of a base puts the message in phase again.

other rather amusing experiment. If a single base were changed in the left-hand end of the B gene, we would expect the gene to remain active, both because this end of the gene seems to be unessential and because the reading of the rest of the message is not shifted. In fact, if the B gene remained active, we would have no way of knowing that a base had been changed. In a few cases, however, we have been able to destroy the activity of the B gene by a base change traceable to the left-hand end of the gene. Presumably the change creates a nonsense triplet. We reasoned that if we could shift the reading so that the message was read in different groups of three, the new reading might not yield a nonsense triplet. We therefore selected a minus and a plus that together allowed the B gene to function, and that were on each side of the presumed nonsense mutation. Sure enough, this combination of three mutants allowed the gene to function [see top illustration on page 190]. In other words, we could abolish the effect of a nonsense triplet by shifting its reading.

All this suggests that the message is read from a fixed point, probably from one end. Here the question arises of how one gene ends and another begins,

since in our picture there is nothing on the backbone of the long DNA molecule to separate them. Yet the two genes A and B are quite distinct. It is possible to measure their function separately, and Benzer has shown that no matter what mutation is put into the A gene, the B function is not affected, provided that the mutation is wholly within the A gene. In the same way changes in the B gene do not affect the function of the A gene.

The Space between the Genes

It therefore seems reasonable to imagine that there is something about the DNA between the two genes that isolates them from each other. This idea can be tested by experiments with a mutant T4 in which part of the rII region is deleted. The mutant, known as T4 1589, has lost a large part of the right end of the A gene and a smaller part of the left end of the B gene. Surprisingly the B gene still shows some function; in fact this is why we believe this part of the B gene is not too important.

Although we describe this mutation as a deletion, since genetic mapping shows that a large piece of the genetic

information in the region is missing, it does not mean that physically there is a gap. It seems more likely that DNA is all one piece but that a stretch of it has been left out. It is only by comparing it with the complete version—the wild type —that one can see a piece of the message is missing.

We have argued that there must be a small region between the genes that separates them. Consequently one would predict that if this segment of the DNA were missing, the two genes would necessarily be joined. It turns out that it is quite easy to test this prediction, since by genetic methods one can construct double mutants. We therefore combined one of our acridine mutations, which in this case was near the beginning of the A gene, with the deletion 1589. Without the deletion present the acridine mutation had no effect on the B function, which showed that the genes were indeed separate. But when 1589 was there as well, the B function was completely destroyed [see top illustration on page 189]. When the genes were joined, a change far away in the A gene knocked out the B gene completely. This strongly suggests that the reading proceeds from one end.

We tried other mutations in the A

gene combined with 1589. All the acridine mutations we tried knocked out the B function, whether they were plus or minus, but a pair of them (plus with minus) still allowed the B gene to work. On the other hand, in the case of the other type of mutation (which we believe is due to the change of a base and not to one being added or subtracted) about half of the mutations allowed the B gene to work and the other half did not. We surmise that the latter are nonsense mutations, and in fact Benzer has recently been using this test as a definition of nonsense.

Of course, we do not know exactly what is happening in biochemical terms. What we suspect is that the two genes, instead of producing two separate pieces of messenger RNA, produce a single piece, and that this in turn produces a protein with a long polypeptide chain, one end of which has the amino acid sequence of part of the presumed A protein and the other end of which has most of the B protein sequence—enough to give some B function to the combined molecule although the A function has been lost. The concept is illustrated schematically at the bottom of this page. Eventually it should be possible to check the prediction experimentally.

How the Message Is Read

So far all the evidence has fitted very well into the general idea that the message is read off in groups of three, starting at one end. We should have got the same results, however, if the message had been read off in groups of four, or indeed in groups of any larger size. To test this we put not just two of our acridine mutations into one gene but three of them. In particular we put in three with the same sign, such as plus with plus with plus, and we put them fairly close together. Taken either singly or in pairs, these mutations will destroy the function of the B gene. But when all three are placed in the same gene, the B function reappears. This is clearly a remarkable result: two blacks will not make a white but three will. Moreover, we have obtained the same result with several different combinations of this type and with several of the type minus with minus with minus.

The explanation, in terms of the ideas described here, is obvious. One plus will put the reading out of phase. A second plus will give the other wrong reading. But if the code is a triplet code, a third plus will bring the message back into phase again, and from then on to the end it will be read correctly. Only between

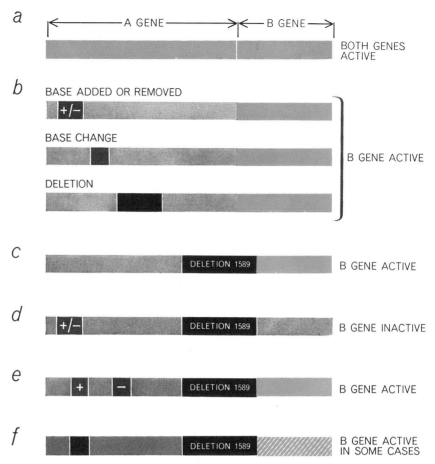

DELETION JOINING TWO GENES makes the B gene vulnerable to mutations in the A gene. The messages in two wild-type genes (*a*) are read independently, beginning at the left end of each gene. Regardless of the kind of mutation in A, the B gene remains active (*b*). The deletion known as 1589 inactivates the A gene but leaves the B gene active (*c*). But now alterations in the A gene will often inactivate the B gene, showing that the two genes have been joined in some way and are read as if they were a single gene (*d, e, f*).

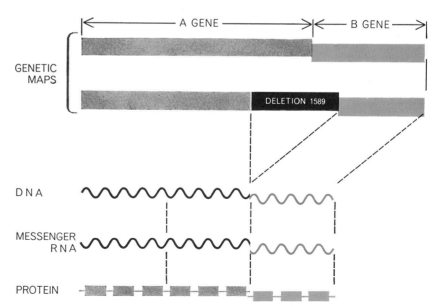

PROBABLE EFFECT OF DELETION 1589 is to produce a mixed protein with little or no A-gene activity but substantial B activity. Although the conventional genetic map shows the deletion as a gap, the DNA molecule itself is presumably continuous but shortened. In virus replication the genetic message in DNA is transcribed into a molecule of ribonucleic acid, called messenger RNA. This molecule carries the message to cellular particles known as ribosomes, where protein is synthesized, following instructions coded in the DNA.

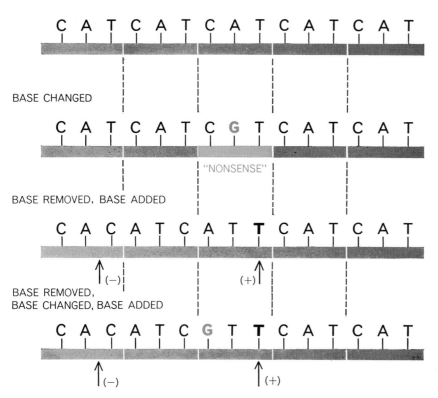

WILD-TYPE GENE

C A T C A T C A T C A T C A T

BASE CHANGED

C A T C A T C G T C A T C A T

"NONSENSE"

BASE REMOVED, BASE ADDED

C A C A T C A T T C A T C A T

↑(−)   ↑(+)

BASE REMOVED,
BASE CHANGED, BASE ADDED

C A C A T C G T T C A T C A T

↑(−)   ↑(+)

NONSENSE MUTATION is one creating a code group that evidently does not represent any of the 20 amino acids found in proteins. Thus it makes the gene inactive. In this hypothetical case a nonsense triplet, CGT, results when an A in the wild-type gene is changed to G. The nonsense triplet can be eliminated if the reading is shifted to put the G in a different triplet. This is done by recombining the inactive gene with one containing a minus-with-plus combination. In spite of three mutations, the resulting gene is active.

the pluses will the message be wrong [see illustration below].

Notice that it does not matter if plus is really one extra base and minus is one fewer; the conclusions would be the same if they were the other way around. In fact, even if some of the plus mutations were indeed a single extra base, others might be two fewer bases; in other words, a plus might really be minus minus. Similarly, some of the minus mutations might actually be plus

plus. Even so they would still fit into our scheme.

Although the most likely explanation is that the message is read three bases at a time, this is not completely certain. The reading could be in multiples of three. Suppose, for example, that the message is actually read six bases at a time. In that case the only change needed in our interpretation of the facts is to assume that all our mutants have been changed by an even number of

bases. We have some weak experimental evidence that this is unlikely. For instance, we can combine the mutant 1589 (which joins the genes) with medium-sized deletions in the A cistron. Now, if deletions were random in length, we should expect about a third of them to allow the B function to be expressed if the message is indeed read three bases at a time, since those deletions that had lost an exact multiple of three bases should allow the B gene to function. By the same reasoning only a sixth of them should work (when combined with 1589) if the reading proceeds six at a time. Actually we find that the B gene is active in a little more than a third. Taking all the evidence together, however, we find that although three is the most likely coding unit, we cannot completely rule out multiples of three.

There is one other general conclusion we can draw about the genetic code. If we make a rough guess as to the actual size of the B gene (by comparing it with another gene whose size is known approximately), we can estimate how many bases can lie between a plus with minus combination and still allow the B gene to function. Knowing also the frequency with which nonsense triplets are created in the misread region between the plus and minus, we can get some idea whether there are many such triplets or only a few. Our calculation suggests that nonsense triplets are not too common. It seems, in other words, that most of the 64 possible triplets, or codons, are not nonsense, and therefore they stand for amino acids. This implies that probably more than one codon can stand for one amino acid. In the jargon of the trade, a code in which this is true is "degenerate."

In summary, then, we have arrived at three general conclusions about the genetic code:

1. The message is read in nonover-

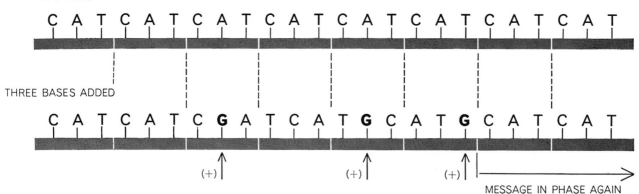

WILD-TYPE GENE

C A T C A T C A T C A T C A T C A T C A T C A T

THREE BASES ADDED

C A T C A T C G A T C A T G C A T G C A T C A T

(+)↑   (+)↑   (+)↑ MESSAGE IN PHASE AGAIN →

TRIPLE MUTATION in which three bases are added fairly close together spoils the genetic message over a short stretch of the gene but leaves the rest of the message unaffected. The same result can be achieved by the deletion of three neighboring bases.

lapping groups from a fixed point, probably from one end. The starting point determines that the message is read correctly into groups.

2. The message is read in groups of a fixed size that is probably three, although multiples of three are not completely ruled out.

3. There is very little nonsense in the code. Most triplets appear to allow the gene to function and therefore probably represent an amino acid. Thus in general more than one triplet will stand for each amino acid.

It is difficult to see how to get around our first conclusion, provided that the B gene really does code a polypeptide chain, as we have assumed. The second conclusion is also difficult to avoid. The third conclusion, however, is much more indirect and could be wrong.

Finally, we must ask what further evidence would really clinch the theory we have presented here. We are continuing to collect genetic data, but I doubt that this will make the story much more convincing. What we need is to obtain a protein, for example one produced by a double mutation of the form plus with minus, and then examine its amino acid sequence. According to conventional theory, because the gene is altered in only two places the amino acid sequences also should differ only in the two corresponding places. According to our theory it should be altered not only at these two places but also at all places in between. In other words, a whole string of amino acids should be changed. There is one protein, the lysozyme of the T4 phage, that is favorable for such an approach, and we hope that before long workers in the U.S. who have been studying phage lysozyme will confirm our theory in this way.

The same experiment should also be useful for checking the particular code schemes worked out by Nirenberg and Matthaei and by Ochoa and his colleagues. The phage lysozyme made by the wild-type gene should differ over only a short stretch from that made by the plus-with-minus mutant. Over this stretch the amino acid sequence of the two lysozyme variants should correspond to the same sequence of bases on the DNA but should be read in different groups of three.

If this part of the amino acid sequence of both the wild-type and the altered lysozyme could be established, one could check whether or not the codons assigned to the various amino acids did indeed predict similar sequences for that part of the DNA between the base added and the base removed.

18

THE GENETIC CODE: III

F. H. C. CRICK
October 1966

The central theme of molecular biology is confirmed by detailed knowledge of how the four-letter language embodied in molecules of nucleic acid controls the 20-letter language of the proteins

The hypothesis that the genes of the living cell contain all the information needed for the cell to reproduce itself is now more than 50 years old. Implicit in the hypothesis is the idea that the genes bear in coded form the detailed specifications for the thousands of kinds of protein molecules the cell requires for its moment-to-moment existence: for extracting energy from molecules assimilated as food and for repairing itself as well as for replication. It is only within the past 15 years, however, that insight has been gained into the chemical nature of the genetic material and how its molecular structure can embody coded instructions that can be "read" by the machinery in the cell responsible for synthesizing protein molecules. As the result of intensive work by many investigators the story

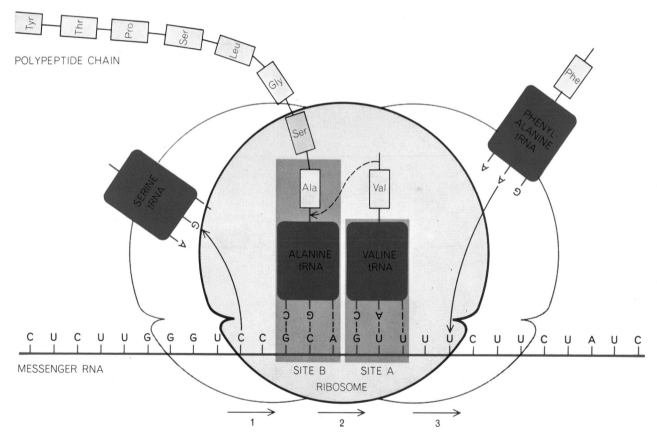

SYNTHESIS OF PROTEIN MOLECULES is accomplished by the intracellular particles called ribosomes. The coded instructions for making the protein molecule are carried to the ribosome by a form of ribonucleic acid (RNA) known as "messenger" RNA. The RNA code "letters" are four bases: uracil (U), cytosine (C), adenine (A) and guanine (G). A sequence of three bases, called a codon, is required to specify each of the 20 kinds of amino acid, identified here by their abbreviations. (A list of the 20 amino acids and their abbreviations appears on the next page.) When linked end to end, these amino acids form the polypeptide chains of which proteins are composed. Each type of amino acid is transported to the ribosome by a particular form of "transfer" RNA (tRNA), which carries an anticodon that can form a temporary bond with one of the codons in messenger RNA. Here the ribosome is shown moving along the chain of messenger RNA, "reading off" the codons in sequence. It appears that the ribosome has two binding sites for molecules of tRNA: one site (*A*) for positioning a newly arrived tRNA molecule and another (*B*) for holding the growing polypeptide chain.

of the genetic code is now essentially complete. One can trace the transmission of the coded message from its original site in the genetic material to the finished protein molecule.

The genetic material of the living cell is the chainlike molecule of deoxyribonucleic acid (DNA). The cells of many bacteria have only a single chain; the cells of mammals have dozens clustered together in chromosomes. The DNA molecules have a very long backbone made up of repeating groups of phosphate and a five-carbon sugar. To this backbone the side groups called bases are attached at regular intervals. There are four standard bases: adenine (A), guanine (G), thymine (T) and cytosine (C). They are the four "letters" used to spell out the genetic message. The exact sequence of bases along a length of the DNA molecule determines the structure of a particular protein molecule.

Proteins are synthesized from a standard set of 20 amino acids, uniform throughout nature, that are joined end to end to form the long polypeptide chains of protein molecules [*see illustration at right*]. Each protein has its own characteristic sequence of amino acids. The number of amino acids in a polypeptide chain ranges typically from 100 to 300 or more.

The genetic code is not the message itself but the "dictionary" used by the cell to translate from the four-letter language of nucleic acid to the 20-letter language of protein. The machinery of the cell can translate in one direction only: from nucleic acid to protein but not from protein to nucleic acid. In making this translation the cell employs a variety of accessory molecules and mechanisms. The message contained in DNA is first transcribed into the similar molecule called "messenger" ribonucleic acid—messenger RNA. (In many viruses—the tobacco mosaic virus, for example—the genetic material is simply RNA.) RNA too has four kinds of bases as side groups; three are identical with those found in DNA (adenine, guanine and cytosine) but the fourth is uracil (U) instead of thymine. In this first transcription of the genetic message the code letters A, G, T and C in DNA give rise respectively to U, C, A and G. In other words, wherever A appears in DNA, U appears in the RNA transcription; wherever G appears in DNA, C appears in the transcription, and so on. As it is usually presented the dictionary of the genetic code employs the letters found in RNA (U, C, A, G) rather than those found in DNA (A, G, T, C).

The genetic code could be broken easily if one could determine both the amino acid sequence of a protein and the base sequence of the piece of nucleic acid that codes it. A simple comparison of the two sequences would yield the code. Unfortunately the determination of the base sequence of a long nucleic acid molecule is, for a variety of reasons, still extremely difficult. More indirect approaches must be used.

Most of the genetic code first became known early in 1965. Since then additional evidence has proved that almost all of it is correct, although a few features remain uncertain. This article describes how the code was discovered and some of the work that supports it.

Scientific American has already presented a number of articles on the genetic code. In one of them ["The Genetic Code," beginning on page 184] I explained that the experimental evidence (mainly indirect) suggested that the code was a triplet code: that the bases on the messenger RNA were read three at a time and that each group corresponded to a particular amino acid. Such a group is called a codon. Using four symbols in groups of three, one can form 64 distinct triplets. The evidence indicated that most of these stood for one amino acid or another, implying that an amino acid was usually represented by several codons. Adjacent amino acids were coded by adjacent codons, which did not overlap.

In a sequel to that article ["The Genetic Code: II," Offprint 153] Marshall W. Nirenberg of the National Institutes of Health explained how the composition of many of the 64 triplets had been determined by actual experiment. The technique was to synthesize polypeptide chains in a cell-free system, which was made by breaking open cells of the colon bacillus (*Escherichia coli*) and extracting from them the machinery for protein synthesis. Then the system was provided with an energy supply, 20 amino acids and one or another of several types of synthetic RNA. Although the exact sequence of bases in each type was random, the proportion of bases was known. It was found that each type of synthetic messenger RNA directed the incorporation of certain amino acids only.

By means of this method, used in a quantitative way, the *composition* of many of the codons was obtained, but the *order* of bases in any triplet could not be determined. Codons rich in G were difficult to study, and in addition a few mistakes crept in. Of the 40 codon compositions listed by Nirenberg

AMINO ACID	ABBREVIATION
ALANINE	Ala
ARGININE	Arg
ASPARAGINE	AspN
ASPARTIC ACID	Asp
CYSTEINE	Cys
GLUTAMIC ACID	Glu
GLUTAMINE	GluN
GLYCINE	Gly
HISTIDINE	His
ISOLEUCINE	Ileu
LEUCINE	Leu
LYSINE	Lys
METHIONINE	Met
PHENYLALANINE	Phe
PROLINE	Pro
SERINE	Ser
THREONINE	Thr
TRYPTOPHAN	Tryp
TYROSINE	Tyr
VALINE	Val

TWENTY AMINO ACIDS constitute the standard set found in all proteins. A few other amino acids occur infrequently in proteins but it is suspected in each case that they originate as one of the standard set and become chemically modified after they have been incorporated into a polypeptide chain.

in his article we now know that 35 were correct.

The Triplet Code

The main outlines of the genetic code were elucidated by another technique invented by Nirenberg and Philip Leder. In this method no protein synthesis occurs. Instead one triplet at a time is used to bind together parts of the machinery of protein synthesis.

Protein synthesis takes place on the comparatively large intracellular structures known as ribosomes. These bodies travel along the chain of messenger RNA, reading off its triplets one after another and synthesizing the polypeptide chain of the protein, starting at the amino end (NH_2). The amino acids do not diffuse to the ribosomes by themselves. Each amino acid is joined chemically by a special enzyme to one of the codon-recognizing molecules known both as soluble RNA (sRNA) and transfer RNA (tRNA). (I prefer the latter designation.) Each tRNA molecule has its own triplet of bases, called an anticodon, that recognizes the relevant codon on the messenger RNA by pairing bases with it [*see illustration on page 192*].

Leder and Nirenberg studied which amino acid, joined to its tRNA molecules, was bound to the ribosomes in the presence of a particular triplet, that is, by a "message" with just three letters. They did so by the neat trick of passing the mixture over a nitrocellulose filter that retained the ribosomes. All the tRNA molecules passed through the filter except the ones specifically bound to the ribosomes by the triplet. Which they were could easily be decided by using mixtures of amino acids

in which one kind of amino acid had been made artificially radioactive, and determining the amount of radioactivity absorbed by the filter.

For example, the triplet GUU retained the tRNA for the amino acid valine, whereas the triplets UGU and UUG did not. (Here GUU actually stands for the trinucleoside diphosphate GpUpU.) Further experiments showed that UGU coded for cysteine and UUG for leucine.

Nirenberg and his colleagues synthesized all 64 triplets and tested them for their coding properties. Similar results have been obtained by H. Gobind Khorana and his co-workers at the University of Wisconsin. Various other

groups have checked a smaller number of codon assignments.

Close to 50 of the 64 triplets give a clearly unambiguous answer in the binding test. Of the remainder some evince only weak binding and some bind more than one kind of amino acid. Other results I shall describe later suggest that the multiple binding is often an artifact of the binding method. In short, the binding test gives the meaning of the majority of the triplets but it does not firmly establish all of them.

The genetic code obtained in this way, with a few additions secured by other methods, is shown in the table below. The 64 possible triplets are set out in a regular array, following a plan

SECOND LETTER

FIRST LETTER		U	C	A	G		THIRD LETTER
	U	UUU UUC } Phe UUA UUG } Leu	UCU UCC UCA UCG } Ser	UAU UAC } Tyr UAA OCHRE UAG AMBER	UGU UGC } Cys UGA ? UGG Tryp		U C A G
	C	CUU CUC CUA CUG } Leu	CCU CCC CCA CCG } Pro	CAU CAC } His CAA CAG } GluN	CGU CGC CGA CGG } Arg		U C A G
	A	AUU AUC } Ileu AUA AUG Met	ACU ACC ACA ACG } Thr	AAU AAC } AspN AAA AAG } Lys	AGU AGC } Ser AGA AGG } Arg		U C A G
	G	GUU GUC GUA GUG } Val	GCU GCC GCA GCG } Ala	GAU GAC } Asp GAA GAG } Glu	GGU GGC GGA GGG } Gly		U C A G

GENETIC CODE, consisting of 64 triplet combinations and their corresponding amino acids, is shown in its most likely version. The importance of the first two letters in each triplet is readily apparent. Some of the allocations are still not completely certain, particularly for organisms other than the colon bacillus (*Escherichia coli*). "Amber" and "ochre" are terms that referred originally to certain mutant strains of bacteria. They designate two triplets, UAA and UAG, that may act as signals for terminating polypeptide chains.

that clarifies the relations between them.

Inspection of the table will show that the triplets coding for the same amino acid are often rather similar. For example, all four of the triplets starting with the doublet AC code for threonine. This pattern also holds for seven of the other amino acids. In every case the triplets XYU and XYC code for the same amino acid, and in many cases XYA and XYG are the same (methionine and tryptophan may be exceptions). Thus an amino acid is largely selected by the first two bases of the triplet. Given that a triplet codes for, say, valine, we know that the first two bases are GU, whatever the third may be. This pattern is true for all but three of the amino acids. Leucine can start with UU or CU, serine with UC or AG and arginine with CG or AG. In all other cases the amino acid is uniquely related to the first two bases of the triplet. Of course, the converse is often not true. Given that a triplet starts with, say, CA, it may code for either histidine or glutamine.

Synthetic Messenger RNA's

Probably the most direct way to confirm the genetic code is to synthesize a messenger RNA molecule with a strictly defined base sequence and then find the amino acid sequence of the polypeptide produced under its influence. The most extensive work of this nature has been done by Khorana and his colleagues. By a brilliant combination of ordinary chemical synthesis and synthesis catalyzed by enzymes, they have made long RNA molecules with various repeating sequences of bases. As an example, one RNA molecule they have synthesized has the sequence UGUG-UGUGUGUG.... When the biochemical machinery reads this as triplets the message is UGU–GUG–UGU–GUG.... Thus we expect that a polypeptide will be produced with an alternating sequence of two amino acids. In fact, it was found that the product is Cys–Val–Cys–Val.... This evidence alone would not tell us which triplet goes with which amino acid, but given the results of the binding test one has no hesitation in concluding that UGU codes for cysteine and GUG for valine.

In the same way Khorana has made chains with repeating sequences of the type XYZ... and also XXYZ.... The type XYZ... would be expected to give a "homopolypeptide" containing one amino acid corresponding to the triplet XYZ. Because the starting point is not clearly defined, however, the homo-polypeptides corresponding to YZX... and ZXY... will also be produced. Thus poly-AUC makes polyisoleucine, polyserine and polyhistidine. This confirms that AUC codes for isoleucine, UCA for serine and CAU for histidine. A repeating sequence of four bases will yield a single type of polypeptide with a repeating sequence of four amino acids. The general patterns to be expected in each case are set forth in the table on this page. The results to date have amply demonstrated by a direct biochemical method that the code is indeed a triplet code.

Khorana and his colleagues have so far confirmed about 25 triplets by this method, including several that were quite doubtful on the basis of the binding test. They plan to synthesize other sequences, so that eventually most of the triplets will be checked in this way.

The Use of Mutations

The two methods described so far are open to the objection that since they do not involve intact cells there may be some danger of false results. This objection can be met by two other methods of checking the code in which the act of protein synthesis takes place inside the cell. Both involve the effects of genetic mutations on the amino acid sequence of a protein.

It is now known that small mutations are normally of two types: "base substitution" mutants and "phase shift" mutants. In the first type one base is changed into another base but the total number of bases remains the same. In the second, one or a small number of bases are added to the message or subtracted from it.

There are now extensive data on base-substitution mutants, mainly from studies of three rather convenient proteins: human hemoglobin, the protein of tobacco mosaic virus and the A protein of the enzyme tryptophan synthetase obtained from the colon bacillus. At least 36 abnormal types of human hemoglobin have now been investigated by many different workers. More than 40 mutant forms of the protein of the tobacco mosaic virus have been examined by Hans Wittmann of the Max Planck Institute for Molecular Genetics in Tübingen and by Akita Tsugita and Heinz Fraenkel-Conrat of the University of California at Berkeley [see "The Genetic Code of a Virus," by Heinz Fraenkel-Conrat; SCIENTIFIC AMERICAN Offprint 193]. Charles Yanofsky and his group at Stanford University have characterized about 25 different mutations of the A protein of tryptophan synthetase.

RNA BASE SEQUENCE	READ AS			AMINO ACID SEQUENCE EXPECTED
(XY)$_n$ . . .	X Y X	Y X Y	X Y X	Y X Y . . .	αβαβ
(XYZ)$_n$ . . .	X Y Z	X Y Z	X Y Z . . .		ααα
. . .	Y Z X	Y Z X	Y Z X . . .		βββ
. . .	Z X Y	Z X Y	Z X Y . . .		γγγ
(XXYZ)$_n$ . . .	X X Y Z	X X Y Z	X X Y Z . . .		αβγδαβγδ
(XYXZ)$_n$ . . .	X Y X Z	X Y X Z	X Y X Z . . .		αβγδαβγδ

VARIETY OF SYNTHETIC RNA's with repeating sequences of bases have been produced by H. Gobind Khorana and his colleagues at the University of Wisconsin. They contain two or three different bases (X, Y, Z) in groups of two, three or four. When introduced into cell-free systems containing the machinery for protein synthesis, the base sequences are read off as triplets (middle) and yield the amino acid sequences indicated at the right.

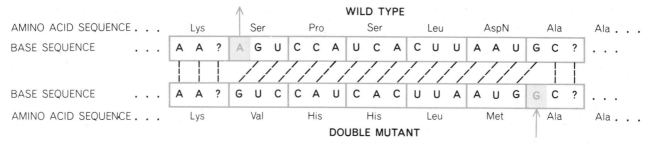

"PHASE SHIFT" MUTATIONS help to establish the actual codons used by organisms in the synthesis of protein. The two partial amino acid sequences shown here were determined by George Streisinger and his colleagues at the University of Oregon. The sequences are from a protein, a type of lysozyme, produced by the bacterial virus T4. A pair of phase-shift mutations evidently removed one base, A, and inserted another, G, about 15 bases farther on. The base sequence was deduced theoretically from the genetic code.

The remarkable fact has emerged that in every case but one the genetic code shows that the change of an amino acid in a polypeptide chain could have been caused by the alteration of a single base in the relevant nucleic acid. For example, the first observed change of an amino acid by mutation (in the hemoglobin of a person suffering from sickle-cell anemia) was from glutamic acid to valine. From the genetic code dictionary on page 194 we see that this could have resulted from a mutation that changed either GAA to GUA or GAG to GUG. In either case the change involved a single base in the several hundred needed to code for one of the two kinds of chain in hemoglobin.

The one exception so far to the rule that all amino acid changes could be caused by single base changes has been found by Yanofsky. In this one case glutamic acid was replaced by methionine. It can be seen from the genetic code dictionary that this can be accomplished only by a change of *two* bases, since glutamic acid is encoded by either GAA or GAG and methionine is encoded only by AUG. This mutation has occurred only once, however, and of all the mutations studied by Yanofsky it is the only one not to back-mutate, or revert to "wild type." It is thus almost certainly the rare case of a double change. All the other cases fit the hypothesis that base-substitution mutations are normally caused by a single base change. Examination of the code shows that only about 40 percent of all the possible amino acid interchanges can be brought about by single base substitutions, and it is only these changes that are found in experiments. Therefore the study of actual mutations has provided strong confirmation of many features of the genetic code.

Because in general several codons stand for one amino acid it is not possible, knowing the amino acid sequence, to write down the exact RNA base sequence that encoded it. This is unfortunate. If we know which amino acid is changed into another by mutation, however, we can often, given the code, work out what that base change must have been. As an example, glutamic acid can be encoded by GAA or GAG and valine by GUU, GUC, GUA or GUG. If a mutation substitutes valine for glutamic acid, one can assume that only a single base change was involved. The only such change that could lead to the desired result would be a change from A to U in the middle position, and this would be true whether GAA became GUA or GAG became GUG.

It is thus possible in many cases (not in all) to compare the nature of the base change with the chemical mutagen used to produce the change. If RNA is treated with nitrous acid, C is changed to U and A is effectively changed to G. On the other hand, if double-strand DNA is treated under the right conditions with hydroxylamine, the mutagen acts only on C. As a result some C's are changed to T's (the DNA equivalent of U's), and thus G's, which are normally paired with C's in double-strand DNA, are replaced by A's.

If 2-aminopurine, a "base analogue" mutagen, is added when double-strand DNA is undergoing replication, it produces only "transitions." These are the same changes as those produced by hydroxylamine—plus the reverse changes. In almost all these different cases (the exceptions are unimportant) the changes observed are those expected from our knowledge of the genetic code.

Note the remarkable fact that, although the code was deduced mainly from studies of the colon bacillus, it appears to apply equally to human beings and tobacco plants. This, together with more fragmentary evidence, suggests that the genetic code is either the same or very similar in most organisms.

The second method of checking the code using intact cells depends on phase-shift mutations such as the addition of a single base to the message. Phase-shift mutations probably result from errors produced during genetic recombination or when the DNA molecule is being duplicated. Such errors have the effect of putting out of phase the reading of the message from that point on. This hypothesis leads to the prediction that the phase can be corrected if at some subsequent point a nucleotide is deleted. The pair of alterations would be expected not only to change two amino acids but also to alter all those encoded by bases lying between the two affected sites. The reason is that the intervening bases would be read out of phase and therefore grouped into triplets different from those contained in the normal message.

This expectation has recently been confirmed by George Streisinger and his colleagues at the University of Oregon. They have studied mutations in the protein lysozyme that were produced by the T4 virus, which infects the colon bacillus. One phase-shift mutation involved the amino acid sequence ...Lys—Ser—Pro—Ser—Leu—AspN—Ala—Ala—Lys.... They were then able to construct by genetic methods a double phase-shift mutant in which the corresponding sequence was ...Lys—Val—His—His—Leu—Met—Ala—Ala—Lys....

Given these two sequences, the reader should be able, using the genetic code dictionary on page 194, to decipher uniquely a short length of the nucleic acid message for both the original protein and the double mutant and thus deduce the changes produced by each of the phase-shift mutations. The correct result is presented in the illustration above. The result not only confirms several rather doubtful codons, such as UUA for leucine and AGU for serine, but also shows which codons are actually involved in a genetic message. Since the technique is difficult, however, it may not find wide application.

Streisinger's work also demonstrates what has so far been only tacitly assumed: that the two languages, both of which are written down in a certain direction according to convention, are in fact translated by the cell in the same direction and not in opposite directions. This fact had previously been established, with more direct chemical methods, by Severo Ochoa and his colleagues at the New York University School of Medicine. In the convention, which was adopted by chance, proteins are written with the amino (NH_2) end on the left. Nucleic acids are written with the end of the molecule containing a "5 prime" carbon atom at the left. (The "5 prime" refers to a particular carbon atom in the 5-carbon ring of ribose sugar or deoxyribose sugar.)

Finding the Anticodons

Still another method of checking the genetic code is to discover the three bases making up the anticodon in some particular variety of transfer RNA. The first tRNA to have its entire sequence worked out was alanine tRNA, a job done by Robert W. Holley and his collaborators at Cornell University [see the article "The Nucleotide Sequence of a Nucleic Acid," by Robert W. Holley, beginning on page 174]. Alanine tRNA, obtained from yeast, contains 77 bases. A possible anticodon found near the middle of the molecule has the sequence IGC, where I stands for inosine, a base closely resembling guanine. Since then Hans Zachau and his colleagues at the University of Cologne have established the sequences of two closely related serine tRNA's from yeast, and James Madison and his group at

the U.S. Plant, Soil and Nutrition Laboratory at Ithaca, N.Y., have worked out the sequence of a tyrosine tRNA, also from yeast.

A detailed comparison of these three sequences makes it almost certain that the anticodons are alanine–IGC, serine–IGA and tyrosine–GΨA. (Ψ stands for pseudo-uridylic acid, which can form the same base pairs as the base uracil.) In addition there is preliminary evidence from other workers that an anticodon for valine is IAC and an anticodon for phenylalanine is GAA.

All these results would fit the rule that the codon and anticodon pair in an antiparallel manner, and that the pairing in the first two positions of the codon is of the standard type, that is, A pairs with U and G pairs with C. The pairing in the third position of the codon is more complicated. There is now good experimental evidence from both Nirenberg and Khorana and their co-workers that one tRNA can recognize several codons, provided that they differ only in the last place in the codon. Thus Holley's alanine tRNA appears to recognize GCU, GCC and GCA. If it recognizes GCG, it does so only very weakly.

The "Wobble" Hypothesis

I have suggested that this is because of a "wobble" in the pairing in the third place and have shown that a reasonable theoretical model will explain many of the observed results. The suggested rules for the pairing in the third position of the anticodon are presented in the table at the top of this page, but

ANTICODON	CODON
U	A G
C	G
A	U
G	U C
I	U C A

"WOBBLE" HYPOTHESIS has been proposed by the author to provide rules for the pairing of codon and anticodon at the *third* position of the codon. There is evidence, for example, that the anticodon base I, which stands for inosine, may pair with as many as three different bases: U, C and A. Inosine closely resembles the base guanine (G) and so would ordinarily be expected to pair with cytosine (C). Structural diagrams for standard base pairings and wobble base pairings are illustrated at the bottom of this page.

this theory is still speculative. The rules for the first two places of the codon seem reasonably secure, however, and can be used as partial confirmation of the genetic code. The likely codon-anticodon pairings for valine, serine, tyrosine, alanine and phenylalanine satisfy the standard base pairings in the first two places and the wobble hypothesis in the third place [see illustration on page 198].

Several points about the genetic code remain to be cleared up. For example, the triplet UGA has still to be allocated.

STANDARD AND WOBBLE BASE PAIRINGS both involve the formation of hydrogen bonds when certain bases are brought into close proximity. In the standard guanine-cytosine pairing (*left*) it is believed three hydrogen bonds are formed. The bases are shown as they exist in the RNA molecule, where they are attached to 5-car-

bon rings of ribose sugar. In the proposed wobble pairing (*right*) guanine is linked to uracil by only two hydrogen bonds. The base inosine (I) has a single hydrogen atom where guanine has an amino (NH_2) group (*broken circle*). In the author's wobble hypothesis inosine can pair with U as well as with C and A (*not shown*).

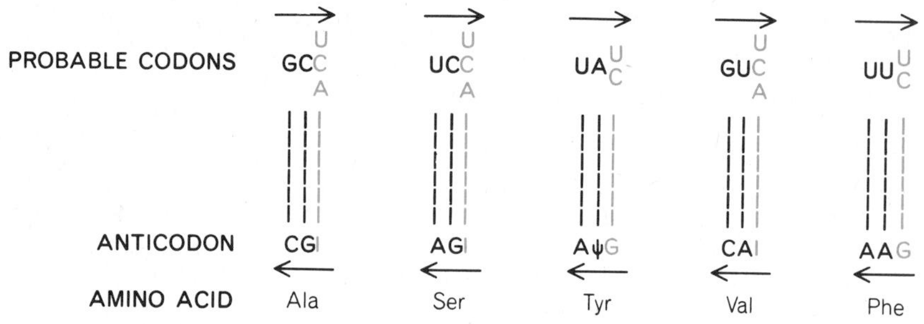

PROBABLE CODONS	GCC̲ᴬᵁ	UCC̲ᴬᵁ	UAᵁ꜀	GUC̲ᴬᵁ	UUᵁ꜀
	‖‖‖	‖‖‖	‖‖‖	‖‖‖	‖‖‖
ANTICODON	CGI	AGI	AψG	CAI	AAG
AMINO ACID	Ala	Ser	Tyr	Val	Phe

CODON-ANTICODON PAIRINGS take place in an antiparallel direction. Thus the anticodons are shown here written backward, as opposed to the way they appear in the text. The five anticodons are those tentatively identified in the transfer RNA's for alanine, serine, tyrosine, valine and phenylalanine. Color indicates where wobble pairings may occur.

The punctuation marks—the signals for "begin chain" and "end chain"—are only partly understood. It seems likely that both the triplet UAA (called "ochre") and UAG (called "amber") can terminate the polypeptide chain, but which triplet is normally found at the end of a gene is still uncertain.

The picturesque terms for these two triplets originated when it was discovered in studies of the colon bacillus some years ago that mutations in other genes (mutations that in fact cause errors in chain termination) could "suppress" the action of certain mutant codons, now identified as either UAA or UAG. The terms "ochre" and "amber" are simply invented designations and have no reference to color.

A mechanism for chain initiation was discovered fairly recently. In the colon bacillus it seems certain that formyl-methionine, carried by a special tRNA, can initiate chains, although it is not clear if all chains have to start in this way, or what the mechanism is in mammals and other species. The formyl group (CHO) is not normally found on finished proteins, suggesting that it is probably removed by a special enzyme. It seems likely that sometimes the methionine is removed as well.

It is unfortunately possible that a few codons may be ambiguous, that is, may code for more than one amino acid. This is certainly not true of most codons. The present evidence for a small amount of ambiguity is suggestive but not conclusive. It will make the code more difficult to establish correctly if ambiguity can occur.

Problems for the Future

From what has been said it is clear that, although the entire genetic code is not known with complete certainty, it is highly likely that most of it is correct. Further work will surely clear up the doubtful codons, clarify the punctuation marks, delimit ambiguity and extend the code to many other species. Although the code lists the codons that *may* be used, we still have to determine if alternative codons are used equally. Some preliminary work suggests they may not be. There is also still much to be discovered about the machinery of protein synthesis. How many types of tRNA are there? What is the structure of the ribosome? How does it work, and why is it in two parts? In addition there are many questions concerning the control of the rate of protein synthesis that we are still a long way from answering.

When such questions have been answered, the major unsolved problem will be the structure of the genetic code. Is the present code merely the result of a series of evolutionary accidents, so that the allocations of triplets to amino acids is to some extent arbitrary? Or are there profound structural reasons why phenylalanine has to be coded by UUU and UUC and by no other triplets? Such questions will be difficult to decide, since the genetic code originated at least three billion years ago, and it may be impossible to reconstruct the sequence of events that took place at such a remote period. The origin of the code is very close to the origin of life. Unless we are lucky it is likely that much of the evidence we should like to have has long since disappeared.

Nevertheless, the genetic code is a major milestone on the long road of molecular biology. In showing in detail how the four-letter language of nucleic acid controls the 20-letter language of protein it confirms the central theme of molecular biology that genetic information can be stored as a one-dimensional message on nucleic acid and be expressed as the one-dimensional amino acid sequence of a protein. Many problems remain, but this knowledge is now secure.

III

MACROMOLECULAR COMPLEXES
AND THEIR ASSEMBLY

Functional Associations
of Nucleic Acids and Proteins

An example of one of the very simplest types of living cells is the bacterium *Escherichia coli*; yet this single cell organism contains over a million protein molecules of about 3000 different types. Most of these proteins are enzymes, the essential catalysts of nearly all the chemical reactions in cells; others serve structural roles in cellular assembly. *E. coli* can duplicate its cell substance and divide into identical daughter cells every forty minutes at 37°C when immersed in a growth medium containing only inorganic salts and the simple sugar glucose. In such a medium glucose is the only source of carbon for the molecules that must be synthesized by the cell, and as it is broken down, it also provides the chemical energy to drive the synthetic processes necessary for growth. Somehow, the cell must be able to coordinate thousands of chemical reactions in space and in time so that the finished products are faithful copies of the original cell.

This assembly line is many times more complex than that required in the construction of an automobile. Over the years certain rules of efficiency in the design of assembly plants have evolved by trial and error. The building of an automobile involves the operation of a number of converging subassembly lines in which various components are manufactured and then brought together to produce the final product. Regulation of the rates of manufacture of the various components and adequate control on their quality are essential for the smooth operation of an assembly line. Imagine as an alternative that all of the raw materials for an automobile were placed in one pile in the center of the factory floor and that several thousand specialized mechanics were then to converge upon this pile en masse to construct an automobile! It is not at all surprising that living systems have developed very efficient assembly lines in the course of evolution and that the complex series of chemical reactions that lead to the synthesis of a cell do not occur simultaneously in one seething vat. The enzymes involved in a particular sequence of biochemical reactions are often localized in the cell and sometimes even physically connected to one another. Such operations as protein synthesis and the replication of DNA are carried out by specific protein complexes. The membranes of cells also serve important roles in the organization of enzymes into functional units. For example, as shown in the article by Efraim Racker in Part I, the processes of oxidative phosphorylation and electron transport in eucaryotes occur in complexes associated with mitochondrial membranes. Membranes serve the additional role of partitioning cells into functional compartments and of providing a selective barrier between the cell and its environment. The mitochondria and the nucleus in eucaryotic cells have their own membrane systems to compartmentalize at least two specialized elements of cell machinery, oxidative energy transformations and chromosome replication, respectively. Procaryotes (bacteria and blue-green algae) are characterized by the fact that intracellular compartmentalization is less evident. Thus, *E. coli* has no well-defined nucleus or mitochondria, and the enzymes of oxidative metabolism are found instead on the plasma membrane. (In fact, most bacteria are roughly the size of the mitochondria found in eucaryotes.) In addition to the plasma membrane, bacteria possess a cell wall which provides structural rigidity to the organism and which also permits compartmentalization of certain enzymes in the *periplasmic space* between the membrane and the wall.

Consider now in more detail the general principles of efficient design that operate in the construction of cells. Two important aspects are evident at the molecular level—*subunit construction* and *self-assembly*. Subunit construction is exemplified in the process of building polymers from monomer subunits. The cell contains relatively few types of polymers, the principal ones being proteins, nucleic acids, and carbohydrates; but each of these polymers uses a unique type of monomer and a characteristic system for linking the monomers together. A vast amount of diversity is possible with even a relatively small number of monomers; thus, as described in the introduction to part I, one can construct 20^{200} different protein molecules using the twenty different amino acids in all possible sequences in proteins of 200 amino acids in length. Since essentially all proteins are synthesized in the same way, (i.e., on particles called ribosomes), a minimum amount of specialized information is needed in the genome for their individual assembly. Subunit construction is also a very useful quality control device—defective units can be rejected at different stages in the assembly process so that the survival of the cell is not dependent upon every amino acid or even every protein being perfect.

The second aspect of efficient design in the economy of the cell is self-assembly. To the extent that functional proteins and other elements can organize themselves into their appropriate complexes spontaneously, the genome is relieved of the responsibility for providing additional information for subcellular assembly. The information for the assembly of a complex organelle such as a ribosome is contained in the amino acid sequences of the individual proteins that constitute the organelle. This fact was illustrated most dramatically by Nomura and co-workers when they succeeded in reconstructing a ribosome from its separated components in vitro.

In this section we consider the nature of some subcellular complexes and some of the problems of their assembly. The ribosome is in many ways the most crucial organelle in the cell. It is the workbench upon which proteins are made and thereby the site at which the genetic code is translated. However, it is much more than a workbench, both in its complexity and in its active contribution to the process of protein synthesis. It is composed of many proteins and several distinct species of RNA molecules. In his article on ribosomes M. Nomura describes the way in which this organelle performs its task and also the way in which its protein and nucleic acid components combine sequentially to form the functional structure. The ribosome assembly process proceeds spontaneously, but it is also cooperative in the sense that certain elements must be present in proper sequence to facilitate the association of other units. We do not know whether the RNA components serve information-carrying roles although it is clear from Nomura's work that not just any RNA of the appropriate size will do. The RNA might have a structural role and help with the organization of the protein components—and it may also be concerned with the proper orientation of the incoming charged transfer RNA molecules. One of the recognition sites in the transfer RNA might combine by base pairing with some complementary region of ribosomal RNA in order to hold the charged transfer RNA in place as the attached amino acid is affixed to the growing polypeptide chain. The active participation of the ribosome in the decoding process is illustrated by the finding that the antibiotic streptomycin interacts with one of the protein subunits and causes mistakes in translation as well as general inhibition of protein synthesis (see Luigi Gorini, *Scientific American* April 1966; Offprint 1041).

Next we consider the construction of viruses, which are the simplest macromolecular structures that possess sufficient genetic information for their own duplication. This definition does not imply, however, that a free virus in an appropriate growth medium is able to reproduce itself. The virus may be regarded as a kind of "regressed parasite" that must take advantage of the biochemical machinery of the host cell within which it reproduces. All viruses utilize the ribosomes of the host cell to synthesize their own proteins. Not only is a virus unable to reproduce outside its host, but, unlike other kinds of parasites, it does not even "metabolize" in the free state; outside its host cell the virus particle is biologically inert. For this reason, some virologists argue that the virus may properly be regarded as a cell component that has become independent enough that it can pass from cell to cell. Viewed in this light, viruses more than fulfill their early promise of being the simplest systems for the study of isolated genetic elements.

Viruses consist of a nucleic acid core (that may be either DNA or RNA) and, usually, a protein coat. The nucleic acid directs the synthesis of the protein coat and additional specialized proteins necessary for the duplication of the virus. The simplest known virus contains information in its RNA genome for its coat protein and an RNA polymerase; all other molecular machinery involved in its reproduction must come from its host. In contrast, some of the more complex viruses contain over one hundred genes, some of which are biochemically equivalent to the genes contained in their hosts. Thus, the T2 bacteriophage (a bacterial virus) specifies enzymes for the synthesis of thymidine and can provide this information even if the host bacterium is a mutant that can no longer synthesize its own thymidine. (Of course, the host does not derive any particular benefit from this viral contribution since it simply enhances the virus takeover of the cell.)

The "phage group" begun by Max Delbrück at Caltech was largely responsible for introducing to the field of molecular biology the strategy that the best way to gain a fundamental understanding of a complex biological phenomenon is to study it first in the simplest system in which it can be found. The T-series of bacteriophage were isolated and initially characterized by Delbrück and his students. The life cycle of the T2 bacteriophage is described in the article by Gunther S. Stent on the multiplication of bacterial viruses. (It should be noted that this article was written just before the presentation of the Watson-Crick model for DNA.) Stent emphasized the application of radioisotopic tracers to study the fate of parental virus particles in the course of their life cycle, an approach that led to the conclusion that the nucleic acid core rather than the protein coat enters the host cell to initiate the process of infection. This result constituted early proof of the genetic role of nucleic acid. It has since been shown that the infecting phage does inject a small amount of protein, generally those enzymes needed to initiate the process of taking over the metabolism of the host. The apparent result that only a portion of the original infecting DNA molecules end up intact in progeny phage is now explained by the large amount of genetic recombination that occurs among the replicating viral DNA molecules within the host cell. (Genetic recombination involves initially a physical exchange of nucleic acid between the recombining molecules.) This article, the earliest in our collection, also presents an electron micrograph of bacteriophage T2. Compare the level of sophistication with this technique in 1953 to that now possible as shown in the most recent article in this collection written by Oscar L. Miller, Jr., some twenty years later.

Viruses occur in a comprehensive array of regular geometrical forms. The article by R. W. Horne on the structure of viruses displays the variety of form that was revealed as the result of a number of important technical improvements in sample preparation for electron microscopy. It is worth emphasizing that the simplest biological structures possess a high degree of symmetry—indeed, most viruses can be crystallized. The protein coat components represent most of the mass of a typical virus particle. That being the case, then how is it possible for the limited information in the viral genome to specify enough coat protein to envelope that genome? The answer is that the virus coat typically contains a regular symmetrical array of many *identical* proteins, again illustrating the principle of subunit construction. Furthermore, the principle of self-assembly also applies in that the aggregation of these coat protein subunits normally occurs spontaneously. In the tobacco mosaic virus (TMV) 2130 identical protein subunits are assembled in a helical array around the RNA core. Even without the RNA, these proteins combine spontaneously to form long cylinders of random lengths. However, the presence of the RNA core limits the association of the protein subunits to virus-length cylinders. In this example the physical size of the genome as well as its nucleotide sequence is important to the completed virus structure.

The T4 bacteriophage is much more complex than TMV, but even here most of the events in its assembly occur spontaneously. W. B. Wood and R. S. Edgar describe our current understanding of the assembly process and its genetic control in their article on building a bacterial virus. The study of the functions required for the synthesis of the T4 bacteriophage involved the mapping of mutations in the T4 genome. Many mutant classes of T4 have been isolated and the locations of the genes responsible for the affected functions have been determined. In the article by Seymour Benzer in Part II we have already learned how mutations may be mapped in the T4 genome. However, a problem arises when the mutation alters some critical function in the virus life cycle, such as the synthesis of the enzymes responsible for the actual duplication of the viral DNA. Such mutations are normally lethal and would be impossible to study because the virus simply could not reproduce. How can the effects of specific mutations be studied when there are no progeny? The solution to this problem was solved largely through the development of several important classes of "conditional lethal" mutations such as the temperature-sensitive mutations. A temperature-sensitive conditional lethal mutation would be one in which the altered gene affects the amino acid sequence of an essential enzyme that is then active at low temperature but whose configuration is changed such that it is rendered inactive at a higher temperature. Clearly the expression of a temperature-sensitive mutation may be controlled simply by changing the temperature.

An important fact that has emerged from the mapping of the viral genes is that their arrangement in the DNA molecule is not random: Genes that control similar functions are often arranged in clusters. Not only have the positions of genes that specify the various protein components of the T4 bacteriophage been determined, but for this complex virus a number of genes have been shown to be required for the assembly process itself. That is, enzymes must be produced to catalyze certain steps in the assembly of the viral proteins into the final phage particle. Thus, we begin to see how an understanding of the sequence of nucleotides in a DNA molecule provides the necessary information for the synthesis and assembly of a complex nucleoprotein structure.

In the second half of this section we consider surface phenomena and complexes involving lipids and carbohydrates as well as proteins. As we have mentioned above, many important biochemical reaction sequences are organized on membranes in the cell. The study of membranes is now one of the most exciting areas of research in molecular biology. Although the morphology of these structures has been studied by electron microscopy for some time, the chemistry of membrane complexes has been elusive and difficult to study, largely because membranes are insoluble in aqueous solvents and also because lipoprotein complexes have tended to be notoriously unstable. The article by Efraim Racker in Part I emphasizes some of these difficulties in the study of the mitochondrial membrane. The last three articles in this section deal with the layers that separate the inside of a bacterium from its environment. *E. coli* and certain other bacteria possess an outer lipopolysaccaride envelope. Richard Losick and Phillips W. Robbins continue the discussion of T4 bacteriophage in their article on the receptor site for a bacterial virus which examines the chemistry and function of this outer cell envelope. The attachment of a virus is highly specific for certain hosts; in *E. coli* this outer membrane provides the specific "brand" that identifies the host as appropriate for T4. The article also considers some of the effects that viral infection may have on the surface properties of the host cell and how such alterations sometimes accompany the transformation of normal mammalian cells to malignancy upon infection by oncogenic viruses.

Inside the lipopolysaccharide sheath we find a rigid cell wall that gives *E. coli* its characteristic rod shape. The bacterial cell wall described by Nathan Sharon is much more complex than the simple cellulose cell wall that surrounds most plant cells. (Animal cells do not have any cell wall at all.) It is possible to remove the bacterial cell wall without rupturing the inner plasma membrane by keeping the cells in a sucrose solution of appropriate osmotic strength. One of the methods for doing this utilizes the enzyme, lysozyme, described in Phillips's article in Part II. Another method requires that the cells grow in the presence of the antibiotic penicillin, which inhibits cell wall synthesis. The cells essentially grow out of their skins and then exist as membrane-bounded "spheroplasts" that are still able to carry out many of the functions of the intact living cell. However, many enzymes that normally reside in the periplasmic space between the wall and the plasma membrane are released when the wall is removed.

This inhibitory action of penicillin on cell wall synthesis forms the basis for a very simple method for selecting certain types of bacterial mutants. Suppose there exists in a normal cell population a mutant that cannot synthesize the amino acid arginine. How might this mutant be isolated from the other, say, 10^8 normal cells in the population? If all the cells are placed in a medium lacking arginine but containing penicillin, the normal cells will grow and will eventually burst as they grow out of their walls. The mutant cells will remain dormant until later when the penicillin is removed and the required arginine supplement is provided in the growth medium; then the mutants will grow in large numbers and can be easily established as an independent cell line.

Nearly one-quarter of all of the proteins in the cell are associated with membranes. C. Fred Fox considers the structure of cell membranes with particular emphasis on the bacterial plasma membrane and the various experimental approaches to its study. He considers the question of why certain proteins are found in membranes and how they differ basically from those found in the internal aqueous environment of the cell. The membrane turns out to be a very dynamic organelle with a structure that is highly flexible and fluid. The fluidity depends upon the unsaturated fatty acid content of the membrane, and it is possible to vary the composition of membranes by feeding the organism different fatty acids. Another important problem is how the proteins find their respective sites on the membrane: Are they placed there as the membrane is assembled, or are they built in later? There is evidence that the proteins involved in transport functions are added after the membrane has been constructed but that they are guided by specific recognition of attachment sites in the membrane.

19

RIBOSOMES

MASAYASU NOMURA

October 1969

They are the organelles that conduct the synthesis of proteins in the living cell. Their structure and functioning are studied by taking them apart and seeing how they reassemble themselves

It is one thing to discover the basic principles of a life process and quite another to know in detail the chemical mechanisms that underlie it. In order to genuinely understand a cellular function one must study the machinery that performs it, and in many cases that means studying a highly organized cellular element, or organelle, that provides the machinery. One must first determine the organelle's structure and learn how it operates and then find out how the organelle itself is generated in the cell. In this article I shall relate how the structural and functional description of the ribosome, the organelle that conducts protein synthesis, has been attempted and is even now being achieved.

The story of the ribosome goes back to the discovery some years ago that the capacity of various types of cell to synthesize proteins was correlated with the cells' content of ribonucleic acid (RNA), and that most of the cellular RNA was in the form of small particles (then known as microsomes) in the cytoplasm of the cell. This suggested that the particles must play some role in protein synthesis, but the real importance of ribosomes emerged only after intensive biochemical investigation.

The pioneer work was done by Paul C. Zamecnik and his collaborators at the Massachusetts General Hospital in the 1950's [see "The Microsome," by Paul C. Zamecnik; SCIENTIFIC AMERICAN Offprint 52]. They homogenized rat-liver cells, added amino acids labeled with atoms of the radioactive isotope carbon 14, fortified the homogenate with adenosine triphosphate (ATP) to provide chemical energy and were able to detect the formation of small amounts of protein. By a process of elimination they established that several cellular organelles, including the nucleus and the mitochondrion, were not necessary for protein synthesis but that the microsomes were essential. They were able to identify other cellular components required for protein synthesis, including the small RNA molecules called transfer RNA and enzymes that attach amino acids to transfer-RNA molecules. These early test-tube assembly systems, however, made only very small amounts of protein. Then in 1961 Marshall W. Nirenberg and J. Heinrich Matthaei of the National Institutes of Health found that in order to obtain intensive synthesis of protein in cell-free extracts of the bacterium *Escherichia coli* it was necessary to include a third type of RNA, called messenger RNA, that had been postulated by François Jacob and Jacques Monod of the Pasteur Institute in Paris.

Once a complete cell-free protein-synthesizing system could be assembled it was possible to study the functioning of its several components. One of the most interesting of these was the ribosome. It was now clear that this particle coordinates the translation of the genetic information in the sequence of nucleotide bases in the messenger RNA (transcribed from the DNA molecule, the gene) to the sequence of amino acids in each protein manufactured by the cell [*see illustration on page 209*]. The first systematic studies of ribosomes were initiated about 1957 by several groups, notably one at Harvard University led by Alfred Tissières and James D. Watson and one in the Carnegie Institution of Washington that included Ellis T. Bolton, Roy Britten and Richard B. Roberts. (I should add that my association with Watson's group at that time, although it was brief, had a great influence on my later research on the ribosome.) The initial studies were done mainly on *E. coli* ribosomes, which consist of two subunits of unequal size that are designated 30S and 50S. The size is determined by the rate, measured in Svedberg units (S), at which a particle sediments when it is spun at high speed in an ultracentrifuge. Together these particles constitute the functional unit in protein synthesis: the 70S ribosome. (The reason the two S values are not additive is that the shape of a particle influences its rate of sedimentation.) In each of these subunits proteins represent about a third of the total mass; the rest is RNA. The 50S ribosome subunit contains a 23S RNA molecule and a 5S RNA molecule. The 30S ribosome subunit incorporates one 16S RNA molecule. In 1961 J. P. Waller and J. I. Harris of Harvard observed that the ribosome contains different kinds of protein molecules, indicating that its structure must be quite complex. Subsequent experiments conducted by many workers, including those in my own group at the University of Wisconsin, show that the 30S ribosomal subunit includes either 19 or 20 different protein molecules and that the 50S subunit apparently has more than 30 protein molecules.

As the early work on the structure of the ribosome was proceeding, a general picture of its functional properties had begun to emerge. The existence of specific ribosomal binding sites for transfer and messenger RNA was demonstrated, forcing the conclusion that the ribosome plays an active role in protein synthesis and is not merely an inert workbench on whose surface amino acids are assembled. The observed physical complexity of the ribosome must therefore reflect the complexity of its function. What we needed to establish was the relation between structure and function. Yet as far as the actual roles of the RNA and proteins and their critical interrelations were concerned, the ribosome was still a mysterious "black box." How was one to understand this complicated piece of machinery? One could take it apart and try to reassemble it, but what tools were delicate enough to avoid destroying the machine in the process?

In 1961 Jacob, Sidney Brenner and Matthew S. Meselson, in a paper describing the classic experiments that proved the messenger-RNA theory, noted the presence of two kinds of ribonucleoprotein particle in mixtures that were centrifuged in a solution of the salt cesium chloride, which forms a density gradient in the centrifuge cell. (Density-gradient centrifugation, originally developed by Meselson, Franklin W. Stahl and Jerome R. Vinograd at the California Institute of Technology, separates large molecules on the basis of their different buoyancy in such a solution.) When they centrifuged bacterial extracts containing ribosomes, they observed two bands containing ribosomal particles. The lighter band (the B band), corre-

sponding to a density of 1.61, contained messenger RNA as well as proteins being synthesized; the heavier band (the A band), corresponding to the density of 1.65, did not.

The presence of the A band was not relevant to the main theme of the paper, and no reason was given for its presence. At the time the paper was published I was working at the University of Osaka on both ribosomes and messenger RNA, and I was quite curious about this phenomenon. We knew then that the 30S and 50S subunits and their aggregate, the 70S ribosome, all have the same chemical composition: 65 percent RNA and 35 percent protein. Since the buoyancy of complex molecules in solution usually reflects their chemical composition, why should density-gradient centrifugation reveal two kinds of ribonucleoprotein particle?

In the summer of 1962 I had an opportunity to visit Meselson's laboratory at Harvard to look into the question.

When we recovered the particles from the two bands, we found that the B band contained undegraded 50S and 30S ribosomal subunits. The denser A band, however, consisted of a mixture of smaller 40S and 23S "core" particles that had been created from the usual ribosomal subunits by the splitting off of about 40 percent of the protein during the density-gradient centrifugation; the split proteins could be found in a protein fraction at the top of the gradient. The explanation for the B band was apparently that in crude bacterial extracts some of the ribosomes are resistant to this splitting, perhaps because they are stabilized by messenger RNA and growing protein chains.

On returning to Osaka I continued experiments with Robert K. Fujimura to characterize these core particles. Initially we prepared 40S and 23S particles (the latter are not to be confused with 23S RNA molecules) by respectively centrifuging purified 50S and 30S ribo-

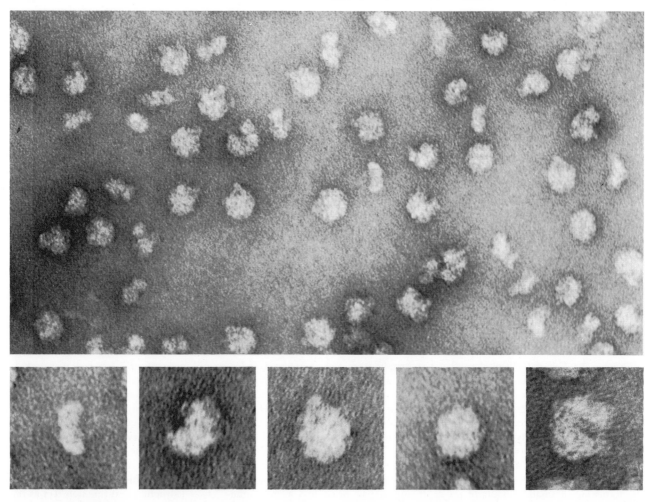

RIBOSOMAL SUBUNITS are enlarged 450,000 diameters in an electron micrograph (top) made by Martin Lubin of the Dartmouth Medical School. To prepare them, intact (70S) ribosomes were dissociated in solutions with a low magnesium concentration and the subunits were negatively stained with uranyl formate. Individual particles are enlarged 800,000 diameters (bottom). The smaller of the two subunits is the 30S (left). The larger is the 50S, seen in three different views characterized by a kidney shape (second from left), a "nose" (third from left) and a groove (fourth from left). The two subunits join to form a 70S ribosome (right).

somal particles in cold cesium chloride solution for 36 hours and then recovering the core particles from the band in the middle of the centrifuge cell [*see middle illustration on following page*]. This procedure was troublesome, however, and unsuitable for large-scale preparation of the particles. We therefore tried omitting the centrifuge step. Reasoning that it was surely the particular salt solution and not the physical centrifugation that disrupted the subunits, we simply kept 50S subunits in the solution for 36 hours, expecting that irreversible splitting of the protein would

take place, yielding the 40S core particles we wanted to study.

When we removed the cesium chloride and examined the products, however, we found to our surprise that the recovered particles behaved just like the original 50S ribosomal subunits. Why had there been no splitting of the proteins? We immediately realized the important implication of this experimental observation: The splitting of the 50S unit is reversible; the reaction is pushed in the direction of dissociation only by separation in the centrifuge. To test this

supposition we prepared core particles and split proteins by the usual centrifugation method. Then we mixed them together and removed the cesium chloride. We found complete conversion of the core particles to intact ribosomal particles. In this way we succeeded in reconstituting the 50S ribosomal subunit from the 40S core and split proteins that had been derived from the 50S, and also in reconstituting the 30S subunit from the 23S core and the homologous split proteins.

In order to prove that the reconstituted ribosomal particles really had the

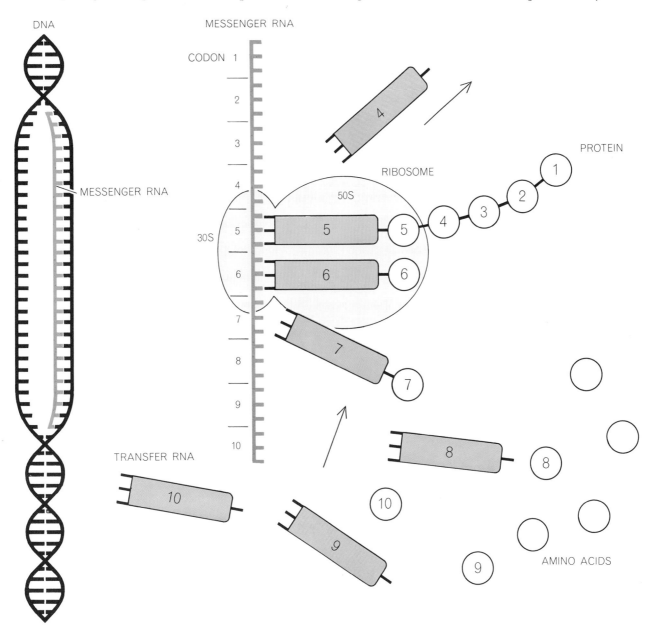

RIBOSOMES conduct protein synthesis. Genetic information is encoded in the sequence of bases (*horizontal elements*) in the double helix of DNA (*left*). This information is transcribed into a complementary sequence of RNA bases to form messenger RNA (*dark color*). Each group of three bases in the messenger RNA constitutes a codon, which specifies a particular amino acid and is recognized by a complementary anticodon on a transfer-RNA

molecule (*lighter color*) that has previously been charged with that amino acid. Here amino acid No. 6, specified by the sixth codon, has just been bound to its site on the ribosome by the corresponding transfer RNA. It will bond to amino acid No. 5, thus extending the growing peptide chain. Then the ribosome will move along the messenger RNA the length of one codon and so come into position to bind transfer RNA No. 7 with its amino acid.

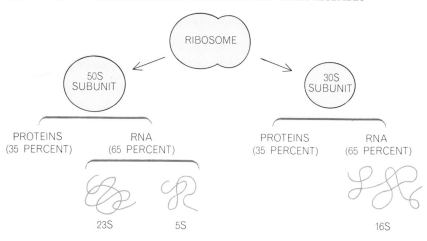

TWO SUBUNITS of a ribosome can be separated by spinning ribosomes in a centrifuge because the subunits are different sizes and move through the centrifuge cell at different rates. Both subunits are about 35 percent protein and 65 percent RNA. The 50S subunit contains a 23S and a 5S RNA molecule and the smaller 30S subunit has a 16S RNA molecule.

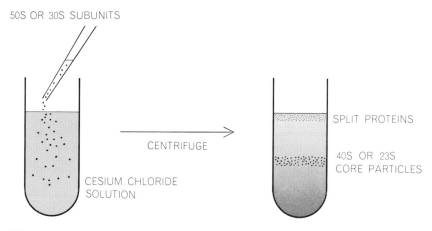

FURTHER CENTRIFUGATION of the subunits breaks them down. The subunits are added to a cesium chloride solution (left). Centrifugation establishes a stable density gradient in the solution (right), within which the subunit components form layers according to density. Some proteins split off, leaving "core" particles of RNA and other proteins.

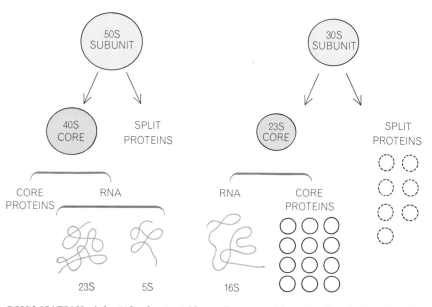

DISSOCIATION of the 50S subunit yields a 40S core particle and split proteins, dissociation of the 30S subunit a 23S core particle and split proteins. In the case of the 30S subunit there appear to be seven split proteins, with about 12 proteins remaining in the core.

same specific structure as the original ones, we had to demonstrate the functional integrity of the reconstituted particles. Before we could succeed in such experiments I left Osaka and moved to the University of Wisconsin. There Keiichi Hosokawa and I were able to show in 1965 that, whereas neither the 23S nor the 40S cores have any activity in a cell-free protein-synthesizing system, reconstituted 30S and 50S particles have activities comparable to the original intact 30S and 50S ribosomal subunits. At the same time Theophil Staehelin and Meselson, who were taking a similar approach at Harvard, independently succeeded in demonstrating the reconstitution of the ribosomal subunits from core particles and split proteins. The functional capabilities of the reconstituted particles can be assayed in various ways. For example, the function of 30S particles is usually assayed by measuring the rate of protein synthesis directed by messenger RNA in the presence of intact 50S subunits and other necessary components. One can also test the subunits' ability to bind several different transfer RNA's in the presence of various messenger RNA's and the ability to bind messenger RNA itself.

The success in reconstitution, although it involved the dissociation and reassociation of only some of the ribosomal proteins (the split proteins), had several important implications. First, the experiment showed that at least part of the ribosome assembly in the test tube is spontaneous; no extraribosomal template or enzyme is required. Second, it provided a system in which the functional roles of individual split proteins could be analyzed. The 30S split-protein fraction consists of seven proteins. By column chromatography we separated the proteins into five pure protein components and one fraction containing two proteins, and we showed that the five proteins differ from one another in amino acid composition. Then we determined the functional need for each of the purified proteins by omitting one of them at a time in reconstitution experiments. We found that three of the five purified proteins are essential for reconstitution, and that the omission of either one of the others has only a partial effect. From this type of experiment we could conclude that all five of the purified proteins are chemically and functionally distinct, and that some of them are absolutely required whereas others are not (although they are required for full activity in protein synthesis).

The partial-reconstitution system was a first step toward the functional analy-

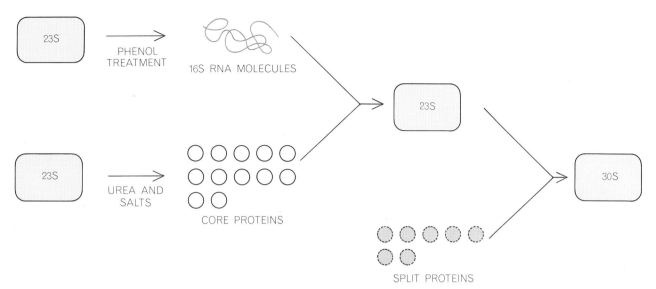

RECONSTITUTION of subunits was first accomplished as shown here. Core particles were treated with phenol to prepare 16S RNA and with urea and salts to yield the core proteins. The RNA and proteins were combined at 37 degrees Celsius to form 23S core particles. Then the reaction was completed by the addition of the split proteins, and in this way the 30S subunits were formed.

sis of the ribosome, but the information that could be provided by such a system was limited. To accomplish complete analysis we needed a way to reconstitute ribosomes entirely from free RNA and completely separated proteins. In 1967 Peter Traub and I began systematic attempts at complete reconstitution with the 30S subunit.

We assumed that we must do the reconstitution in two steps, first making 23S cores from 16S RNA and proteins and then making complete 30S subunits from the 23S cores and the split proteins. We therefore prepared 16S RNA by treating 23S cores with phenol. We separated proteins from other 23S cores by treating the core particles with urea and a high concentration of lithium chloride. We mixed these core proteins with the 16S RNA under several differ-

ent conditions, hoping to obtain 23S core particles. Then we added the split proteins to the reaction mixture, recovered the particles by centrifugation and assayed the activity of the recovered particles in a cell-free protein-synthesizing system. As typical enzyme chemists, we felt it was essential to protect the ribosomal proteins and any sensitive intermediates from inactivation by heat, and so we performed all these operations in a cold room and kept everything on ice.

Our initial attempts were failures. We could find only very slight protein-synthesizing activity. It seemed that the reaction might not be possible. Then we realized that living E. coli cells multiply most rapidly at 37 degrees Celsius (body temperature) and not at all at freezing temperatures, and that multiplying cells must certainly be assembling

ribosomes quite efficiently. We also recognized that the cytoplasm of living E. coli contains a rather high concentration of salts; the salts might discourage nonspecific RNA-protein aggregation and thereby promote the specific RNA-protein assembly reaction. We therefore attempted the reconstitution at 37 degrees and with a high concentration of a salt, potassium chloride. Success! The 30S ribosome could be self-assembled [see illustration above]. We found, indeed, that the reconstitution of 30S subunits from RNA and proteins is independent of the order of addition of proteins, whether they are core proteins or split proteins. By simply mixing all the proteins prepared directly from the 30S subunit with 16S RNA and incubating them in an optimal ionic environment at about 40 degrees C. for 10 minutes we were able to convert almost all

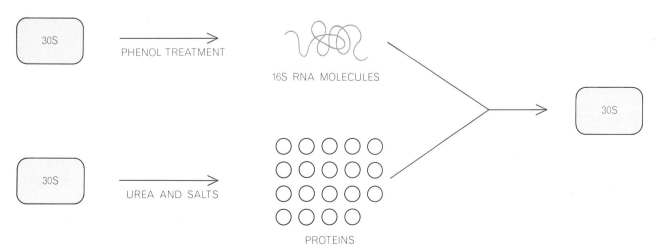

CORE AND SPLIT PROTEINS need not be assembled separately. Reconstitution can be accomplished by mixing all ribosomal proteins from 30S subunits with 16S RNA in the correct ionic environment and incubating at 37 to 40 degrees C. for 10 to 20 minutes.

the 16S RNA in the mixture into 30S particles [*see bottom illustration on page 211*]. In protein-synthesizing activity, in protein content and in sedimentation behavior the reconstituted particles were almost identical with the original 30S particles.

From this point one could proceed in many directions. One basic goal was to determine the functional role of the 16S RNA and of each 30S ribosomal protein. With regard to the RNA, we first considered the specificity requirements. For example, is an RNA molecule that is merely similar in size to *E. coli* 16S RNA competent to reconstitute a physically and functionally intact 30S subunit? No. Neither 16S ribosomal RNA from yeast cells nor *E. coli* 23S RNA degraded to a 16S-sized fragment was active in reconstitutions with *E. coli* 30S proteins. In fact, the inactive products of such combinations did not even resemble 30S ribosomes physically, judging by their sedimentation behavior. This finding certainly came as no sur-

prise; one would expect the requirements for a functional ribosome to be stringent. How stringent? When we performed the reconstitution with 16S RNA from one bacterial species and 30S ribosomal proteins from another species that is distantly related, we found that in many cases such "artificial" ribosomes were as active as the respective homologous RNA-protein combinations. We conclude from this that although there is definitely a specificity requirement for the RNA, the requirement is not absolute.

In determining the roles of the various ribosomal proteins our basic approach was to perform the reconstitution with one component omitted or specifically modified and then see if physically intact 30S particles were formed and, if so, whether or not they were functionally active. We first had to separate the 30S protein mixture into each of its 19 components. The fractionation of 30S ribosomal proteins had already been achieved by research groups at the University of Geneva, the Max Planck Institute for

Molecular Genetics in Berlin, the University of Wisconsin and the University of Illinois. We employed methods similar to theirs, relying mainly on various types of column chromatography.

We then did the reconstitution with 16S RNA and 19 purified proteins rather than with the unfractionated protein mixture used in our earlier experiments. The extent of the reconstitution was not as good as it was with unfractionated proteins (and we therefore could not exclude the possibility that there are some ribosomal components other than the RNA and 19 protein molecules), but the reasonably high efficiency of the reconstitution made it possible to undertake the functional analysis of the separated protein components.

The first protein we studied in detail was the one responsible for sensitivity or resistance to the antibiotic streptomycin. Earlier studies had indicated that the drug's primary site of action is the bacterial ribosome, specifically the 30S subunit. When streptomycin is added to a cell-free system containing ribosomes

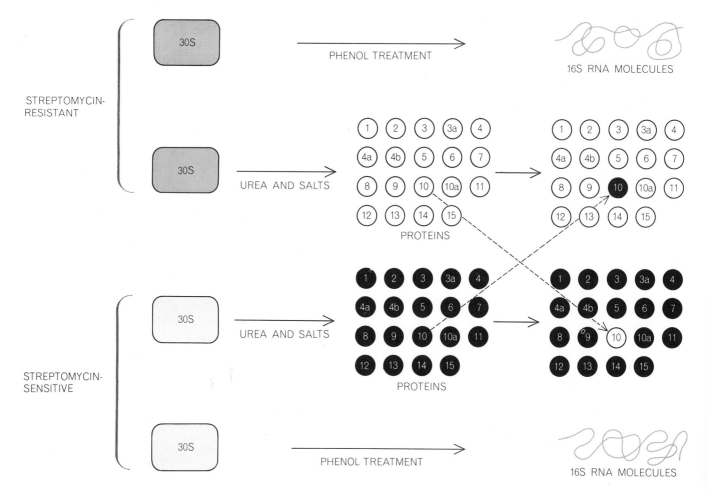

ONE PROTEIN in the 30S subunit is responsible for the effect of streptomycin on ribosome functioning. Proteins and RNA are separated from subunits derived from bacterial cells that are resistant to streptomycin and from other cells that are susceptible to its

effect. The proteins are isolated. When the protein designated *P*10 from susceptible cells is combined with all the other proteins from resistant cells and with the RNA from resistant cells, the 30S ribosomes that result turn out to be susceptible to streptomycin. On the from a streptomycin-sensitive strain, it inhibits protein synthesis. Streptomycin also causes the misreading of certain synthetic messenger RNA's, that is, it induces the incorporation into proteins of amino acids other than the ones dictated by the genetic code. This misreading effect of streptomycin was discovered first by Julian E. Davies, Walter Gilbert and Luigi Gorini at Harvard [see "Antibiotics and the Genetic Code," by Luigi Gorini; SCIENTIFIC AMERICAN Offprint 1041].

Bacteria can become resistant to streptomycin through mutation, and streptomycin does not inhibit synthesis or cause misreading of messages in a cell-free system if 30S ribosomal particles from streptomycin-resistant mutants are used. Traub and I showed that the component altered by the mutation was not the RNA but had to be in the protein fraction; 30S particles reconstituted from the protein of a resistant mutant and the RNA of a susceptible strain were resistant to streptomycin in cell-free protein-synthesizing systems,

whereas the reverse combination produced 30S particles that were susceptible to streptomycin. Makoto Ozaki and Shoji Mizushima took over the job of identifying the altered protein. We purified 30S ribosomal proteins from both susceptible and resistant bacteria, systematically substituted single proteins from a resistant strain in a mixture of proteins from a susceptible strain [see bottom illustration at left] and assayed the reconstituted ribosomes for their response to streptomycin. In this way we established that a single protein, one we had designated P10, determines the susceptibility of the entire 30S ribosomal particle to the inhibitory action of streptomycin, its susceptibility to streptomycin-induced misreading and its ability to bind the antibiotic.

Having learned how an alteration in a given protein can affect the function of the ribosome, we investigated what happens to the ribosome when this protein is simply left out. Is the ribosome still able to assemble itself and, if so, how are its functional capabilities altered? We found that in the absence of P10, RNA and the other ribosomal proteins can still assemble into particles that sediment at 30S. These P10-deficient particles have several interesting properties, however. Under the conditions of the assays these particles show high activity when a synthetic messenger RNA is used as a template, but their activity is weak when directed by RNA from a natural source. It is known that a special mechanism for initiating protein synthesis is needed in the system directed by natural messenger RNA's but not in the system directed by synthetic messenger RNA's [see the article "How Proteins Start," by Brian F. C. Clark and Kjeld A. Marcker, beginning on page 288]; the P10-deficient particles cannot carry out this special initiating function.

The other interesting finding is that the frequency of translation errors with the P10-deficient particles is much reduced not only in the presence of streptomycin but also in the presence of certain other antibiotics, of ethyl alcohol or of high concentrations of magnesium ions, all of which are known to induce translation errors. In fact, the deficient particles read synthetic messenger RNA more accurately in the presence of error-inducing agents than normal ribosomes do even in the absence of such agents. In other words, it appears that protein P10 plays a role in increasing the frequency of errors in the translation of the genetic message. The inherent ability of

ribosomes to make mistakes may be advantageous to bacterial cells, as Gorini suggested, since it can suppress the effects of harmful mutations. On the other hand, the property may simply be an unavoidable consequence of the complexity of the machinery.

Although we have not yet completed similar detailed analyses of all the proteins, preliminary experiments with them lead to certain general conclusions. Omitting any one of several proteins affects a number of known 30S ribosomal functions. Conversely, several different 30S ribosomal functions are affected by the omission of any of a number of proteins. That is, these functions seem to require the presence of more than one protein component, and so one can say that the 30S ribosomal proteins function cooperatively. We have also found that the omission of some proteins drastically affects physical assembly. Particles formed in the absence of one of these proteins are deficient in several other proteins, including some that were present in the reconstitution mixture. In other words, the presence of certain proteins is essential for other proteins to be bound. In this sense the assembly process itself is cooperative.

One of the most effective tools that are available for the study of a reaction is chemical kinetics, the study of the rate at which a reaction proceeds. Most chemical reactions have a distinct kinetic mode, or "order," which is determined experimentally from the reaction rate's dependence on the concentration of the reactants. Since the 30S reconstitution reaction involves at least 20 components, one might expect the rate to have a very high order of dependence on their concentration—so that doubling the concentration would increase the rate by as much as 2^{20}, or more than a million times! To be sure, this would be an absurdity, since it would mean that it was necessary for all the components to collide simultaneously in order to form a complete subunit, and subunit formation would be an incredibly rare event. As a matter of fact, most chemical reactions, even those involving a number of reactants, turn out to have a first-order or second-order dependence on reactant concentration (that is, doubling the concentration doubles or quadruples the reaction rate). The reason is that most complex reactions proceed in steps, with one unimolecular rearrangement or bimolecular interaction being slower than the others and hence determining the overall rate.

30S   STREPTOMYCIN-SENSITIVE
10

30S   STREPTOMYCIN-RESISTANT
10

other hand, when P10 is derived from resistant cells and all the other components are taken from susceptible cells, the resulting subunit is resistant to streptomycin.

This generalization, however, has in the past applied to reactions involving far fewer components than are needed for reconstitution, and so we expected to find a somewhat higher order of reaction. We were rather surprised to observe that our assembly was in fact a first-order reaction; whether we double or halve the concentration of the reactants, the time it takes for all the components to assemble themselves into completed ribosomes is roughly the same—about five minutes under optimal conditions in the test tube. When one observes first-order kinetics in a reaction involving more than a single component, one can conclude that there is an intermediate step, involving a relatively slow rearrangement of a single component, that must take place before the reaction can be completed. This is exactly what we observe in the reconstitution reaction. The slow step in the assembly process may be occurring at any time before the binding of the 20 proteins to the RNA molecule is complete—after none of them or only some of them are bound—or after all are bound; in any case, our observation would be the same.

We had noted with interest that the reaction is extremely dependent on temperature. It turned out that a considerable amount of heat energy is required to effect the rearrangement of the unknown intermediate product—about 40,-000 calories per mole of ribosomes. On the other hand, many of the proteins attach themselves in an ice-cold solution. We were therefore able to isolate the intermediate, activate it by warming the solution and then observe the almost instantaneous binding of the rest of the proteins to form a completed ribosomal subunit. And so we can describe the general nature of the pathway of self-assembly: a rapid binding of some of the proteins to the RNA, a slow structural rearrangement of this intermediate that requires thermal energy and then a rapid binding of the rest of the proteins.

After obtaining all this information on assembly in the test tube, one comes to an obvious question: Do the same principles operate in the living cell? The problem of ribosome synthesis in living cells was being attacked long before test-tube assembly was even seriously considered. The early work was done by the group including Britten, Brian J. McCarthy and Roberts at the Carnegie Institution, and they were followed by a number of groups, notably Shozo Osawa's in Japan and David Schlessinger's at Washington University. In a series of intricate experiments they delivered short pulses of radioactive components to growing bacterial cells and monitored the flow of the labeled components into ribosomes. It was possible in this way to postulate the presence of several classes of precursor particles, but it was difficult to isolate and analyze them. The obvious limitations of such an experimental approach encouraged investigators to seek other directions. One was genetics. Genetics has been a powerful tool for identifying the flow of intermediates in numerous biosynthetic pathways because one of the easiest ways to find out how something works is to see what happens when it does not work. For example, it should be possible to isolate mutants that are defective in a specific step in the biosynthesis of ribosomes; the step reveals itself by the accumulation of the precursor whose conversion is blocked by the mutational defect.

What was needed was a systematic method of isolating mutants defective in ribosome assembly. The trouble is that since ribosomes are essential for growth such mutants are ordinarily inviable and cannot be cultured. They can be isolated only as "conditionally lethal mutants": cells with defects such that the organism is inviable under one condition but functions normally under some other condition. As I have mentioned, in our detailed study of the test-tube reaction we had been struck by the remarkable dependence of the reaction rate on temperature. If the same principle operated in living cells, we reasoned, then many mutational defects in ribosomes or in related components should manifest themselves more severely at lower temperatures, and so some assembly-defective mutants should be conditionally lethal—viable at high temperatures but inviable at lower temperatures. They could therefore be isolated as cold-sensitive mutants.

Our reasoning proved to be correct. Christine Guthrie, Hiroko Nashimoto and I have isolated a large number of cold-sensitive mutants of E. coli, a significant fraction of which appear to be defective specifically in ribosome assembly. (Independently John L. Ingraham and his co-workers at the University of California at Davis have found abnormal ribosome biosynthesis at low temperatures in cold-sensitive mutants of the related bacterium Salmonella typhimurium.) By sucrose-gradient sedimentation we have already identified three distinct classes of particles, from three different mutants, that accumulate in cells grown at 20 degrees C. Two of these particles appear to be precursors of 50S subunits; the third appears to be a 30S precursor. While proceeding with the biochemical characterization of these particles we are also conducting genetic analyses of the various mutants in the hope of obtaining information on the genetic organization and genetic control of the ribosome and of ribosome assembly. We hope that through the coupling of genetic techniques with the biochemical techniques of test-tube reconstitution this sophisticated and complex cellular instrument will soon be understood on a truly molecular level.

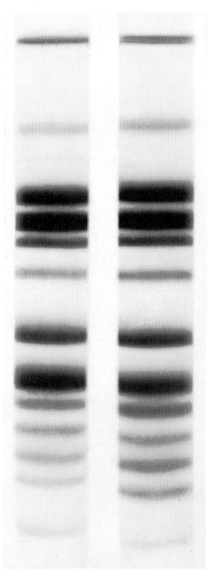

ALL THE PROTEINS normally found in a 30S subunit are found in a reconstituted 30S subunit, as shown by a comparison of electrophoresis results for the natural (left) and the reconstituted (right) particles. The protein mixtures are layered onto the top of a polyacrylamide gel column. When an electric current is applied, the proteins migrate down the column, each protein forming a band that moves at a different rate depending on the charge and the size of the molecule. Staining visualizes each protein.

20

THE MULTIPLICATION OF BACTERIAL VIRUSES

GUNTHER S. STENT

May 1953

The organisms that infect bacteria provide a means of studying the mechanism of heredity. Some new tracer experiments reveal that their reproduction has several rather unexpected features

THE PROCESS of heredity—how like begets like—is one of the most fascinating mysteries in biology, and all over the world biologists are investigating it with enthusiasm and ingenuity. Of the many angles from which they are attacking the problem, none is more exciting than the experiments on bacterial viruses. Here is an organism that reproduces its own kind in a simple and dramatic way. A virus attaches itself to a bacterium and quickly slips inside. Twenty-four minutes later the bacterium pops open like a burst balloon, and out come about 200 new viruses, each an exact copy of the original invader. What is the trick by which the virus manages to make all these living replicas of itself from the hodgepodge of materials at hand? What happens in the host cell in those critical 24 minutes?

Within the past few years studies with radioactive tracers have made it possible to begin to answer these questions. By labeling with radioactive atoms the substances of the virus or of the medium in which it multiplies, experimenters can follow these materials and trace the events that lead to the construction of a new virus. This article will tell about some of the experiments and the facts learned from them.

The bacterial virus, a tiny organism only seven millionths of an inch long, is a nucleoprotein: that is, a particle made up half of protein and half of nucleic acid. The latter is desoxyribonucleic acid—the well-known DNA which is a basic stuff of all cell nuclei [see "The Chemistry of Heredity," by A. E. Mirsky; SCIENTIFIC AMERICAN Offprint 28]. We are interested in the respective roles of the two parts of the virus molecule: the protein and the DNA. We are also interested in where the various materials

come from when a virus synthesizes replicas of itself inside the bacteria growing in a culture medium.

First let us consider the tracer technique. Suppose we wish to label the DNA part of the virus particles. Since an important constituent of DNA is its phosphate links, we shall label the element phosphorus with the radioactive isotope phosphorus 32. We begin with the medium in which we are growing bacteria that are to be infected by the virus. The culture contains inorganic phosphate as the source of phosphorus for the bacteria. To this medium we add a little radiophosphorus, so that there is one radioactive atom for every billion atoms of ordinary, non-radioactive phosphorus. The bacteria will take up the same proportion of radioactive and ordinary phosphorus. We can tell how much phosphate the bacteria contain simply by counting the radiophosphorus

atoms with a Geiger counter: the total amount of phosphorus is a billion times that.

Now if we infect the culture of bacteria with viruses, the virus progeny also will have the same proportion of radiophosphorus. But to measure their phosphorus we must isolate them, for the culture contains a great deal of phosphorus not incorporated in them. We can separate the viruses in three ways: (1) by a series of centrifuging operations that remove the other materials through their differences in weight; (2) by adding non-radioactive bacteria, on which the viruses become fixed and which can then be removed by low-speed centrifugation; or (3) by adding a serum (developed in rabbits) which contains antibodies that combine with the viruses and precipitate them from the culture.

Two radioactive isotopes are used in

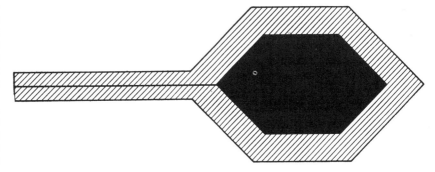

BACTERIAL VIRUS of the T2 strain, which infects the bacterium *Escherichia coli,* has a hexagonal head and a tail and is approximately seven millionths of an inch long. In this schematic drawing the virus is divided into two parts. Its outer layer (*diagonal lines*) is composed of protein which has the ability to attach itself to the surface of a bacterium of the appropriate species and to react with antivirus serum. Its core (*black*) is made up of nucleic acid, which is protected by the layer of protein.

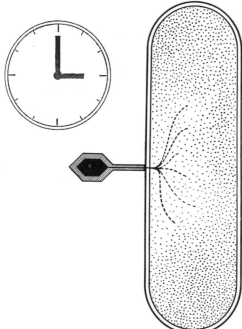

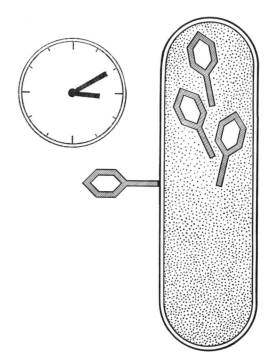

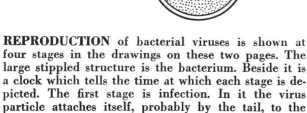

REPRODUCTION of bacterial viruses is shown at four stages in the drawings on these two pages. The large stippled structure is the bacterium. Beside it is a clock which tells the time at which each stage is depicted. The first stage is infection. In it the virus particle attaches itself, probably by the tail, to the surface of the bacterium. The nucleic acid core of the virus empties into the bacterial cell; the protein coat of the virus remains outside. The second stage, called the "dark period," is shown about 10 minutes later. The virus nucleic acid has begun to multiply within the bacterial cell, and has induced the formation of

the bacterial virus work: phosphorus 32 to label phosphate and the DNA part of the virus, sulfur 35 to label the protein part of the virus. Now let us look at the experiments.

ALL OF THESE experiments were done on bacterial viruses of the strain called T2, which infects the common bacterium *Escherichia coli.* Some years ago two investigators—Thomas F. Anderson of the University of Pennsylvania and Roger M. Herriott of The Johns Hopkins University—observed that something curious happened to bacterial viruses when they were exposed to "osmotic shock," namely, a sudden change in osmotic pressure effected by adding distilled water to the liquid in which they were suspended. These viruses could still attack and kill bacteria. But they had lost their ability to reproduce. Under the electron microscope they looked like sacs that had been emptied of their contents, and a chemical analysis indicated that they had lost all their DNA.

Recently A. D. Hershey and M. W. Chase, working at the Carnegie Institution of Washington genetics laboratory in Cold Spring Harbor, N. Y., repeated and confirmed these experiments with the help of radioactive tracers. The DNA, labeled with radiophosphorus, was indeed removed from the virus by osmotic shock. It remained as DNA in the solution, but it was easily broken

down by an enzyme—an indication that it had lost the protection of the protein "coat" of the virus. As for the protein shell of the virus, when separated from the solution and placed in a culture of bacteria it showed all its old power to seize upon and kill the bacteria. It also retained its ability to react with antivirus serum.

This looked very much as if the two parts of the virus had specialized functions. Apparently the virus' ability to attach itself to and kill a bacterium resided in its protein "coat." Did its power to reproduce and build hereditary images of itself reside in its DNA core? Other investigators had found that DNA did control hereditary continuity in bacteria. Hershey and Chase proceeded to investigate the question in their viruses.

They first put viruses in cultures of bacteria that had been killed by heat. The viruses attached themselves to the dead bacteria and apparently poured out their DNA, for the DNA (labeled with radiophosphorus) was easily broken down by the enzyme desoxyribonuclease, just as when it was spilled out from viruses after osmotic shock. Similarly, when bacteria were killed by heat after viruses had infected them, the enzyme again broke down the DNA. The enzyme had no effect, however, on DNA discharged into *living* bacteria. It seems that the living membrane of a bacterium protects DNA from the enzyme, but when the bacterium is killed, its mem-

brane becomes permeable and lets the enzyme through.

WHAT HAPPENS to the protein coat of the virus after it has emptied its DNA into the bacterium? Hershey and Chase infected living bacteria with virus, this time labeling the protein with radiosulfur. Then they shook up the suspension of infected bacteria in a Waring blender—the device used for stirring laboratory mixtures and for making milk shakes. The shearing force of the mixer stripped more than 80 per cent of the labeled protein off the bacteria. On the other hand, it did not remove any significant amount of DNA or interfere with the reproduction of viruses within the bacteria. The experiment showed that the virus protein stays outside the bacterium, and its job is finished as soon as it enables the DNA to gain entry into the cell. By the same token, it indicated strongly that the DNA is responsible for reproduction.

Once inside the host, the task of the nucleic acid is to reproduce itself 200-fold. It must also stimulate the production of 200 protein coats exactly like the one it has just shed. Where do the raw materials come from, and how are they put together?

In 1946 Seymour S. Cohen of the University of Pennsylvania, the first investigator to study bacterial virus reproduction with radioactive tracers, conceived

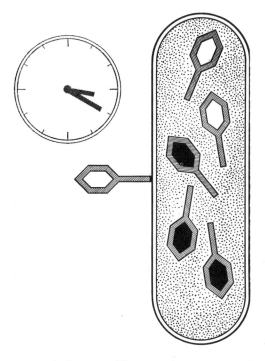

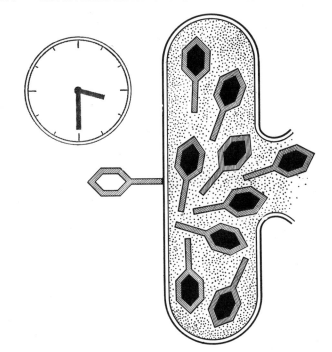

new protein coats. The protein coats contain no nucleic acid; there are no infective particles, not even the particle that caused the infection. The third stage, called the "rise period," is shown about 20 minutes after infection. Now some of the protein coats contain nucleic acid; the first infective particles of the new generation have made their appearance within the bacterial cell. The final stage is shown about 30 minutes after the first. The infected bacterium bursts and releases the new generation of virus particles into the surrounding medium. In the final drawing only a few of the 200 particles in the new generation of bacterial viruses are depicted.

an experiment directed to this question. He wished to find out whether the needed raw materials, particularly the phosphorus, came from the bacterial cell itself or from the medium surrounding it. He grew two cultures of bacteria, one in a medium containing radiophosphorus, the other in a non-radioactive medium. Then he removed the bacterial cells from the liquid in the two test tubes and switched them, putting the non-radioactive bacteria in the radioactive medium and *vice versa*. Now he infected both cultures with viruses. When the bacteria burst and the new viruses emerged, he isolated the viruses and measured their radioactivity. The viruses that came out of the non-radioactive bacteria transferred to the radioactive medium were radioactive: they had two thirds as high a concentration of radiophosphorus as the medium in which the bacteria had been immersed. On the other hand, the viruses that came from the radioactive bacteria in the non-radioactive medium had only one third as much radiophosphorus as the bacteria. Cohen therefore concluded that the new generations of viruses had obtained two thirds of their phosphorus from the growth medium while they were being formed and only one third from their host bacteria. This was a great surprise to those bacteriologists who had long supposed that bacterial viruses were formed from ready-made structures already present in the host cell.

At the University of Chicago Frank W. Putnam and Lloyd M. Kozloff, making similar studies with nitrogen 15 as the tracer, have found that the protein of viruses, like their DNA, is derived mostly from substances assimilated from the growth medium.

COHEN'S EXPERIMENT had covered the state of the system at just one stage: the next step was to follow the whole history of the conversion of inorganic phosphorus into virus DNA, from the moment bacteria began to grow in the medium until the newborn viruses finally emerged. At the State Serum Institute of Denmark Ole Maaloe and I extended Cohen's experiment with radiophosphorus, making the switch of radioactive bacteria to a non-radioactive medium and *vice versa* at many different stages in the development of the culture, both before and after infection of the bacteria with virus. In this way we were able to determine just how much of the phosphorus that the bacteria eventually donated to the new viruses was assimilated by them from the medium during the various periods of development. Before they were infected with virus, the bacteria took up that phosphorus at the rate of their own growth, which means that they were using the phosphorus to make their own DNA. But after infection, their assimilation of phosphorus that they were to donate to the viruses increased sharply. Most of the phos-

phorus the bacteria were now taking up was going directly into the synthesis of new viruses.

We also observed that it takes at least 12 minutes to convert inorganic phosphorus into virus DNA. Hence any phosphorus that is to go into the making of the new viruses must have been assimilated by the bacteria by the end of the first 12 minutes of the 24-minute period during which the viruses are synthesized in the cell. As a matter of fact, A. H. Doermann has found that the 24-minute latent period divides into two 12-minute phases. In experiments at Cold Spring Harbor he opened infected bacteria at various stages. During the first half of the latent period there were no fully formed viruses with infective power within the bacterial cell; even the original invader had disappeared. Then, after 12 minutes, the first infective particle appeared, and more followed until there were 200 just before the cell burst. The explanation is clear. The original invading virus had shed its protein coat on entering the cell and therefore was no longer an infective unit. No virus could appear in the cell until at least one new protein coat had been manufactured and coupled with a unit of DNA. Apparently this proceeding takes some 12 minutes.

It seems that the manufacture of protein and of DNA goes on side by side within the cell. In experiments with radiosulfur as the label, Maaloe and Neville Symonds of the California Insti-

THREE BACTERIAL VIRUSES of the T2 strain are shown in this electron micrograph made by Robley C. Williams and Dean Fraser of the Virus Laboratory at the University of California. At the upper right are two intact viruses; their tails and hexagonal heads are clearly visible. At the lower left is a virus from which the nucleic acid has been removed; as a consequence the head of the virus has collapsed. This reproduction of the micrograph magnifies viruses approximately 100,000 diameters.

tute of Technology have recently shown that by the time the first new infective virus appears, there is already enough virus protein in the cell to form about 60 viruses. On the other hand, in similar experiments with radiophosphorus as the label we have found indications that completed units of DNA do not unite with protein units until the last moment; the particle then becomes infective.

THE DNA of the original invading virus is responsible, as we have seen, for reproduction within the cell, both of DNA itself and of protein. How does it go about its job? Putnam and Kozloff labeled viruses with radiophosphorus

and followed the radioactivity to see what happened to the phosphorus after the viruses infected bacteria. They found that about 40 per cent of the labeled phosphorus showed up in the viruses' progeny, the rest being discarded in the debris. Experiments with radiocarbon have shown that the same is true of other constituents of the DNA. In other words, about 40 per cent of the DNA of the parent viruses is passed on to the descendants.

How is the old DNA passed on? Is the parent's DNA handed on intact to a single individual virus offspring in each bacterial cell in a random 40 per cent of the cases, or is it distributed generally

among the descendants? At Washington University of St. Louis Hershey, Martin D. Kamen, Howard Gest and J. W. Kennedy examined this question. They infected bacteria in a highly radioactive medium (one in every 1,000 phosphorus atoms was radioactive) with non-radioactive viruses. The DNA of the parent viruses was stable; not containing any radiophosphorus, it would not decay by radioactivity. Hence if it was passed on intact, a recognizable number of the viruses' descendants also should have stable DNA. But this was not the case. The descendant population steadily lost its infectivity, due to radioactive decay of its phosphorus atoms, until fewer than one tenth of 1 per cent of the descendants were infective.

Is it possible that the hereditary continuity of the virus resides in a fraction consisting of 40 per cent of the DNA, and that the rest of the DNA does not participate in reproduction at all? To answer this question Maaloe and James D. Watson at the State Serum Institute of Denmark produced three generations of virus. The first had its DNA labeled with radiophosphorus. A single virus of this generation then produced generation II, and passed on to it 40 per cent of its radiophosphorus. Now if the radiophosphorus transmitted from generation I to generation II was carried in a special reproductive fraction of the DNA, all of it should have been passed on to generation III. Actually it was found that generation III received only the usual 40 per cent. One must therefore conclude that the parent DNA material is not handed on in intact fractions but rather is distributed in a general fashion over the structures of the descendants.

SUMMING UP, the tracer studies so far have given us the following picture of how bacterial viruses reproduce themselves. By means of some property residing in its protein coat, a virus is able to attach itself to the surface of a bacterial cell. The contact immediately uncorks the virus, and it pours its DNA into the cell. The emptied protein coat is left outside the cell and thereafter plays no further part. Inside the cell the virus DNA begins to make replicas of itself, using as raw materials the nucleic acids of the bacterium and fresh substances absorbed by the bacterium from the medium surrounding it. About 40 per cent of the parent virus DNA itself is conserved and will reappear in the descendants. The virus DNA also induces the synthesis of new protein in the cell. Finally units of the protein combine with the DNA replicas to form 200 exact copies of the parent virus.

The facts discovered so far give us only an outline of the process, but they seem a good start on the road to solving the mystery of how organisms build structural copies of themselves and pass on their heredity from generation to generation.

21

THE STRUCTURE OF VIRUSES

R. W. HORNE

January 1963

The electron microscope reveals that these infectious particles possess three principal types of symmetry. Each species of virus is ingeniously assembled from just a few kinds of building block

When the smaller members of the virus family are enlarged several hundred thousand times in the electron microscope, they are found to possess an extremely high degree of structural symmetry. In such viruses it is probable that the subunits visible in electron micrographs are individual protein molecules, often identical in kind, packed together to form a simple geometric structure. In the larger viruses the geometry is usually more complex, and a certain degree of structural flexibility begins to appear. Viewing the micrographs one has the impression of being shown how the inanimate world of atoms and molecules shades imperceptibly into the world of forms possessing some of the attributes of life.

Viruses are the smallest biological structures that embody all the information needed for their own reproduction. Essentially they consist of a shell of protein enclosing a core of nucleic acid—either ribonucleic acid (RNA) or deoxyribonucleic acid (DNA). The shell serves as a protective jacket and in some instances as a means for breaching the walls of those living cells that the virus is capable of attacking. The nucleic acid core enters the cell and redirects the cell machinery toward the production of scores of complete virus particles. When the job is done, the cell ruptures and the viruses spill out.

Most viruses fall in a size range between 10 and 200 millimicrons; in other terms, between a fortieth of a wavelength and half a wavelength of violet light. Since objects smaller than the wavelength of light cannot be seen in an ordinary microscope, viruses can be observed directly only with the aid of the electron microscope. These instruments employ a beam of electrons whose wavelength is much smaller than the dimensions of a virus. Viruses can also be studied indirectly by placing crystals of a pure virus preparation in an X-ray beam and recording the diffraction patterns produced when the X rays are reflected from the planes of atoms in the crystal. Analysis of such X-ray diffraction patterns suggested that the protein subunits forming the virus shell were arranged symmetrically. The tobacco mosaic virus, for example, showed up in early electron micrographs as a slender rod without visible subunits. When the virus was examined by X-ray diffraction, however, one could see patterns suggesting that the subunits were arranged in a helix. On the other hand, most small viruses, which looked spherical in electron micrographs, gave rise to X-ray patterns indicating that they had a cubic symmetry. This suggested that they were regular polyhedrons and also members of the group of Platonic solids: solids with four, six, eight, 12 and 20 sides.

In the light of the X-ray results, and arguing from general principles, F. H. C. Crick and James D. Watson proposed in 1956 and 1957 that the amount of nucleic acid present in the small viruses was limited, and that the information it carried would be sufficient to code for only a few kinds of protein. They suggested, therefore, that the shells of small "spherical" viruses were probably built from a number of identical protein subunits packed symmetrically. The most likely way for identical units to be packed on the surface of a sphere, Crick and Watson pointed out, would be in some pattern having cubic symmetry.

Some of the predictions of Crick and Watson were subsequently confirmed by electron micrography. There was a period, however, when the design and development of the electron microscope outpaced methods of preparing virus specimens for observation. Dehydrated virus particles are essentially transparent to an electron beam. Various techniques have had to be devised to make the particles visible. One of the earliest and simplest methods was to create "shadows" by allowing a stream of heavy-metal atoms to fall on the virus particles at an angle. This was done by placing the specimen of virus particles in a vacuum chamber and evaporating the metal atoms from a source toward the side of the chamber. The metal atoms that accumulated on the virus particle itself would block the passage of electrons, whereas electrons could pass freely through the shadows where metal atoms had not been deposited. In this way it was possible to discern the overall shape of the virus particle but not all the fine details of its surface structure.

Within the past few years a new and simple method of "staining" isolated particles such as viruses and large protein molecules has been even more successful than shadowing for revealing fine detail at the high magnifications now available in electron microscopes. It consists of surrounding the particles to be examined by an electron-dense material: potassium phosphotungstate. This is achieved by mixing the virus suspension with a solution of the phosphotungstate and spraying the mixture or depositing droplets on the specimen mounts. Since the phosphotungstate method produces images that are reversed compared with those obtained with the normal preparation procedures, it is called "negative staining" or "negative contrast." Application of this method to a large number of viruses has shown that they fall into three main symmetry groups: those with cubic symmetry, those with helical symmetry and those with complex symmetry or combined symmetries.

The class of polyhedrons that have

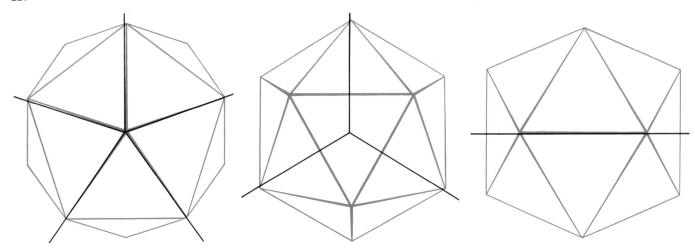

AXES OF SYMMETRY are shown for a regular icosahedron, a figure with 12 corners, 20 faces and 30 edges. Viewed along an axis at any corner, the figure can be rotated in five positions without changing its appearance (*left*). Rotated around any face axis, a regular icosahedron exhibits threefold symmetry (*middle*). Rotated around any edge axis, the figure shows twofold symmetry (*right*).

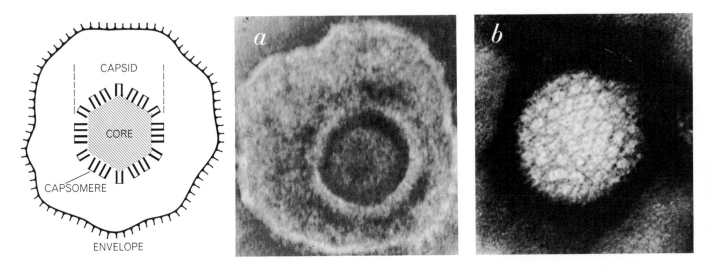

VIRUS NOMENCLATURE covers principal features observed in electron micrographs.

HERPES VIRUS sometimes has an envelope (*a*). The magnification is 310,000 diameters. The capsid (*b*) is composed of 162 capsomeres. Negative staining (*c*) indicates

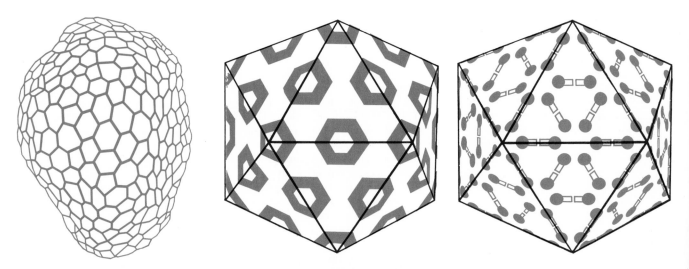

RADIOLARIANS, small marine organisms, have skeletons built of pentagons and hexagons.

ALTERNATIVE SCHEMES show how a regular icosahedron containing 42 pentagonal and hexagonal capsomeres (*left*) could be built up from 120 (or 240) small subunits.

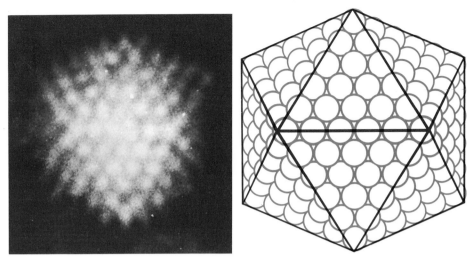

ADENOVIRUS is shown embedded in phosphotungstate, magnified about one million diameters (*left*). The drawing shows how the particle's 252 surface subunits, or capsomeres, are arranged with icosahedral symmetry. There are 12 on corners, 240 on faces or edges.

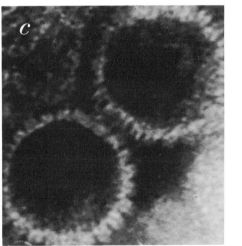

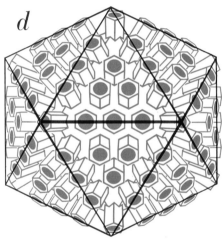

that they are hollow. The drawing (*d*) shows icosahedral arrangement. Micrographs are by P. Wildy and W. C. Russell of the Institute of Virology in Glasgow and the author.

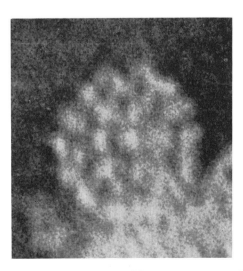

POLYOMA VIRUS is magnified one million diameters in micrograph by Wildy, M. G. P. Stoker and I. A. Macpherson of the Institute of Virology. It has 42 capsomeres (*right*).

cubic symmetry includes the regular tetrahedron (four faces), dodecahedron (12 faces) and icosahedron (20 faces). Shadowed preparations of the tipula iridescent virus, which causes a disease in the larvae of several insects, showed it to have the shape of a regular icosahedron, and the symmetry was self-evident [*see bottom illustration on page 223*]. Smaller viruses, on the other hand, do not reveal symmetry unless they are examined at very high magnification, and this requires the use of negative phosphotungstate staining.

Consider the symmetry properties of a regular icosahedron, in which each face is an equilateral triangle. If spokes are projected from the center of the icosahedron through the corners of the triangles, the spokes will represent one axis of rotational symmetry. Spokes projected from the center of the solid through the center of each face will represent a second axis. And spokes projected from the center through the midpoint of each edge will represent a third axis. (There will be 12 corner spokes, 20 face spokes and 30 edge spokes.) If the icosahedron is viewed along the spoke at any corner, one finds that the body can be rotated in five positions without changing its appearance [*see top illustration on opposite page*]. If the icosahedron is viewed along the spoke at any face, the body can be rotated in three positions without changing its appearance. And if the icosahedron is viewed along an edge spoke, it can be rotated in two positions without change of appearance. The regular icosahedron is thus said to have 5.3.2. symmetry.

Let us see now what implication this symmetry pattern has for a particle of adenovirus, which is associated with respiratory disease in man. The electron microscope shows that the surface of the particle is composed of regularly arranged structural units resembling tiny balls. Moreover, these balls are seen on the vertexes, faces and edges of an icosahedron [*see top illustration on this page*]. One can identify certain balls surrounded by five neighbors, which indicates that they are located on vertexes and therefore on axes of fivefold symmetry. Balls surrounded by six neighbors must lie on faces or edges and thus must occupy axes of either threefold or twofold symmetry. Along each edge there are six balls, including two balls occupying vertexes. To calculate the total number of balls covering the entire icosahedron one applies the simple formula $10(n - 1)^2 + 2$, where n is the number of balls along one edge. Substituting 6

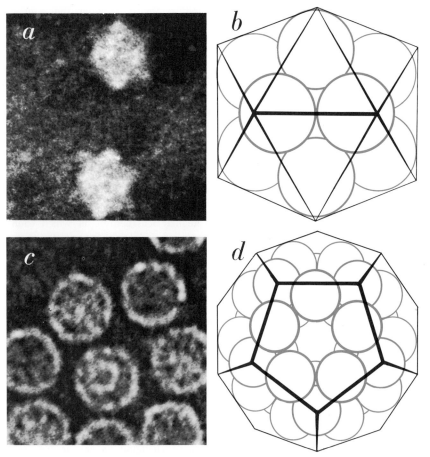

BACTERIOPHAGE ΦX174, magnified 750,000 diameters in *a*, appears to consist of 12 capsomeres arranged in icosahedral symmetry as shown in *b*. In other micrographs (*c*) smaller subunits seem to be arranged in ringlike structures. Each capsomere might actually be formed from five subunits as shown in *d*. Thirty such subunits would form a dodecahedron.

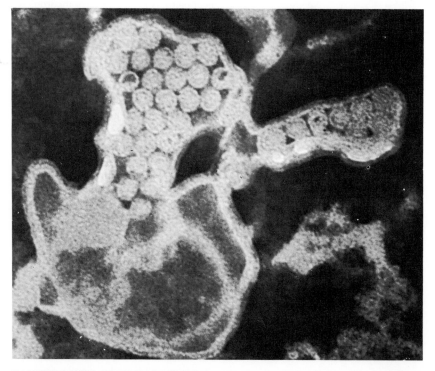

POLIOMYELITIS VIRUS PARTICLES are shown inside a fragment of an infected cell. The particles, magnified 250,000 diameters, appear to be composed of subunits smaller than the typical capsomere. The structural arrangement is not established. Electron micrograph is by Jack Nagington of the Public Health Laboratory in Cambridge and the author.

for *n* yields 252 as the number of morphological units composing the shell of the adenovirus particle.

For purposes of description (and to avoid the term "subunit," which can be applied to morphological, structural or chemical features) I shall adopt the recent terminology suggested for the various viral components [*see illustration at middle left, page 220*]. The morphological units composing the shell have been given the name "capsomeres." The shell itself is the "capsid." The region inside the capsid is the "core." The outer membrane, seen surrounding the capsid of some viruses, is the "envelope."

One merit of the negative-staining technique is that the electron-dense material is capable of penetrating into extremely small regions between, and even within, the capsomeres. A striking instance of such penetration can be seen in the electron micrograph of the herpes virus shown at the middle left on the preceding page. (In man the herpes virus causes, among other things, "cold sores.") Electron micrographs of the shadowed particle had indicated that it had the same external shape and symmetry as the adenovirus. When the two viruses were negatively stained and still further magnified, however, it could be seen on close examination that the capsomeres of the herpes virus, unlike those of the adenovirus, were elongated hollow prisms, some hexagonal in cross section and others pentagonal. In a number of particles the phosphotungstate penetrated into the central region, or core, normally containing the nucleic acid. In these "empty" particles the elongated capsomeres stand out clearly in profile at the periphery of the virus, and one can see their hollow form and the precision of their radial arrangement.

From the micrographs the number of capsomeres located on each edge was estimated to be five, giving a total of 162 capsomeres for the herpes virus. Of the 162 capsomeres, 12 are pentagonal prisms and 150 are hexagonal prisms. To satisfy the packing arrangement in accordance with icosahedral symmetry, the 12 pentagonal prisms would have to be placed at the corners and the 150 hexagonal prisms located on the edges or faces of the particle [*see drawing at middle right on preceding page*].

The need for pentagonal units goes deeper than the simple need to satisfy icosahedral symmetry. As early geometers observed, there is no way to arrange a system of hexagons so that they will enclose space. But if pentagonal units are included with hexagons, it is possible to enclose space in an almost

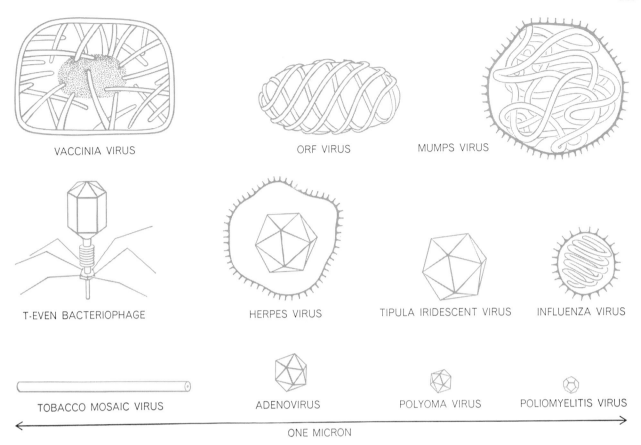

VACCINIA VIRUS

ORF VIRUS

MUMPS VIRUS

T-EVEN BACTERIOPHAGE

HERPES VIRUS

TIPULA IRIDESCENT VIRUS

INFLUENZA VIRUS

TOBACCO MOSAIC VIRUS

ADENOVIRUS

POLYOMA VIRUS

POLIOMYELITIS VIRUS

ONE MICRON

RELATIVE SIZES OF VIRUSES are shown in this chart. A micron, used as a measuring stick, is a thousandth of a millimeter; it is enlarged 175,000 times. The five viruses with polyhedral structures possess cubic symmetry. The tobacco mosaic virus and the internal components of influenza and mumps virus have helical symmetry. The remaining viruses exhibit complex symmetry.

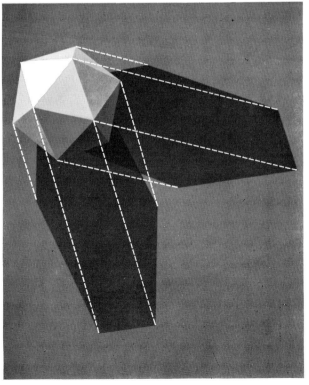

TIPULA IRIDESCENT VIRUS, an insect virus, is so large that its geometrically regular structure shows up clearly when specimens are shadowed with atoms of a heavy metal and enlarged in the electron microscope. In the doubly shadowed micrograph (left) the virus particles are enlarged about 58,000 diameters. The shadows indicate that each particle is a regular icosahedron (right). The micrograph was made by Kenneth Smith of the University of Cambridge and Robley C. Williams of the University of California.

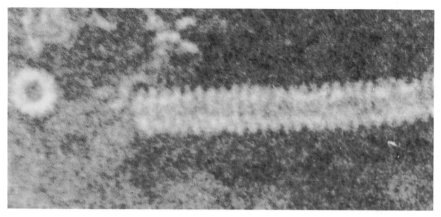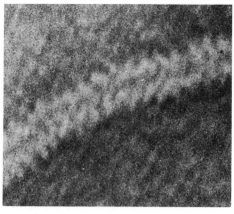

HELICAL SYMMETRY is shown in the electron micrographs of the rodlike tobacco mosaic virus, magnified 800,000 diameters, at left. The second electron micrograph, of the same magnification, shows an internal thread from a disrupted member of the myxovirus group. It too seems to possess helical symmetry. (Intact myxovirus particles are shown directly below.) The tobacco mosaic virus has

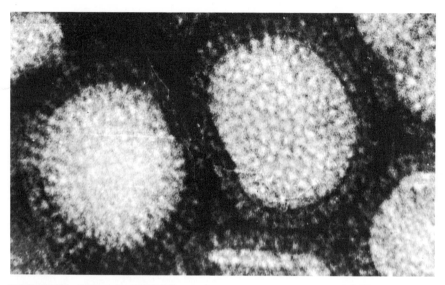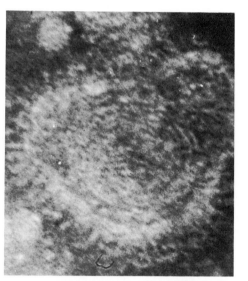

INFLUENZA VIRUS PARTICLES, members of the myxovirus family, are magnified 700,000 diameters in the electron micrograph at left. Although the particles are irregular in both size and shape, they appear to bristle with regularly spaced surface projections. In the second micrograph, which has a magnification of 600,000 diameters, phosphotungstate has penetrated the core of a particle, reveal-

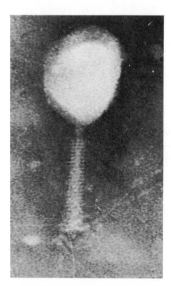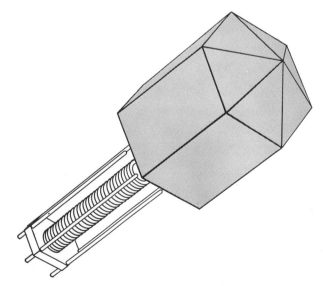

COMPLEX SYMMETRY is displayed by the T2 bacteriophage and other members of the "T even" family. Electron micrographs, in which the particle is magnified 300,000 diameters, clearly show that T2 exists in "untriggered" and "triggered" forms. The untriggered form is shown in the first pair of illustrations. The head of the phage is a bipyramidal hexagonal prism. The tail is a tube-like structure surrounded by a helical sheath. An end plate carries six tail fibers. When triggered, as shown in the second pair of illus-

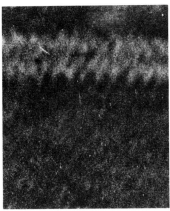

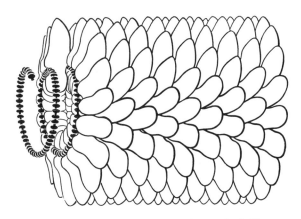

2,130 elongated capsomeres, consisting of protein molecules, arranged around a hollow core, as shown in the diagram at far right. The helical coil embedded in the capsomeres represents viral nucleic acid. The micrographs are by Nagington, A. P. Waterson and the author.

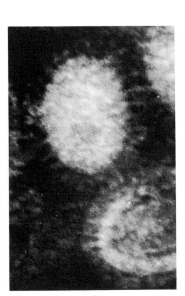

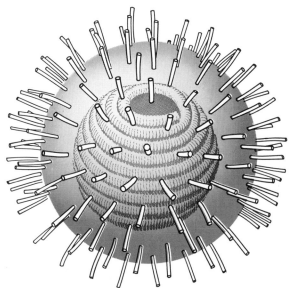

ing a coiled structure inside. The diagram at right shows a possible arrangement of the components in a typical myxovirus. The diagram follows a model built by L. Hoyle, Waterson and the author. The micrographs are by Waterson, Wildy, A. E. Farnham and the author.

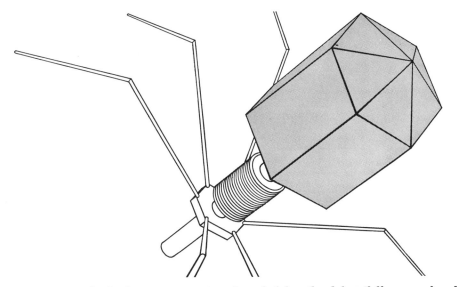

trations, the sheath contracts away from the end of the tail and the tail fibers are released. Presumably this coincides with the ejection of the DNA core (*not shown*), which previously had been coiled up in the head of the particle. The micrographs are by Sydney Brenner, George Streisinger, S. Champe, Leslie Barnett, Seymour Benzer, M. Rees and the author.

endless variety of ways, with forms both regular and irregular. The radiolarians, a group of marine protozoa, provide a fascinating example of varied structures assembed from pentagonal and hexagonal units [*see illustration at bottom left on page 220*].

Viruses smaller than the herpes virus usually have fewer capsomeres, but the relation between size and capsomere number is somewhat variable. The polyoma virus, which produces tumors in rodents and has stimulated a search for viruses in human cancer, appears to be almost spherical when examined by the shadowing technique. Nevertheless, negative staining shows that the outer shell is probably composed of 42 elongated angular capsomeres arranged in icosahedral symmetry [*see bottom illustration, page 221*]. Such a shell can be constructed by placing 12 pentagonal prisms at the corners of an icosahedron and 30 hexagonal prisms on the 30 edges. In this case the 20 faces have no capsomeres of their own, which helps to explain the nearly spherical appearance of the virus.

In the electron microscope the turnip yellow mosaic virus, which causes a disease of the leaves in the turnip and-related plants, appears to have 32 capsomeres arranged in accordance with cubic symmetry. Crystals of the same virus studied by X-ray diffraction also show cubic symmetry, but this method indicates that there are 60 subunits instead of 32. Strictly speaking, neither number can be used to construct an icosahedron. But both numbers of subunits can be disposed symmetrically on the surface of an icosahedron. The smaller number can be distributed by placing 12 subunits on corners, 20 on faces and none on edges. (The 32 capsomeres could also be placed on the 32 vertexes of a pentakis dodecahedron or a rhombic triacontahedron.) The larger number can be distributed according to strict icosahedral symmetry by placing two subunits on each of the 30 edges and none on corners or faces. It is evident that if the two figures were transparent, one could be fitted over the other and the subunits of one would fall precisely in between the subunits of the other without overlapping. This suggests that the 60 subunits inferred from X-ray diffraction patterns may combine in some fashion to give the appearance of 32 subunits when the virus particle is observed in the electron microscope.

It has therefore been suggested that in the small spherical viruses the morphological features resolved as pentagons

VIRUS	SYMMETRY	NUMBER OF CAPSOMERES	SIZE OF CAPSID (ANGSTROM UNITS)	NUCLEIC ACID
TIPULA IRIDESCENT	CUBIC	812	1,300	DNA
ADENOVIRUS	CUBIC	252	700–750	?
GAL (GALLUS ADENO-LIKE)	CUBIC	252	950–1,000	?
INFECTIOUS CANINE HEPATITIS	CUBIC	252	820	?
HERPES SIMPLEX	CUBIC	162	1,000	DNA
WOUND TUMOR	CUBIC	92	?	RNA
POLYOMA	CUBIC	42	450	DNA
WARTS	CUBIC	42	500	?
TURNIP YELLOW MOSAIC	CUBIC	32	280–300	RNA
ΦX174	CUBIC	12	230–250	DNA
TOBACCO MOSAIC	HELICAL	2,130	3,000 × 170	RNA
MUMPS	HELICAL	—	170 (DIAMETER)	RNA
NEWCASTLE DISEASE	HELICAL	—	170 (DIAMETER)	RNA
SENDAI	HELICAL	—	170 (DIAMETER)	?
INFLUENZA	HELICAL	—	90–100 (DIAMETER)	RNA
T-EVEN BACTERIOPHAGE	COMPLEX	—	1,000 × 800 (HEAD)	DNA
CONTAGIOUS PUSTULAR DERMATITIS (ORF)	COMPLEX	—	2,600 × 1,600	?
VACCINIA	COMPLEX	—	3,030 × 2,400	DNA

TABLE OF VIRUSES shows the symmetry classification, number of capsomeres and capsid size of some of the principal families. (An angstrom unit is a ten-millionth of a millimeter; the wavelength of violet light is 4,000 angstrom units.) Nucleic acid (*column at far right*) is the genetic material of the virus. DNA is deoxyribonucleic acid; RNA, ribonucleic acid.

and hexagons may actually be built up from smaller structural subunits. These subunits may not all be identical, but they may be of two or three different molecular species. The diagram at the bottom right on page 220 indicates how such subunits might be assembled to produce pentagonal and hexagonal units, in strict accordance with icosahedral symmetry. The arrangement illustrated, one of several possible combinations, was proposed by A. Klug, D. L. D. Caspar and J. Finch of the University of London. It shows how 42 capsomeres could be formed from 120 (or 240) smaller subunits. Recent evidence suggests that the capsomeres in some of the larger viruses are linked together by small structures that may well correspond to the subunits.

High-resolution electron micrographs have revealed that structures originally identified as capsomeres in one very small virus are indeed composed of still smaller subunits. The virus, known as φX174, has been intensively studied because it contains an unusual single-stranded form of DNA [see "Single-stranded DNA," by Robert L. Sinsheimer; SCIENTIFIC AMERICAN Offprint 128]. When first examined in the elec-

tron microscope, the virus appeared to have a shell composed of 12 spherical capsomeres, the minimum number needed for icosahedral symmetry. More recent electron micrographs indicate that each capsomere is formed from five subunits, but since each capsomere may be shared with a neighbor, the number of subunits is 30 [see top illustration on page 222]. If they are not shared and each capsomere is composed of five subunits, the total would be 60 and the shape would be that of a dodecahedron. Similar subunits smaller than capsomeres have been observed in electron micrographs of the virus of poliomyelitis, but it has not yet been possible to count them accurately [see bottom illustration on page 222].

The second broad group of viruses I shall discuss are those that have helical symmetry. Far and away the best known of this group is the virus that causes the mosaic disease of tobacco. Its helical structure was originally inferred from X-ray diffraction data. These data, combined with evidence from other physical and chemical observations, have led to a detailed knowledge of the tobacco mosaic virus' architecture. The

subunits appear to be elongated structures so arranged that about 16 subunits form one turn of a helix. The subunits project from a central axial hole that runs the entire length of the virus. The nucleic acid of the virus does not occupy the hole, as might be expected, but is deeply embedded in the protein subunits and describes a helix of its own. The virus is composed of 2,130 identical protein subunits. Each subunit is a large molecule formed by the joining together of 168 amino acid molecules. The diagram of the virus' structure at the top of this page is based on a model by R. E. Franklin, Klug, Caspar and K. Holmes of the University of London.

Until recently helical symmetry was observed only in plant viruses. Now it has also been found in the complex animal viruses that are members of the influenza, or myxovirus, group. The group includes the viruses of mumps, Newcastle disease (a respiratory ailment of fowl), fowl plague and Sendai disease (a form of influenza). Electron micrographs produced by the shadow-casting technique showed these viruses to be of various shapes and sizes. Some were roughly spherical, some were filaments and others were complex and irregular. Thin sections of purified virus and particles seen at the surface of infected cells suggested the existence of an internal component in the form of ringlike structures surrounded by an outer membrane.

Recent studies using the negative staining method have shown that the internal component, or capsid, has the same dimensions and appearance as the rods of tobacco mosaic virus but is more flexible. This is particularly evident in electron micrographs of mumps virus, which show that the helical capsid forms coils or loops after being released. The particles of influenza and fowl plague are more structurally compact than the mumps virus and, unless subjected to special chemical treatment, are rarely observed releasing their internal components.

The envelopes of influenza virus and fowl plague virus carry surface projections that evidently contain the protein known as hemagglutinin, so named because it causes red blood cells to agglutinate. If these two viruses are treated with ether, the internal helix is released and can be separated from the hemagglutinin in a centrifuge. When this inner component is studied by electron microscopy, it is found to be of smaller diameter than that in the viruses of mumps, Newcastle disease virus and Sendai disease virus. The precise length of the helical components in the various myxoviruses is not yet known, nor the way they are packed within their envelopes. A possible arrangement for a typical myxovirus is shown in the diagram at the middle on the preceding page.

The last of the three broad groups of viruses are those whose symmetry is complex. This category includes the large bacterial viruses, such as the T2 virus that infects the bacterium *Escherichia coli*, and the large pox viruses. The T2 virus and several of its "T even" relatives are particularly remarkable because they contain some sort of contractile mechanism, a feature that has not been discerned in any other family of viruses. The electron micrographs at the bottom of page 224 show that the T2 virus has a head shaped in the form of a bipyramidal hexagonal prism. Attached to one end of the prism is a tail sructure consisting of a helical contractile sheath surrounding a central hollow core. At the extreme end of the core there is a curious hexagonal plate carrying six slender tail fibers. The plate structure and tail fibers probably make initial contact with the wall of the bacterium that is being attacked. After contact has been made the helical sheath contracts, allowing the nucleic acid core of the virus to enter the bacterium.

The contraction of the T2 sheath raises many fascinating questions. The entire T2 virus appears to contain only a few different kinds of protein molecule. If these are allocated to the construction of the different structures— head, sheath, tail plate and tail fibers— one must conclude that the contractile sheath is composed of only two or at most three different kinds of protein. How can so few kinds of building block produce a sheath with contractile ability? What substances trigger the contraction? And how is the contraction related to the ejection of the long DNA molecule that is tightly packed in the T2 core?

Still larger viruses having complex symmetry are several important members of the pox virus family: the viruses of variola, vaccinia, cowpox and ectromelia. They are among the few viruses large enough to be seen in the light microscope. In early shadowed electron micrographs the vaccinia virus appeared to have a three-dimensional bricklike shape with a spherical dense central region. More detailed studies of the virus seen in infected cells after staining and thin sectioning revealed morphological features not observed in other viruses. The central dense region appeared to be surrounded by a number of layers, or membranes, of varying opacity to the electron beam. In some micrographs tubelike structures could be seen between the outer membranes and the central region. The electron micrographs below illustrate the structural variations that exist between two members of the pox group. In the particles of the virus that causes orf, or contagious pustular dermatitis, the tubular components form a definite crisscross pattern. It is difficult to say whether the tubular structures should be described as capsids or as capsomeres, nor can one say just where the nucleic acid is located in relation to them.

The electron microscope, together with other methods, has greatly contributed to the study of viruses, and it has shown that they come in a surprising variety of mathematically ordered families. It has been understood for many years, of course, that proteins are versatile building blocks and that they account for the tremendous diversity of living forms. But it required the electron microscope to reveal directly what intricate and exquisite structures can be created by putting together only a few kinds of protein molecule.

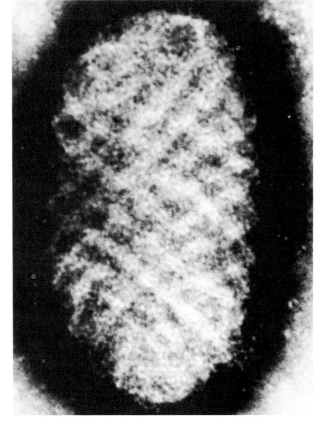

VACCINIA VIRUS, one of the giant pox viruses, is about twice the diameter of the smallest living cells, which are known as pleuro-pneumonia-like organisms. The magnification is 400,000 diameters.

ORF VIRUS, another pox virus, has components wound in a criss-cross pattern. The magnification is 450,000 diameters. Micrographs of the orf and vaccinia viruses are by Nagington and the author.

22

BUILDING A BACTERIAL VIRUS

WILLIAM B. WOOD AND R. S. EDGAR
July 1967

T4 viruses with mutations in certain genes produce unassembled viral components. These particles are combined in the test tube in an effort to learn how the genes of a virus specify its shape

Slice an orange in half, squeeze the juice into a pitcher and then drop in the rind. It comes as no surprise that the orange does not reconstitute itself. If, on the other hand, the components of the virus that causes the mosaic disease of tobacco are gently dissociated and then brought together under the proper conditions, they do reassociate, forming complete, infectious virus particles. The tobacco mosaic virus consists of a single strand of ribonucleic acid with several thousand identical protein subunits assembled around it in a tubular casing. The orange, of course, is a large and complex structure composed of a variety of cell types incorporating many different kinds of proteins and other materials. Yet both orange and virus are examples of biological architecture that must arise as a consequence of the action of genes.

Molecular biologists have now provided a fairly complete picture of how

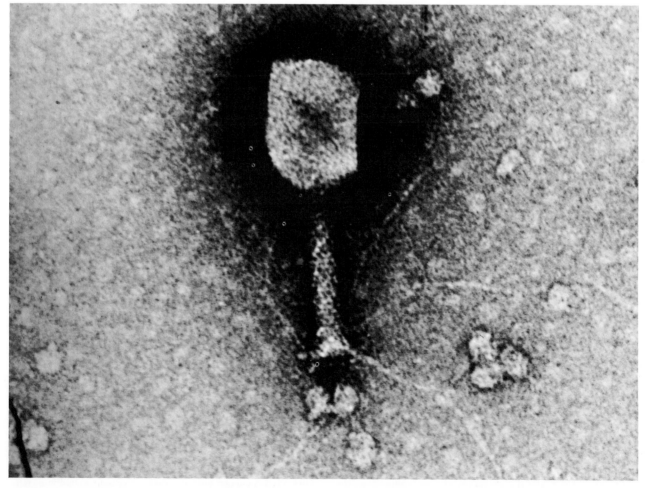

COMPLETE T4 PARTICLE was built by assembling component parts in the test tube. The virus is enlarged about 300,000 diameters in this electron micrograph made, like the ones on the next page, by Jonathan King of the California Institute of Technology.

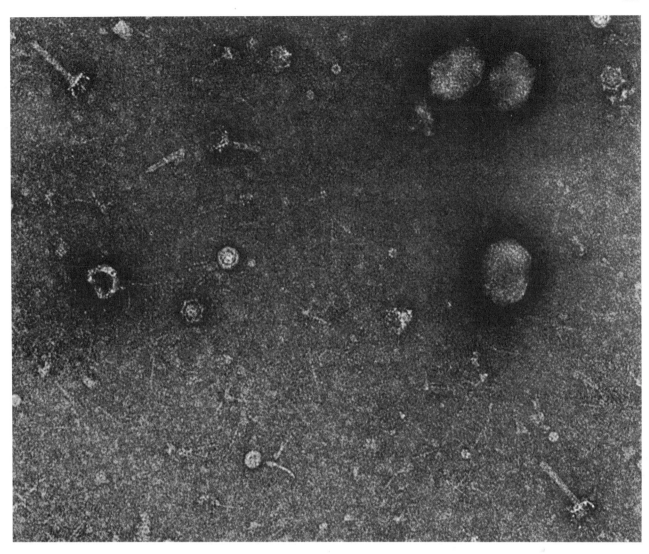

UNASSEMBLED PARTS of the T4 virus are present in this extract. It was prepared by infecting colon bacilli with a mutant virus defective in gene No. 18, which specifies the synthesis of the sheath (*see upper illustration on page 230*). The result is the accumulation of all major components except the sheath: heads, free tail fibers and "naked" tails consisting of cores and end plates.

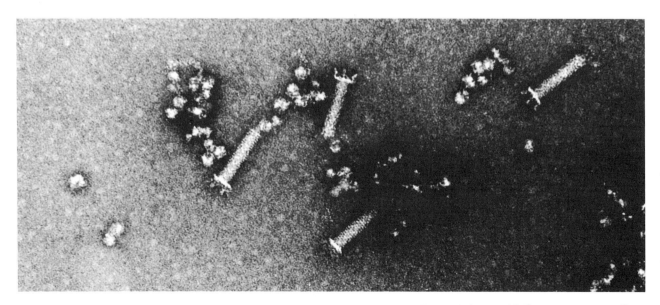

COMPLETE TAILS, enclosed in sheaths, were produced by a different mutant, defective in a gene involved in head formation. The tails were separated from the resulting extract (along with some spherical bacterial ribosomes) by being spun in a centrifuge. If the tails are added to the extract (*top photograph*), they combine with the heads and free fibers in it to form infectious virus.

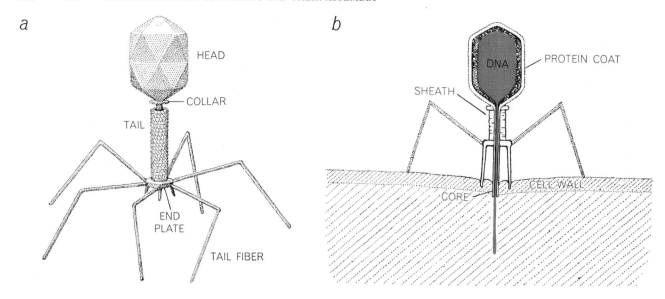

T4 BACTERIAL VIRUS is an assembly of protein components (*a*). The head is a protein membrane, shaped like a kind of prolate icosahedron with 30 facets and filled with deoxyribonucleic acid (DNA). It is attached by a neck to a tail consisting of a hollow core surrounded by a contractile sheath and based on a spiked end plate to which six fibers are attached. The spikes and fibers affix the virus to a bacterial cell wall (*b*). The sheath contracts, driving the core through the wall, and viral DNA enters the cell.

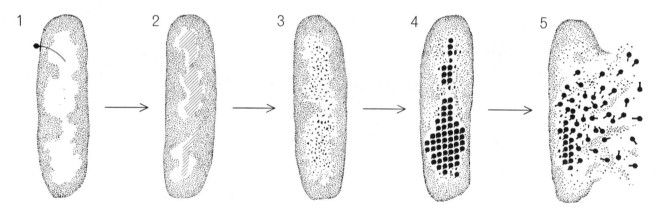

VIRAL INFECTION begins when viral DNA (*color*) enters a bacterium (*1*). Bacterial DNA is disrupted and viral DNA replicated (*2*). Synthesis of viral structural proteins (*3*) and their assembly into virus (*4*) continues until the cell bursts, releasing particles (*5*).

genes carry out their primary function: the specification of protein structure. The segment of nucleic acid (DNA or RNA) that constitutes a single gene specifies the chain of amino acids that comprises a protein molecule. Interactions among the amino acids cause the chain to fold into a unique configuration appropriate to the enzymatic or structural role for which it is destined. In this way the information in one gene determines the three-dimensional structure of a single protein molecule.

Where does the information come from to direct the next step: the assembly of many kinds of protein molecules into more complex structures? To build the relatively simple tobacco mosaic virus no further information is required; the inherent properties of the strand of RNA and the protein subunits cause them to interact in a unique way that results in the formation of virus particles. Clearly such a self-assembly process cannot explain the morphogenesis of an orange. At some intermediate stage on the scale of biological complexity there must be a point at which self-assembly becomes inadequate to the task of directing the building process. Working with a virus that may be just beyond that point, the T4 virus that infects the colon bacillus, we have been trying to learn how genes supply the required additional information.

Although the T4 virus is only a few rungs up the biological ladder from the tobacco mosaic virus, it is considerably more complex. Its DNA, which comprises more than 100 genes (compared with five or six in the tobacco mosaic virus), is coiled tightly inside a protein membrane to form a polyhedral head. Connected to the head by a short neck is a springlike tail consisting of a contractile sheath surrounding a central core and attached to an end plate, or base, from which protrude six short spikes and six long, slender fibers.

The life cycle of the T4 virus begins with its attachment to the surface of a colon bacillus by the tail fibers and spikes on its end plate. The sheath then contracts, driving the tubular core of the tail through the wall of the bacterial cell and providing an entry through which the DNA in the head of the virus can pass into the bacterium. Once inside, the genetic material of the virus quickly takes over the machinery of the cell. The bacterial DNA is broken down, production of bacterial protein stops and within less than a minute the cell has begun to manufacture viral proteins under the control of the injected virus genes. Among the first proteins to be made are the enzymes needed for viral DNA replication, which begins five minutes after infection. Three minutes later a second set of genes starts to direct the synthesis of the structural proteins that will form the head components and the tail components, and the process of viral morpho-

genesis begins. The first completed virus particle materializes 13 minutes after infection. Synthesis of both the DNA and the protein components continues for 12 more minutes until about 200 virus particles have accumulated within the cell. At this point a viral enzyme, lysozyme, attacks the cell wall from the inside to break open the bacterium and liberate the new viruses for a subsequent round of infection.

Additional insight into this process has come from studying strains of T4

carrying mutations—molecular defects that arise randomly and infrequently in the viral DNA during the course of its replication [see "The Genetics of a Bacterial Virus," by R. S. Edgar and R. H. Epstein; SCIENTIFIC AMERICAN Offprint 1004]. When a mutation is present, the protein specified by the mutant gene is synthesized in an altered form. This new protein is often nonfunctional, in which case the development of the virus stops at the point where the protein is required. Normally such a mutation has little experimental use, since the virus in

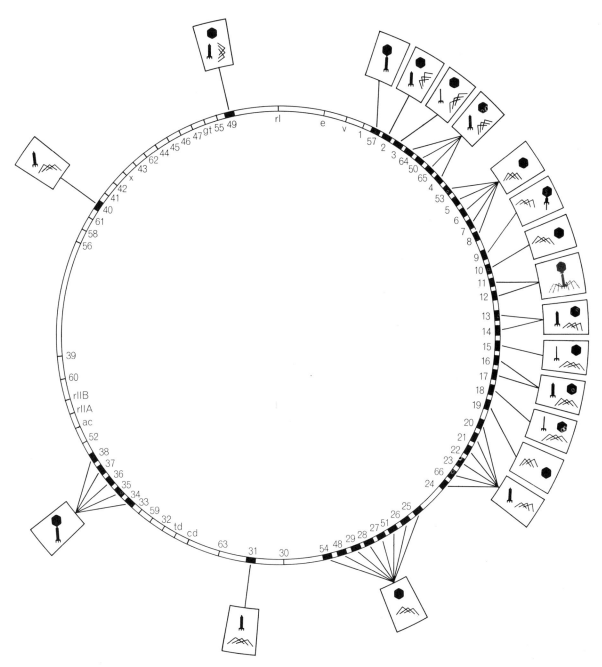

GENETIC MAP of the T4 virus shows the relative positions of more than 75 genes so far identified on the basis of mutations. The solid black segments of the circle indicate genes with morphogenetic functions. The boxed diagrams show which viral components are seen in micrographs of extracts of cells infected by mutants defective in each morphogenetic gene. A defect in gene No. 11 or 12 produces a complete but fragile particle. Heads, all tail parts, sheaths or fibers are the missing components in other extracts.

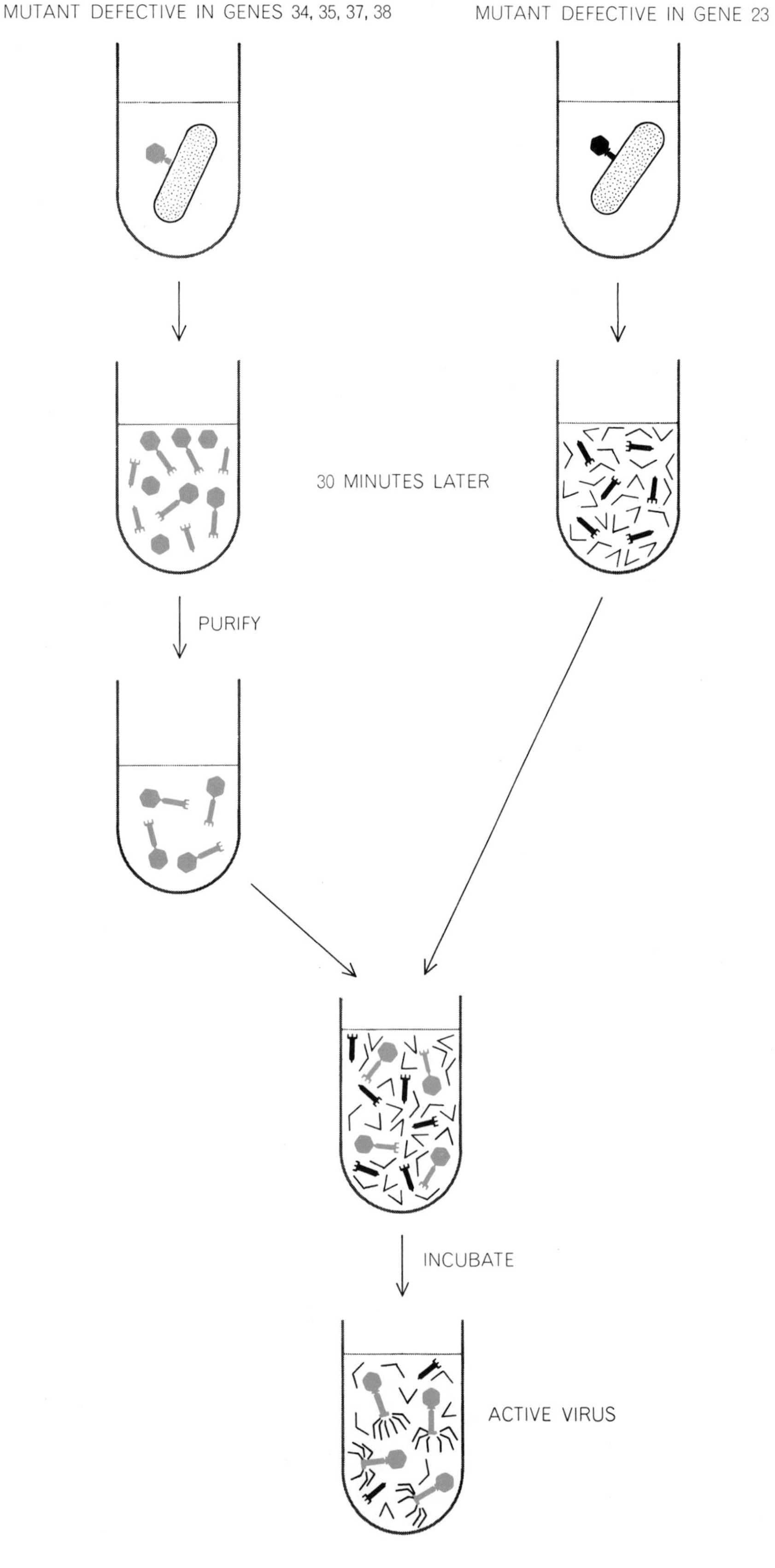

30 MINUTES LATER

PURIFY

INCUBATE

ACTIVE VIRUS

TAIL FIBERS are attached to fiberless particles in the experiment diagrammed here. Cells are infected with a virus (*color*) bearing defective tail-fiber genes. The progeny particles, lacking fibers, are isolated with a centrifuge. A virus with a head-gene mutation (*black*) infects a second bacterial culture, providing an extract containing free tails and fibers. When the two preparations are mixed and incubated at 30 degrees centigrade, the fiberless particles are converted to infectious virus particles by the attachment of the free fibers.

which it arises is dead and hence cannot be recovered for study. Edgar and Epstein, however, found mutations that are only "conditionally lethal": the mutant protein is produced in either a functional or a nonfunctional form, depending on the conditions of growth chosen by the experimenter. Under "permissive" conditions reproduction is normal, so that the mutants can be cultured and crossed for genetic studies. Under "restrictive" conditions, however, viral development comes to a halt at the step where the protein is needed, and by determining the point at which development is blocked the investigator can infer the normal function of the mutated gene. In this way a number of conditionally lethal mutations have been assigned to different genes, have been genetically mapped and have been tested for their effects on viral development under restrictive conditions [*see illustration on page 231*].

In the case of genes that control the later stages of the life cycle, involving the assembly of virus particles, mutations lead to the accumulation of unassembled viral components. These can be identified with the electron microscope. By noting which structures are absent as a result of mutation in a particular gene, we learn about that morphogenetic gene's normal function. For example, genes designated No. 23, No. 27 and No. 34 respectively appear to control steps in the formation of the head, the tail and the tail fibers; these are the structures that are missing from the corresponding mutant-infected cells.

A blockage in the formation of one of these components does not seem to affect the assembly of the other two, which accumulate in the cell as seemingly normal and complete structures. This information alone provides some insight into the assembly process. The virus is apparently not built up the way a sock is knitted—by a process starting at one end and adding subunits sequentially until the other end is reached. Instead, construction seems to follow an assembly-line process, with three major branches that lead independently to the formation of heads, tails and tail fibers. The finished components are combined in subsequent steps to form the virus particle.

A second striking aspect of the genetic map is the large number of genes controlling the morphogenetic process. More than 40 have already been discovered, and a number probably remain to be identified. If all these genes specify proteins that are component parts of the virus, then the virus is considerably more complex than it appears to be. Alternatively, however, some gene products may play directive roles in the assembly process without contributing materially to the virus itself. Studies of seven genes controlling formation of the virus's head support this possibility [see "The Genetic Control of the Shape of a Virus," by Edouard Kellenberger; SCIENTIFIC AMERICAN Offprint 1058].

In order to determine the specific functions of the many gene products involved in morphogenesis, it seemed necessary to seek a way to study individual assembly steps under controlled conditions outside the cell. One of us (Edgar) is a geneticist by training, the other (Wood) a biochemist. The geneticist is inclined to let reproductive processes take their normal course and then, by analyzing the progeny, to deduce the molecular events that must have occurred within the organism. The biochemist is eager to break the organism open and search among the remains for more direct clues to what is going on inside. For our current task a synthesis of these two approaches has proved to be most fruitful. Since it seemed inconceivable that the T4 virus could be built from scratch like the tobacco mosaic virus, starting with nucleic acid and individual protein molecules, we decided to let cells infected with mutants serve as sources of preformed viral components. Then we would break open the cells and, by determining how the free parts could be assembled into complete infectious virus, learn the sequence of steps in assembly, the role of each gene product and perhaps its precise mode of action.

Our first experiment was an attempt to attach tail fibers to the otherwise complete virus particle—a reaction we suspected was the terminal step in morphogenesis. Cells infected with a virus bearing mutations in several tail fiber genes (No. 34, 35, 37 and 38) were broken open, and the resulting particles —complete except for fibers and noninfectious—were isolated by being spun in a high-speed centrifuge. Other cells, infected with a gene No. 23 mutant that was defective in head formation, were similarly disrupted to make an extract containing free fibers and tails but no heads. When a sample of the particles was incubated with the extract, the level of infectious virus in the mixture increased rapidly to 1,000 times its initial value. Electron micrographs of samples taken from the mixture at various times showed that the particles were indeed acquiring tail fibers as the reaction proceeded.

In that first experiment the production

of infectious virus required only one kind of assembly reaction—the attachment of completed fibers to completed particles. We went on to test more demanding mixtures of defective cell extracts. For example, with a mutant blocked in head formation and another one blocked in tail formation we prepared two extracts, one containing tails and free tail fibers but no heads and another containing heads and free tail fibers

but no tails. When a mixture of these two extracts also gave rise to a large number of infectious viruses, we concluded that at least two reactions must have occurred: the attachment of heads to tails and the attachment of fibers to the resulting particles.

By infecting bacilli with mutants bearing defects in different genes con-

TWO ASSEMBLY REACTIONS occur in this experiment: union of heads and tails and attachment of fibers. One virus (*color*), with a defective tail gene, produces heads and fibers. Another (*black*), with a mutation in a head gene, produces tails and fibers. When the two extracts are mixed and incubated, the parts assemble to produce infectious virus.

cerned with assembly, we prepared 40 different extracts containing viral components but no infectious virus. When we tested the extracts by mixing pairs of them in many of the appropriate combinations, some mixtures produced active virus and others showed no detectable activity. The production of infective virus implied that the two extracts were complementing each other in the test tube, that each was supplying a component that was missing or defective in the other and that could be assembled into complete, active virus under our experimental conditions. Lack of activity,

on the other hand, suggested that both extracts were deficient in the same viral component—a component being defined as a subassembly unit that functions in our experimental system. By analyzing the pattern of positive and negative results we could find out how many functional components we were dealing with.

It developed that there are at least 13 such components. That is, analysis of our pair combinations produced 13 complementation groups, the members of which did not complement one another but did complement any member of any other group. Two of these groups were

quite large [see illustration below]. Since one gene produces one protein and since each extract has a different defective gene product, a mixture of any two extracts should include all the proteins required for building the virus. The fact that members of these large groups do not complement one another must mean that our experimental system is not as efficient as an infected cell; whatever the gene products that are missing in each of these extracts do, they cannot do it in the test tube.

The idea that a complementation group consisted of extracts deficient in

EXTRACT GROUP	MUTANT GENES	COMPONENTS PRESENT	INFERRED DEFECT
I	5, 6, 7, 8, 10, 25, 26, 27, 28, 29, 48, 51, 53		TAIL
II	20, 21, 22, 23, 24, 31		HEAD (FORMATION)
	2, 4, 16, 17, 49, 50, 64, 65		HEAD (COMPLETION)
III	54		TAIL CORE
IV	13, 14		?
V	15		
VI	18		
VII	9		?
VIII	11		?
IX	12		
X	37, 38		TAIL FIBERS
XI	36		
XII	35		
XIII	34		

COMPLEMENTATION TESTS defined 13 groups of defective extracts, as described in the text. Mixing any two extracts in a single group fails to produce infectious virus in the test tube, but mixing any two members of different groups yields infectious virus. Apparently each group represents the genes concerned with the synthesis of a component that is functional under experimental conditions. The precise nature of the defect in some extracts, and hence the function of the missing gene product, could not be identified on the basis of the structures recognized in electron micrographs and remained to be determined by additional experiments.

the same functional component could be checked against the earlier electron micrograph results. Micrographs of the 12 defective extracts of Group I, for example, all show virus heads and tail fibers but no tails. Each of these extracts must therefore be deficient in a gene product that has to do with a stage of tail formation that cannot be carried out in our extracts. The second large complementation group appeared at first to be anomalous in terms of electron micrography: some extracts contained only tails and tail fibers, whereas others contained heads as well. Tests against extracts known to contain active tails revealed, however, that these heads—although they looked whole—could not combine to produce active virus in the test tube. In other words, heads, like tails, must be nearly completed within an infected cell before they become active for complementation. The early stages of head formation are still inaccessible to study in mixed extracts.

The remaining defective extracts gave rise to active virus in almost all possible pair combinations, segregating into another 11 complementation groups. With a total of 13 groups, there must be at least 12 assembly steps that can occur in mixtures of extracts. The defects recognizable in micrographs suggest what some of these steps must be: the completion and union of heads and tails, the assembly of tail fibers and the attachment of fibers to head-tail particles. These, then, are the steps that can be studied further in our present experimental system. We have in effect a virus-building kit, some of whose more intricate parts have been preassembled at the cellular factory.

Our next experiments were designed to determine the normal sequence of assembly reactions and further characterize those whose nature remained ambiguous. Examples of the latter were the steps controlled by genes No. 13, 14, 15 and 18. Defects in the corresponding gene products resulted in the accumulation of free heads and tails, suggesting that they are somehow involved in head-tail union. It was unclear, however, whether these gene products are required for the attachment process itself or for completion of the head or the tail before attachment. We could distinguish the alternatives by complementation tests using complete heads and tails. These we isolated from the appropriate extracts in the centrifuge, taking advantage of their large size in relation to the other materials present. On the basis of the evidence for the independent assembly of heads and tails, we assumed that

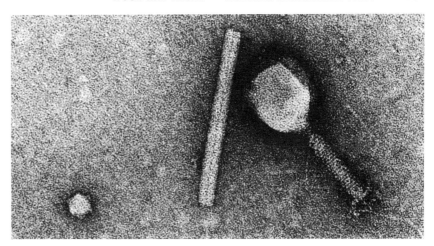

TWO SIMPLER VIRUSES are shown with the T4 in an electron micrograph made by Fred Eiserling of the University of California at Los Angeles. The icosahedral ΦX174 virus (*left*) infects the colon bacillus, as does the T4. The rod-shaped tobacco mosaic virus reassembles itself in the test tube after dissociation. The enlargement is 200,000 diameters.

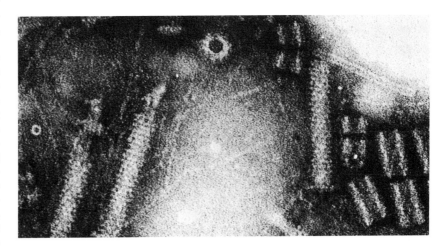

COMPLEX STRUCTURE of the T4 tail is shown in an electron micrograph made by E. Boy de la Tour of the University of Geneva. The parts were obtained by breaking down virus particles, not by synthesis, which is why fibers are attached to tails. The hollow interiors of the free core (*top right*) and pieces of sheath are delineated by dark stain that has flowed into them. There are end-on views of pieces of core (*left*) and sheath (*top center*).

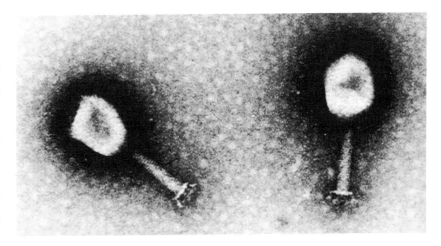

FIBERLESS PARTICLES, otherwise complete, are the products of infection by a mutant defective in one of the fiber-forming genes. Heads, tails and fibers are each formed by a subassembly line (*see illustration on page 237*). The electron micrograph was made by King.

the heads we isolated from a tail-defective extract would be complete, as would tails isolated from a head-defective extract.

The results of the tests were unambiguous. The addition of isolated heads to extracts lacking the products of gene No. 13 or 14 resulted in virus production, whereas the addition of tails did not. We could therefore conclude that the components missing from these extracts normally affect the head structure, and that genes No. 13 and 14 control head completion rather than tail completion or head-tail union. The remaining two of the four extracts gave the opposite result; these were active with added tails but not with added heads, indicating that genes No. 15 and 18 are involved in the completion of the tail. All four of these steps must precede the attachment of heads to tails, since defects in any of

the corresponding genes block head-tail union.

By manipulating extracts blocked at other stages we worked out the remaining steps in the assembly process with the help of Jonathan King and Jeffrey Flatgaard. The various reactions were characterized and their sequence determined by many experiments similar to those described above. In addition, more detailed electron micrographs of defective components helped to clarify the nature of some individual steps. For example, knowing that genes No. 15 and 18 were concerned with tail completion, we went on to find just what each one did. Electron micrographs showed that in the absence of the No. 18 product no contractile sheaths were made. If No. 18 was functional but No. 15 was defective, the sheath units were assembled on the core but were unstable and could fall

away. The addition of the product of gene No. 15 (and of No. 3 also, as it turned out) supplied a kind of "button" at the upper end of the core and thus apparently stabilized the sheath.

The results to date of this line of investigation can be summarized in the form of a morphogenetic pathway [see illustration on page 237]. As we had thought, it consists of three principal independent branches that lead respectively to the formation of the head, the tail and the tail fibers.

The earliest stages of head morphogenesis are controlled by six genes. These genes direct the formation of a precursor that is identifiable as a head in electron micrographs but is not yet functional in extract-complementation experiments. Eight more gene products must act on this precursor to produce a

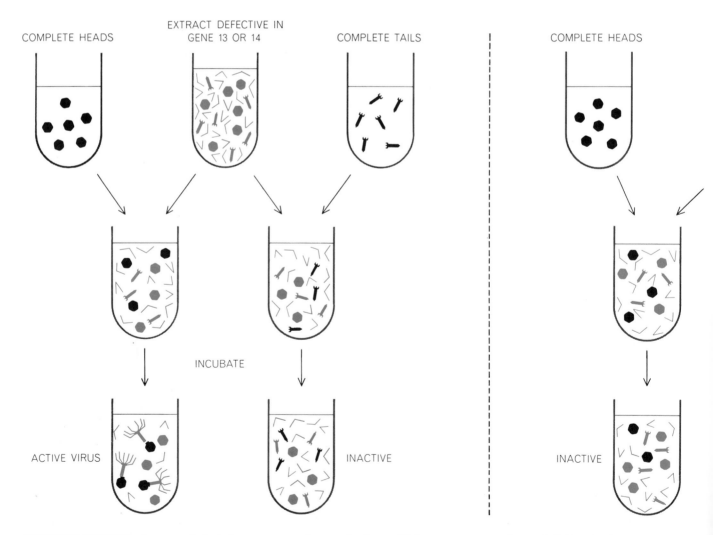

ASSEMBLY DEFECTS of mutants (*color*) that seem to produce complete heads, tails and fibers are identified, using isolated complete heads and tails (*black*) as test reagents. When complete heads are added to some extracts to be tested (*left*), infectious virus is produced, but the addition of complete tails is ineffective. This indicates that the tails made by these mutants must be functional, head structure that is active in complementation experiments. This active structure undergoes the terminal step in head formation (the only one so far demonstrated in the test tube): conversion to the complete head that is able to unite with the tail. The nature of this conversion, which is controlled by genes No. 13 and 14, remains unclear. A likely possibility would be that these genes control the formation of the upper neck and collar, but evidence on this point is lacking. The attachment of head structures to tails has never been observed in extracts prepared with mutants defective in gene No. 13 or 14, or with any of the preceding class of eight genes. It therefore appears that completion of the head is a prerequisite for the union of heads and tails.

The earliest structure so far identified in the morphogenesis of the tail is the

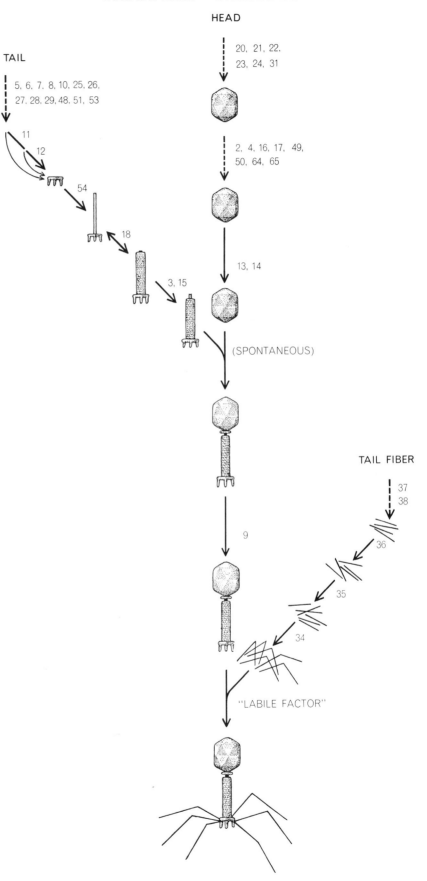

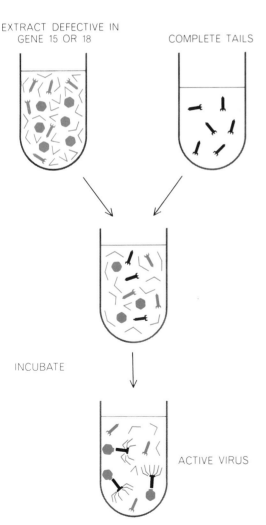

implying that the heads must be defective. In the case of other mutants (*right*), such tests indicate that the tails must be defective.

MORPHOGENETIC PATHWAY has three principal branches leading independently to the formation of heads, tails and tail fibers, which then combine to form complete virus particles. The numbers refer to the gene product or products involved at each step. The solid portions of the arrows indicate the steps that have been shown to occur in extracts.

end plate. It is apparently an intricate bit of machinery, since 15 different gene products participate in its formation. All the subsequent steps in tail formation can be demonstrated in the test tube. The core is assembled on the end plate under the control of the products of gene No. 54 and probably No. 19; the resulting structure appears as a tail without a sheath. The product of gene No. 18 is the principal structural component of the sheath, which is somehow stabilized by the products of genes No. 3 and 15. Tails without sheaths do not attach themselves to head structures, indicating that the tail as well as the head must be completed before head-tail union can occur. Moreover, unattached tail structures are never fitted with fibers, suggesting that these can be added only at a later stage of assembly.

Completed heads and tails unite spontaneously, in the absence of any additional factors, to produce a precursor particle that interacts in a still undetermined manner with the product of gene No. 9, resulting in the complete head-plus-tail particle. It is only at this point that tail fibers can become attached to the end plate.

At least five gene products participate in the formation of the tail fiber. In the first step, which has not yet been demonstrated in extracts, the products of genes No. 37 and 38 combine to form a precursor corresponding in dimensions to one segment of the finished fiber. This precursor then interacts sequentially with the products of genes No. 36, 35 and 34 to produce the complete structure. Again the completion of a major component—in this case the tail fiber—

appears to be a prerequisite for its attachment, since we have never seen the short segments linked to particles.

The final step in building the virus is the attachment of completed tail fibers to the otherwise finished particle. We have studied this process in reaction mixtures consisting of purified particles and a defective extract containing complete tail fibers but no heads or tails. When we divided the extract into various fractions, we found that it supplies two components, both of which are necessary for the production of active virus. One of these of course is the tail fiber. The other is a factor whose properties suggest that it might be an enzyme. For one thing, the rate at which fibers are attached depends on the level of this factor present in the reaction mixture, and yet the factor does not appear to be used up in the process. Moreover, the rate of attachment depends on the temperature of incubation—increasing by a factor of about two with every rise in temperature of 10 degrees centigrade. These characteristics suggest that the factor could be catalyzing the formation of bonds between the fibers and the tail end plate. At the moment we can only speculate on its possible mechanism of action, since the chemical nature of these bonds is not yet known; we call it simply a "labile factor," not an enzyme. Although no gene controlling the factor has yet been discovered, we assume that its synthesis must be directed by the virus, since it is not found in extracts of uninfected bacteria.

The T4 assembly steps so far accomplished and studied in the test tube

represent only a fraction of the total number. Already, however, it is apparent that there is a high degree of sequential order in the assembly process; restrictions are somehow imposed at each step that prevent its occurrence until the preceding step has been completed. Only two exceptions to this rule have been discovered. The steps controlled by genes No. 11 and 12, which normally occur early in the tail pathway, can be bypassed when these gene products are lacking. In that case the tail is completed, attaches itself to a head and acquires tail fibers, but the result is a fragile, defective particle. The particle can, however, be converted to a normal active virus by exposure to an extract containing the missing gene products. These are the only components whose point of action in the pathway appears to be unimportant.

The problem has now reached a tantalizing stage. A partial sequence of gene-controlled assembly steps can be written, but the manner in which the corresponding gene products contribute to the process remains unclear, and the questions posed at the beginning of this article cannot yet be answered definitively. There is the suggestion that the attachment of tail fibers is catalyzed by a virus-induced enzyme. If this finding is substantiated, it would overthrow the notion that T4 morphogenesis is entirely a self-assembly process. Continued investigation of this reaction and the assembly steps that precede it can be expected to provide further insight into how genes control the building of biological structures.

Functional Associations of Lipids, Carbohydrates, and Proteins

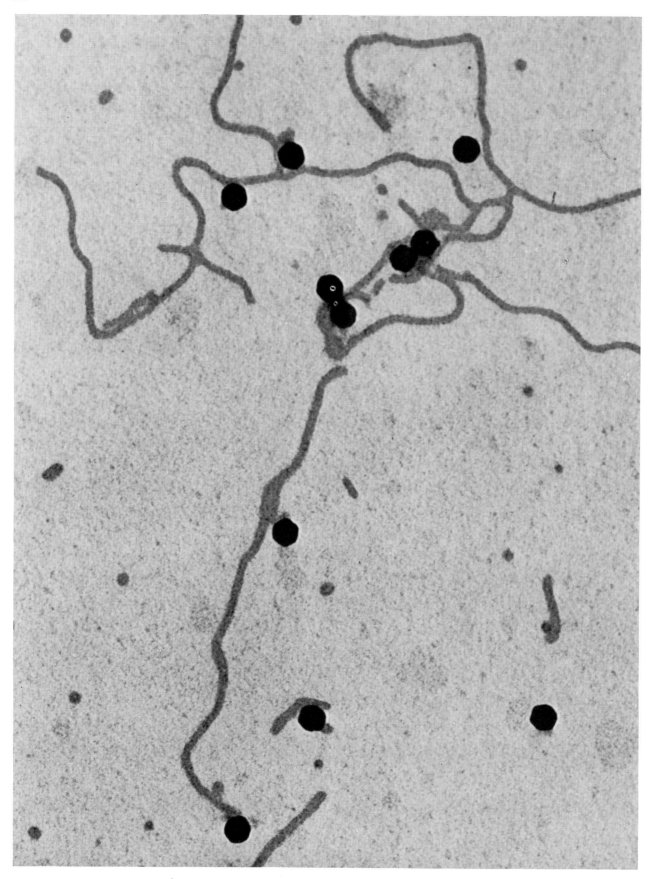

BACTERIAL VIRUSES, or phages, are adsorbed to the surface of cells they infect at specific receptor sites that have a particular chemical structure. In the case of phage ε^{15} the site is in the lipopolysaccharide (fats and sugars) component of the surface of the bacterium *Salmonella anatum*. In this electron micrograph, made by Cecil E. Hall of the Massachusetts Institute of Technology, ε^{15} particles, enlarged about 185,000 diameters, have become bound to ribbons of lipopolysaccharide isolated from *S. anatum* cells. Such particles do not become bound to lipopolysaccharide isolated from other bacterial strains that are not subject to infection by this virus.

23

THE RECEPTOR SITE FOR A BACTERIAL VIRUS

RICHARD LOSICK AND PHILLIPS W. ROBBINS

November 1969

A virus infects a bacterium by attaching itself to a specific site on the bacterium's surface. The chemistry of such a site is studied by seeing how it changes after dormant infection by the same virus

The critical first step in the infection of a bacterial cell by a phage, or bacterial virus, is the attachment of the virus to the surface of the cell. The precise sequence of events in that process has been most clearly outlined in the case of the phage designated T4 and its host, the bacterium *Escherichia coli*. Phage T4 is a complex virus consisting of a head that contains the genetic material, a tail through which the material is injected into the cell and six long tail fibers. Electron micrographic studies, such as those conducted by Lee Simon and Thomas F. Anderson of the Institute for Cancer Research in Philadelphia, show that the phage's tail fibers first make contact with the cell and then bring a pronged plate at the base of the tail into close proximity with the surface. After the base plate is fixed to the cell surface the tail contracts and the phage genetic material is injected into the cell, leaving the empty virus particle behind [*see top illustration on next page*].

Bacterial viruses vary in structure, and many of them are quite different from T4; no doubt their modes of attachment differ too. One important generalization can be made, however, about phage adsorption to the surface of a cell: It is a highly specific process. A particular phage generally attaches itself only to a limited range of bacterial strains. Moreover, a bacterium that adsorbs many kinds of phages may mutate, giving rise to a new strain that no longer adsorbs one of these phages but still serves readily as a host for others. The surface of a bacterium is apparently a mosaic of different receptor sites, each of which is able to form a highly specific bond with a particular phage.

It is clear that receptor sites are located on and characteristic of the surface of bacteria. T4, for example, is adsorbed by empty cell walls of *E. coli* from which the cellular contents have leaked. Some strains of phage are adsorbed by particular sites on appendages of the cell surface. Phage f1 attaches itself to the tips of hairlike structures known as F-pili on the surface of male bacteria; phage R17 attaches itself to the sides of these pili and phage χ attaches itself to the whiplike flagella of certain other bacteria.

The nature of these various receptor sites has remained obscure in spite of considerable investigation ever since the early days of phage genetics. Much was learned about the details of phage infection, and the specificity and diversity of the sites was amply demonstrated, but it was not possible to describe the architecture of even one such site at the molecular level. Then in 1952 Shoei Iseki and Tatsuo Saki of Gunma University in Japan discovered a new bacterial virus, ε^{15}, that is adsorbed by only a few bacteria. One of them is *Salmonella anatum*. In the past 10 years a great deal has been learned about the structure of a particular component of *S. anatum*'s surface. In our laboratory at the Massachusetts Institute of Technology we have been able, by studying the relation between ε^{15} and this component, to determine some structural details of the ε^{15} receptor site. At the same time, and from another point of view, we have made use of the process of phage infection to analyze the chemistry of the bacterial surface.

The cell envelope, the outer portion of the bacterial cell, is a complex structure consisting of an inner plasma membrane and a rigid mucopeptide layer, the cell wall proper, that confers strength and shape [see the article "The Bacterial Cell Wall," by Nathan Sharon, beginning on page 245]. The gram-negative bacteria, such as *E. coli* and the salmonellae, have in addition an outer membrane, composed in part of lipopolysaccharide, a combination of fats and sugars. It was found that ε^{15} particles are bound to lipopolysaccharide extracted from *S. anatum* but not to the same kind of material extracted from other bacteria that do not adsorb ε^{15} [*see illustration on opposite page*]. This demonstrated that the lipopolysaccharide fraction of the outer membrane contains the receptor site.

One component of this fraction is a long polymer, or molecular chain, composed of the sugars galactose, rhamnose and mannose repeated in sequence. This polysaccharide chain, known as O-antigen, was shown to be a necessary part of the receptor site for ε^{15} because mutants of *S. anatum* that do not synthesize O-antigen also fail to adsorb ε^{15}. O-antigen has two structural features that turn out to be important in receptor-site chemistry. Sugars are joined to one another by either of two linkages, known as alpha and beta; the galactose units in O-antigen are joined to the adjacent mannose units by an alpha linkage, in which the linkage to mannose is below the plane of the galactose unit. And the galactose units in O-antigen have a substituent, an acetyl group, attached to them [*see bottom illustration on next page*].

Working with one of us (Robbins), Andrew Wright and Marcello Dankert unraveled the biochemical steps involved in the synthesis in cells of the O-antigen chain. They linked radioactively labeled sugars by means of enzymes taken from *S. anatum* and found that in a cell-free laboratory system the galactose, rhamnose and mannose are not immediately joined to one another in repeating chains. There is an intermediate step. A carrier molecule—a lipid, or fat—serves as a site on which the trisaccharide galactose-rhamnose-mannose is assembled. Then the trisaccharide

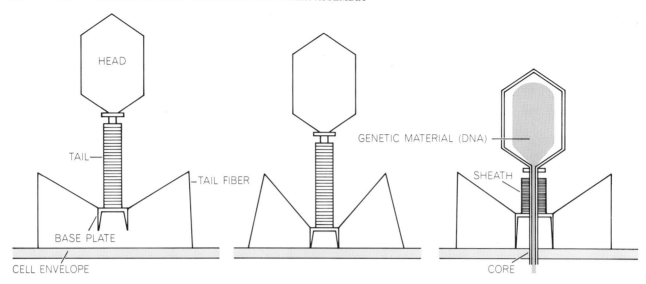

PHAGE T4 attaches itself to the surface of a bacterial cell in a series of steps. First the long tail fibers become affixed to the surface of the cell (*left*). Then the base plate is brought into contact with the surface (*middle*). Finally the tail sheath contracts, driving the phage's tubular core through the cell envelope, and the genetic material in the head enters the cell through the core (*right*).

units are polymerized into long polysaccharide chains by the enzyme alpha-polymerase, the galactose of one trisaccharide being attached to the mannose of another unit in an alpha linkage. (In a mutant that lacks alpha-polymerase the trisaccharide units, linked to their lipid carriers, simply accumulate in the cell envelope and no O-antigen is formed.) After polymerization another enzyme, transacetylase, attaches acetyl groups to the galactose units. Finally the

O-antigen polymers are released from their lipid carriers and become attached to a new site on the outer surface of the cell [*see top illustration on opposite page*].

There is a special form of virus infection of a bacterium in which the phage genetic material, instead of subverting the machinery of the cell to produce hundreds of virus particles and kill the cell, enters into a dormant state and is integrated into the bacterial chromo-

some. This dormant form of infection is called lysogeny [see "Viruses and Genes," by François Jacob and Elie L. Wollman; SCIENTIFIC AMERICAN Offprint 89]. In 1956 Hisao Uetake of Kyoto University made a remarkable observation. He noticed that cells that were lysogenic for ε^{15} (cells dormantly infected by ε^{15} and therefore containing its genetic material) would not adsorb new ε^{15} particles; lysogenic infection by the phage apparently modified the bacterium's sur-

O-ANTIGEN, the component of *S. anatum* lipopolysaccharide to which ε^{15} attaches itself, is a repeating chain of three sugars: mannose, rhamnose and galactose. In O-antigen from uninfected cells the galactose units carry an acetyl group (CH₃CO) and trisac-

charide (three-sugar) units are joined by alpha linkages, in which the linkage to the mannose is below the plane of galactose (*top*). In O-antigen from cells that are lysogenic for ε^{15} the acetyl is missing and the linkage is beta; the linkage is above the galactose (*bottom*).

face and made subsequent adsorption impossible. Six years later one of us (Robbins) and Takahiro Uchida were able to learn the chemical basis for this observation: O-antigen that is synthesized after infection by ε^{15} has a different structure from normal O-antigen. The polysaccharide chains lack the acetyl substituent. And the galactose-mannose linkages are beta linkages rather than alpha; they can be cleaved experimentally by the enzyme beta-galactosidase and not by alpha-galactosidase. In other words, lysogeny by ε^{15} changes the chemical structure of its own receptor site.

With John Keller and Dennis Bray, we undertook genetic and biochemical experiments in an effort to determine the events that accomplish this change in structure. We had reason to believe these events were mediated by specific phage genes, since Uetake and Uchida had in 1959 isolated ε^{15} mutants that do not alter receptor-site structure in the usual way. We were able to track each change in the receptor to a specific phage gene incorporated in the bacterial chromosome. One gene, called gene a, blocks the synthesis of S. anatum transacetylase and thus eliminates the acetyl groups on O-antigen polymer that is synthesized after infection. Another gene, b, makes a new enzyme, beta-polymerase, that replaces the S. anatum alpha-polymerase; like the normal enzyme, the new enzyme links up trisaccharide units, but it does so by generating beta linkages, in which the linkage to mannose is above the plane of the galactose instead of below it. Finally, another ε^{15} gene inactivates the host's alpha-polymerase by making a protein that inhibits it [see bottom illustration at right]. The protein has been purified in our laboratory, and it inhibits alpha-polymerase in a cell-free system.

It is interesting to note that two different mechanisms of inactivation are employed here. The transacetylase is repressed, that is, its synthesis is blocked; this is shown by the fact that after infection there is no increase in transacetylase activity but enzyme that was previously synthesized continues to show activity. The alpha-polymerase, on the other hand, continues to be synthesized by the cell but is inactivated by an inhibitor; this is shown by the fact that after infection its activity decreases or ceases altogether.

Since ε^{15} alters the structure of O-antigen both by preventing the attachment of acetyl groups and by changing the linkage, the question arises: Which of these two structural features is essential

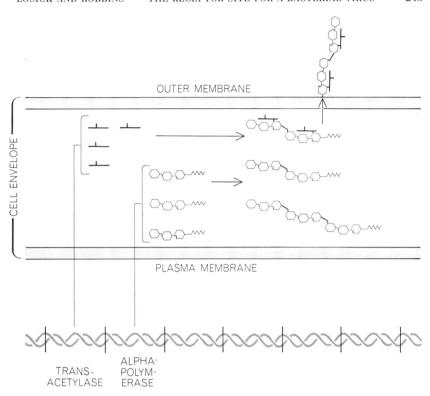

BIOSYNTHESIS of O-antigen takes place in the cell envelope. Genes in the bacterial chromosome (gray helix) encode instructions for two enzymes involved in the synthesis. One, alpha-polymerase, links individual trisaccharides on lipid carriers into long polysaccharides, forming the alpha linkage. The other, transacetylase, attaches acetyl groups (black bars) to galactose units. Then the O-antigen chains are transferred to the cell surface.

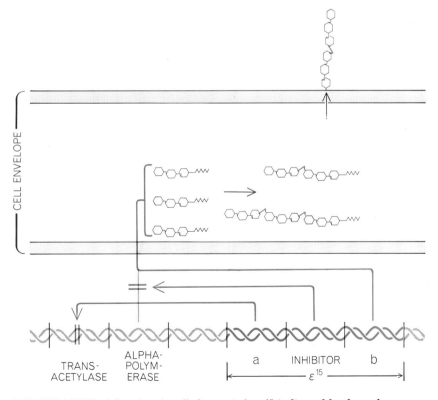

MODIFICATION of O-antigen in cells lysogenic for ε^{15} is directed by three phage genes (colored helix) integrated into the bacterial chromosome. Host-cell alpha-polymerase is inactivated by a protein made by the inhibitor gene. Gene a is responsible for repressing the synthesis of host transacetylase. Gene b makes beta-polymerase, which generates an O-antigen with beta linkages that replaces the normal alpha-linked O-antigen on the cell surface.

for adsorption—or are both of them essential? In order to answer this question Bray isolated mutants of *S. anatum* that synthesize an alpha-linked polymer without acetyl groups. He found that ε^{15} particles are adsorbed by these transacetylase mutants, and so the acetyl groups must not be prerequisites for attachment. The next step was to test an O-antigen that retains acetyl groups but has the beta linkage. *S. anatum* was made to synthesize such a polymer by infecting it with a mutant phage, $\varepsilon^{15}a$, that changes the linkage without removing the acetyl groups. Bacteria so lysogenized fail to adsorb ε^{15}, showing that the alpha linkage is the critical feature of the receptor site.

We soon found that alteration of the receptor site is not the only consequence of lysogeny by ε^{15}. The very changes that prevent the attachment of more ε^{15} particles make possible the attachment of a different virus, ε^{34}. In fact, ε^{34} will infect only cells that are lysogenic for ε^{15}; it requires a receptor site without acetyl groups and with beta linkages. Thus one phage makes it possible for another phage to infect the same bacterial cell.

Phage ε^{34}, in turn, brings about a further modification of the cell surface. It adds glucose to the galactose units of the O-antigen. Wright determined the steps that occur in this modification [*see illustration below*]. By working with

various mutants of ε^{34} he found that one phage gene is responsible for linking glucose to a carrier molecule, a lipid similar to the one involved in the polymerization of O-antigen. Another gene is responsible for transferring glucose from the carrier to the polysaccharide chain. The presence of the glucose makes it impossible, in turn, for new ε^{34} particles to be adsorbed by cells that have been lysogenized by ε^{34}.

There may be significant similarities between the way ε^{15} or ε^{34} alters the surface of bacterial cells and the action of tumor viruses on animal cells. Tumor viruses such as simian virus 40 and the polyoma virus transform normal cells into cancer cells. Renato Dulbecco and his colleagues at the Salk Institute for Biological Studies have shown that the transformation, the process of "cancerization," is induced by viral genetic material [see "The Induction of Cancer by Viruses," by Renato Dulbecco; SCIENTIFIC AMERICAN Offprint 1069]. Dulbecco's group has recently reported evidence that the chromosome of tumor viruses sometimes persists in the cancerized cell in a dormant state, integrated into the host cell's chromosome. This dormant state may be analogous to the lysogenic state of such phages as ε^{15} and ε^{34}. Moreover, cancerized cells appear to have altered surface characteristics. A

new, virus-specific antigen that can be detected by immunological techniques usually appears on the cancer cell's surface. Contact inhibition, the process whereby the persistent undulating motion of normal cells is stopped when one cell touches another, is ineffective in a cancer-cell culture; the cells keep on moving even after touching, suggesting again that their surface has somehow been altered. It may be that a specific gene of the tumor virus is responsible for each change in the cell surface just as specific ε^{15} genes bring about changes in the surface of *S. anatum*.

The analogy is nice, but it should be treated with caution. Normal cells can be transformed into cancer cells by chemical agents or radiation without any apparent viral influence at all, implying that cancerization does not always require specific virus genes. Moreover, even when a virus is the agent, it may be that cancerization is incidental to the entire process of infection and integration into the host chromosome and does not involve specific genes. As work on phage infection and on cancerization proceeds, it will be interesting and perhaps helpful to watch for further similarities and differences between the mechanism by which a phage alters the surface of its bacterial host and the more complex mechanisms by which a tumor virus affects animal cells.

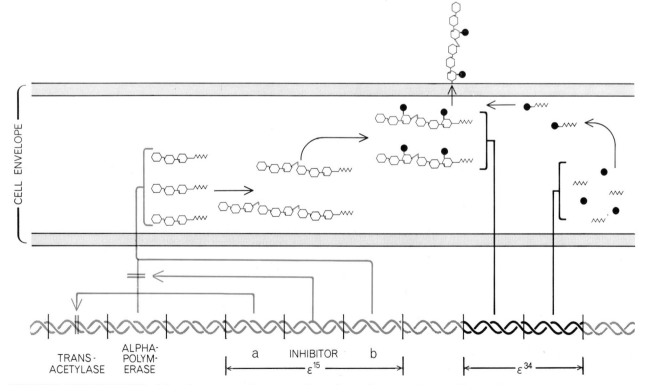

FURTHER MODIFICATION of O-antigen occurs when a second phage, ε^{34}, infects cells that are already lysogenic for ε^{15}. One gene of the new phage's genetic material (*black helix*) is responsible for the attachment of glucose (*black disks*) to a lipid carrier. A second phage gene transfers the glucose to the galactose units of the beta-linked O-antigen chains, which are then transferred to the surface.

24

THE BACTERIAL CELL WALL

NATHAN SHARON

May 1969

*The tough, inflexible envelope that surrounds a
bacterium is a single giant molecule made up of amino
acid and sugar units. Its formation is blocked by drugs
such as penicillin*

The pioneer microscopist Anton van
Leeuwenhoek, who in the 17th
century was the first to observe the
one-celled organisms later called bacteria, made sketches showing their shapes.
In this way he identified the major
classes from which bacteria derive their
names: cocci (spherical), bacilli (rodlike) and spirilla (helical). It is evident
from van Leeuwenhoek's writings that
he realized bacteria have a structure of
some kind that holds the organism together and preserves its shape.

This structure is known to us as the
bacterial cell wall. It surrounds the fragile plasma membrane of the bacterial cell
and protects the membrane and the cytoplasm within from adverse influences in
the environment. The rigid wall structure is characteristic of bacteria; the cells
of animals are enclosed only by the plasma membrane. Plant cells have a rigid
or semirigid envelope, but the structural
material of plant cell walls (with the exception of a few seaweeds) is cellulose.
The bacterial cell wall is made up of
a different material whose composition
and structure are far more complex than
those of cellulose. Indeed, studies of the
bacterial cell wall's molecular architecture have revealed substances and structures found only in bacteria and not in
plants and animals.

The composition of the cell-wall material is of particular interest because
the cell wall and certain substances that
are attached to it are largely responsible
for the virulence of bacteria. The symp-
toms of a broad spectrum of bacterial
diseases can be produced by the injection into animals of cell walls rather
than whole bacteria. Similarly, animals
can be immunized against certain bacteria (for example enteric bacteria, bru-
cellae and pasteurellae) by means of the
bacterial cell walls rather than whole
killed cells.

The bacterial cell wall became the
subject of intensive study about 10 years
ago, when it was discovered that certain
antibiotics, notably penicillin, kill bacteria by interfering with the synthesis of
the cell-wall material. Bacteria that lack
cell walls can be grown in the presence

of penicillin under certain conditions.
Such naked cells, called protoplasts, can
also be prepared by means of enzymes,
such as lysozyme, that normally kill susceptible bacteria by lysing, or dissolving, their walls. The substances that are
stripped from bacteria to expose the na-

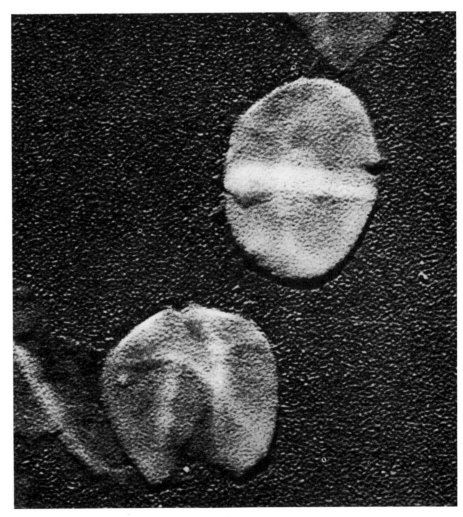

ROUND CELL WALLS were isolated from the spherical bacteria known as *Micrococcus
lysodeikticus*. The wall structure is responsible for the characteristic shapes of bacteria.

ked cell have been investigated in order to learn the composition of the cell wall and the process by which bacteria synthesize the wall material. Thanks to these and other studies that I shall discuss, it is now possible to understand the selective action that makes penicillin such an effective antibacterial agent.

The foundations for the investigation of the bacterial cell wall were laid by Milton R J. Salton at the University of Cambridge in the early 1950's, when he succeeded in isolating cell walls and examining them in the electron microscope. At a large magnification the isolated walls look like empty bags or deflated balloons that have been spread out on a table [*see illustrations on pages 245 and 246*]. Salton found that, as one might have expected, the cell walls had the same shapes as the bacteria they had formerly housed. The cell walls of cocci are round and the walls of bacilli are elongated.

Although the investigation of these structures has flourished during the past decade, before that time progress was slow. The main difficulty was the dimensions of bacteria. Most bacterial cells are between .5 and one micron wide and between one and five microns long. (A micron is a thousandth of a millimeter.) The investigation of such tiny cells was impeded not only by technical difficulties but also by the widely held premise that cells of such small dimensions must be structureless.

The work that has elucidated the fine structure of bacteria came about as the result both of a radical change in the conception of the living cell and the development of new techniques for its study. An essential tool in investigating bacteria is the electron microscope, whose resolving power is more than 100 times that of the optical microscope. Another important method is paper chromatography, which enables one to examine the chemical composition of cellwall material and conduct experiments on its structure even when only minute

quantities of the material are available.

It may appear to be contradictory that whereas bacteria are killed by agents that disrupt their cell walls, the microorganisms can be viable when the cell wall is removed. To clarify this apparent contradiction, let us compare the behavior of bacteria with the behavior of animal cells under certain conditions. Animal cells such as red blood corpuscles will remain intact in a solution that has about the same concentration of salt as the concentration inside the cell (an isotonic solution). On the other hand, if red blood corpuscles are placed in a hypotonic solution (distilled water, for instance) whose osmotic pressure is lower than the pressure of the cell contents, water will enter the cells in accordance with the elementary principles of osmosis. In a short time the corpuscles swell up, and eventually they burst. If the corpuscles are suspended in a hypertonic solution, so that the osmotic pressure is higher than the pressure inside the cell, the corpuscles shrivel up to an unrecog-

ELONGATED CELL WALL retains the shape of the rodlike bacterium *Bacillus licheniformis* from which it was taken. In this electron micrograph and the one on preceding page, both of which were made by the author, the cell walls are enlarged 53,000 diameters.

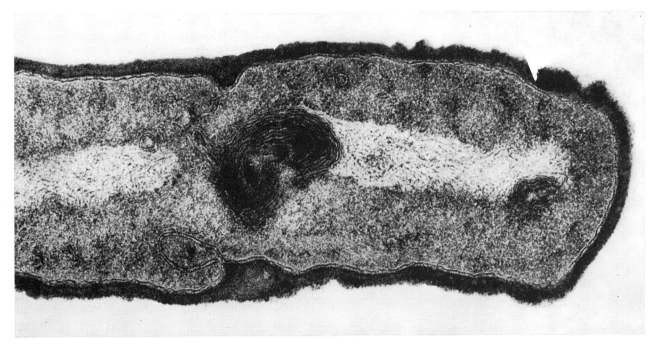

SECTIONAL VIEW of *Bacillus subtilis* shows the cell wall as a dark layer around the edge of the cell. Lining the cell wall is the thin layer of the plasma membrane. Various subcellular particles are visible in the cytoplasm of the bacillus. The bacillus is seen in longitudinal section in this electron micrograph, which was made by Elizabeth H. Leduc, now at Brown University, and Philippe Granboulan at the Institut de Recherches Scientifiques sur le Cancer. The bacillus is enlarged approximately 80,000 diameters.

nizable shape as liquid flows out of them.

In contrast to this behavior, bacteria do not change in their outward appearance when they are in a hypotonic or hypertonic solution. This is because of the tough and inflexible cell wall, whose shape and volume are nearly unchangeable. In bacteria, as in animal cells, the plasma membrane has the property of selective permeability and thus performs the function of "pumping" solutes from the environment into the cell. Indeed, the membrane is so efficient at concentrating metabolites that the osmotic pressure inside the cell can reach 20 atmospheres (300 pounds per square inch). The plasma membrane by itself cannot withstand such pressure; it remains intact only because it transmits the pressure to the cell wall. Hence if the wall of a bacterium is damaged or removed, the membrane bursts.

Weakening or destroying the bacterial cell wall by treating it with lysozyme (or inhibiting the wall's synthesis by means of substances such as penicillin) kills bacteria by causing them to burst. A bacterium that lacks a cell wall will survive, however, if its plasma membrane is preserved intact, which can be done by keeping the microorganism in an isotonic solution. Accordingly one method of preparing the naked protoplasts is to treat bacteria with lysozyme when they are in an isotonic solution. (A relatively concentrated solution of sucrose is used.)

It is interesting that although bacteria have a variety of shapes, their protoplasts are always spherical. The reason is that the weak, flexible plasma membrane assumes the spherical shape of maximum stability. As long as protoplasts remain in an isotonic solution, they take up oxygen and eliminate carbon dioxide, conduct biosynthesis and under special circumstances will even reproduce. In the form of protoplasts, bacteria with flagella retain these whiplike appendages. The flagella do not, however, propel the protoplasts as they do the bacteria, because the solid fulcrum the firm wall structure affords is lacking. Protoplasts from which the cell wall has been entirely removed (some wall-stripping agents do not do a complete job) are not attacked by bacteriophages, the viruses that parasitize bacteria.

If the osmotic pressure of a solution in which protoplasts are suspended is re-

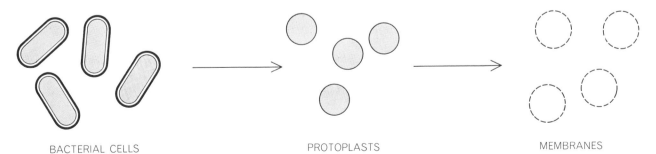

BACTERIAL CELLS   PROTOPLASTS   MEMBRANES

CELL WALLS ARE REMOVED to expose protoplasts (viable naked cells) by treating bacteria with penicillin or with the enzyme lysozyme, which lyses, or dissolves, the wall substance. During the process the bacteria must be kept in a solution that balances the osmotic pressure across the plasma membrane. Bacteria without a wall assume a spherical shape. Plasma membranes of bacteria are prepared by reducing the osmotic pressure of the solution. The protoplasts then swell and burst, releasing their contents.

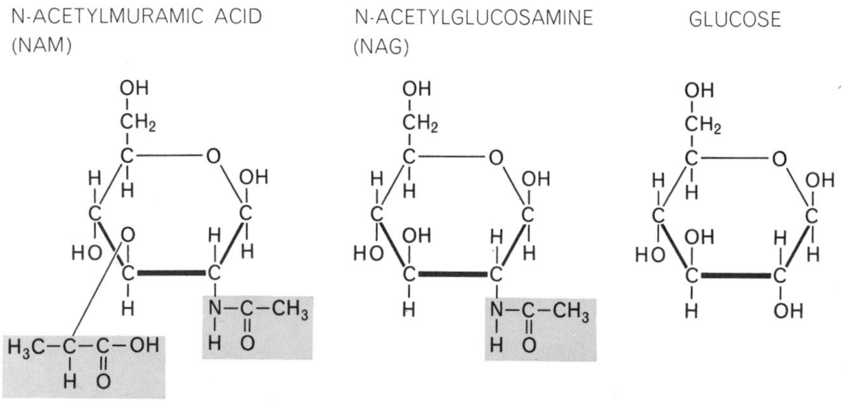

N-ACETYLMURAMIC ACID
(NAM)

N-ACETYLGLUCOSAMINE
(NAG)

GLUCOSE

SACCHARIDE MOLECULES indicate (*color*) where the amino sugars N-acetylmuramic acid (NAM) and N-acetylglucosamine (NAG) differ from glucose. NAM and NAG are components of the glycopeptide that forms the foundation in cell walls of many bacteria. Cellulose, the structural material in plant cell walls, is composed of linked glucose units.

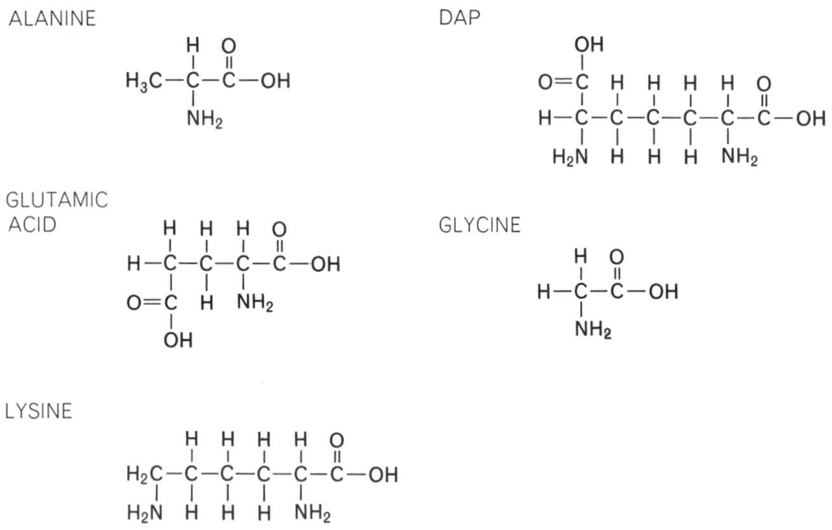

ALANINE

DAP

GLUTAMIC
ACID

GLYCINE

LYSINE

AMINO ACID MOLECULES are the other components of the glycopeptide that gives the cell wall its shape. Alanine and glutamic acid have a configuration different from the one they have in other natural materials (*see illustration below*). In cell walls of many bacteria, lysine is replaced by diaminopimelic acid (DAP). DAP is peculiar to the cell wall.

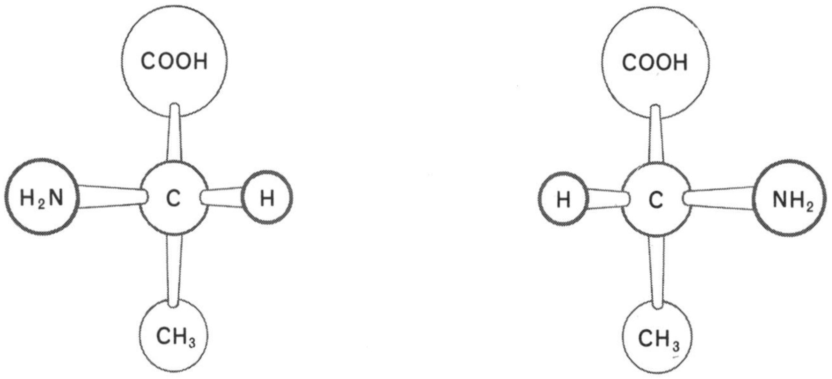

TWO CONFIGURATIONS OF ALANINE are found (in equal amounts) in the glycopeptide of the cell wall. When in solution, one configuration, or stereoisomer, designated *L*-alanine (*left*) rotates plane-polarized light in one direction; its mirror image, *D*-alanine (*right*), rotates polarized light in the opposite direction. *L*-amino acids are normally found in natural materials. Only *D*-glutamic acid appears in cell-wall glycopeptide.

duced, the cells burst and their contents escape, leaving the plasma membrane an empty sac or in fragments. This is the basis of a technique that is used to prepare bacterial plasma membranes, which are now under study in many laboratories.

In investigating the cell wall of bacteria two basic approaches can be followed. In one the wall is studied as part of the intact cell. Thus the examination of very thin slices of bacterium in the electron microscope reveals that the thickness of the cell wall is usually between 15 and 20 millimicrons, which is about 1 percent of the overall thickness of an average bacterial cell. The wall appears to envelop the cell completely; it is also possible to observe changes in the wall during cell division.

The second approach to studying cell walls is to isolate them from bacteria, as one must do in order to analyze their chemical composition. A simple and effective method for the preparation of cell walls is to shake a thick suspension of bacteria and water together with tiny glass beads. The cells are punctured, their contents pour out into the water and the cell walls, which are insoluble, can be readily separated from the denser glass beads and the water-soluble contents of the bacteria. After several washings a white precipitate is obtained that consists solely of cell walls, as can be shown by examination under the electron microscope.

Early investigators were surprised to find that the bacterial wall accounts for 20 to 30 percent of the cell's dry weight. Although the wall is much smaller in volume than the cell cytoplasm, it is mostly solid material whereas the cytoplasm is a relatively dilute solution.

The chemical composition of the cell wall was first analyzed by Salton. His analyses indicated that the wall substance includes almost all the kinds of material that are found in living cells, with the exception of the nucleic acids. He found sugars linked in chains (polysaccharides), amino acids linked in the long chains of proteins and the shorter ones of peptides, and also lipids (fats). As more bacteria were examined, however, it became clear that all cell walls have a few common components: they are two simple sugars and three or four amino acids. These form the "basal structure" of the wall. In the instances where large amounts of other substances were found, it was shown that they came from substances linked to the wall, such as the substances that form the gummy or slimy capsule encasing many bacteria. Treatment with the appropriate solvents (for example salt solutions or organic liquids) was found to remove these substances, leaving an insoluble material that, when it was examined in the electron microscope, still retained the characteristic wall shape. It is therefore this material that provides the cell wall's rigid foundation.

When this material is decomposed by prolonged heating in acid solution, one obtains its constituent saccharides and amino acids. In some bacteria the cell wall is composed almost exclusively of these components. The two sugars are simple ones related to glucose. One is glucosamine, an amino sugar (so named because an amino group–NH_2–is attached to the sugar molecule). Glucosamine is a common constituent of natural polymers. For instance, chitin, the principal structural material of the external skeleton of insects, which among natural polymers is probably second only to cellulose in abundance, is made up solely of glucosamine units. This amino sugar is also found in animal connective tissues and in many protein-carbohydrate compounds. It is usually in the form of N-acetylglucosamine (NAG), and it is in this form in the bacterial cell wall as well.

The other saccharide of the bacterial cell wall is a lactic acid derivative of glucosamine. It was named muramic acid (from *murus,* the Latin word for wall). Muramic acid has been found only in bacteria and microorganisms that resemble bacteria, such as the blue-green algae. Like glucosamine, it is usually in the form of the N-acetyl derivative.

Only a limited number of the 20 amino acids that form protein are found in the cell-wall material. In the walls of many bacteria the amino acids are glutamic acid, glycine, lysine and alanine. Two of the amino acids, however, are found in an "unnatural" configuration, that is, certain of the atoms or groups of atoms in their molecules are arranged differently from the way they are in the amino acids of other natural substances. This configuration, which is designated dextro (*D*), is a mirror image of the "normal" configuration, levo (*L*). Glutamic acid appears as a *D*-stereoisomer in bacterial walls whereas lysine has the usual *L* configuration. Alanine in the cell wall is found as a *D*-stereoisomer and also as an *L* one [*see bottom illustration on opposite page*]. In many bacteria there is no lysine in the cell wall; instead the wall contains diaminopimelic acid, whose

structure resembles that of lysine. Diaminopimelic acid is peculiar to the bacterial cell wall; it is not found in any protein.

How are the basic constituents of cell-wall material linked to form its rigid structure? In answering this question lysozyme proved a particularly valuable tool. The enzyme was discovered by Sir Alexander Fleming in 1922, six years before his discovery of penicillin. Fleming observed that tears, saliva, nasal mucus and other secretions of the body are capable of dissolving certain bacteria, and he attributed this activity to a cell-lysing enzyme (which he accordingly named lysozyme). He hoped that lysozyme might be useful in fighting pathogenic bacteria. This hope, however, has never been realized.

Lysozyme is also found abundantly in

the white of hen's eggs, and it was from this source that lysozyme was isolated and crystallized by other investigators. Fleming found that the bacterium most susceptible to dissolution by lysozyme is *Microccus lysodeikticus.* (*Lysodeikticus* means "lysis meter.") This bacterium is still widely used in investigations of lysozyme and enzymes that resemble it. If a tear or a drop of egg white is added to a suspension of *M. lysodeikticus,* in only a minute or so one can see with the unaided eye the turbid suspension clear up. Microscopic examination shows that the cells have simply disappeared.

Until Salton had investigated this phenomenon it was not clear exactly what lysozyme did to the cells, although some information on its action had been gathered by Karl Meyer at Columbia University and by other workers. Salton

CELLULOSE

CHITIN

GLYCOPEPTIDE

STRUCTURAL POLYSACCHARIDES in plant cell walls (*top*), the external skeleton of insects (*middle*) and bacterial cell walls (*bottom*) are similar. The manner in which the chitin and cell-wall polysaccharides differ from cellulose is indicated in color. The polysaccharide of bacterial walls consists of alternating units of NAM and NAG. By cleaving bonds between these units lysozyme dissolves the cell wall. *R* stands for an OH group or a peptide.

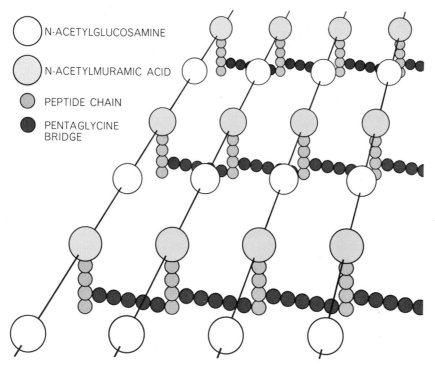

○ N-ACETYLGLUCOSAMINE

◯ N-ACETYLMURAMIC ACID

● PEPTIDE CHAIN

● PENTAGLYCINE BRIDGE

STRUCTURE OF CELL-WALL GLYCOPEPTIDE is represented diagrammatically. Attached to the polysaccharide chains are chains of amino acid units (peptides). The peptides, linked to one another by pentaglycine bridges, cross-link the polysaccharide chains. By inhibiting cross-linking, penicillin prevents cell-wall formation in growing bacteria.

showed that lysozyme dissolves the cell wall by cleaving the chemical bonds between its amino sugar subunits.

In investigating the composition of cell walls dissolved by lysozyme, Salton discovered a two-sugar substance composed of acetylglucosamine and acetylmuramic acid. The detailed structure of this disaccharide was worked out in 1963 by Roger W. Jeanloz, Harold M. Flowers and myself in the Laboratory for Carbohydrate Research at the Massachusetts General Hospital. Other investigations have served to clarify the cell wall's architecture, notably those by Jack L. Strominger of the University of Wisconsin School of Medicine, by Jean Marie Ghuysen of the University of Liège, by Howard J. Rogers and H. R. Perkins of the National Institute for Medical Research in London and also those by my colleagues and myself at the Weizmann Institute of Science in Israel. A glycopeptide that appears to represent the fundamental repeating unit of most bacterial cell walls was recently isolated and characterized in our laboratory by David Mirelman.

These investigations have led to the conclusion that the walls of bacteria are made up of two types of polymer, one composed of saccharide subunits and one of amino acid subunits. The saccharide portion of the wall consists of long chains of alternating units of acetylglucosamine and acetylmuramic acid, linked by bridges that include an oxygen atom (glycosidic linkages). This polysaccharide closely resembles the cellulose of plant cell walls and the chitin of insects [see illustration on preceding page].

The cell wall of a bacterium ought to be insoluble by virtue of its polysaccharide alone, as the walls of plants and the external skeletons of insects are. In the bacterial wall, however, there is an additional factor that contributes to the insolubility and also to the strength and toughness of the material. This factor is provided by peptides, the short chains made up of amino acids. The peptide chains, which are cross-linked to one another, are attached to the polysaccharide chains, so that they serve to join them and form a three-dimensional network [see illustration on this page]. The cross-linking of polymers is well known as a means of achieving toughness and mechanical strength in synthetic materials.

The amino acids of the cell wall are linked by peptide bonds (HN–CO), but they are not susceptible to digestion by enzymes, such as trypsin or pepsin, that act to split such bonds (proteolytic enzymes). This resistance may well be related to the unusual D configuration of the wall components glutamic acid and alanine. Most proteolytic enzymes are unable to split peptide bonds between D-amino acids, and it may be that bacteria are naturally protected from lysis by these enzymes. To be sure, the cell-wall peptides are not totally resistant to all enzymes. If this were so, the world would be full of the cell walls of dead bacteria.

The three-dimensional network of the cell wall can be conceived of in a general way as a gigantic bag-shaped molecule. Such a structure belongs to the general class of glycopeptides; the wall probably is a glycopeptide with a molecular weight of tens of billions. Penicillin, as I have noted, prevents this three-dimensional structure from forming. At what point in the synthesis of the wall glycopeptide does penicillin act?

Fleming himself tried to solve the problem of penicillin's mechanism of action, and the problem has occupied many other workers, particularly during the 1940's, when penicillin came into wide use as a drug. The key to the problem was discovered in 1949 by James T. Park, who is now at the Tufts University Medical School. Park found that when staphylococci are incubated for several hours in a medium containing penicillin, there is an accumulation of substances that at the time were unknown. The detailed structure of these substances was revealed several years later by Park and by Strominger. They were shown to have a resemblance to the glycopeptide of the cell wall, but they had a low molecular weight. There was also present in these substances in combined form a compound that plays an important role in the biosynthesis of polysaccharides, namely uridine diphosphate.

In 1957 Park and Strominger jointly put forward a hypothesis to account for these findings. The substances associated with uridine diphosphate that accumulate under the influence of penicillin were identified as intermediates in cell-wall synthesis; they result from interference with the synthetic process. This hypothesis explained the well-known fact that penicillin kills only multiplying cells: only at the time of reproduction do bacteria need to synthesize new cell-wall material. It is noteworthy that shortly before Park and Strominger published the results of their work, Joshua Lederberg, who was then at the University of Wisconsin, concluded, on the basis of other evidence, that penicillin kills bacteria by inhibiting the synthesis of their walls.

The hypothesis that the cell wall is penicillin's site of action explained the great susceptibility of bacteria to this drug, which is not toxic to animals. (Only a few micrograms of penicillin are needed to kill staphylococci, but humans are usually not affected by doses 10 million times larger.) This response is to be expected, since penicillin interferes with the synthesis of a structure that is vital for bacteria but is not even present in animals.

During the past decade Park and Strominger's hypothesis has been verified experimentally. The details of the biosynthesis of the intermediate substances and the ways these small water-soluble molecules are converted into the gigantic insoluble glycopeptide of the cell wall have been clarified. Today it is possible to synthesize the uridine diphosphate intermediates from the amino sugars and amino acids of the cell wall with the aid of about 12 specific enzymes that have been extracted from bacteria.

The synthesis of intermediates is only the first step in biosynthesis of the cell wall, and penicillin does not interfere with the process at this stage. In the next stage the uridine-linked intermediate is polymerized together with uridine-linked acetylglucosamine to form the polysaccharide backbone with the aid of enzyme bound to the plasma membrane and a special cofactor (a phosphate derivative of a polyisoprenoid alcohol). Penicillin does not influence this process either. Other antibiotics such as vancomycin and bacitracin do, however, inhibit synthesis at this stage. Quite recently Park and Strominger independently discovered that penicillin inhibits the last stage in synthesis of the cell-wall glycopeptide. This is the synthetic step in which adjacent polysaccharide chains are linked by reaction between their peptide side chains.

Thus on a molecular level the two great discoveries made by Fleming, lysozyme and penicillin, have been joined.

Both substances interact with the cell-wall glycopeptide, although at different sites. The investigation of penicillin and lysozyme has not, however, come to an end. Penicillin's mode of action is being studied in further detail in the search for information that might be useful in combating resistant bacteria, for example through the synthesis of new "tailor-made" antibacterial drugs. Recently there has been an upsurge of interest in lysozyme as the first enzyme whose three-dimensional structure has been precisely determined, so that its properties are understood in atomic detail. It is believed that lysozyme, which we too are studying in terms of its detailed effect on the bacterial cell wall, may offer the key to understanding the secrets of enzyme action.

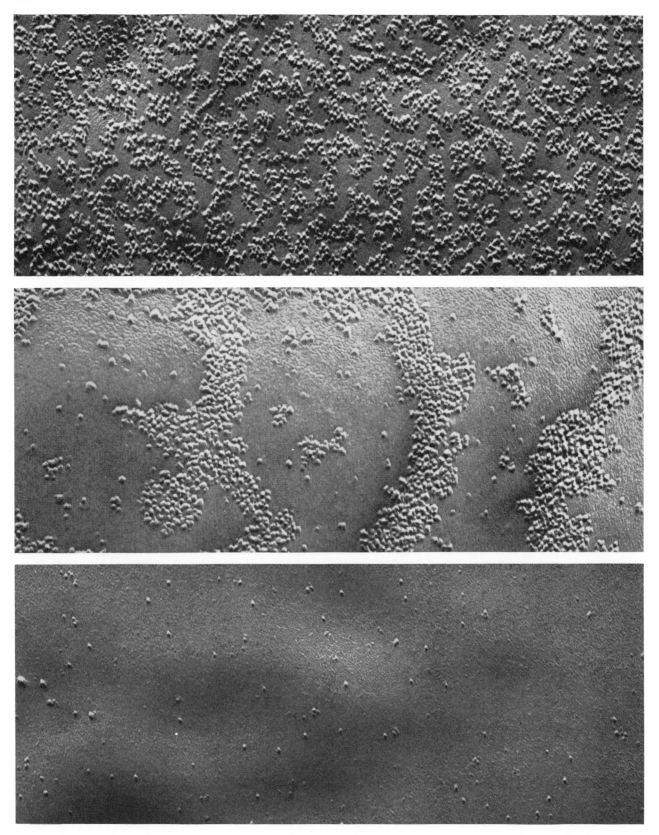

EVIDENCE FOR PROTEINS within the bilayer structure of cell membranes is provided by freeze-etch electron microscopy. A suspension of membranes in water is frozen and then fractured with a sharp blade. The fracture will often split a membrane in the middle along a plane parallel to the surface. After platinum and carbon vapors are deposited along the fracture surface the specimen can be studied in the electron microscope. The micrograph at the top shows many particles 50 to 85 angstroms in diameter embedded in a fractured membrane from rabbit red blood cells. The other two views show how the number of particles is greatly reduced if the membrane is first treated with a proteolytic enzyme that digests 45 percent (*middle*) or 70 percent (*bottom*) of the original membrane protein. The missing particles have presumably been digested by the enzyme. The membrane preparations are enlarged some 95,000 diameters in these micrographs made by L. H. Engstrom in Daniel Branton's laboratory at the University of California at Berkeley.

25

THE STRUCTURE OF CELL MEMBRANES

C. FRED FOX

February 1972

The thin, sturdy envelope of the living cell consists of lipid, phosphate and protein. The proteins act as both gatekeepers and active carriers, determining what passes through the membrane

Every living cell is enclosed by a membrane that serves not only as a sturdy envelope inside which the cell can function but also as a discriminating portal, enabling nutrients and other essential agents to enter and waste products to leave. Called the cytoplasmic membrane, it can also "pump" substances from one side to the other against a "head," that is, it can extract a substance that is in dilute solution on one side and transport it to the opposite side, where the concentration of the substance is many times higher. Thus the cytoplasmic membrane selectively regulates the flux of nutrients and ions between the cell and its external milieu.

The cells of higher organisms have in addition to a cytoplasmic membrane a number of internal membranes that isolate the structures termed organelles, which play various specialized roles. For example, the mitochondria oxidize foodstuffs and provide fuel for the other activities of the cell, and the chloroplasts conduct photosynthesis. Single-cell organisms such as bacteria have only a cytoplasmic membrane, but its structural diversity is sufficient for it to serve some or all of the functions carried out by the membranes of organelles in higher cells. It is clear that any model formulated to describe the structure of membranes must be able to account for an extraordinary range of functions.

Membranes are composed almost entirely of two classes of molecules: proteins and lipids. The proteins serve as enzymes, or biological catalysts, and provide the membrane with its distinctive functional properties. The lipids provide the gross structural properties of the membrane. The simplest lipids found in nature, such as fats and waxes, are insoluble in water. The lipids found in membranes are amphipathic, meaning

that one end of the molecule is hydrophobic, or insoluble in water, and the other end hydrophilic, or water-soluble. The hydrophilic region is described as being polar because it is capable of carrying an ionic (electric) charge; the hydrophobic region is nonpolar.

In most membrane lipids the nonpolar region consists of the hydrocarbon chains of fatty acids: hydrocarbon molecules with a carboxyl group (COOH) at one end. In a typical membrane lipid two fatty-acid molecules are chemically bonded through their carboxyl ends to a backbone of glycerol. The glycerol backbone, in turn, is attached to a polar-head group consisting of phosphate and other groups, which often carry an ionic charge [*see illustration on next page*]. Phosphate-containing lipids of this type are called phospholipids.

When a suspension of phospholipids in water is subjected to high-energy sound under suitable conditions, the phospholipid molecules cluster together to form closed vesicles: small saclike structures called liposomes. The arrangement of phospholipids in the walls of both liposomes and biological membranes has recently been deduced with the help of X-ray diffraction, which can reveal the distance between repeating groups of atoms. An X-ray diffraction analysis by M. F. Wilkins and his associates at King's College in London indicates that two parallel arrays of the polar-head groups of lipids are separated by a distance of approximately 40 angstroms and that the fatty-acid tails are stacked parallel to one another in arrays of 50 or more phospholipid molecules.

The X-ray data suggest a structure for liposomes and membranes in which the phospholipids are arranged in two parallel layers [*see illustrations on page 255*]. The polar heads are arrayed externally

on the bilayer surfaces, and the fatty-acid tails are pointed inward, perpendicular to the plane of the membrane surface. This model of phospholipid structure in membranes is identical with one proposed by James F. Danielli and Hugh Davson in the mid-1930's, when no precise structural data were available. It is also the minimum-energy configuration for a thin film composed of amphipathic molecules, because it maximizes the interaction of the polar groups with water.

Unlike lipids, proteins do not form orderly arrays in membranes, and thus their arrangement cannot be assessed by X-ray diffraction. The absence of order is not surprising. Each particular kind of membrane incorporates a variety of protein molecules that differ widely in molecular weight and in relative numbers; a membrane can incorporate from 10 to 100 times more molecules of one type of protein than of another.

Since little can be learned about the disposition of membrane proteins from a general structural analysis, investigators have chosen instead to study the orientation of one or a few species of the proteins in membranes. In the Danielli-Davson model the proteins are assumed to be entirely external to the lipid bilayer, being attached either to one side of the membrane or to the other. Although information obtained from X-ray diffraction and high-resolution electron microscopy indicates that this is probably true for the bulk of the membrane protein, biochemical studies show that the Danielli-Davson concept is an oversimplification. The evidence for alternative locations has been provided chiefly by Marc Bretscher of the Medical Research Council laboratories in Cambridge and by Theodore L. Steck, G. Franklin and Donald F. H. Wallach of the Harvard Medical School. Their results suggest that certain proteins penetrate the lipid bilayer and that others extend all the way through it.

Bretscher has labeled a major protein of the cytoplasmic membrane of human red blood cells with a radioactive substance that forms chemical bonds with the protein but is unable to penetrate the membrane surface. The protein was labeled in two ways [*see illustration, pages 256 and 257*]. First, intact red blood cells were exposed to the label so that it became attached only to the portion of the protein that is exposed on the outer surface of the membrane. Second, red blood cells were broken up before the radioactive label was added. Under these conditions the label could attach itself to parts of the protein exposed on the internal surface of the membrane as well as to parts on the external surface.

The two batches of membrane, labeled under the two different conditions, were treated separately to isolate the protein. The purified protein from the two separate samples was degraded into definable fragments by treatment with a proteolytic enzyme: an enzyme that cleaves links in the chain of amino acid units that constitutes a protein. A sample from each batch of fragments was now placed on the corner of a square of filter paper for "fingerprinting" analysis. In this technique the fragments are separated by chromatography in one direction on the paper and by electrophoresis in a direction at right angles to the first. In the chromatographic step each type of fragment is separated from the others because it has a characteristic rate of travel across the paper with respect to the rate at which a solvent travels. In the electrophoretic step the fragments are further separated because they have characteristic rates of travel in an imposed electric field.

Once a separation had been achieved the filter paper was laid on a piece of X-ray film so that the radioactively labeled fragments could reveal themselves by exposing the film. When films from the two batches of fragments were developed, they clearly showed that more labeled fragments were present when both the internal and the external surface of the cell membrane had been exposed to the radioactive label than when the outer surface alone had been exposed. This provides strong evidence that the portion of the protein that gives rise to the additional labeled fragments is on the inner surface of the membrane.

Steck and his colleagues obtained similar results with a procedure in which they prepared two types of closed-membrane vesicle, using as a starting material the membranes from red blood cells. In one type of vesicle preparation (right-side-out vesicles) the outer membrane surface is exposed to the external aqueous environment. In the other type of preparation (inside-out vesicles) the inner surface of the membrane is exposed to the external aqueous environment. When the two types of vesicle are treated with a proteolytic enzyme, only those proteins exposed to the external aqueous environment should be degraded. Steck found that some proteins are susceptible to digestion in both the right-side-out and inside-out vesicles, indicating that these proteins are exposed on both membrane surfaces. Other proteins are susceptible to proteolytic digestion in right-side-out vesicles but not in inside-out vesicles. Such proteins are evidently located exclusively on only one side of the membrane. This information lends credence to the concept of sidedness in membranes. Such sidedness had been suspected for many years because the inner and outer surfaces of cellular membranes are thought to have different biological functions. The development of a technique for preparing vesicles with right-side-out and inside-out configurations should be extremely useful in determining on which side of the membrane a given species of protein resides and thus functions.

Daniel Branton and his associates at the University of California at Berkeley have developed and exploited the technique of freeze-etch electron microscopy to study the internal anatomy of membranes. In freeze-etch microscopy a suspension of membranes in water is frozen rapidly and fractured with a sharp blade. Wherever the membrane surface runs parallel to the plane of fracture much of the membrane will be split along the middle of the lipid bilayer. A thin film of platinum and carbon is then evaporated onto the surface of the fracture. This makes it possible to examine the anatomy of structures in the fracture plane by electron microscopy.

The electron micrographs of the fractured membrane reveal many particles, approximately 50 to 85 angstroms in diameter, on the inner surface of the lipid bilayer. These particles are not observed if the membrane samples are first treated with proteolytic enzymes, indicating that the particles probably consist of protein [*see illustration on page 252*]. From quantitative estimates of the number of particles revealed by freeze-etching, Branton and his colleagues have suggested that between 10 and 20 percent of the internal volume of many biological membranes is protein.

Somewhere between a fifth and a quarter of all the protein in a cell is physically associated with membranes. Most of the other proteins are dissolved in the aqueous internal environment of the cell. In order to dissolve membrane proteins in aqueous solvents detergents must be added to promote their dispersion. One might therefore expect membrane proteins to differ considerably from soluble proteins in chemical composition. This, however, is not the case.

The amino acids of which proteins are composed can be classified into two groups: polar and nonpolar. S. A. Rosenberg and Guido Guidotti of Harvard University analyzed the amino acid composition of proteins from a number of membranes and found that they contain about the same percentage of polar and nonpolar amino acids as one finds in the soluble proteins of the common colon

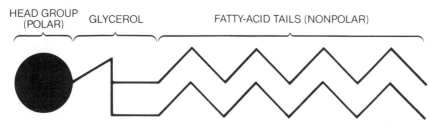

| HEAD GROUP (POLAR) | GLYCEROL | FATTY-ACID TAILS (NONPOLAR) |

TYPICAL MEMBRANE LIPID is a complex molecular structure, one end of which is hydrophilic, or water-soluble, and the other end hydrophobic. Such a substance is termed amphipathic. The hydrophilic, or polar, region consists of phosphate and other constituents attached to a unit of glycerol. The polar-head group, when in contact with water, often carries an electric charge. The glycerol component forms a bridge to the hydrocarbon tails of two fatty acids that constitute the nonpolar region of the lipid. In this highly schematic diagram the zigzag lines represent hydrocarbon chains; each angle is occupied by a carbon atom and two associated hydrogen atoms. The terminal carbon of each chain is bound to three hydrogen atoms. Phosphate-containing amphipathic lipids are called phospholipids.

bacterium *Escherichia coli*. Thus differences in amino acid composition cannot account for the water-insolubility of membrane proteins.

Studies conducted by L. Spatz and Philipp Strittmatter of the University of Connecticut indicate that the most likely explanation for the water-insolubility of membrane proteins is the arrangement of their amino acids. Spatz and Strittmatter subjected membranes of rabbit liver cells to a mild treatment with a proteolytic enzyme. The treatment released the biologically active portion of the membrane protein: cytochrome b_5. In a separate procedure they solubilized and purified the intact cytochrome b_5 and treated it with the proteolytic enzyme. This treatment also released the water-soluble, biologically active portion of the molecule, together with a number of small degradation products that were insoluble in aqueous solution. The biologically active portion of the molecule, whether obtained from the membrane or from the purified protein, was found to be rich in polar amino acids. The protein fragments that were insoluble in water, on the other hand, were rich in nonpolar amino acids. These observations suggest that many membrane proteins may be amphipathic, having a nonpolar region that is embedded in the part of the membrane containing the nonpolar fatty-acid tails of the phospholipids and a polar region that is exposed on the membrane surface.

We are now ready to ask: How do substances pass through membranes? The nonpolar fatty-acid-tail region of a phospholipid bilayer is physically incompatible with small water-soluble substances, such as metal ions, sugars and amino acids, and thus acts as a barrier through which they cannot flow freely. If one measures the rate at which blood sugar (glucose) passes through the phospholipid-bilayer walls of liposomes, one finds that it is far too low to account for the rate at which glucose penetrates biological membranes. Information of this kind has given rise to the concept that entities termed carriers must be present in biological membranes to facilitate the passage of metal ions and small polar molecules through the barrier presented by the phospholipid bilayer.

Experiments with biological membranes indicate that the hypothetical carriers are highly selective. For example, a carrier that facilitates the transport of glucose through a membrane plays no role in the transport of amino acids or other sugars. An interesting experimental

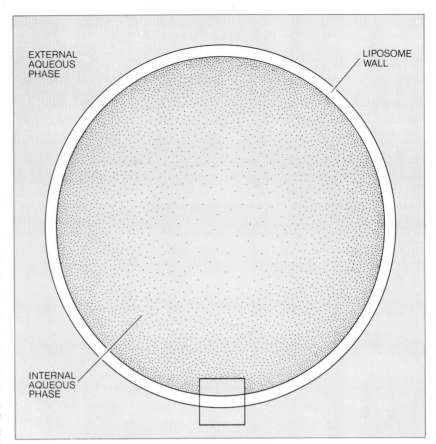

ARTIFICIAL MEMBRANE-ENCLOSED SAC, known as a liposome, is created by subjecting an aqueous suspension of phospholipids to high-energy sound waves. X-ray diffraction shows that the phospholipids in the liposome assume an orderly arrangement resembling what is found in the membranes of actual cells. Area inside the square is enlarged below.

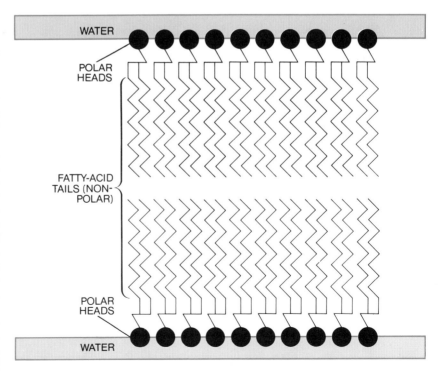

CROSS SECTION OF LIPOSOME WALL shows how the membrane is formed from two layers of lipid molecules. The polar heads of amphipathic lipids face toward the aqueous solution on each side while the nonpolar fatty-acid tails face inward toward one another.

system for measuring selective ion transport was developed by A. D. Bangham, M. M. Standish and J. C. Watkins of the Agricultural Research Council in Cambridge, England, and by J. B. Chappell and A. R. Crofts of the University of Cambridge. As a model carrier they used valinomycin, a nonpolar, fat-soluble antibiotic consisting of a short chain of amino acids (actually 12); such short chains are termed polypeptides to distinguish them from true proteins, which are much larger. Valinomycin combines with phospholipid-bilayer membranes and makes them permeable to potassium ions but not to sodium ions.

The change in permeability is conveniently studied by measuring the change in electrical resistance across a phospholipid bilayer between two chambers containing a potassium salt in aqueous solution. The experiment is performed by introducing a sample of phospholipid into a small hole between the two chambers. The lipid spontaneously thins out until the chambers are separated by only a thin membrane consisting of a phospholipid bilayer. Electrodes are then placed in the two chambers to measure the resistance across the membrane.

The resistance across a phospholipid bilayer in the absence of valinomycin is several orders of magnitude higher than the resistance across a typical biological membrane: 10 million ohms centimeter squared compared with between 10 and 10,000. This indicates that phospholipid-bilayer membranes are essentially impermeable to small hydrophilic ions. If a small amount of valinomycin (10^{-7} gram per milliliter of salt solution) is intro-

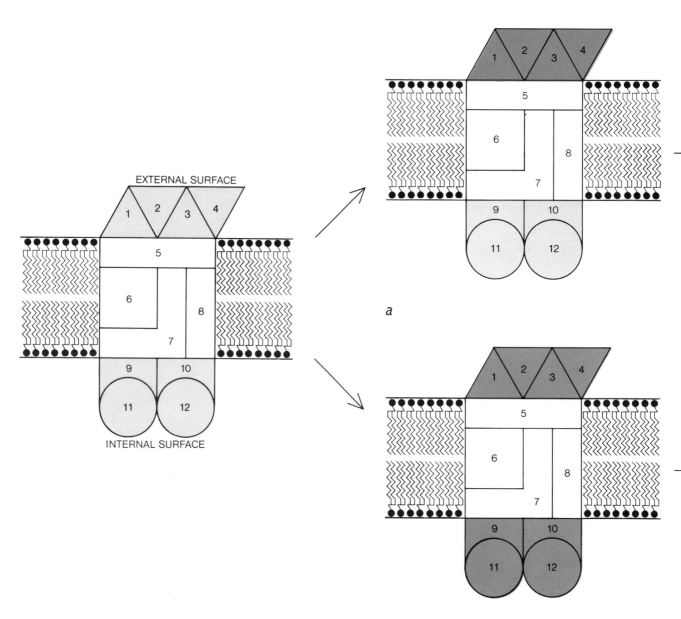

LOCATION OF PROTEINS IN MEMBRANES can be inferred by attaching radioactive labels to the proteins. These diagrams depict an experiment in which a major protein in the membrane of red blood cells was labeled (a). When intact cells (top sequence) are exposed to the radioactive substance, only the portion of the protein on the outside wall picks up the label (color). When the cells are broken before labeling (bottom sequence), the radioactive la- bel is able to reach portions of the protein that are exposed to the internal as well as to the external surfaces of the membrane. This can be demonstrated by isolating and purifying the protein labeled under the two conditions. The protein is then broken up into defined fragments (numbered shapes) by treating it with a proteolytic enzyme (b). Portions of the two batches of fragments are spotted on the corners of filter paper for "fingerprinting" (c). This is a duced into the chambers containing the potassium solution, the resistance falls by five orders of magnitude and the permeability of the phospholipid bilayer to potassium ions rises by a like amount. The permeability of the experimental membrane now essentially duplicates the permeability of biological membranes.

If the experiment is repeated with a sodium chloride solution in the chambers, one finds that the addition of valinomycin causes only a slight change in resistance. Hence valinomycin meets two of the most important criteria for a biological carrier: it enhances permeability and it is highly selective for the transported substance. The question that now arises is: How does valinomycin work?

First of all, valinomycin is nonpolar. Thus it is physically compatible with and can dissolve in the portion of the bilayer that contains the nonpolar fatty-acid tails. Second, valinomycin can evidently diffuse between the two surfaces of the bilayer. S. Krasne, George Eisenman and G. Szabo of the University of California at Los Angeles have shown that the enhancement of potassium-ion transport by valinomycin is interrupted when the bilayer is "frozen" by lowering the temperature. Third, valinomycin must bind potassium ions in such a way that the ionic charge is shielded from the nonpolar region of the membrane. Finally, valinomycin itself must have a selective binding capacity for potassium ions in preference to sodium or other ions.

With valinomycin as a model for carrier-mediated transport, one can postulate three essential steps: recognition of the ion, diffusion of the ion through the membrane, and its release on the other

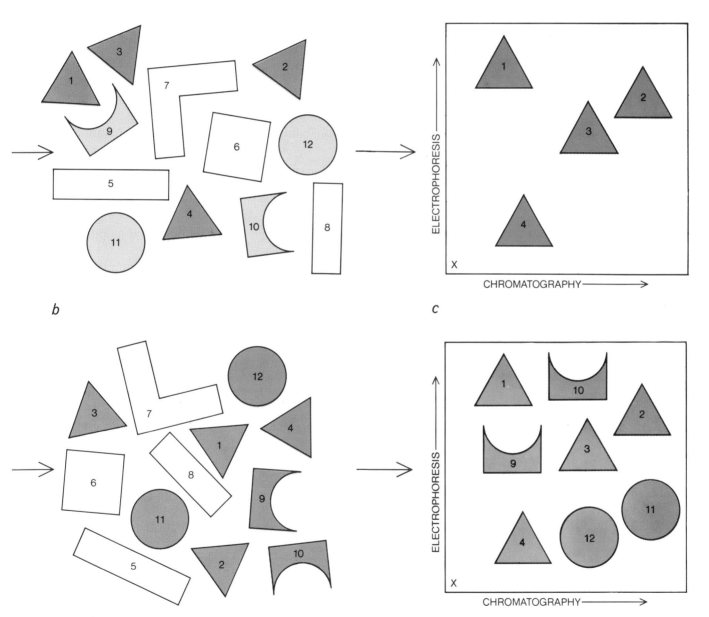

b

c

technique that combines chromatography with electrophoresis. By chromatography alone protein fragments would migrate at different rates depending primarily on their solubility in the solvent system. Electrophoresis involves establishing an electric-potential gradient along one axis of the filter paper. Since various fragments have different densities of electric charge they are further separated. A piece of X-ray film is then placed over each sheet of filter paper. Radiation from the labeled fragments exposes the film and reveals where the various fragments have come to rest. A comparison of the X-ray films produced in the parallel experiments shows that more protein fragments are labeled when the red blood cells are broken before labeling and that the additional fragments (9, 10, 11, 12) must represent portions of the original protein that extend through the membrane and penetrate the inner surface.

side. In the first step some part of the valinomycin molecule, embedded in the membrane, "recognizes" the potassium ion as it approaches the surface of the membrane and captures it. In the second step the complex consisting of valinomycin and the potassium ion diffuses through the membrane. Finally, on reaching the opposite surface of the membrane the potassium ion is dissociated from the complex and is released.

The argument to this point can be summarized in a few words. The fundamental structure of biological membranes is a phospholipid bilayer, the phospholipid bilayer is a permeability barrier and carriers are needed to breach it. In addition, the membrane barrier must often be breached in a directional way. In a normally functioning cell hundreds of kinds of small molecule must be present at a higher concentration inside the cell than outside, and many other small molecules must be present at a lower concentration inside the cell than outside. For example, the concentration of potassium ions in human cells is more than 100 times greater than it is in the blood that bathes them. For sodium ions the concentrations are almost exactly reversed. The maintenance of these differences in concentration is absolutely essential; even slight changes can result in death.

Although the model system based on valinomycin provides considerable insight into the function and selectivity of carriers, it sheds no light on the transport mechanism that can pump a substance from a low concentration on one side of the membrane to a higher concentration on the other. Our understanding of concentrative transport (or, as it is usually termed, active transport) owes much to the pioneering effort of Georges Cohen, Howard Rickenberg, Jacques Monod and their associates at the Pasteur Institute in Paris. The Pasteur group studied the transport of milk sugar (lactose) through the cell membrane of the bacterium *Escherichia coli*. Genetic experiments suggested that the carrier for lactose transport was a protein. Studies of the rate of transport revealed that the transport process behaves like a reaction catalyzed by an enzyme, giving further support to the idea that the carrier is a protein. The Pasteur group also found that the lactose-transport system is capable of active transport, producing a lactose concentration 500 times greater inside the cell than outside. The active-transport process depends on the expenditure of metabolic energy; poisons that block energy metabolism destroy the ability of the cell to concentrate lactose.

A model that accounts for many (but not all) of the properties of the active-transport system that are typified by the lactose system postulates the existence of a carrier protein that can change its shape. The protein is visualized as resembling a revolving door in the membrane wall [*see illustration on opposite page*]. The "door" contains a slot that fits the target substance to be transported. The slot normally faces the cell's external environment. When the target substance enters the slot, the protein changes shape and is thereby enabled to rotate so that the slot faces into the cell. When the target substance has been discharged into the cell, the protein remains with its slot facing inward until the cell expends energy to rotate the protein so that the slot again faces outward.

Working with Eugene P. Kennedy at the Harvard Medical School in 1965, I succeeded in identifying the lactose-transport carrier. We found, as we had expected, that it is a protein with an enzyme-like ability to bind lactose. Since then a number of other transport carriers have been identified, and all turn out to be proteins. The lactose carrier resides in the membrane and is hydrophobic; thus it is physically compatible

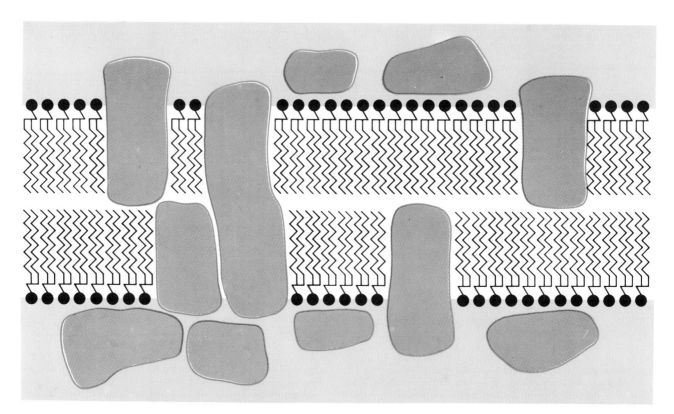

ANATOMY OF BIOLOGICAL MEMBRANE is suggested in this schematic diagram. Phospholipid molecules stacked side by side and back to back provide the basic structure. The gray shapes represent protein molecules. In some cases several proteins (for example the five at the left) are bound into a single functional complex. Proteins can occupy all possible positions with respect to the phospholipid bilayer: they can be entirely outside or inside, they can penetrate either surface or they can extend through the membrane.

with the nonpolar-lipid phase of the membrane.

In 1970 Ron Kaback and his associates at the Roche Institute of Molecular Biology observed that the energy that drives the active transport of lactose and dozens of other low-molecular-weight substances in *E. coli* is directly coupled to the biological oxidation of metabolic intermediates such as D-lactic acid and succinic acid. How energy derived from the oxidation of D-lactic acid can be used to drive active transport is one of the more interesting unsolved problems in membrane biology.

Since transport carriers must be mobile within the membrane in order to move substances from one surface to the other, one might guess that the region of the membrane containing the fatty-acid tails should not have a rigid crystalline structure. X-ray diffraction studies indicate that the fatty acids of membranes in fact do have a "liquid crystalline" structure at physiological temperature, that is, around 37 degrees Celsius. In other words, the fatty acids are not aligned in a rigid crystalline lattice. The techniques of electron paramagnetic resonance and nuclear magnetic resonance can be used to study the flexibility of the fatty-acid side chains in membranes. Several investigators, notably Harden M. McConnell and his associates at Stanford University, have concluded that the fatty acids of membranes are quasi-fluid in character.

Membranes incorporate two classes of fatty acids: saturated molecules, in which all the available carbon bonds carry hydrogen atoms, and unsaturated molecules, in which two or more pairs of hydrogen atoms are absent (with the result that two or more pairs of carbon atoms have double bonds). The fluid character of membranes is largely determined by the structure and relative proportion of the unsaturated fatty acids. In phospholipids consisting only of saturated fatty acids the fatty-acid tails are aligned in a rigidly stacked crystalline array at physiological temperatures. In phospholipids consisting of both saturated and unsaturated fatty acids the fatty acids are packed in a less orderly fashion and thus are more fluid. The double bonds of unsaturated fatty acids give rise to a structural deformation that interrupts the ordered stacking necessary for the formation of a rigid crystalline structure [*see illustration on next page*].

My colleagues and I at the University of Chicago (and later at the University of California at Los Angeles) and Peter Overath and his associates at the Uni-

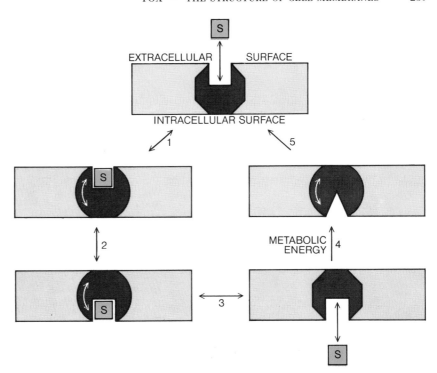

MECHANISM OF "ACTIVE" TRANSPORT may involve a carrier protein (*dark gray*) with the properties of a revolving door. A carrier protein can capture a substance, S, that exists outside the membrane in dilute solution and transport it to the inside of the cell, where the concentration of S is greater than it is outside. When S is bound to the protein, the protein changes shape (*1*), thus enabling it to rotate (*2*). When S becomes detached and enters the cell (*3*), the protein returns to its immobile form. Metabolic energy must be expended (*4*) to alter the protein's shape so that it can rotate and again present its binding site to the cell exterior (*5*). Other protein carriers have the capacity to transport substances from low concentration inside the cell to solutions of higher concentration outside the cell.

versity of Cologne have varied the fatty-acid composition of biological membranes to study the effects of fatty-acid structure on transport. When the membrane lipids are rich in unsaturated fatty acids, transport can proceed at rates up to 20 times faster than it does when the membrane lipids are low in unsaturated fatty acids. These experiments show that normal membrane function depends on the fluidity of the fatty acids.

The temperature at which cells live and grow can have a pronounced effect on the amount of unsaturated fatty acid in their membranes. Bacteria grown at a low temperature have membranes with a greater proportion of unsaturated fatty acid than those grown at a higher temperature. This adjustment in fatty-acid composition is necessary if the membranes are to function normally at low temperature. A similar adjustment can take place in higher organisms. For example, there is a temperature gradient in the legs of the reindeer; the highest temperature is near the body, the lowest is near the hooves. To compensate for this temperature gradient the cells near the hooves have membranes whose lipids are enriched in unsaturated fatty acids.

Although, as we have seen, phospholipids can spontaneously form bilayer films in water, this process only provides a physical rationale as to why the predominant structure in membranes is a phospholipid bilayer. The events leading to the assembly of a biological membrane are far more complex. The cells of higher organisms contain a number of unique membrane structures. They differ widely in lipid composition, and each type of membrane has its own unique complement of proteins. The diversity in protein composition and in the location of proteins within membranes explains the functional diversity of different types of membrane. Rarely does a single species of protein exist in more than one type of membrane.

Since all membrane proteins are synthesized at approximately the same cellular location, what is it that determines that one type of protein will be incorporated only into the cytoplasmic membrane and that another type will turn up only in a mitochondrial membrane? At present this question can be answered only by conjecture tinctured with a few facts. Two general hypotheses for membrane assembly can be offered. One pos-

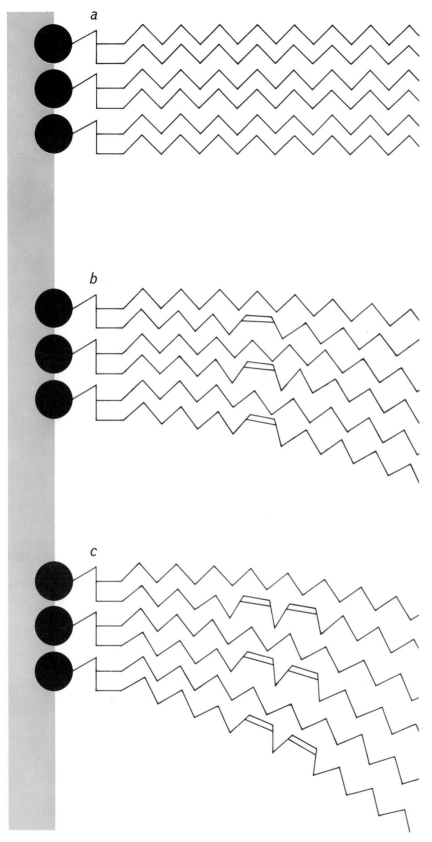

VARIATION IN FATTY-ACID COMPOSITION can disrupt the orderly stacking of phospholipids in a biological membrane. In a lipid layer composed entirely of saturated fatty acids (*a*) the fatty-acid chains contain only single bonds between carbon atoms and thus nest together to form rigid structures. In a lipid layer containing unsaturated fatty acids with one double bond (*b*) the double bonds introduce a deformation that interferes with orderly stacking and makes the fatty-acid region somewhat fluid. When fatty acids with two double bonds are present (*c*), the deformation and the consequent fluidity are greater still.

sibility is that new pieces of membrane are made from scratch by a self-assembly mechanism in which all the components of a new piece of membrane come together spontaneously. This new piece could then be inserted into an existing membrane. A second possibility is that newly made proteins are simply inserted at random into a preexisting membrane.

Recent studies in my laboratory at the University of California at Los Angeles and in the laboratories of Philip Siekevitz and George E. Palade at Rockefeller University support the second hypothesis. That is all well and good, but what determines why a given protein is incorporated only into a given kind of membrane? Although this must be answered by conjecture, it is known that many proteins are specifically bound to other proteins in the same membrane. Such protein-protein interactions are not uncommon; many of the functional entities in membranes are complexes of several proteins. Thus the proteins in a membrane may provide a template that is recognized by a newly synthesized protein and that helps to insert the newly synthesized protein into the membrane. In this way old membrane could act as a template for the assembly of new membrane. This might explain why different membranes incorporate different proteins.

Why, then, do different membranes have different lipid compositions? The answers to this question are even more obscure. In general lipids are synthesized within the membrane; the enzymes that catalyze the synthesis are part of the membrane. Some lipids, however, are made in one membrane and then shuttled to another membrane that has no inherent capacity to synthesize them. Since there is an interchange of lipids between various membranes, it seems unlikely that the variations in lipid composition in different membranes can be explained by dissimilarities in the synthetic capacity of a given membrane for a given type of lipid. There are at least two possible ways of accounting for differences in lipid composition. One possibility is that different membranes may destroy different lipids at different rates; another is that the proteins of one species of membrane may selectively bind one type of lipid, whereas the proteins of another species of membrane may bind a different type of lipid. It is obvious from this discussion that concrete evidence on the subject of membrane assembly is scant but that the problems are well defined.

IV

INFORMATION TRANSFER AND CONTROL

Protein Synthesis and Regulation

IV

INFORMATION
TRANSFER
AND CONTROL

INTRODUCTION

Elaborate control mechanisms have evolved in living cells to coordinate the thousands of chemical reactions necessary for reproduction; our present knowledge of these mechanisms is described in the articles in this final section. We begin with a discussion of protein synthesis since much if not most of the metabolic effort in cells is directed to the synthesis of proteins. An overview of the processes by which proteins are constructed has already been presented in a number of earlier articles in this volume (particularly note Robert W. Holley's article which focuses on the "key" to the genetic code). A sequence of nucleotides in DNA is transcribed into sequence of nucleotides in RNA, and this RNA in turn is translated into a sequence of amino acids in a polypeptide chain which eventually assumes the three-dimensional configuration of a functional protein. In Part II we have examined some of the roles of these proteins as finished products. The process of protein synthesis—the central dogma of molecular biology—is elegant in its simplicity. However, as a molecular event it also has until recently eluded direct visualization. Most of molecular biology deals with inferred conclusions based upon genetic and biochemical techniques applied to systems that we cannot actually see. This is sometimes discomforting to those who are not familiar with these techniques and who may wonder about their validity. (In fact, such approaches are generally no more susceptible to artifact and misinterpretation than are methods that purport to give a direct view of the process under study.) It was with great excitement and perhaps some relief that electron microscopy recently added visual support for currently accepted mechanisms of gene action. These remarkable pictures are presented in the first article in this section by O. L. Miller, Jr. As one examines the electron micrographs, one should appreciate that these are not pictures of static structures but that a frenzy of metabolic activity has been caught in one frame of a high-speed motion picture film. The article points out some important differences in the spatial and temporal relationships between transcription and translation when eucaryotes and procaryotes are compared. The top picture on p. 271 illustrates the close coupling of the processes of transcription and translation in a bacterium.

Different numbers of RNA copies may be produced from different genes. In order to accomplish the high rate of protein synthesis necessary in some eucaryotic cells, such as amphibian oöcytes, many ribosomes must be assembled, and this in turn requires that many copies of the ribosomal RNA components must be made. This is sometimes accomplished by the selective replication of the genes that code for this RNA; such a process is called "gene amplification," and the replicated DNA is found in thousands of copies in an organelle called the nucleolus, which lies adjacent to the chromosome region that contains the "master copy" for ribosomal RNA. Also, in some systems a "tandem duplication" of the genes for ribosomal RNA is found. That is, many copies of the same gene are present in series in the chromosome, separated by short segments of "spacer DNA." In both cases the DNA is transcribed into RNA by many RNA polymerase molecules working sequentially and simultaneously from unique starting points. This process is dramatically and artistically displayed in several of the electron micrographs in Miller's article.

We have already considered the nature of the genetic code and the translation mechanism which requires two key elements in a series, first, an enzyme, an amino acyl synthetase, that is specific for each type of amino acid; these enzymes recognize and attach the "activated" amino acid to an appropriate transfer RNA molecule; and, second, the transfer RNA molecule which contains the anticodon triplet capable of recognizing the complementary codon triplet sequence in messenger RNA. Many other enzymes and structural components are needed in addition to the actual decoding machinery. The article by Nomura in Part III emphasizes the complexity of the assembly bench, the ribosome, upon which proteins are made even though in electron micrographs the details of these complex structures cannot be resolved.

Charles Yanofsky's article on gene structure and protein structure reviews the in vivo evidence that the amino acid sequences in protein do indeed correspond directly to nucleotide sequences in the gene. His article, along with that of Seymour Benzer in Part II, attests to the power and elegance of genetic analysis in attacking biochemical problems. Since that article was written, additional factors required for protein synthesis have been discovered.

Protein synthesis can be conveniently subdivided into three stages: initiation, chain elongation, and chain termination. The initiation stage is discussed in the article by Brian F. C. Clark and Kjeld A. Marcker on how proteins start, and it is now known to require three initiation factors (protein themselves) which serve to form an "initiation complex" on the 30s ribosomal subunit. The formation of this complex in bacteria involves the attachment of a transfer RNA molecule carrying a modified form of methionine and then joining to the 50s ribosomal subunit to form the functional 70s ribosomal complex. This formyl-methionine transfer RNA recognizes an initiating codon, AUG or GUG, and protein chain elongation then begins, reading along the messenger RNA in the 5' to 3' direction. In the initiation stage, energy is required in the form of one molecule of guanosine triphosphate (GTP) which loses one phosphate group and is thereby converted to guanosine diphosphate (GDP). Chain elongation also utilizes energy from GTP, as transfer factor one assists with the binding of each amino acid-tRNA complex to the ribosome. Yet another GTP is split to GDP in the process of "translocation" in which an unloaded tRNA is ejected from the ribosome and the messenger RNA is racheted one more condon along through the ribosome in preparation for acceptance of the next charged tRNA. The initial charging of each tRNA with its appropriate amino acid involves the splitting of an ATP to AMP and pyrophosphate, which is then cleaved by the enzyme pyrophosphatase to orthophosphate. Therefore, the addition of each amino acid in the growing protein chain requires the expenditure of four high-energy phosphate bonds, and a typical protein of 150 amino acids would require the generation of 600 ATP molecules for its synthesis. Nearly twenty glucose molecules must be "burned" to provide this energy.

The third stage of protein synthesis, termination, involves the action of protein release factors (called "rho" factors) that recognize the termination signals (UAA, UAG, or UGA) in the messenger RNA. A given messenger RNA molecule may contain in tandem the information for a number of different proteins that may generally perform related functions. Such a "polycistronic" messenger RNA is transcribed from the DNA as a unit, but it is punctuated with "start" and "stop" signals to separate the different units of translation. The unit of transcription is called an "operon," and this entity is also a unit of regulation, as discussed below.

What determines the relative numbers of different proteins that the cell must produce? In certain growth environments some enzymes are needed in large amounts while others are not needed at all. In general only a part of the genome is expressed as functional proteins at any given time. Cells of the bacterium *E. coli* growing in very simple, chemically defined media with the sugar glucose as the sole source of carbon (and energy) must produce enzymes for synthesizing all of the amino acids. However, if excess amounts of these amino acids are added to the medium, it becomes a waste of effort and energy for the cells to continue to produce enzymes that are no longer needed. A negative feedback mechanism operates at the level of messenger RNA transcription to control enzyme synthesis. The mechanism was postulated by Francois Jacob and Jaques Monod over ten years ago, but only recently were the putative "genetic repressors" isolated by Mark Ptashne and by Walter Gilbert. The control of transcription by the binding of repressor proteins to regulatory genes in an operon is an effective and persistent mechanism for controlling enzyme activities in the cell. However, it is not a very rapidly acting process since the existing enzymes continue to function even when their further synthesis has ceased. A more rapid type of feedback regulation exists at the level of the enzymes themselves, in that the end-product of a biochemical pathway (e.g., a particular amino acid) can interact with and inhibit the activity of one of the enzymes (generally the first one) in the pathway. This mode of regulation is also considered in detail in the article by Jean-Pierre Changeux.

Recently, evidence has been obtained that the molecule "cyclic AMP" plays an important role in the regulation of transcription in some operons. This control is exerted at the level of the "promotor" (which is the attachment site for RNA polymerase), and it supercedes regulation by repressors at the operator site in the operon. Ira Pastan discusses the various regulatory roles of cyclic AMP, including the control of transcription in bacteria. The articles in this section do not document all of the levels at which control of enzyme synthesis and function can occur, and, in fact, we still do not understand the intricacies of many of the regulatory mechanisms that operate in cells. However, we do know that regulation can occur at the following levels: (1) transcription, (2) translation, (3) degradation of messenger RNA, (4) enzyme activation or inhibition, and (5) enzyme degradation. Also, the synthesis of RNA can be enhanced by gene amplification, as described above. In multicellular organisms the problems of regulation are clearly much more complex than those faced by bacteria, with the added elements of differentiation as designated cells must express different groups of genes and communication is established to coordinate the growth and development of different parts of the organism. This latter topic is treated by Gunther S. Stent in the final article of Part IV; it carries the story from the transfer of information via nucleic acid carriers to more general problems of cellular communication.

One commonly thinks of replication and transcription as the major "transactions" carried out by DNA in the cell. As one examines the genetic substance more closely, one finds, however, a number of other activities involving DNA that are equally important to the survival of the cell. Some of these involve natural "labeling" of DNA. Once the DNA in an organism such as *E. coli* has been synthesized, it is "branded" so that certain intracellular degradative enzymes can recognize it. These degradative enzymes might be considered the security police of the cell, and they serve to identify and destroy invading foreign DNA elements lacking the appropriate "brand." This branding process involves the methylation of certain of the bases in the DNA. The degradative enzymes can detect unmarked invading foreign DNA and quickly reduce it to harmless nucleotides. On the other hand, some viruses also degrade the host DNA after infection and must therefore brand their own DNA so that *their* enzymes can tell the difference. The T2 phage hydroxymethylates its cytosine so that it can be "recognized" by its own nucleases; it then further attaches glucose to the hydroxymethyls as insurance against the host nucleases. Nobel laureate Salvador Luria describes the details of the phenomena of host modification and restriction in his article on the recognition of DNA.

In the course of evolution living systems have developed a variety of repair mechanisms to ensure genetic continuity by eliminating defects from their DNA. Some defects are produced as errors in the process of normal replication, in which case the repair mechanisms can serve as quality control elements. Other defects are produced by physical or chemical attack from the environment, for example, by ultraviolet light or some components of smog or other chemical agents. The nature of the best known and most ubiquitous repair process is discussed in our own article on the repair of DNA. Since that article was prepared, we have learned a great deal more about the excision-repair scheme and a related recombinational mechanism for correcting structural defects in DNA. The DNA polymerase I discovered by Kornberg has turned out to have unique capabilities for excision repair. This remarkable enzyme possesses several separate enzymatic functions in one polypeptide chain. In addition to its 3' polymerase activity and a 3' exonuclease activity, there is a 5' exonuclease activity that can perform the excision of defects from DNA. It carries out the "patch and cut" process that was described in our *Scientific American* article in 1967. At that time we proposed that a repair "complex" might traverse the DNA excising the defective nucleotides ahead as it filled in the gaps behind with normal nucleotides. It was no surprise to us when John Cairns showed in 1969 that a mutant *E. coli* strain deficient in DNA polymerase I was also sensitive to ultraviolet light. What did come as a surprise to many was the report that this mutant grew and replicated its DNA normally. It has since been shown that *E. coli* contains at least two additional DNA polymerases, II and III, and that normal semi-conservative DNA replication is specifically dependent upon DNA polymerase III. Studies by Warren Masker at Stanford suggest that DNA polymerase II can "pinch hit" in the excision-repair process in the absence of DNA polymerase I. The fact that at least two DNA polymerases can carry out excision repair further attests to the evolutionary importance of the process. It is likely that the ubiquity of double-stranded, and, hence, genetically redundant, DNA as the primary genetic material is related to the potential for repair that is inherent in such a complementary structure.

One of the early proofs that DNA carried genetic information was the demonstration of genetic transformation in bacteria by Avery, MacLeod, and MacCarty in 1944. Pure DNA from one bacterium may be assimilated by another bacterium, and the donor DNA is sometimes permanently incorporated into the recipient genome through genetic recombination. However, not all bacterial strains can perform this most elementary of sexual processes. Alex Tomasz discusses in his article some of the cellular factors in genetic transformation that render cells "competent." It is not surprising that the cell wall and other surface properties of the cell are crucial factors. The uptake of DNA apparently does not show species specificity in that a competent cell will take up any DNA. However, only "homologous DNA" will participate in the subsequent recombinational event. Genetic recombination between two nucleic acid molecules requires that the double strands separate and that reannealing occur between complementary sequences of the participating molecules. By homologous DNA we mean simply that identical or closely similar sequences of nucleotides exist in the two recombining molecules. The process of reannealing of DNA strands is described in detail in the article by Roy J. Britten and David E. Kohne on repeated segments of DNA. DNA hybridization by reannealing is one of the most powerful techniques of molecular biology since it provides an accurate gauge of the relatedness between two or more selected DNA molecules.

The use of this technique has provided new insights into the phylogenetic relatedness of different organisms. It has also been useful for determining the degree of redundancy of nucleotide sequences in a given collection of DNA molecules. In viruses and procaryotic cell systems there appears to be very little sequence redundancy. However, in eucaryotic systems some 20 to 80 percent of the DNA may exist in multiple copies of the same genes; the classical example is the process of gene amplification discussed above. Some shorter sequences of nucleotides may be present in nearly a million copies, and there is much speculation as to possible roles for these repeated segments of DNA. Many of the suggestions center on possible regulatory roles.

The central dogma of molecular biology as enunciated by Francis H. C. Crick provides that the flow of information is from DNA to RNA to protein in all living systems. RNA viruses posed no exception to this sequence although the idea of RNA *duplication* has been added. The identification of a reverse transcriptase (that promotes DNA synthesis from an RNA template) in mammalian cells infected with certain RNA viruses raised for some the issue of whether this violated the central dogma. Howard M. Temin discusses this novel phenomenon in his article on RNA-directed DNA synthesis and shows how it is in fact not at odds with the central dogma which asserts that information lodged in amino acid sequences is irretrievable only in the living cells. The DNA synthesized by reverse transcriptase can replicate and be transcribed back into RNA (i.e., more virus). It can also undergo "lysogeny" in which it is incorporated physically into the host genome; this property is perhaps the principal justification for the evolution of the reverse transcription mechanism.

In the final article of this section, Gunther S. Stent describes, within the general concept of cellular communication, the several diverse ways in which information is passed from cell to cell both across generational lines (heredity constitutes "genetic communication") and among the cells and tissues of multicellular organisms (hormonal and neural communication). It is interesting to note that the role of environment becomes increasingly significant in modifying cellular communication as one moves from the genetic through the hormonal to the neural systems. The genetic system is to a large extent insulated from the organism's environment in that there is no inheritance of acquired characteristics, and it is only through rare mutational events caused by radiation or other mutagens, or indirectly by selective pressures on phenotypes, that the environment can affect the genetic information that is passed from one generation to the next. Some hormones exert their effects at the level of gene expression in selected target tissues whereas others affect the nervous system and play an important role in mediating organism's behavioral responses to environmental stimuli, as in the "fight or flight" response to stress. Finally, it is as a result of the exquisite elaboration of the nervous system of higher animals that we are able to monitor our environments, store information based on experience both in memory and in writing, and pass this acquired knowledge on to our offspring. The nature of genetic communication has made possible Darwinian or biological evolution; the nature of neural communication has made possible Lamarckian or cultural evolution.

26

THE VISUALIZATION OF GENES IN ACTION

O. L. MILLER, JR.
March 1973

The electron microscope reveals individual genes being transcribed into RNA and their RNA being translated into protein. The pictures look remarkably like diagrams based on genetic and biochemical data

Since the middle 1950's a major objective in biology has been to document and add detail to what is called the central dogma of genetics: that DNA is the hereditary material, that its information is encoded in the sequences of its subunits that constitute the genes and that this information is transcribed into RNA and then translated into protein. By a variety of remarkable biochemical and genetic techniques many of the steps in this process of information transfer and the substances and cellular elements involved in them were identified; the DNA-protein dictionary was worked out, and the effect of different conditions and foreign agents on transcription and translation was determined. The increasingly detailed picture of gene action that emerged was necessarily based largely on indirect and collective evidence, however; there was little direct evidence of how individual genes function.

At the Oak Ridge National Laboratory in 1967 my colleagues and I began attempting to make electron micrographs of individual genes in action. The success we have had indicates that electron microscopy is potentially a valuable tool for the study of cell genetics at the molecular level. Meanwhile one of the gratifying aspects of our work has been that many of our pictures bear an almost uncanny resemblance to diagrams of transcription and translation that have been published over the years in technical journals and in magazines such as *Scientific American*. In other words, the pictures tend to confirm what had been proposed through consideration of painstakingly accumulated quantitative data.

The transfer of information from the deoxyribonucleic acid (DNA) of the genes into protein is accomplished in a series of steps involving the transcription of DNA into three kinds of ribonucleic acid (RNA). One is messenger RNA, which serves as the template on which amino acid subunits are assembled into proteins in the translation step. Another is ribosomal RNA, a constituent of the structures called ribosomes on which translation is accomplished. The third is transfer RNA, which carries the amino acid subunits of protein to their proper site along the messenger-RNA template.

The steps involved in the transcription of DNA into RNA and the translation of RNA into protein are essentially the same in all living cells, but the temporal and spatial relations of the steps in eukaryotic cells, whose DNA is confined within a membrane-bounded nucleus during most of the cell cycle, are different from the steps in prokaryotic cells, which lack such a nucleus. In prokaryotes, which include the many species of bacteria, transcription and translation of messenger RNA occur at the same time and place. In eukaryotic cells, which range from single-celled organisms such as the paramecium to the cells of higher vertebrates, the two processes are separated in time and space: transcription of DNA into the various RNA's that accomplish protein synthesis occurs in the nucleus; then the RNA's migrate through the membranous nuclear envelope into the cytoplasm, the main body of the cell, where the machinery for protein synthesis is located [*see illustration on page 271*].

We decided to begin by looking at the genetic machinery of a eukaryote: the oöcyte, or female reproductive cell, of amphibians such as frogs and salamanders. This type of oöcyte is a convenient cell with which to study genetic activity for a number of reasons; what was most important to us was that it would give us a good chance of seeing a great many identifiable genes, those that code for the two large molecules of RNA that are found in ribosomes. In most other eukaryotic cells these genes are found in clusters localized in nucleoli: dense bodies located at specific sites on certain chromosomes. In amphibian oöcytes large numbers of copies are made of the genes for ribosomal RNA, and these copies are in hundreds of nucleoli floating free and unattached to any chromosome. The result of this "amplification" of ribosomal-RNA genes is that a single oöcyte nucleus is equivalent to many hundreds of typical cell nuclei with respect to an identifiable genetic locus, and the amplified genes are not confusingly mingled with the other genes in the chromosomes.

There are two other advantages in studying amphibian oöcytes. One is that while they are growing they are actively synthesizing messenger RNA, and their chromosomes are in a greatly enlarged and uncoiled "lampbrush" stage where we might be able to see structural details of DNA being transcribed into messenger RNA. The final advantage is that these oöcytes are remarkably large. In some species they are as much as two millimeters in diameter at maturity and have nuclei as much as a millimeter in diameter. With a low-power microscope and jeweler's forceps one can readily isolate individual nuclei and manipulate their contents to prepare them for electron microscopy.

Barbara Beatty and I found that if the contents of an oöcyte nucleus were isolated and then put quickly into distilled water, the granular outer layers of the extrachromosomal nucleoli rapidly dispersed, allowing the compact fibrous

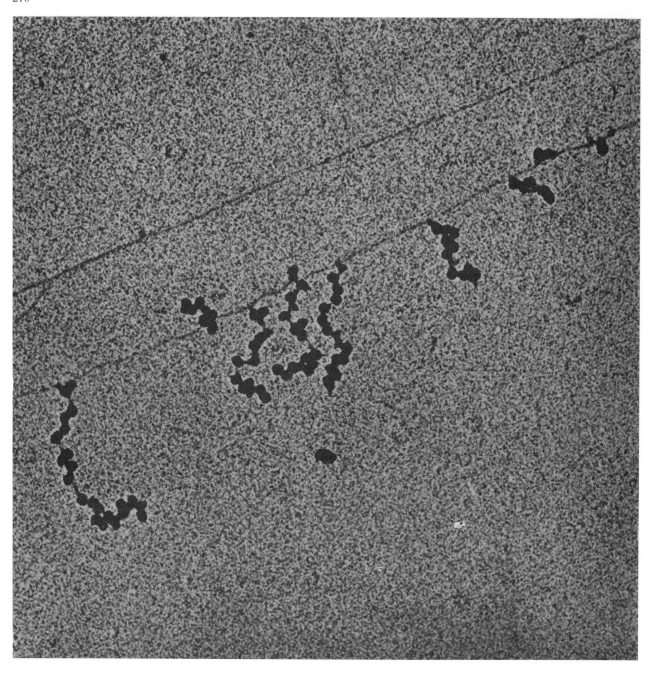

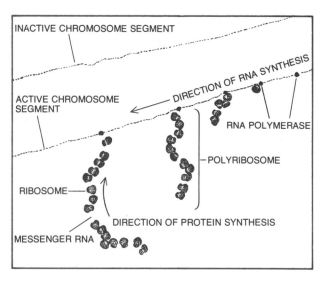

INACTIVE CHROMOSOME SEGMENT

ACTIVE CHROMOSOME
SEGMENT

DIRECTION OF RNA SYNTHESIS

RNA POLYMERASE

POLYRIBOSOME

RIBOSOME

DIRECTION OF PROTEIN SYNTHESIS

MESSENGER RNA

BACTERIAL GENE IN ACTION is enlarged 150,000 diameters in an electron micrograph made, like the others illustrating this article, by the author and his colleagues. The micrograph is interpreted in the somewhat simplified drawing at the right. One sees two segments of the chromosome of the bacterium *Escherichia coli*. The lower segment is active, that is, its DNA is being transcribed into messenger RNA and the RNA is being translated into protein. In the micrograph one can see molecules of RNA polymerase, the enzyme that catalyzes the transcription of DNA into RNA; one, at the far right, is at the approximate initiation site, where the transcription of each molecule of RNA begins. Successively transcribed messenger-RNA strands (*color in drawing*) in effect peel off toward the left; the longest one, now at the left of the micrograph, was the first to have been synthesized. As each RNA strand lengthens, ribosomes attach themselves and move along it, toward the chromosome, translating the RNA into protein (*not shown*).

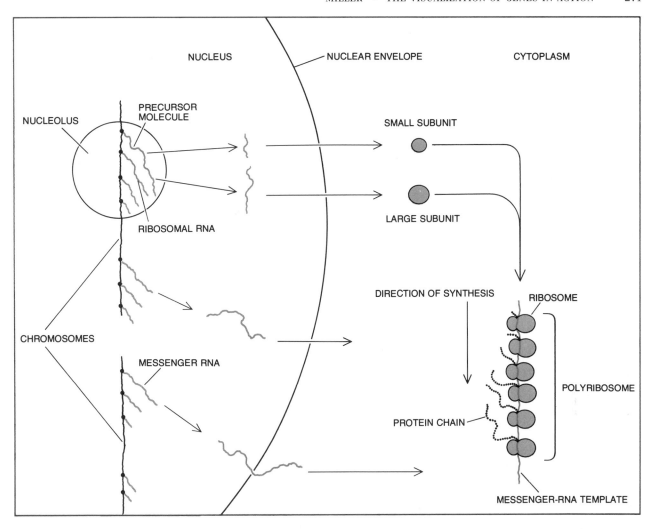

TRANSCRIPTION AND TRANSLATION occur respectively in the nucleus and the cytoplasm of eukaryotic cells. Ribosomal RNA (*light color*) transcribed from genes in the nucleolus forms a precursor molecule that is cleaved within the nucleolus to form the two large RNA molecules found in the two units of a ribosome. (RNA's from the same precursor are actually unlikely to end up in the same ribosome.) Messenger RNA's (*dark color*) transcribed from other genes serve as the templates on which polyribosomes assemble amino acid subunits into protein chains. Both kinds of RNA are complexed with protein (*not shown*) as they are being synthesized. In amphibian oöcytes a large number of copies of ribosomal-RNA genes are also present in extrachromosomal nucleoli.

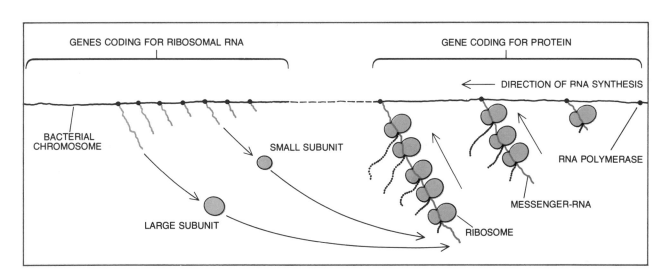

SIMULTANEOUS transcription and translation take place in prokaryotic cells such as bacteria, which have no true nucleus. Ribosomal RNA (*light color*) is transcribed from two genes, apparently contiguous, on certain segments of the single bacterial chromosome. Messenger RNA (*dark color*) is transcribed from genes at other sites and immediately translated into protein by ribosomes.

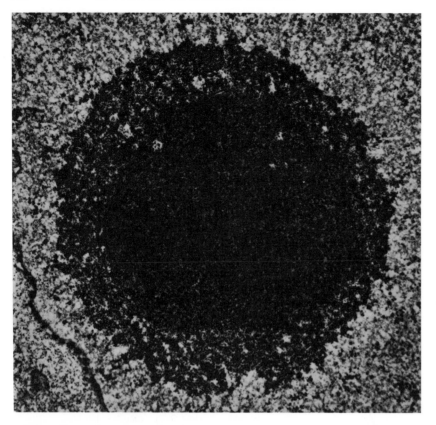

EXTRACHROMOSOMAL NUCLEOLUS in an oöcyte of the spotted newt, *Triturus viridescens*, is seen in thin section, enlarged 15,000 diameters. The nucleolus has a fibrous core and a granular cortex. Part of the nuclear envelope is visible (*lower left*), with the cytoplasm outside it. The core contains many copies of the genes for ribosomal RNA.

cores—which we hoped contained the genes we were looking for—to expand. We suspended such cores in a formalin solution to fix them and then spun them in a centrifuge tube where they were thrown against a thin carbon film on a wire-mesh specimen grid for an electron microscope. The problem was to spread the cores well without distorting them and destroying the pattern of gene activity we were looking for. We did that by treating the carbon films to make them hydrophilic, or water-attracting, and by rinsing the specimen grid in water containing an agent to reduce surface tension before drying it. Then we stained the specimen with a heavy metal to give it more contrast in the electron beam and made electron micrographs of the unwound cores.

Each unwound core appears as a long fiber in the form of a tangled, collapsed circle, along which there is a series of repeating shapes that we call matrix units [*see illustration on next page*]. Each of these units is about 2.4 microns long, and between each unit there is a segment of bare fiber that is usually somewhat shorter. Each matrix unit consists of a set of fibrils extending laterally from the long fiber. The length of the fibrils increases regularly along the fiber from one end of the matrix unit to the other, forming a structure with the outline of an arrowhead. All the matrix units on a single core fiber have the same polarity, that is, all the arrowheads point in the same direction. By treating the specimens with specific stains and enzymes we could identify the long axial fiber as a complex of DNA and protein and the lateral fibrils as a complex of RNA and protein. By means of autoradiography with radioactively labeled constituents of RNA we found that the matrix units were actively synthesizing RNA; the spacer segments between the matrix units were apparently not being transcribed. The nascent RNA molecules in the fibrils were complexed with protein that was made in the cell cytoplasm and that migrated into the nucleus and associated with the RNA molecules as they were being synthesized.

At about the same time we were making these observations some very pertinent biochemical information became available. Several investigators reported that the DNA in the extrachromosomal nucleoli of amphibian oöcytes contained subunit sequences that were complementary to the ribosomal RNA of ribosomes in oöcyte cytoplasm, thus confirming that the nucleolar DNA was the source of information for ribosomal RNA. The steps in the transfer of that information were also being elucidated in many laboratories. It turned out that in both mammals and amphibians the two large RNA molecules that are found in ribosomes are not transcribed from DNA separately but are snipped out of a single precursor molecule. (Subsequent studies have shown that this is probably true in all eukaryotes.) The precursor molecule has a molecular weight of about 2.7 million in amphibians, and about 2.7 microns of double-strand DNA would be necessary to code for such a molecule. The matrix units we were observing were about 2.4 microns long, a figure reasonably close to the 2.7 microns of DNA. Since extrachromosomal nucleoli had been shown to contain DNA coding for ribosomal RNA, and since our observations localized RNA synthesis in the matrix units of nucleolar cores, we concluded that the DNA fiber in each matrix unit was indeed a gene coding for a precursor molecule of ribosomal RNA, and that the lateral fibrils contained such molecules in the process of being synthesized.

On some fibers that were slightly stretched during preparation we could see something more: we could resolve the individual molecules of RNA polymerase, the enzyme that catalyzes the reading of DNA, that were active at each transcription site [*see illustration, page 274*]. Our success in visualizing nucleolar genes was due to the fact that when such genes are active, some 80 to 100 polymerase molecules are simultaneously transcribing each gene, forming fibrils of ribosomal-RNA precursor molecules (plus protein) in a graded set of lengths at successive stages of synthesis. The slight discrepancy between the observed 2.4-micron length of the matrix units and the 2.7-micron length expected for the genes can probably be explained by the fact that the DNA is unwound for transcription at each of the many transcription points; any bellying out of the DNA double helix would tend to shorten the overall length of the gene.

When we prepared amphibian oöcyte chromosomes in the same way that we had prepared the extrachromosomal nucleoli, we found that the fine detail of RNA synthesis could similarly be observed on the long lateral loops characteristic of the lampbrush stage [*see illustrations on page 275*]. Again we see the RNA polymerase molecules closely spaced along the active portions of the

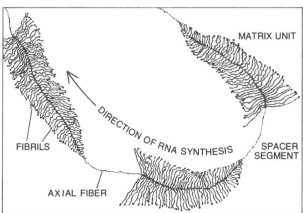

ACTIVE GENES for ribosomal RNA are arrayed within an unwound nucleolar core that was stained with tungstic acid and enlarged 26,000 diameters in the electron micrograph, elements of which are interpreted in the simplified drawing (*left*). One sees a long axial fiber in the form of a collapsed circle, with a series of arrowhead-shaped "matrix units" along it. Each unit is composed of a set of fine lateral fibrils of graded lengths. By staining and enzyme testing the axial fiber is identified as a complex of DNA and protein, the fibrils as complexes of RNA and protein. The segment of DNA within each matrix unit is a gene for ribosomal RNA; each fibril is a protein-complexed ribosomal-RNA precursor molecule that was being transcribed from the gene in the living oöcyte. The stretches of the fiber between matrix units are inactive "spacer" segments the function of which is not yet known.

fiber, each at the base of one of the nascent RNA-protein fibrils that form a continuous gradient along the loop. The fibrils grow much longer than those on the nucleolar genes, however, and so they must contain very large RNA molecules. We do not know the subsequent function of the fibrils produced by the oöcyte chromosomes; presumably their RNA gives rise to messenger RNA's for the synthesis of specific proteins, but so far we cannot assign any specific messenger to any one loop.

With my help, Aimée H. Bakken has recently extended our work with eukaryotes to a mammalian cell: the human tumor cell line called HeLa, which has been established in laboratory cultures for 20 years and in which the synthesis of RNA and proteins has been intensively studied by chemical techniques. The fact that these cells are much smaller than oöcytes (only some 30 microns or less in diameter) called for different methods of isolating and dispersing the contents of their nuclei. In an attempt to avoid tedious micromanipulation we sought to develop a chemical method for disrupting cells and nuclei so that we could spread out the nucleoli and chromosomes without degrading the structures we hoped to observe. After many trials with various agents we attained success by treating the cells with low concentrations of Joy, an ordinary household dishwashing detergent.

We have isolated from HeLa cells (presumably from the nucleoli) active chromosome segments that we have identified tentatively as genes for ribosomal RNA [see top illustration on page 276]. The identification is based primarily on their striking structural similarity to the ribosomal matrix units of oöcytes, whose identity is on firm ground. In both cases the active segments of DNA are fully loaded with RNA polymerase molecules, with as many as 150 such polymerases per gene in HeLa cells; small, dense granules appear at the free ends of the growing fibrils, and neighboring active segments are separated by inactive spacer segments, which in the HeLa cells are about the same length as the putative ribosomal-RNA genes. The active segments are longer in HeLa cells than they are in the oöcyte but shorter than would be expected on the basis of the molecular weight of the ribosomal-RNA precursor in mammals: the segments are about 3.5 microns long instead of an expected 4.5 microns. Again the discrepancy is probably due to some unwinding of the DNA at each of the numerous closely spaced transcription sites.

In considering non-ribosomal-RNA synthesis in HeLa cells, we knew from work done in other laboratories that 95 percent or so of the RNA synthesized in a HeLa cell at a given time does not survive to reach the cytoplasm but is degraded in the nucleus. This RNA is called heterogeneous nuclear RNA and its molecules range from about a tenth the size of the HeLa ribosomal-RNA-precursor molecule to many times that size. There is good evidence that small portions of the heterogeneous-nuclear-RNA molecules are not degraded and are precursors of the messenger RNA's found in the HeLa cytoplasm.

In addition to the distinctive ribosomal-RNA-precursor fibrils we do see RNA-protein fibrils of various sizes at scattered sites along dispersed HeLa chromosomes, but none of these fibrils are arrayed in a graded set of lengths. This fact indicates that RNA synthesis is initiated much less frequently, and that the level of polymerase activity is therefore much lower, on active sites of HeLa chromosomes than it is on the active

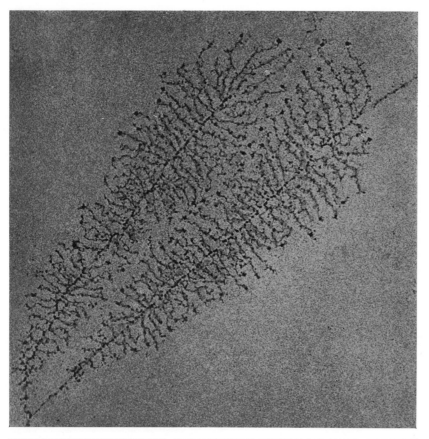

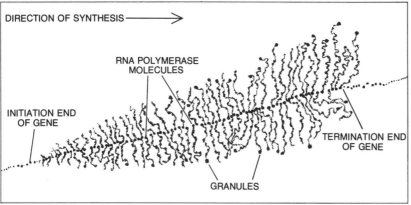

RNA POLYMERASE MOLECULES that are active at each transcription site are resolved on two nucleolar genes that were stretched in the process of preparation and are enlarged 48,000 diameters in the electron micrograph; one of the genes is reproduced in the drawing. The polymerase molecules are visible as dense granules at the base of each fibril; each polymerase was moving from left to right as it transcribed the gene. This micrograph also shows the granules, whose function is unknown, that are seen at the tip of maturing fibrils.

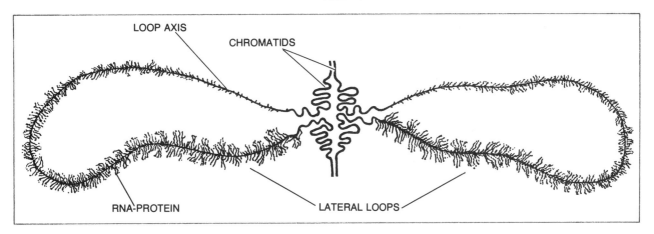

LOOP AXIS

CHROMATIDS

RNA-PROTEIN

LATERAL LOOPS

LAMPBRUSH CHROMOSOMES are a special form seen in oöcyte nuclei during such cells' long growth period. At the lampbrush stage the active portions of the sister chromatids that constitute each chromosome axis are greatly extended in lateral loops; the DNA axis of each loop is transcribed into RNA that is complexed with protein. The increasing length of the RNA molecules accounts for the increasing thickness of the loop from one point of juncture with the chromosome axis around to the next juncture. The colored rectangle shows approximately the segment of a lateral loop that is seen in the micrograph at the bottom of the page.

lateral loops of amphibian lampbrush chromosomes. Since heterogeneous nuclear RNA constitutes 95 percent or more of the RNA being synthesized in the HeLa nucleus, it seems likely that most of the dispersed RNA-protein fibrils we see attached to HeLa chromosomes contain nascent RNA of that kind. As in the case of the lampbrush chromosomes, however, no specific genetic role can yet be assigned to any of the observed RNA synthesis.

After our first success with amphibian oöcytes I collaborated with Charles A. Thomas, Jr., and Barbara A. Hamkalo at the Harvard Medical School in adapting the preparative techniques to a prokaryotic cell: the bacterium *Escherichia coli.* Later Hamkalo continued the work with me at Oak Ridge. Since there is a broad background of biochemical and genetic studies on the transcription and translation of various genes in *E. coli* and other bacteria, we could predict what active bacterial genes should look like if we could take their picture. The first problem, however, was to weaken the tough outer wall of the bacterial cells enough to open them up. We found that rapid cooling of growing *E. coli* followed by judicious application of the enzyme lysozyme removed enough of the wall to leave us with osmotically sensitive protoplasts: naked cells that could be burst open by osmotic shock when we diluted them in water. We then centrifuged the

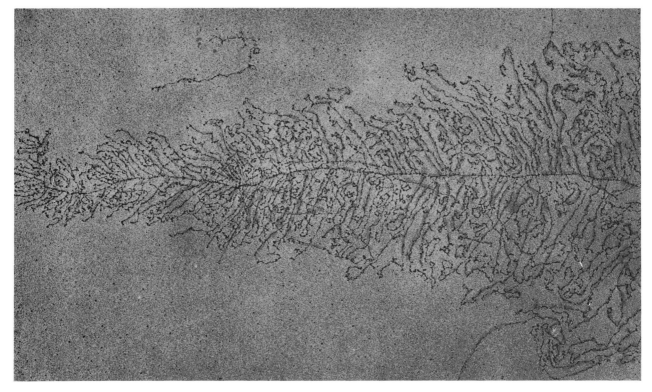

LATERAL LOOP of a lampbrush chromosome from a newt oöcyte is enlarged 22,000 diameters. The thin end, near the juncture with the chromosome axis, is at the left. RNA synthesis begins there, as shown by the increasing length of the fibrils from left to right. RNA polymerase molecules are visible at the base of each fibril. No specific genetic function can yet be assigned to these RNA's.

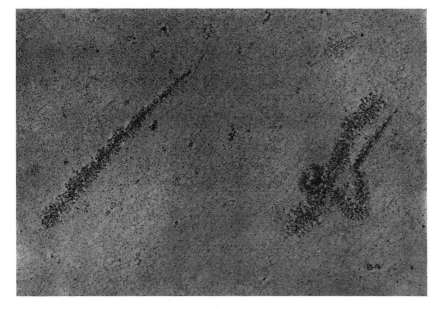

MATRIX UNITS isolated from the nucleus of a HeLa cell, a human tumor cell line, are identified as ribosomal-RNA genes on the basis of their similarity to the nucleolar genes of amphibian oöcytes: the graded set of closely spaced fibrils with a polymerase at the base of each, the dense granules at the tips of the fibrils and the occurrence of spacer segments (not resolved here) between the active genes. The enlargement is 16,000 diameters.

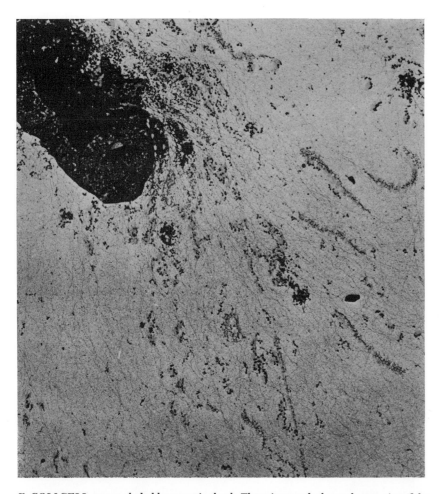

E. COLI CELL was exploded by osmotic shock. The micrograph shows the remains of the cell wall (upper left) and extruded cellular contents consisting of fine fibers with fibrillar segments and attached strings of granules. The fibers are portions of bacterial chromosome, the strings of granules are polyribosomes and the fibrillar segments are ribosomal genes. The enlargement is 22,000; the micrograph on page 270 further enlarges similar material.

burst cells much as we had the oöcyte cores and made electron micrographs of their extruded contents.

We knew that in bacteria, unlike in eukaryotes, nascent messenger RNA is normally complexed with translating ribosomes. As soon as a newly initiated strand of messenger RNA is long enough, a ribosome attaches itself to the strand and translation of the RNA into protein begins. As the synthesis of the messenger RNA proceeds, giving rise to a longer strand of RNA, more ribosomes attach themselves to the strand. They form a string called a polyribosome, which continues the translation of the elongating messenger RNA. Any active genes coding for protein synthesis in our specimens should therefore have a graded set of strings of polyribosomes attached along their length.

Micrographs of the extruded E. coli contents show masses of thin fibers, some of them with attached strings of granules [see bottom illustration at left]. If we treat a specimen with the enzyme deoxyribonuclease, the fibers are destroyed; if we treat one with ribonuclease, the granular strings are removed from the fibers. We conclude that the fibers are portions of the bacterial chromosome and that the granules are ribosomes that were translating messenger RNA. The fact that there are no attached polyribosomes at all along large stretches of the chromosomes is consistent with reports that a large part of the E. coli chromosome either codes for rare species of RNA or is never transcribed. Also, our bacteria are grown under optimal conditions, so that many of the biosynthetic pathways may be inactive and the genes coding for their enzymes may be repressed.

At higher magnification the detailed configuration of the ribosome-coated regions—the genes coding for the synthesis of protein—is revealed [see illustration, page 270]. The gene is visible as a thin axial fiber. Along it there is a graded set of strings of ribosomes. At the junction point of each string of ribosomes with the gene we can resolve the active RNA polymerase molecule that is catalyzing the transcription. With negative staining we can see that the enzymes are smaller and more irregular in shape than they are in eukaryotic cells [see top illustration on next page]. Unfortunately the pictures do not show the growing protein chains that are presumably associated with each of the ribosomes; the amino acids that make up the chain are simply too small for our procedures to resolve.

As for ribosomal-RNA genes in E. coli,

we could predict that they should look very different from the genes for protein synthesis. Whereas the messenger RNA's of genes coding for proteins are contained within polyribosomes, the nascent ribosomal RNA should be complexed with some protein but not mature ribosomes. Quantitative studies of ribosomal-RNA synthesis in rapidly growing *E. coli*, combined with estimates that within each *E. coli* chromosome there are no more than seven segments coding for ribosomal-RNA synthesis, suggested that these segments should be relatively scarce in our preparations and should be densely packed with active RNA polymerase molecules. There was also evidence that the two large RNA molecules that are found in prokaryote ribosomes are synthesized separately rather than being cleaved from a single precursor as they are in eukaryotes; the genes for these two RNA's were thought to be contiguous, and so they should show up as two adjacent matrix units of fibrils. The molecular weights of the two bacterial ribosomal RNA's indicated that the matrix unit of one should be twice as long as the matrix unit of the other and that the total length of the two units should be close to 1.7 microns.

We have identified in our specimens segments of *E. coli* chromosome that meet these specifications rather precisely [*see bottom illustration, right*]. The segments have 60 to 70 attached fibrils arranged in two consecutive matrix units. As in the amphibian cells and the HeLa cells, these segments are somewhat shorter than had been predicted, again presumably because of the unwinding of DNA. The ribosomal-RNA genes of *E. coli* appear not to be close to one another on the chromosome as they are in eukaryotes. In fact, they are separated by chromosome segments that show the polyribosomal activity of protein-synthesis genes. We have not yet determined the minimum length of these intervening segments.

The fact that rather simple techniques of isolation and preparation have enabled us to see such fine structural details of genetic activity in several types of cell gives us confidence that electron microscopy will become an increasingly important tool for the study of cell genetics. Refinement of the techniques may make it possible to observe directly active genes from almost any kind of cell, and eventually to identify those genes with their products and learn how they function at different times in the cell's life cycle, how they are affected by varying conditions and how they relate to specific cell activities.

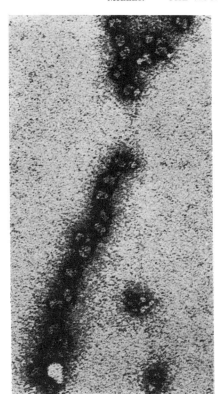
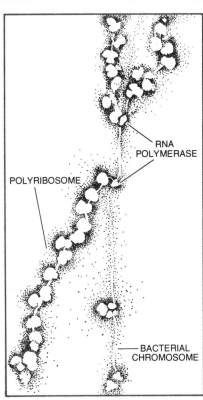

NEGATIVE STAINING (with uranyl acetate) outlines material from an *E. coli* cell, showing that the RNA polymerase molecules (*see drawing at right*) are smaller and more irregular than in eukaryotes. The staining reveals stretches of what is presumably messenger RNA but does not define the chromosome well. The enlargement is 150,000 diameters.

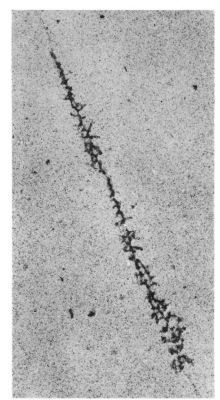
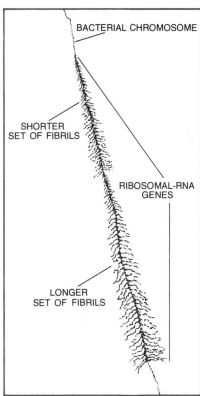

TWO CONTIGUOUS SETS of fibrils reflect transcription activity on genes for the two large RNA molecules of *E. coli* ribosomes. One matrix unit is about twice as long as the other and together they measure about 1.5 microns. The enlargement is 64,000 diameters.

27

GENE STRUCTURE AND PROTEIN STRUCTURE

CHARLES YANOFSKY

May 1967

A linear correspondence between these two chainlike molecules was postulated more than a dozen years ago. Here is how the correspondence was finally demonstrated

The present molecular theory of genetics, known irreverently as "the central dogma," is now 14 years old. Implicit in the theory from the outset was the notion that genetic information is coded in linear sequence in molecules of deoxyribonucleic acid (DNA) and that the sequence directly determines the linear sequence of amino acid units in molecules of protein. In other words, one expected the two molecules to be colinear. The problem was to prove that they were.

Over the same 14 years, as a consequence of an international effort, most of the predictions of the central dogma have been verified one by one. The results were recently summarized in these pages by F. H. C. Crick, who together with James D. Watson proposed the helical, two-strand structure for DNA on

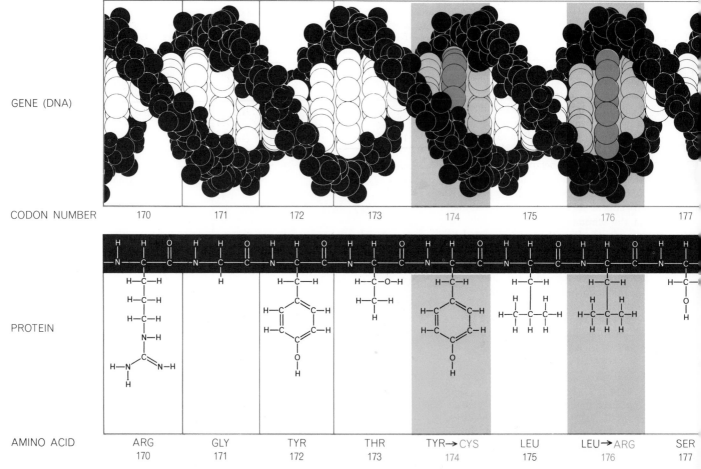

GENE (DNA)

CODON NUMBER   170   171   172   173   174   175   176   177

PROTEIN

AMINO ACID   ARG   GLY   TYR   THR   TYR→CYS   LEU   LEU→ARG   SER
             170   171   172   173   174   175   176   177

STRUCTURES OF GENE AND PROTEIN have been shown to bear a direct linear correspondence by the author and his colleagues at Stanford University. They demonstrated that a particular sequence of coding units (codons) in the genetic molecule deoxyribonucleic acid, or DNA (*top*), specifies a corresponding sequence of amino acid units in the structure of a protein molecule (*bottom*).

In the DNA molecule depicted here the black spheres represent repeating units of deoxyribose sugar and phosphate, which form the helical backbones of the two-strand molecule. The white spheres connecting the two strands represent complementary pairs of the four kinds of base that provide the "letters" in which the genetic message is written. A sequence of three bases attached to which the central dogma is based [see the article "The Genetic Code: III," by F. H. C. Crick, beginning on page 192]. Here I shall describe in somewhat more detail how our studies at Stanford University demonstrated the colinearity of genetic structure (as embodied in DNA) and protein structure.

Let me begin with a brief review. The molecular subunits that provide the "letters" of the code alphabet in DNA are the four nitrogenous bases adenine (A), guanine (G), cytosine (C) and thymine (T). If the four letters were taken in pairs, they would provide only 16 different code words—too few to specify the 20 different amino acids commonly found in protein molecules. If they are taken in triplets, however, the four letters can provide 64 different code words, which would seem too many for the efficient specification of the 20 amino acids. Accordingly it was conceivable that the cell might employ fewer than the 64 possible triplets. We now know that na-

ture not only has selected the triplet code but also makes use of most (if not all) of the 64 triplets, which are called codons. Each amino acid but two (tryptophan and methionine) are specified by at least two different codons, and a few amino acids are specified by as many as six codons. It is becoming clear that the living cell exploits this redundancy in subtle ways. Of the 64 codons, 61 have been shown to specify one or another of the 20 amino acids. The remaining three can act as "chain terminators," which signal the end of a genetic message.

A genetic message is defined as the amount of information in one gene; it is the information needed to specify the complete amino acid sequence in one polypeptide chain. This relation, which underlies the central dogma, is sometimes expressed as the one-gene-one-enzyme hypothesis. It was first clearly enunciated by George W. Beadle and Edward L. Tatum, as a result of their studies with the red bread mold *Neurospora crassa* around 1940. In some cases

a single polypeptide chain constitutes a complete protein molecule, which often acts as an enzyme, or biological catalyst. Frequently, however, two or more polypeptide chains must join together in order to form an active protein. For example, tryptophan synthetase, the enzyme we used in our colinearity studies, consists of four polypeptide chains: two alpha chains and two beta chains.

How might one establish the colinearity of codons in DNA and amino acid units in a polypeptide chain? The most direct approach would be to separate the two strands of DNA obtained from some organism and determine the base sequence of that portion of a strand which is presumed to be colinear with the amino acid sequence of a particular protein. If the amino acid sequence of the protein were not already known, it too would have to be established. One could then write the two sequences in adjacent columns and see if the same codon (or its synonym) always appeared adjacent to a particular amino acid. If it

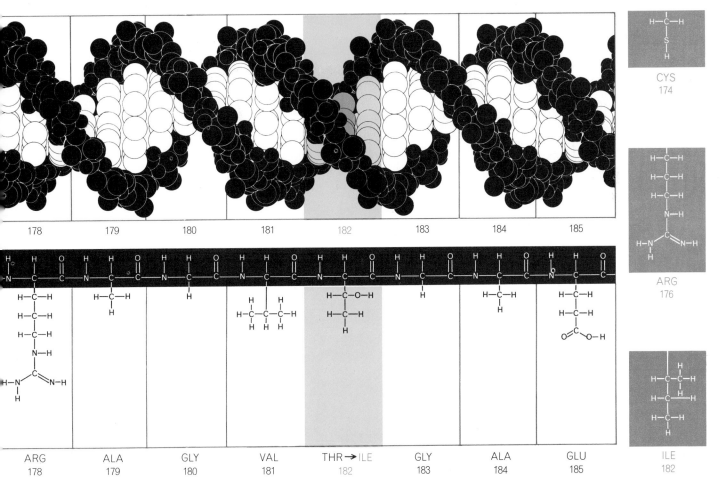

one strand of DNA is a codon and specifies one amino acid. The amino acid sequence illustrated here is the region from position 170 through 185 in the *A* protein of the enzyme tryptophan synthetase produced by the bacterium *Escherichia coli*. It was found that mutations in the *A* gene of *E. coli* altered the amino acids at three places (*174, 176 and 182*) in this region of the *A* protein. (A key to the amino acid abbreviations can be found on page 281). The three amino acids that replace the three normal ones as a result of mutation are shown at the extreme right. Each replacement is produced by a mutation at one site (*dark color*) in the DNA of the *A* gene. In all, the author and his associates correlated mutations at eight sites in the *A* gene with alterations in the *A* protein.

did, a colinear relation would be established. Unfortunately this direct approach cannot be taken because so far it has not been possible to isolate and identify individual genes. Even if one could isolate a single gene that specified a polypeptide made up of 150 amino acids (and not many polypeptides are that

small), one would have to determine the sequence of units in a DNA strand consisting of some 450 bases.

It was necessary, therefore, to consider a more feasible way of attacking the problem. An approach that immediately suggests itself to a geneticist is to construct a genetic map, which is a

representation of the information contained in the gene, and see if the map can be related to protein structure. A genetic map is constructed solely on the basis of information obtained by crossing individual organisms that differ in two or more hereditary respects (a refinement of the technique originally

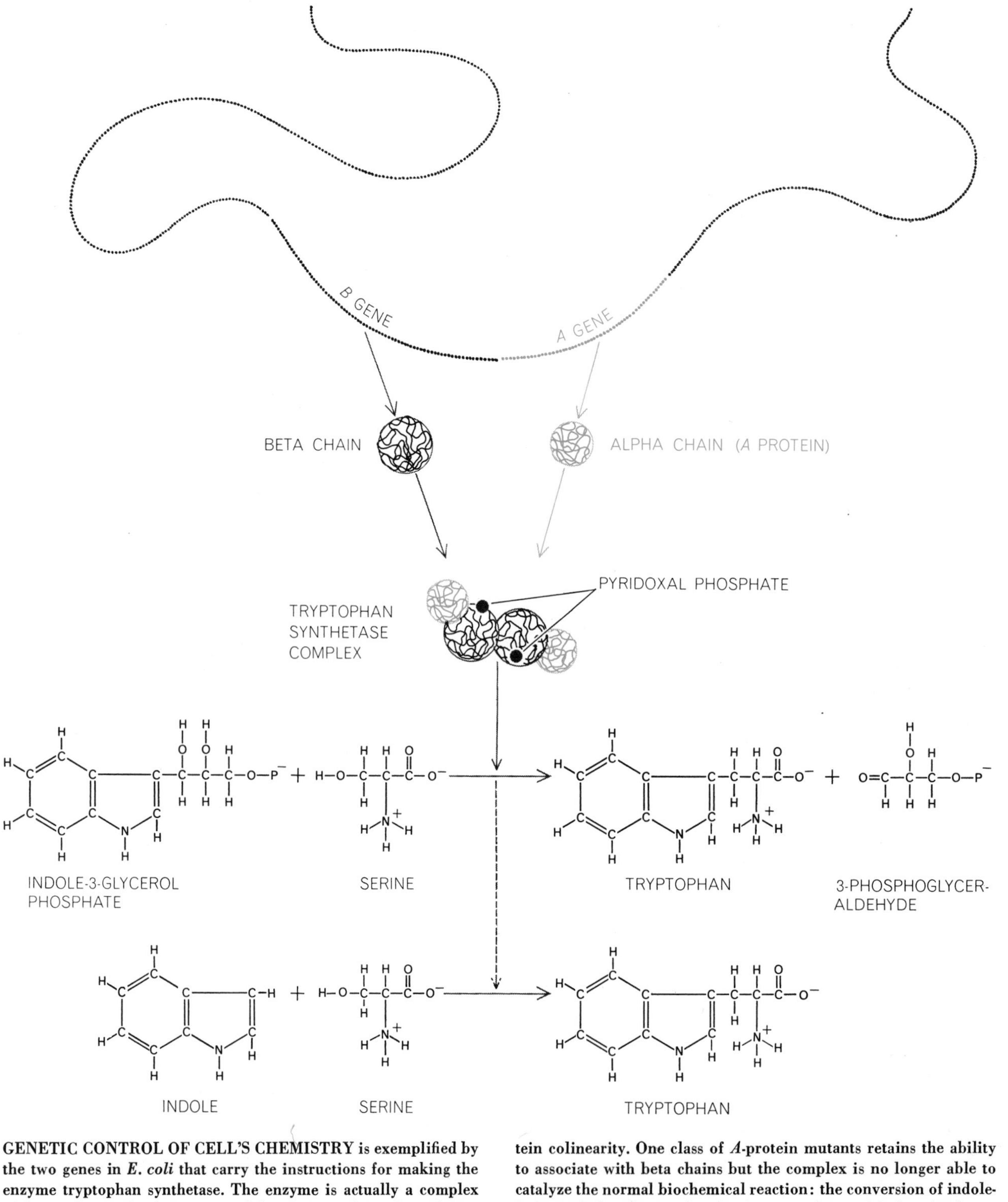

GENETIC CONTROL OF CELL'S CHEMISTRY is exemplified by the two genes in *E. coli* that carry the instructions for making the enzyme tryptophan synthetase. The enzyme is actually a complex of four polypeptide chains: two alpha chains and two beta chains. The alpha chain is the *A* protein in which changes produced by mutations in the *A* gene have provided the evidence for gene-protein colinearity. One class of *A*-protein mutants retains the ability to associate with beta chains but the complex is no longer able to catalyze the normal biochemical reaction: the conversion of indole-3-glycerol phosphate and serine to tryptophan and 3-phosphoglyceraldehyde. But the complex can still catalyze a simpler nonphysiological reaction: the conversion of indole and serine to tryptophan.

used by Gregor Mendel to demonstrate how characteristics are inherited).

By using bacteria and bacterial viruses in such studies one can catalogue the results of crosses involving millions of individual organisms and thereby deduce the actual distances separating the sites of mutational changes in a single gene. The distances are inferred from the frequency with which parent organisms, each with at least one mutation in the same gene, give rise to offspring in which neither mutation is present. As a result of the recombination of genetic material the offspring can inherit a gene that is assembled from the mutation-free portions of each parental gene. If the mutational markers lie far apart on the parental genes, recombination will frequently produce mutation-free progeny. If the markers are close together, mutation-free progeny will be rare [see bottom illustration on next page].

In his elegant studies with the "rII" region of the chromosome of the bacterial virus designated T4, Seymour Benzer, then at Purdue University, showed that the number of genetically distinguishable mutation sites on the map of the gene approaches the estimated number of base pairs in the DNA molecule corresponding to that gene. (Mutations involve pairs of bases because the bases in each of the two entwined strands of the DNA molecule are paired with and are complementary to the bases in the other strand. If a mutation alters one base in the DNA molecule, its partner is eventually changed too during DNA replication.) Benzer also showed that the only type of genetic map consistent with his data is a map on which the sites altered by mutation are arranged linearly. Subsequently A. D. Kaiser and David Hogness of Stanford University demonstrated with another bacterial virus that there is a linear correspondence between the sites on a genetic map and the altered regions of a DNA molecule isolated from the virus. Thus there is direct experimental evidence indicating that the genetic map is a valid representation of DNA structure and that the map can be employed as a substitute for information about base sequence.

This, then, provided the basis of our approach. We would pick a suitable organism and isolate a large number of mutant individuals with mutations in the same gene. From recombination studies we would make a fine-structure genetic map relating the sites of the mutations. In addition we would have to be able to isolate the protein specified by that gene and determine its amino acid sequence. Finally we would have to analyze the protein produced by each mutant (assuming a protein were still produced) in order to find the position of the amino acid change brought about in its amino acid sequence by the mutation. If gene structure and protein structure were colinear, the positions at which amino acid changes occur in the protein should be in the same order as the positions of the corresponding mutationally altered sites on the genetic map. Although this approach to the question of colinearity would require a great deal of work and much luck, it was logical and experimentally feasible. Several research groups besides our own set out to find a suitable system for a study of this kind.

The essential requirement of a suitable system was that a genetically

ALA	ALANINE	GLY	GLYCINE	PRO	PROLINE
ARG	ARGININE	HIS	HISTIDINE	SER	SERINE
ASN	ASPARAGINE	ILE	ISOLEUCINE	THR	THREONINE
ASP	ASPARTIC ACID	LEU	LEUCINE	TRP	TRYPTOPHAN
CYS	CYSTEINE	LYS	LYSINE	TYR	TYROSINE
GLN	GLUTAMINE	MET	METHIONINE	VAL	VALINE
GLU	GLUTAMIC ACID	PHE	PHENYLALANINE		

AMINO ACID ABBREVIATIONS identify the 20 amino acids commonly found in all proteins. Each amino acid is specified by a triplet codon in the DNA molecule (see below).

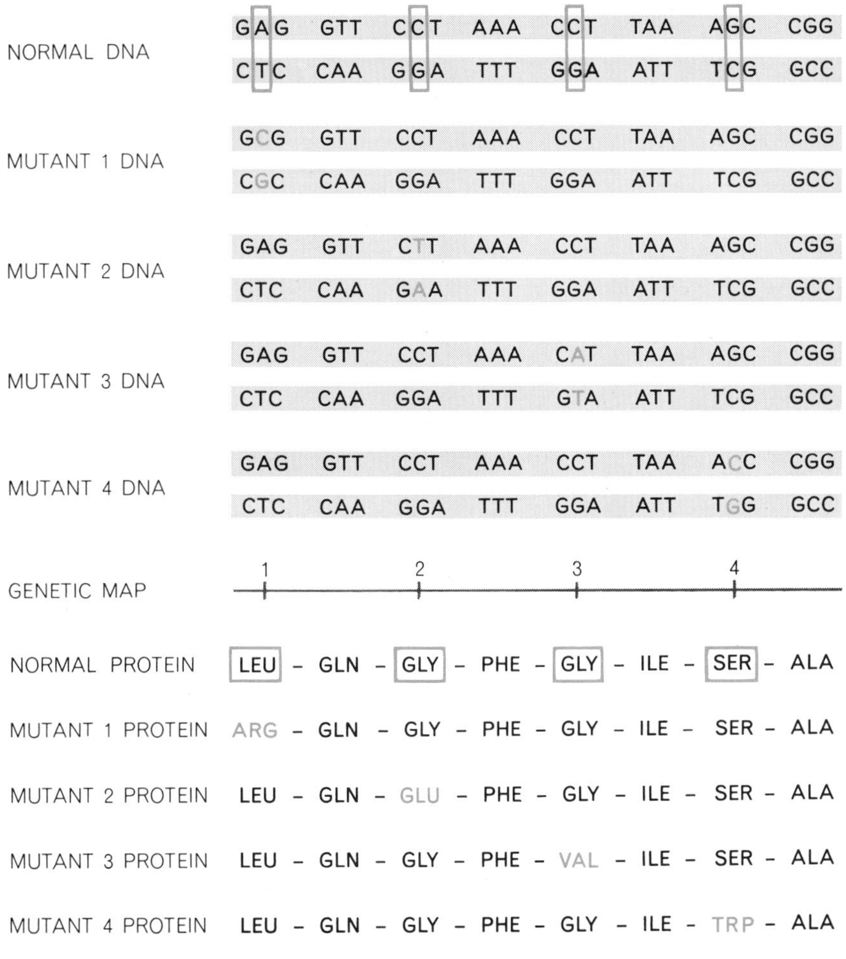

NORMAL DNA	GAG	GTT	CCT	AAA	CCT	TAA	AGC	CGG
	CTC	CAA	GGA	TTT	GGA	ATT	TCG	GCC
MUTANT 1 DNA	GCG	GTT	CCT	AAA	CCT	TAA	AGC	CGG
	CGC	CAA	GGA	TTT	GGA	ATT	TCG	GCC
MUTANT 2 DNA	GAG	GTT	CTT	AAA	CCT	TAA	AGC	CGG
	CTC	CAA	GAA	TTT	GGA	ATT	TCG	GCC
MUTANT 3 DNA	GAG	GTT	CCT	AAA	CAT	TAA	AGC	CGG
	CTC	CAA	GGA	TTT	GTA	ATT	TCG	GCC
MUTANT 4 DNA	GAG	GTT	CCT	AAA	CCT	TAA	ACC	CGG
	CTC	CAA	GGA	TTT	GGA	ATT	TGG	GCC

GENETIC MAP   1   2   3   4

NORMAL PROTEIN   LEU – GLN – GLY – PHE – GLY – ILE – SER – ALA

MUTANT 1 PROTEIN   ARG – GLN – GLY – PHE – GLY – ILE – SER – ALA

MUTANT 2 PROTEIN   LEU – GLN – GLU – PHE – GLY – ILE – SER – ALA

MUTANT 3 PROTEIN   LEU – GLN – GLY – PHE – VAL – ILE – SER – ALA

MUTANT 4 PROTEIN   LEU – GLN – GLY – PHE – GLY – ILE – TRP – ALA

GENETIC MUTATIONS can result from the alteration of a single base in a DNA codon. The letters stand for the four bases: adenine (A), thymine (T), guanine (G) and cytosine (C). Since the DNA molecule consists of two complementary strands, a base change in one strand involves a complementary change in the second strand. In the four mutant DNA sequences shown here (top) a pair of bases (color) is different from that in the normal sequence. By genetic studies one can map the sequence and approximate spacing of the four mutations (middle). By chemical studies of the proteins produced by the normal and mutant DNA sequences (bottom) one can establish the corresponding amino acid changes.

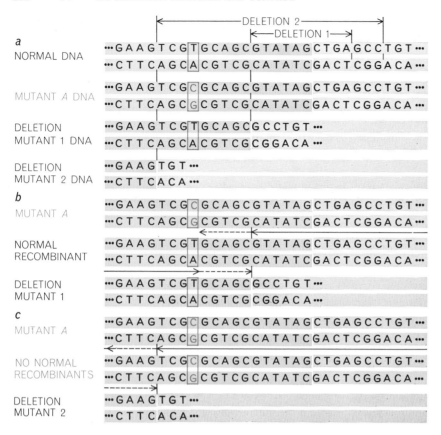

"DELETION" MUTANTS provide one approach to making a genetic map. Here (a) normal DNA and mutant A differ by only one base pair (C–G has replaced T–A) in a certain portion of the A gene (colored area). In deletion mutant 1 a sequence of 10 base pairs, including six pairs from the A gene, has been spontaneously deleted. In deletion mutant 2, 22 base pairs, including 15 pairs from the A gene, have been deleted. By crossing mutant A with the two different deletion mutants in separate experiments (b, c), one can tell whether the mutated site (C–G) in the A gene falls inside or outside the deleted regions. A normal-type recombinant will appear (b) only if the altered base pair falls outside the deleted region.

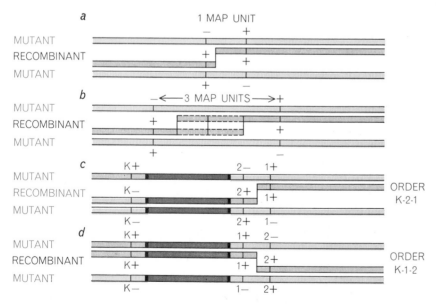

OTHER MAPPING METHODS involve determination of recombination frequency (a, b) and the distribution of outside markers (c, d). The site of a mutational alteration is indicated by "−," the corresponding unaltered site by "+." If the altered sites are widely spaced (b), normal recombinants will appear more often than if the altered sites are close together (a). In the second method the mutants are linked to another gene that is either normal (K+) or mutated (K−). Recombinant strains that contain 1+ and 2+ will carry the K− gene if the correct order is K–2–1. They will carry the K+ gene if the order is K–1–2.

mappable gene should specify a protein whose amino acid sequence could be determined. Since no such system was known we had to gamble on a choice of our own. Fortunately we were studying at the time how the bacterium *Escherichia coli* synthesizes the amino acid tryptophan. Irving Crawford and I observed that the enzyme that catalyzed the last step in tryptophan synthesis could be readily separated into two different protein species, or subunits, one of which could be clearly isolated from the thousands of other proteins synthesized by *E. coli*. This protein, called the tryptophan synthetase A protein, had a molecular weight indicating that it had slightly fewer than 300 amino acid units. Furthermore, we already knew how to force *E. coli* to produce comparatively large amounts of the protein—up to 2 percent of the total cell protein—and we also had a collection of mutants in which the activity of the tryptophan synthetase A protein was lacking. Finally, the bacterial strain we were using was one for which genetic procedures for preparing fine-structure maps had already been developed. Thus we could hope to map the A gene that presumably controlled the structure of the A protein.

To accomplish the mapping we needed a set of bacterial mutants with mutational alterations at many different sites on the A gene. If we could determine the amino acid change in the A protein of each of these mutants, and discover its position in the linear sequence of amino acids in the protein, we could test the concept of colinearity. Here again we were fortunate in the nature of the complex of subunits represented by tryptophan synthetase.

The normal complex consists of two A-protein subunits (the alpha chains) and one subunit consisting of two beta chains. Within the bacterial cell the complex acts as an enzyme to catalyze the reaction of indole-3-glycerol phosphate and serine to produce tryptophan and 3-phosphoglyceraldehyde [*see illustration, page 280*]. If the A protein undergoes certain kinds of mutations, it is still able to form a complex with the beta chains, but the complex loses the ability to catalyze the reaction. It retains the ability, however, to catalyze a simpler reaction when it is tested outside the cell: it will convert indole and serine to tryptophan. There are still other kinds of A-gene mutants that evidently lack the ability to form an A protein that can combine with beta chains; thus these strains are not able to catalyze even the simpler reaction. The first class of mutants—those that produce an A protein that is still able to combine with beta chains and exhibit catalytic activity when they are tested outside the cell—proved to be the most important for our study.

A fine-structure map of the *A* gene was constructed on the basis of genetic crosses performed by the process called transduction. This employs a particular bacterial virus known as transducing phage *P1kc*. When this virus multiplies in a bacterium, it occasionally incorporates a segment of the bacterial DNA within its own coat of protein. When the virus progeny infect other bacteria, genetic material of the donor bacteria is introduced into some of the recipient cells. A fraction of the recip-

ients survive the infection. In these survivors segments of the bacterium's own genetic material pair with like segments of the "foreign" genetic material and recombination between the two takes place. As a result the offspring of an infected bacterium can contain characteristics inherited from its remote parent as well as from its immediate one.

In order to establish the order of mutationally altered sites in the *A* gene we have relied partly on a set of mutant bacteria in which one end of a deleted segment of DNA lies within the *A* gene. In each of these "deletion" mutants a segment of the genetic material of the bacterium was deleted spontaneously.

Thus each deletion mutant in the set retains a different segment of the *A* gene. This set of mutants can now be crossed with any other mutant in which the *A* gene is altered at only a single site. Recombination can give rise to a normal gene only if the altered site does not fall within the region of the *A* gene that is missing in the deletion mutant [*see top illustration on opposite page*]. By crossing many *A*-protein mutants with the set of deletion mutants one can establish the linear order of many of the mutated sites in the *A* gene. The ordering is limited only by the number of deletion mutants at one's disposal.

A second method, which more closely

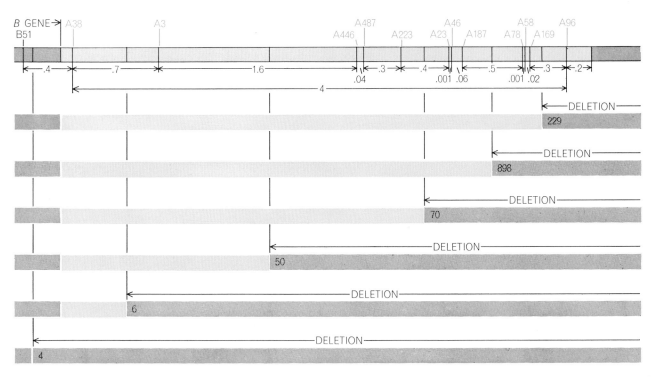

MAP OF *A* GENE shows the location of mutationally altered sites, drawn to scale, as determined by the three genetic-mapping methods illustrated on the opposite page. The total length of the *A* gene is slightly over four map units (probably 4.2). Below map are six deletion mutants that made it possible to assign each of the 12 *A*-gene mutants to one of six regions within the gene. The more sensitive mapping methods were employed to establish the order of mutations and the distance between mutation sites within each region.

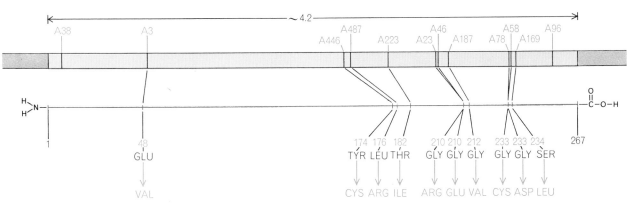

COLINEARITY OF GENE AND PROTEIN can be inferred by comparing the *A*-gene map (*top*) with the various amino acid changes in the *A* protein (*bottom*), both drawn to scale. The amino acid changes associated with 10 of the 12 mutations are also shown.

MET – GLN – ARG – TYR – GLU – SER – LEU – PHE – ALA – GLN – LEU – LYS – GLU – ARG – LYS – GLU – GLY – ALA – PHE – VAL –
1   20

PRO – PHE – VAL – THR – LEU – GLY – ASP – PRO – GLY – ILE – GLU – GLN – SER – LEU – LYS – ILE – ASP – THR – LEU – ILE –
21   40

A3

GLU – ALA – GLY – ALA – ASP – ALA – LEU – GLU – LEU – GLY – ILE – PRO – PHE – SER – ASP – PRO – LEU – ALA – ASP – GLY –
41   ↓VAL   60

PRO – THR – ILE – GLN – ASN – ALA – THR – LEU – ARG – ALA – PHE – ALA – ALA – GLY – VAL – THR – PRO – ALA – GLN – CYS –
61   80

PHE – GLU – MET – LEU – ALA – LEU – ILE – ARG – GLN – LYS – HIS – PRO – THR – ILE – PRO – ILE – GLY – LEU – LEU – MET –
71   100

TYR – ALA – ASN – LEU – VAL – PHE – ASN – LYS – GLY – ILE – ASP – GLU – PHE – TYR – ALA – GLN – CYS – GLU – LYS – VAL –
101   120

GLY – VAL – ASP – SER – VAL – LEU – VAL – ALA – ASP – VAL – PRO – VAL – GLN – GLU – SER – ALA – PRO – PHE – ARG – GLN –
121   140

ALA – ALA – LEU – ARG – HIS – ASN – VAL – ALA – PRO – ILE – PHE – ILE – CYS – PRO – PRO – ASP – ALA – ASP – ASP – ASP –
141   160

A446   A487

LEU – LEU – ARG – GLN – ILE – ALA – SER – TYR – GLY – ARG – GLY – TYR – THR – TYR – LEU – LEU – SER – ARG – ALA – GLY –
161   ↓CYS   ↓ARG   180

A223

VAL – THR – GLY – ALA – GLU – ASN – ARG – ALA – ALA – LEU – PRO – LEU – ASN – HIS – LEU – VAL – ALA – LYS – LEU – LYS –
181 ↓ILE   200

A23 A46   A187

GLU – TYR – ASN – ALA – ALA – PRO – PRO – LEU – GLN – GLY – PHE – GLY – ILE – SER – ALA – PRO – ASP – GLN – VAL – LYS –
201   ↓↓   ↓   220
                                     ARG GLU   VAL

A78 A58 A169

ALA – ALA – ILE – ASP – ALA – GLY – ALA – ALA – GLY – ALA – ILE – SER – GLY – SER – ALA – ILE – VAL – LYS – ILE – ILE –
221   ↓↓   ↓   240
                                                         CYS ASP LEU

GLU – GLN – HIS – ASN – ILE – GLU – PRO – GLU – LYS – MET – LEU – ALA – ALA – LEU – LYS – VAL – PHE – VAL – GLN – PRO –
241   260

MET – LYS – ALA – ALA – THR – ARG – SER
261   267

AMINO ACID SEQUENCE OF *A* PROTEIN is shown side by side with a ribbon representing the DNA of the *A* gene. It can be seen that 10 different mutations in the gene produced alterations in the amino acids at only eight different places in the *A* protein. The explanation is that at two of them, 210 and 233, there were a total of four alterations. Thus at No. 210 the mutation designated A23 changed glycine to arginine, whereas mutation A46 changed glycine to glutamic acid. At No. 233 glycine was changed to cysteine by one mutation (A78) and to aspartic acid by another mutation (A58). On the genetic map A23 and A46, like A78 and A58, are very close.

resembles traditional genetic procedures, relies on recombination frequencies to establish the order of the mutationally altered sites in the *A* gene with respect to one another. By this method one can assign relative distances—map distances—to the regions between altered sites. The method is often of little use, however, when the distances are very close.

In such cases we have used a third method that involves a mutationally altered gene, or genetic marker, close to the *A* gene. This marker produces a recognizable genetic trait unrelated to the *A* protein. What this does, in effect, is provide a reading direction so that one can tell whether two closely spaced mutants, say No. 58 and No. 78, lie in the order 58–78, reading from the left on the map, or vice versa [*see bottom illustration on page 282*].

With these procedures we were able to construct a genetic map relating the altered sites in a group of mutants responsible for altered *A* proteins that could themselves be isolated for study. Some of the sites were very close together, whereas others were far apart [*see upper illustration, page 283*]. The next step was to determine the nature of the amino acid changes in each of the mutationally altered proteins.

It was expected that each mutant of the *A* protein would have a localized change, probably involving only one amino acid. Before we could hope to identify such a specific change we would have to know the sequence of amino acids in the unmutated *A* protein. This was determined by John R. Guest, Gabriel R. Drapeau, Bruce C. Carlton and me, by means of a well-established procedure. The procedure involves breaking the protein molecule into many short fragments by digesting it with a suitable enzyme. Since any particular protein rarely has repeating sequences of amino acids, each digested fragment is likely to be unique. Moreover, the fragments are short enough—typically between two and two dozen amino acids in length—so that careful further treatments can release one amino acid at a time for analysis. In this way one can identify all the amino acids in all the fragments, but the sequential order of the fragments is still unknown. This can be established by digesting the complete protein molecule with a different enzyme that cleaves it into a uniquely different set of fragments. These are again analyzed in detail. With two fully analyzed sets of fragments in hand, it is not difficult to

SEGMENT OF PROTEIN	MUTANT										NOR-MAL
	H11	C140	B17	B272	H32	B278	C137	H36	A489	C208	
I	+	+	+	+	+	+	+	+	+	+	+
II	−	+	+	+	+	+	+	+	+	+	+
III	−	−	+	+	+	+	+	+	+	+	+
IV	−	−	−	+	+	+	+	+	+	+	+
V	−	−	−	−	+	+	+	+	+	+	+
VI	−	−	−	−	−	+	+	+	+	+	+
VII	−	−	−	−	−	−	+	+	+	+	+
VIII	−	−	−	−	−	−	−	+	+	+	+
IX	−	−	−	−	−	−	−	−	+	+	+
X	−	−	−	−	−	−	−	−	−	+	+
XI	−	−	−	−	−	−	−	−	−	−	+

GENETIC MAP   H11 C140 B17 B272 H32   B278   C137   H36   A489 C208

INDEPENDENT EVIDENCE FOR COLINEARITY of gene and protein structure has been obtained from studies of the protein that forms the head of the bacterial virus T4D. Sydney Brenner and his co-workers at the University of Cambridge have found that mutations in the gene for the head protein alter the length of head-protein fragments. In the table "+" indicates that a given segment of the head protein is produced by a particular mutant; "−" indicates that the segment is not produced. When the genetic map was plotted, it was found that the farther to the right a mutation appears, the longer the fragment of head protein.

find short sequences of amino acids that are grouped together in the fragment of one set but that are divided between two fragments in the other. This provides the clue for putting the two sets of fragments in order. In this way we ultimately determined the identity and location of each of the 267 amino acids in the unmutated *A* protein of tryptophan synthetase.

Simultaneously my colleagues and I were examining the mutants of the *A* protein to identify the specific sites of mutational changes. For this work we used a procedure first developed by Vernon M. Ingram, now at the Massachusetts Institute of Technology, in his studies of naturally occurring abnormal forms of human hemoglobin. This procedure also uses an enzyme (trypsin) to break the protein chain into peptides, or polypeptide fragments. If the peptides are placed on filter paper wetted with certain solvents, they will migrate across

the paper at different rates; if an electric potential is applied across the paper, the peptides will be dispersed even more, depending on whether they are negatively charged, positively charged or uncharged under controlled conditions of acidity. The former separation process is chromatography; the latter, electrophoresis. When they are employed in combination, they produce a unique "fingerprint" for each set of peptides obtained by digesting the *A* protein from a particular mutant bacterium. The positions of the peptides are located by spraying the filter paper with a solution of ninhydrin and heating it for a few minutes at about 70 degrees centigrade. Each peptide reacts to yield a characteristic shade of yellow, gray or blue.

When the fingerprints of mutationally altered *A* proteins were compared with the fingerprint of the unmutated protein, they were found to be remarkably similar. In each case, however, there was a difference. The mutant fingerprint usually lacked one peptide spot that appears in the nonmutant fingerprint and exhibited a spot that the nonmutant fingerprint lacks. The two peptides would presumably be related to each other with the exception of the change resulting from the mutational event. One can isolate each of the peptides and compare their amino acid composition. Guest, Drapeau, Carlton and I, together with D. R. Helinski and U. Henning, identified the amino acid substitutions in each of a variety of altered A proteins.

The final step was to compare the locations of these changes in the A protein with the genetic map of the mutationally altered sites. There could be no doubt that the amino acid sequence of the A protein and the map of the A gene are in fact colinear [*see lower illustration on page 283*].

One can also see that the distances between mutational sites on the map of the A gene correspond quite closely to the distances separating the corresponding amino acid changes in the A protein. In two instances two separate mutational changes, so close as to be almost at the same point on the genetic map, led to changes of the same amino acid in the unmutated protein. This is to be expected if a codon of three bases in DNA is required to specify a single amino acid in a protein. Evidently the most closely spaced mutational sites in our genetic map represent alterations in two bases within a single codon.

Thus our studies have shown that each

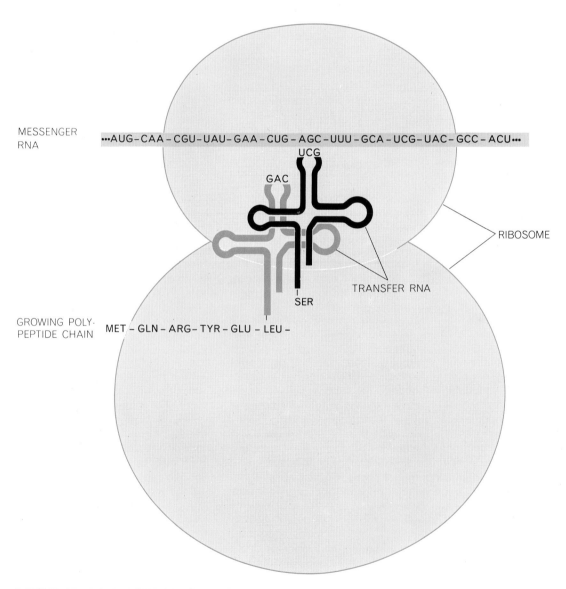

SCHEME OF PROTEIN SYNTHESIS, according to the current view, involves the following steps. Genetic information is transcribed from double-strand DNA into single-strand messenger ribonucleic acid (RNA), which becomes associated with a ribosome. Amino acids are delivered to the ribosome by molecules of transfer RNA, which embody codons complementary to the codons in messenger RNA. The next to the last molecule of transfer RNA to arrive (*color*) holds the growing polypeptide chain while the arriving molecule of transfer RNA (*black*) delivers the amino acid that is to be added to the chain next (serine in this example). The completed polypeptide chain, either alone or in association with other chains, is the protein whose specification was originally embodied in DNA.

```
     AGY                          AGX AGY                          AGY
  CGZ GGZ UAX ACZ [UAX] XUZ [CUZ] UCZ CGZ GCZ GGZ GUZ [ACW] GGZ GCZ GAY AAX CGZ GCZ GCZ XUZ
 -ARG-GLY-TYR-THR-[TYR]-LEU-[LEU]-SER-ARG-ALA-GLY-VAL-[THR]-GLY-ALA-GLU-ASN-ARG-ALA-ALA-LEU-
  170                                                                                    190

  CCZ XUZ AAX CAX XUZ GUZ GCZ AAY XUZ AAY GAY UAX AAX GCZ GCZ CCZ CCZ XUZ CAY [GGA]
  PRO-LEU-ASN-HIS-LEU-VAL-ALA-LYS-LEU-LYS-GLU-TYR-ASN-ALA-ALA-PRO-PRO-LEU-GLN-[GLY]-
  191                                                                        210

          AGX
  UUX [GGZ] AUW UCZ GCZ CCZ GAX CAY GUZ AAY GCZ GCZ AUW GAX GCZ GGZ GCZ GCZ GGZ GCZ
  PHE-[GLY]-ILE-SER-ALA-PRO-ASP-GLN-VAL-LYS-ALA-ALA-ILE-ASP-ALA-GLY-ALA-ALA-GLY-ALA-
  211                                                                        230

      AGX
  AUW UCZ [GGX] [UCZ] GCZ AUW GUZ AAY AUW AUW GAY CAY CAX AAX AUW GAY CCZ GAY AAY AUG
  ILE-SER-[GLY]-[SER]-ALA-ILE-VAL-LYS-ILE-ILE-GLU-GLN-HIS-ASN-ILE-GLU-PRO-GLU-LYS-MET-
  231                                                                        250
```

W = U, C or A   X = U or C   Y = A or G   Z = U, C, A or G

PROBABLE CODONS IN MESSENGER RNA that determines the sequence of amino acids in the *A* protein are shown for 81 of the protein's 267 amino acid units. The region includes seven of the eight mutationally altered positions (*colored boxes*) in the *A* protein. The codons were selected from those assigned to the amino acids by Marshall Nirenberg and his associates at the National Institutes of Health and by H. Gobind Khorana and his associates at the University of Wisconsin. Codons for the remaining 186 amino acids in the *A* protein can be supplied similarly. In most cases the last base in the codon cannot be specified because there are usually several synonymous codons for each amino acid. With a few exceptions the synonyms differ from each other only in the third position.

unique sequence of bases in DNA—a sequence constituting a gene—is ultimately translated into a corresponding unique linear sequence of amino acids— a sequence constituting a polypeptide chain. Such chains, either by themselves or in conjunction with other chains, fold into the three-dimensional structures we recognize as protein molecules. In the great majority of cases these proteins act as biological catalysts and are therefore classed as enzymes.

The colinear relation between a genetic map and the corresponding protein has also been convincingly demonstrated by Sydney Brenner and his co-workers at the University of Cambridge. The protein they studied was not an enzyme but a protein that forms the head of the bacterial virus T4. One class of mutants of this virus produces fragments of the head protein that are related to one another in a curious way: much of their amino acid sequence appears to be identical, but the fragments are of various lengths. Brenner and his group found that when the chemically similar regions in fragments produced by many mutants were matched, the fragments could be arranged in order of increasing length. When they made a genetic map of the mutants that produced these fragments, they found that the mutationally altered sites on the genetic map were in the same order as the termination points in the protein fragments. Thus the length of the fragment of the head protein produced by a mutant increased as the site of mutation was displaced farther from one end of the genetic map [*see illustration on page 285*].

The details of how the living cell translates information coded in gene structure into protein structure are now reasonably well known. The base sequence of one strand of DNA is transcribed into a single-strand molecule of messenger ribonucleic acid (RNA), in which each base is complementary to one in DNA. Each strand of messenger RNA corresponds to relatively few genes; hence there are a great many different messenger molecules in each cell. These messengers become associated with the small cellular bodies called ribosomes, which are the actual site of protein synthesis [*see illustration on page 286*]. In the ribosome the bases on the messenger RNA are read in groups of three and translated into the appropriate amino acid, which is attached to the growing polypeptide chain. The messenger also contains in code a precise starting point and stopping point for each polypeptide.

From the studies of Marshall Nirenberg and his colleagues at the National Institutes of Health and of H. Gobind Khorana and his group at the University of Wisconsin the RNA codons corresponding to each of the amino acids are known. By using their genetic code dictionary we can indicate approximately two-thirds of the bases in the messenger RNA that specifies the structure of the A-protein molecule. The remaining third cannot be filled in because synonyms in the code make it impossible, in most cases, to know which of two or more bases is the actual base in the third position of a given codon [*see illustration*

above]. This ambiguity is removed, however, in two cases where the amino acid change directed by a mutation narrows down the assignment of probable codons. Thus at amino acid position 48 in the A-protein molecule, where a mutation changes the amino acid glutamic acid to valine, one can deduce from the many known changes at this position that of the two possible codons for glutamic acid, GAA and GAG, GAG is the correct one. In other words, GAG (specifying glutamic acid) is changed to GUG (specifying valine). The other position for which the codon assignment can be made definite in this way is No. 210. This position is affected by two different mutations: the amino acid glycine is replaced by arginine in one case and by glutamic acid in the other. Here one can infer from the observed amino acid changes that of the four possible codons for glycine, only one—GGA—can yield by a single base change either arginine (AGA) or glutamic acid (GAA).

Knowledge of the bases in the messenger RNA for the A protein can be translated, of course, into knowledge of the base pairs in the A gene, since each base pair in DNA corresponds to one of the bases in the RNA messenger. When the ambiguity in the third position of most of the codons is resolved, and when we can distinguish between two quite different sets of codons for arginine, leucine and serine, we shall be able to write down the complete base sequence of the A gene—the base sequence that specifies the sequence of the 267 amino acids in the A protein of the enzyme tryptophan synthetase.

28

HOW PROTEINS START

BRIAN F. C. CLARK AND KJELD A. MARCKER
January 1968

*The chain of amino acid units that constitutes a
protein molecule begins to grow when a variant of one
of the standard amino acids is delivered to the site of
synthesis by a specific transfer agent*

Over the past 15 years a tremendous amount of information has been amassed on how the living cell makes protein molecules. Step by step investigators in laboratories all over the world are clarifying the architecture of specific proteins, the nature of the genetic material that incorporates the instructions for building them, the code in which the instructions are written and the processes that translate the instructions into the work of construction. With the information now available experimenters have already synthesized a number of protein-like molecules from cell-free materials, and the day seems not far off when we shall be able to describe, and perhaps control, every step in the making of a protein.

How is the building of a protein initiated? Until recently this question seemed to create no special problems. Given a supply of the amino acids from which a protein is made, the cell assembles them into a polypeptide chain that grows into a protein molecule, and it did not appear that the cell used any special machinery to start the construction of the chain. We have now learned, however, that the cell does indeed possess a starting mechanism. With the discovery of this mechanism it has become possible to study in detail the first step in the production of a protein molecule.

In order to discuss this new development we must first review the general features of protein synthesis by the cell. Proteins are made up of some 20 varieties of amino acid. A protein molecule consists of a long chain of amino acid units, typically from 100 to 500 or more of them, linked together in a specific sequence. The instructions for the particular order in each protein (the cell manufactures hundreds of different proteins) reside in the chainlike molecule of deoxyribonucleic acid (DNA). The DNA molecule consists of units called nucleotides; each nucleotide contains a side group of atoms called a base, and the sequence of bases along the DNA chain specifies the sequence for amino acids in the protein. There are four different bases in DNA: adenine (A), guanine (G), thymine (T) and cytosine (C). A "triplet" (a sequence of three bases) constitutes the "codon" that specifies a particular amino acid. The four bases taken three at a time in various sequences provide 64 possible codons; thus the four-letter language of DNA provides a vocabulary that is more than sufficient to designate the 20 amino acids. (In fact, some amino acids can be indicated by more than one codon.)

DNA does not guide the construction of the protein directly. Its message is first transcribed into the daughter molecule called messenger ribonucleic acid (mRNA). Messenger RNA also has four bases; three of them (A, G and C) are the same as in DNA, but the fourth, taking the place of thymine, is uracil (U). The RNA molecule is generated from DNA by a coupling process based on the fact that U couples to A and G couples to C. Thus during the transcription of DNA into RNA the four bases A, G, T and C in DNA give rise respectively to U, C, A and G in RNA [see *illustration on opposite page*].

The coded message is then read off the messenger RNA and translated into the construction of a protein molecule. This process takes place on the cell particles known as ribosomes, and it requires the assistance of smaller RNA molecules called transfer RNA (tRNA) that bring amino acids to the indicated sites. Each transfer RNA is specific for a particular amino acid, to which it attaches itself with the aid of an enzyme. It possesses an "anticodon" corresponding to a particular codon on the messenger RNA molecule. The ribosome moves along the messenger RNA molecule, reading off each codon in succession, and in this way it mediates the placement of the appropriate amino acids as they are delivered. As the amino acids join the chain they are linked together through peptide bonds formed by means of enzymes.

The decipherment of the genetic code for protein synthesis began in 1961 when Marshall W. Nirenberg and J. Heinrich Matthaei of the National Institutes of Health synthesized a simplified form of messenger RNA, composed of just one type of nucleotide, and found that it could generate the formation of a protein-like chain molecule made up of one variety of amino acid. Their artificial messenger RNA was the polynucleotide called "poly-U," containing uracil as the base. When it was added to a mixture of amino acids, extracts from cells of the bacterium *Escherichia coli* and energy-supplying compounds, it caused the synthesis of a polypeptide chain composed of the amino acid phenylalanine. Thus the poly-U codon (UUU) was found to specify phenylalanine.

This breakthrough quickly led to the identification of the codons for a number of other amino acids by means of the same device: using synthetic forms of messenger RNA. The experiments suggested that the initiation of synthesis of a protein was a perfectly straightforward matter. It appeared that the first codon in the messenger RNA chain simply called forth the delivery and placement of the specified amino acid and that no special starting signal was required. In 1964, however, Frederick Sanger and one of the authors of this article (Marcker) discovered a peculiar form of an amino acid, in combination with its transfer RNA, that threw entirely new light on the situation.

Using extracts from the *E. coli* bacterium, we were studying the chemical characteristics of the combination of the amino acid methionine with its specific tRNA. In the course of this study we decided to investigate the breakdown of the compound by pancreatic ribonu-

clease, an enzyme known to split RNA chains at certain specific bonds [*see top illustration on page 291*]. In order to facilitate identification of the products we labeled the methionine in advance with radioactive sulfur, and after treatment of the methionine-tRNA compound with

●	CARBON	A	ADENINE
○	OXYGEN	T	THYMINE
∙	HYDROGEN	G	GUANINE
◑	NITROGEN	C	CYTOSINE
○	SULFUR	U	URACIL
●	PHOSPHORUS		

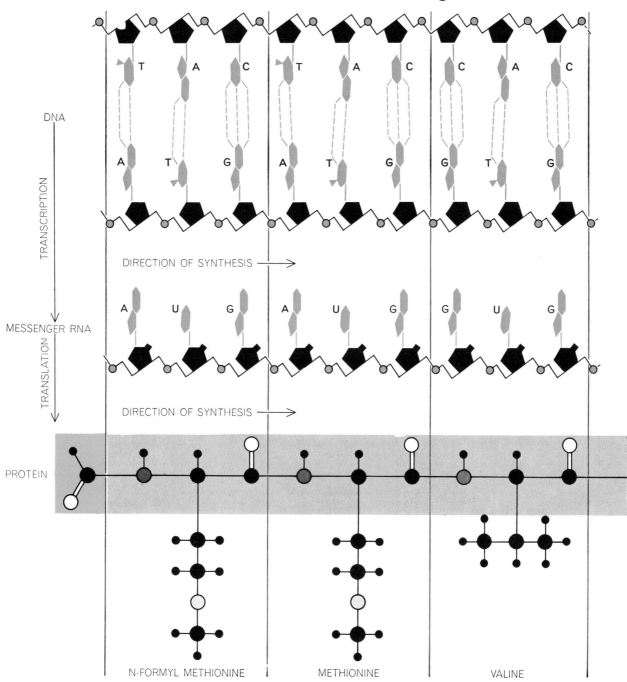

DNA

TRANSCRIPTION

DIRECTION OF SYNTHESIS ⟶

MESSENGER RNA

TRANSLATION

DIRECTION OF SYNTHESIS ⟶

PROTEIN

N-FORMYL METHIONINE   METHIONINE   VALINE

TRANSMISSION OF GENETIC INFORMATION takes place in two main steps. First the linear code specifying a particular protein is transcribed from deoxyribonucleic acid (DNA) into messenger ribonucleic acid (RNA). The code letters in DNA are the four bases adenine (A), thymine (T), guanine (G) and cytosine (C). Hydrogen bonds (*broken lines*) between the complementary bases A–T and G–C hold the two strands of the DNA molecule together. The strands, which run antiparallel, consist of alternating units of deoxyribose sugar (*pentagons*) and phosphate (PO_3H). The code letters in messenger RNA duplicate those attached to one strand of the DNA except that uracil (U) replaces thymine. In RNA the sugar is ribose. In the second step of the process messenger RNA is translated into protein. The code letters in RNA are read in triplets, or codons, each of which specifies one (or sometimes more) of the 20 amino acids that form protein molecules. It has now been found that the codon AUG can specify a modification of methionine known as formyl methionine, which signals the start of a protein chain. Inside the chain AUG specifies ordinary methionine. The codon GUG, which codes for valine inside the chain, can also specify formyl methionine and initiate chain synthesis.

the enzyme we separated the products by means of electrophoresis, the technique that segregates electrically charged molecules according to their charge, size and shape. As was to be expected, one of the products was the compound known as methionyl-adenosine, a combination of methionine with the terminal adenosine portion of the tRNA molecule. But we also found, to our surprise, that the products included a considerable amount of a formylated variety of this compound, that is, a variation in which a formyl group (CHO) replaced a hydrogen atom in the amino group

(NH₂) of the molecule. It turned out that this was by no means an artifact of the treatment to which the original compound had been subjected; growing cells proved to contain a high proportion of formylated methionine tRNA.

It was immediately evident that formylated methionine must occupy a special position in the protein molecule. The attachment of the formyl group to the amino group would prevent the amino group from forming a peptide bond [*see illustration below*]. Consequently the formylated amino acid must be an end unit in the protein molecule. Since an

amino group forms the "front" end of protein molecules when they are being assembled, formylated methionine must constitute the initial unit of the molecule.

We were able to separate the methionine tRNA of *E. coli* into two distinct species, and found that only one can be formylated. The formylatable species constitutes about 70 percent of the bacterium's methionine tRNA [*see bottom illustration on opposite page*]. Recent work in our laboratory at the Medical Research Council in Cambridge has established that the compound is formylated (at methionine's amino group) only after the amino acid has become attached to the tRNA molecule. The donor of the formyl group is 10-formyl tetrahydrofolic acid, and the reaction is catalyzed by a specific enzyme that acts exclusively on the combination of methionine with the formylatable species of tRNA.

Our laboratory and others have proceeded to analyze the initiation of protein formation by several experimental techniques. We began by testing a number of different synthetic messenger RNA's for their ability to bring about synthesis of a polypeptide incorporating methionine. Only two of the synthetic polynucleotides we tried proved to be capable of doing this. One contained the bases uracil, adenine and guanine (poly-UAG); the other had only uracil and guanine (poly-UG). We found that in a mixture of amino acids and other cell-free materials where only the formylatable species of methionine tRNA was present, either poly-UAG or poly-UG would cause the synthesis of a polypeptide with methionine in the starting position—and only in that position. Surprisingly, this was true even when no formyl group was attached to the methionine-tRNA compound. We had to conclude that the formylatable version of the tRNA for methionine possessed a special adaptation that helped it to function as a polypeptide-chain initiator.

A thorough search was made for formylated varieties of other tRNA's: that is, of tRNA's for amino acids other than methionine. None were found. This raised an interesting question. In the proteins produced by *E. coli* cells the amino acid at the "front" end of the protein molecule is not always methionine; often it is alanine or serine. These amino acids are never found to be formylated. How, then, does either of them become the initial member of the protein chain?

Experiments with natural messenger RNA's (rather than synthetic polynucleotides) have suggested an explanation. Jerry Adams and Mario Capecchi, work-

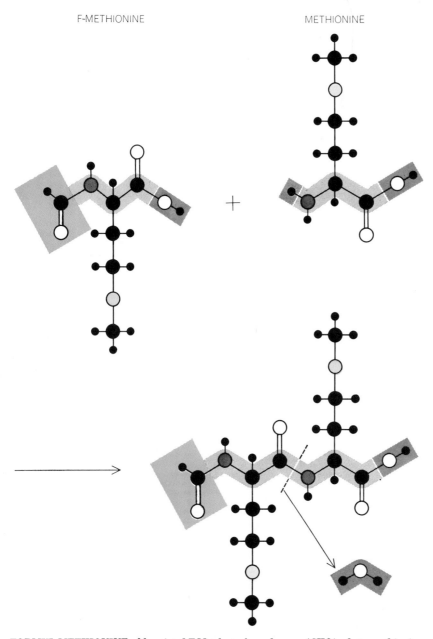

F-METHIONINE   METHIONINE

FORMYL METHIONINE, abbreviated F-Met, has a formyl group (CHO) where methionine (Met) has a hydrogen atom as part of a terminal amino (NH₂) group. When an amino acid enters a protein chain, one of the hydrogens from the amino end of one molecule combines with an OH group from the carboxyl (COOH) end of another molecule to form a molecule of water. The two molecules are then linked by a peptide bond. The formyl group prevents this reaction, hence F-Met can appear only at the beginning of a protein chain.

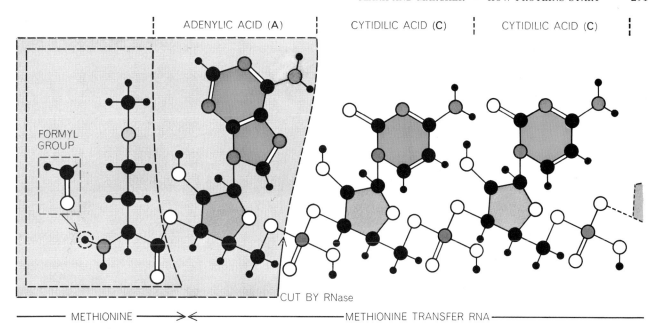

ADENYLIC ACID (A)   CYTIDILIC ACID (C)   CYTIDILIC ACID (C)

FORMYL GROUP

CUT BY RNase

——— METHIONINE ———>< ——————— METHIONINE TRANSFER RNA ———————

TRANSFER OF AMINO ACID to the site of protein synthesis is accomplished by molecules of transfer RNA (tRNA). There is at least one species of transfer RNA for each amino acid. All transfer RNA molecules contain the base sequence CCA at the terminal that holds the amino acid. Such a terminal is diagrammed here and shown coupled to methionine. Methionine that subsequently can be converted to formyl methionine is transferred by a different tRNA. When treated with the enzyme ribonuclease (RNase), the final base (adenine) and its coupled amino acid are split off from the rest of the transfer RNA. The fragment is called an aminoacyl adenosine.

ing in the laboratory of James D. Watson at Harvard University, and Norton D. Zinder and his collaborators at Rockefeller University have used messenger RNA's extracted from bacterial viruses. These RNA's direct the synthesis of the proteins that form the coat of the virus. The experimenters in Watson's and Zin-

der's laboratories found that when such an RNA was added to cell-free materials in the test tube, formylated methionine turned up at the starting end of the coat proteins that were synthesized. This was most surprising, because normally in living systems the initial amino acid of the viruses' coat protein is alanine. A signifi-

cant clue was found, however, in the fact that the coat proteins synthesized in the cell-free systems invariably had an alanine in the second position, following the formyl methionine. From this it seems reasonable to deduce that in living systems, as in the cell-free system, the formation of the protein starts with formyl

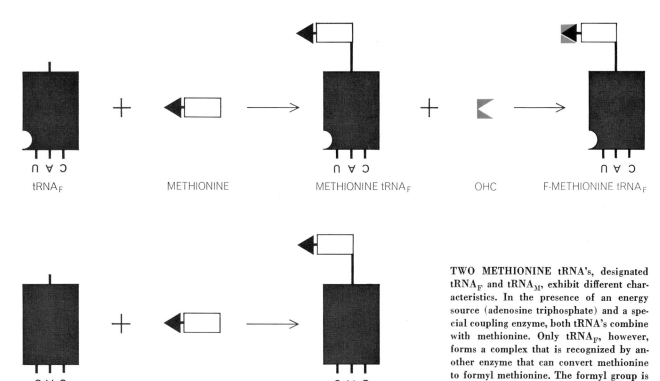

tRNA$_F$ METHIONINE   METHIONINE tRNA$_F$ OHC   F-METHIONINE tRNA$_F$

tRNA$_M$ METHIONINE   METHIONINE tRNA$_M$

TWO METHIONINE tRNA's, designated tRNA$_F$ and tRNA$_M$, exhibit different characteristics. In the presence of an energy source (adenosine triphosphate) and a special coupling enzyme, both tRNA's combine with methionine. Only tRNA$_F$, however, forms a complex that is recognized by another enzyme that can convert methionine to formyl methionine. The formyl group is provided by 10-formyl tetrahydrofolic acid. The complex that results is F-Met-tRNA$_F$.

tRNA	CODONS
MET-tRNA$_M$	AUG
MET-tRNA$_F$	AUG
F-MET-tRNA$_F$	GUG

CODON ASSIGNMENTS show the bases in messenger RNA that cause the two Met-tRNA's to deliver methionine or formyl methionine for insertion in a protein chain.

methionine, and that the bacterial cells supply an enzyme that chops off the formyl methionine later, leaving alanine in the first position.

Experiments with *E. coli* RNA in our laboratory and others have produced similar results. Messenger RNA extracted from these bacteria, like that extracted from bacterial viruses, causes cell-free systems to synthesize proteins with formyl methionine in the first position. On the other hand, the proteins extracted from living *E. coli* cells usually have unformylated methionine or alanine or serine in the lead position. It therefore seems likely that the living cells remove the formyl group from methionine or split off the entire formyl methionine unit after synthesis of the protein chain has got under way. The significance of the frequent appearance of alanine and serine at the front end of *E. coli* proteins is not clear; no satisfactory explanation has yet been found for the cell's selection of alanine and serine to follow formyl methionine. At all events, what does seem plausible now is that in *E. coli* the synthesis of all proteins starts with formyl methionine as the first unit.

How does the messenger RNA convey the message calling for formyl methionine as the starting unit? Does it use a special codon addressed specifically to the formylatable variety of methionine tRNA? We tested various codons for their ability to bring about the delivery of formyl methionine to the protein-synthesizing ribosomes. A codon for methionine was already known: it is AUG. We found that AUG was "read" by both varieties of methionine tRNA—the formylatable and the unformylatable. Either variety of tRNA delivered and bound methionine to the ribosome in response to AUG. We found that the formylatable tRNA (but not the other variety) also recognized and responded to another codon: GUG.

These findings were consistent with our earlier observation that either poly-UAG or poly-UG could effect the incorporation of methionine into a polypeptide in a cell-free system. Poly-UAG, of course, can contain the codons AUG and GUG, depending on the sequence in which the bases happen to be arranged in this polynucleotide; poly-UG provides the codon GUG. That both AUG and GUG can initiate the synthesis of a methionine polypeptide was confirmed and clearly spelled out in detail by experiments in the laboratory of H. Gobind Khorana at the University of Wisconsin. Using synthetic messenger RNA's in which the bases were arranged only in these triplet sequences (AUG and GUG), Khorana's group showed that both codons led to the formation of a chain with formyl methionine in the starting position. AUG also placed methionine in internal positions in the chain, but GUG, which can code only for the formylatable version of the tRNA, incorporated methionine only at the starting end [see illustration below].

Investigators in the laboratories of Severo Ochoa at New York University and Paul M. Doty at Harvard obtained the same results. They also noted that both codons possess a certain versatility as signals, depending on their location in the messenger RNA. Located at or near the beginning of the messenger RNA chain, the AUG triplet is recognized by the formylatable variety of tRNA and leads to the placement of formylated methionine at the starting end of the polypeptide; farther on in the messenger RNA chain the same triplet is recognized by unformylatable tRNA and causes the placement of unformylated methionine in the internal part of the polypeptide. In short, at the "front" end of the RNA message the AUG codon says to the cell's synthesizing machinery, "Start the formation of a protein"; when it is located internally in the message, AUG simply says, "Place a methionine here." Similarly, the codon GUG was found to have two possible meanings: located at the beginning of the message, it orders the initiation of a protein with formylated methionine; in an internal position in the message it is the code word for placement not of methionine but of the amino acid valine.

How is it that each of these codons signifies a starting signal in one position and has a different meaning in another? Obviously this question will have to be answered in order to clarify the language of the protein-starting mechanism. Indeed, we cannot be sure that a codon in itself constitutes the entire message for the initiation of a protein. The signaling mechanism may be more complex than one might assume from the findings developed so far. Those findings are based almost entirely on work done with artificial messenger RNA's, and it is possible that the messages they provide are only approximations—meaningful enough to stimulate the cell machinery but not the full story.

When we consider how important the

SYNTHETIC MESSENGER	SOURCE OF METHIONINE		POSITION OF METHIONINE IN POLYPEPTIDE		CODONS USED
	MET-tRNA$_M$	MET-tRNA$_F$ / F-MET-tRNA$_F$	INTERNAL	N-TERMINAL	
RANDOM POLY-UG	−	+	−	+	GUG
RANDOM POLY-AUG	+	+	+	+	AUG, GUG
POLY-(UG)$_n$	−	+	−	+	GUG
POLY-(AUG)$_n$	+	+	+	+	AUG

INCORPORATION OF METHIONINE in protein-like chains has been studied with synthetic messenger RNA's and the two species of methionine transfer RNA: tRNA$_M$ and tRNA$_F$. The plus sign indicates combinations that lead to incorporation. In random poly-UG and random poly-AUG the bases can occur in any sequence, but presumably the only effective sequences are GUG and AUG. Poly-(UG)$_n$ and poly-(AUG)$_n$ are synthetic chains of RNA consisting of 30 or more repetitions of the base sequences indicated.

codons AUG and GUG are in initiating the synthesis of polypeptides, it is certainly odd that a synthetic messenger RNA such as poly-U, which of course cannot supply those codons, nevertheless manages to cause the ribosomes to produce a polypeptide. We can only conclude that they do so by mistake, so to speak, that is, by acting in a way not entirely specified by the available information. (It is ironic that the genetic code was broken because artificial systems were able to make the right kind of mistake!) Are there circumstances that tend to assist these systems in accomplishing proper mistakes? One influential factor has been found. It is the concentration of magnesium in the cell-free system of building materials. A high magnesium concentration makes it possible for many kinds of synthetic messenger RNA to generate polypeptides; when the magnesium concentration is lowered, only the RNA's that contain AUG or GUG succeed in doing so. What magnesium may have to do with polypeptide initiation is still unclear.

Let us come back to the placement of the initial methionine as the normal first step in the construction of a protein. We have noted that the methionine-tRNA complex that places the amino acid in the initial position does not necessarily contain a formyl group. Evidently under conditions of a relatively high concentration of magnesium the formyl group of itself plays no essential role in the installation of the amino acid. What seems to be important is the character of the tRNA: only the formylatable variety of methionine tRNA can initiate the synthesis, and it can do so even when it is not formylated. What, then, are the specific properties that account for its role as an initiator?

A reasonable supposition is that this variety of tRNA has a special shape or configuration that helps it to fit into a particular site on the ribosome. As a matter of fact there is evidence that ribosomes possess two kinds of site for the attachment of tRNA's. One kind, called an amino acid site, simply receives and positions the tRNA when it arrives with its amino acid; the other kind, called a peptide site, holds the tRNA while a peptide bond is formed between its amino acid and an adjacent neighbor [see illustration at right]. It is therefore plausible to suppose that the formylatable variety of the tRNA for methionine may have a shape that helps it to fit into a peptide site on the ribosome and thus be in a position to start the linking together of amino acids.

Evidence in support of this hypothesis has been obtained in our laboratory by Mark S. Bretscher and one of us (Marcker) and by Philip Leder and his associates at the National Institutes of Health in experiments using the antibiotic puromycin. The structure of puromycin is similar to that of the end of a tRNA molecule that attaches to an amino acid [see illustration on next page]. Because it has an NH₂ group, puromycin can form a peptide bond with an amino acid, but since it lacks the free carboxyl (COOH) group of a normal amino acid it cannot form a second peptide bond.

Thus it cannot participate in chain elongation. Various experiments indicate that puromycin will add on to—and terminate—a growing polypeptide chain only when the tRNA holding the chain is bound in the peptide site.

In other experiments it has been found that the formylatable variety of methionine tRNA, when bound to a ribosome, will combine with puromycin; the unformylatable variety of the tRNA, on the other hand, will not react with puromycin. The experimental results therefore indicate that there are indeed two kinds of ribosomal site or state: one where a

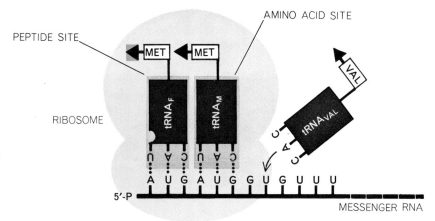

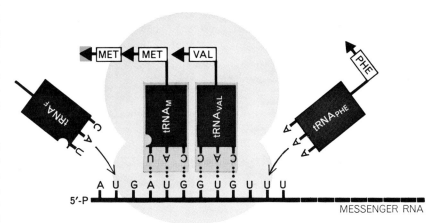

PROTEIN SYNTHESIS takes place on cellular particles called ribosomes, which travel along the "instruction tape" of messenger RNA, reading off the genetic message. The ribosome evidently has two sites for accommodating molecules of transfer RNA: a peptide site and an amino acid site. It appears that the structure of tRNA_F enables it to go directly to the peptide site, thereby initiating the protein chain. This special structure is symbolized by a notch in tRNA_F. Other tRNA's may acquire the configuration needed for the peptide site after first occupying the amino acid site. In step 1 (top) the codon AUG at the front end (5'-phosphate end) of messenger RNA pairs with the anticodon CAU that is believed to exist on tRNA_F, which delivers a molecule of formyl methionine to start the protein chain. The codon AUG in the second position is paired with the CAU anticodon of tRNA_M, which delivers a molecule of ordinary methionine. In step 2 (bottom) the tRNA_F molecule has moved away and the peptide site has been occupied by tRNA_M, which is now coupled to the growing protein chain. Valine transfer RNA has moved into the amino acid site.

peptide bond cannot be formed between the peptide chain and puromycin and one where it can. Most likely the latter is the peptide site. Furthermore, the experiments have strengthened the suspicion that the formylatable tRNA possesses a unique structure that somehow helps it to move into the peptide site on a ribosome. Apparently the structure of the formylatable tRNA has been particularly tailor-made for its function as a chain initiator.

The question therefore arises: What is the precise role of the formyl group? If the formyl group per se has nothing to do with placing methionine in the starting position, what function does it have? Our earlier experiments, in which we used a relatively high magnesium concentration, suggested that the formyl group is involved somehow in the formation of the first peptide bond, which launches the building of the polypeptide chain. When the methionine tRNA complex is formylated, synthesis of the polypeptide proceeds much faster than when it is not. This effect can be ascribed to the fact that the presence of the formyl group somehow facilitates the entry into the peptide site. It still remains to be determined just how the formyl group helps to promote such an effect.

Further light has been shed on the problem of protein-chain initiation in the past year by the work of several laboratories, including our own. Special protein agents, still poorly defined, have been implicated together with a cofactor in the formation of the initiation complex on the ribosome. When these new components are present and the supply of magnesium is low, the formyl group is necessary if the formylatable methionine tRNA is to be attached to the ribosomal peptide site by a messenger. Quite recently the cofactor has been identified as being a nucleotide derivative: guanosine triphosphate. Hence we are coming to the view that the conditions prevailing within the living cell are approached by these low-magnesium conditions, where there is strict specificity for forming the initiation complex and for unambiguous polypeptide formation. In our present state of knowledge, however, it is still unclear how these new components help to ensure the placement of the formylated methionine tRNA in the peptide site on the ribosome.

The specific findings concerning the initiation of protein synthesis that we have discussed in this article apply only to bacterial cells. So far no such form of tRNA (containing the formyl group or any other blocking agent) has been found in the cells of mammals. Accordingly the mechanism of protein-chain initiation is possibly different in mammalian cells from the mechanism discussed here. The process of polypeptide initiation in the cells of higher organisms is currently under study in several laboratories.

Meanwhile the investigation of the *E. coli* system is being pursued with experiments that promise to yield further insights. The way in which the vaguely characterized protein agents and guanosine triphosphate are involved in the initiation of a polypeptide chain is being explored. Much work is under way on analyzing the sequence of nucleotides in natural messenger RNA's, with a view to determining whether or not AUG or GUG constitutes a complete coding signal for protein initiation. We are searching for differences between the formylatable and unformylatable varieties of methionine tRNA, in their nucleotide sequences and in their three-dimensional structures, that may throw light on their respective interactions with the ribosomes.

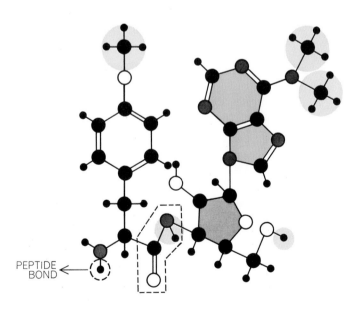

PEPTIDE BOND ←

PEPTIDE BOND

tRNA CHAIN

TYROSINE tRNA

PEPTIDE BOND ←

PUROMYCIN

PROTEIN-CHAIN TERMINATION can be induced by adding puromycin, an antibiotic, to a protein-synthesizing system. The structure of puromycin closely resembles the structure formed by the amino acid tyrosine and the terminal base of tRNA. Colored disks mark the atomic differences. Tyrosine can be inserted in a protein chain because it can form two peptide bonds. Puromycin can form only one peptide bond because the —CONH— linkage (*inside broken line*) is less reactive than the —COO— linkage in tyrosine tRNA.

29

THE CONTROL OF BIOCHEMICAL REACTIONS

JEAN-PIERRE CHANGEUX

April 1965

The cell is a factory and enzymes are its machines. Two feedback systems control production, one regulating synthesis of enzymes, another their activity. Models of the two systems are described

The analogy between a living organism and a machine holds true to a remarkable extent at all levels at which it is investigated. To be sure, living things are machines with exceptional powers, set apart from other machines by their ability to adapt to the environment and to reproduce themselves. Yet in all their functions they seem to obey mechanistic laws. An organism can be compared to an automatic factory. Its various structures work in unison, not independently; they respond quantitatively to given commands or stimuli; the system regulates itself by means of automatic controls consisting of specific feedback circuits.

These principles have long been recognized in the behavior of living organisms at the physiological level. In response to the tissues' need for more oxygen during exercise the heart speeds up its pumping of blood; in response to a rise in the blood-sugar level the pancreas increases its secretion of insulin. Now analogous systems have been discovered at work within the living cell. The new findings of molecular biology show that the cell is a mechanical microcosm: a chemical machine in which the various structures are interdependent and controlled by feedback systems quite similar to the systems devised by engineers who specialize in control theory. In this article we shall survey the experimental findings and hypotheses that have developed from the viewpoint that the cell is a self-regulating machine.

We can think of the cell as a completely automatic chemical factory designed to make the most economical use of the energy available to it. It manufactures certain products—for example proteins—by means of series of reactions that constitute its production lines, and most of the energy goes to power these processes. Regulating the production lines are control circuits that themselves require very little energy. Typically they consist of small, mobile molecules that act as "signals" and large molecules that act as "receptors" and translate the signals into biological activity.

The elementary machines of the cellular factory are the biological catalysts known as enzymes. The synthesis of any product (for example a specific protein) entails a series of steps, each of which calls for a specific enzyme. Obviously there are two possible ways in which the cell can control its output of a given product: (1) it may change the number of machines (enzyme molecules) available for some step in the chain or (2) it may change their rate of operation. Therefore in order to reduce the output of the product in question the cell may cut down the number of enzyme molecules or inhibit some of them or do both.

An excellent demonstration of such control has been obtained in experiments with the common bacterium *Escherichia coli*. The experiments involved the bacterial cell's production of the amino acid L-isoleucine, which it uses, along with other amino acids, to make proteins. Would the cell go on synthesizing this amino acid if it already had more than it needed for building proteins? L-isoleucine labeled with radioactive atoms was added to the medium in which the bacteria were growing; the experiments showed that when the substance was present in excess, the bacteria ceased to produce it. The amount of the amino acid in the cell in this case serves as the signal controlling its synthesis: if the amount is below a certain level, the cell produces more L-isoleucine; if it rises above that level, the cell stops producing L-isoleucine. Like the temperature level in a house with a thermostatically regulated heating system, the level of L-isoleucine in the cell exerts negative-feedback control on its own production.

How is the control carried out? H. Edwin Umbarger and his colleagues, working in the laboratory of the Long Island Biological Association, found that the presence of an excess of L-isoleucine has two effects on the cell: it inhibits the activity of the enzyme (L-threonine deaminase) needed for the first step in the chain of synthesizing reactions, and it stops production by the cell of all the enzymes (including L-threonine deaminase) required for L-isoleucine synthesis. Curiously it turned out that the two control mechanisms are independent of each other. By experiments with mutant strains of *E. coli* it was found that one mutation deprived the cell of the ability represented by the inhibition of L-threonine deaminase by L-isoleucine; another mutation deprived it of the ability to halt production of the entire set of enzymes. The two mutations were located at different places on the bacterial chromosome. Therefore it is clear that the two control mechanisms are completely separate.

Let us first examine the type of mechanism that controls the manufacture of enzymes. It was Jacques Monod and Germaine Cohen-Bazire of the Pasteur Institute in Paris who discovered the phenomenon of repression: the inhibition of enzyme synthesis by the presence of the product, the product serving as a signal that the enzymes are not needed. The signal substance

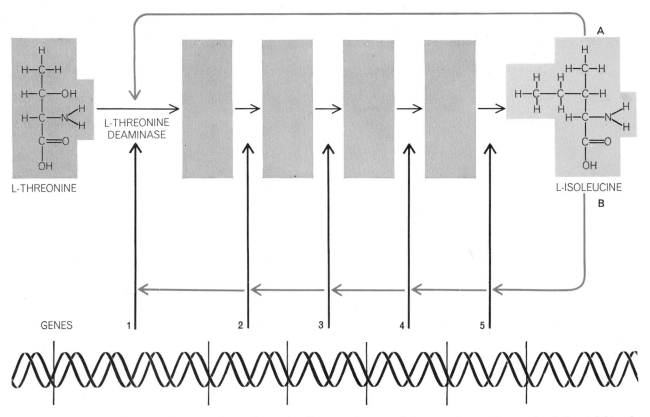

TWO FEEDBACK SYSTEMS control the biosynthesis of cell products, as shown here for the synthesis of the amino acid L-iso-leucine in the bacterium *Escherichia coli*. The end product of the synthesizing chain acts as a regulatory signal that inhibits the activity of the first enzyme in the chain, L-threonine deaminase (*A*), and also represses the synthesis of all the enzymes (*B*).

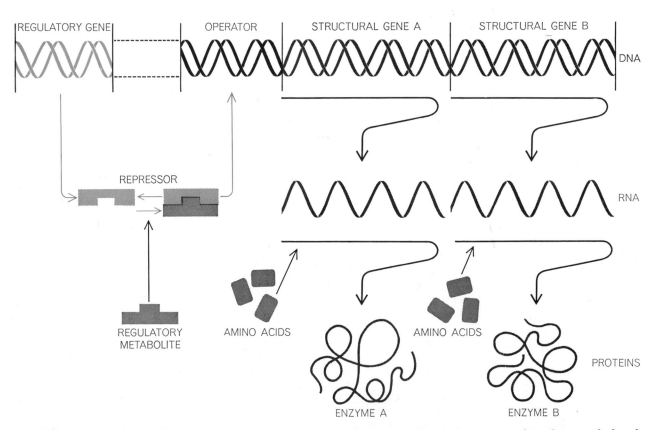

CONTROL OF PROTEIN SYNTHESIS by a genetic "repressor" was proposed by François Jacob and Jacques Monod. A regulatory gene directs the synthesis of a molecule, the repressor, that binds a metabolite acting as a regulatory signal. This binding either activates or inactivates the repressor, depending on whether the system is "repressible" or "inducible." In its active state the repressor binds the genetic "operator," thereby causing it to switch off the structural genes that direct the synthesis of the enzymes.

in their experiments was the amino acid tryptophan. They found that when the medium in which *E. coli* cells were growing contained an abundance of tryptophan, the cells stopped producing tryptophan synthetase, the enzyme required for the synthesis of the amino acid. This efficient behavior has since been demonstrated in many cells, not only bacteria but also the cells of higher organisms. The addition of an essential product to the cells' growth medium results in a negative-feedback signal that causes them to stop synthesizing enzymes they do not need.

In other systems the response of the cell is not negative but positive. We have been considering signals that repress the synthesis of enzymes; the cell can also respond to signals calling on it to produce enzymes. An example of such a situation is that the cell is confronted with a compound it must break down into substances it requires for growth.

The "induction" of enzyme synthesis in cells was discovered at the turn of the century by Frédéric Dienert of the Agronomical Institute in France. He was studying the effect of a yeast (*Saccharomyces ludwigii*) in fermenting the milk sugar lactose. He found that strains of the yeast that had been grown for several generations in a medium containing lactose would begin to work on the sugar immediately, causing it to start fermenting within an hour. These cells had a high level of lactase, an enzyme that specifically breaks down lactose. Yeast cells that had not been grown in lactose lacked this enzyme, and not surprisingly they failed to ferment lactose on being introduced to the sugar. After 14 hours, however, fermentation of the sugar did get under way; it developed that the presence of the lactose had induced the yeast to produce the enzyme lactase. The adaptation was quite specific: only lactose caused the yeast to synthesize this enzyme; other sugars failed to do so.

In recent years Monod and François Jacob of the Pasteur Institute have worked out some of the basic mechanisms of enzymatic adaptation by the cell, in both the repression and induction aspects. First they discovered that a single mutation in *E. coli* could eliminate the control of lactase synthesis by lactose: the mutant cells produced lactase just as well in the absence of lactose as in its presence. In these cells only the triggering effect was changed; the enzyme they produced was exactly the same as that synthesized by nonmutant strains. In other words, it ap-

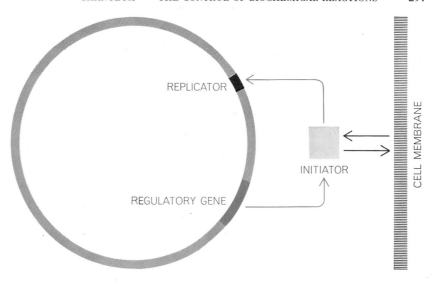

REPLICATION OF DNA of a bacterial chromosome may be under a control like that of protein synthesis. A regulatory gene directs the synthesis of an "initiator," which receives a signal (perhaps from the cell membrane) that makes it act on the "replicator."

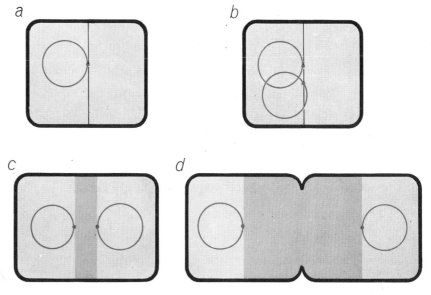

ROLE OF CELL MEMBRANE in replication is suggested by the fact that a bacterial chromosome is attached to a point on the membrane (*a*). It could be a signal from the membrane that initiates the formation of daughter chromosomes (*b*). Then the membrane begins to grow, separating the points of attachment (*c*) until the cell is ready to divide (*d*).

peared that the rate of production of the enzyme was controlled by one gene and that the structure of the enzyme was determined by quite another gene. This was confirmed by genetic experiments that showed that the "regulatory gene" and the "structural gene" were indeed in separate positions on the bacterial chromosome.

How does the regulatory gene work? Arthur B. Pardee, Jacob and Monod found that it causes the cell to produce a "repressor" molecule that controls the functioning of the structural gene. In the absence of lactose the repressor molecule prevents the structural gene from directing the synthesis of lactase molecules. The repressor does not act on the structural gene directly; it binds itself to a special structure that is closely linked on the chromosome with the structural gene for the enzyme and with several other genes involved in lactose metabolism. This special genetic structure is called an "operator." The binding of the repressor to the operator causes the latter to switch off the activity of the adjacent structural genes, and in this way it blocks the complex series of events that would lead to synthesis of the enzyme.

Jacob and Monod have shown that this scheme of control applies to any category of "adaptive" enzymes [*see bottom illustration on page 296*]. The repression and induction of enzymes can be regarded as opposite sides of the same coin. In a repressible system the binding of the regulatory signal on the repressor activates the repressor so that it blocks the synthesis of the enzyme. In an inducible system, on the other hand, the binding of the inducing signal on the repressor *inactivates* the repressor, thus releasing the cell machinery to synthesize the enzyme. Mutant cells that lose the repressive machinery need no inducer: they synthesize the enzyme almost limitlessly without requiring any induction signal.

In brief, the various repressors in the cell are specialized receptors, each capable of recognizing a specific signal. And within its chromosomes a cell possesses instructions for synthesizing a wide variety of enzymes, each of which can be evoked simply by the presenta-

tion of the appropriate signal to the appropriate repressor.

The cell's selection of chromosomal records for transcription is so efficient as to seem almost "conscious." Actually, however, the responses of the cell are automatic, and like any other automatic mechanism they can be "tricked." It is as though a vending machine were made to work by a false coin: certain artificial compounds closely resembling lactose are excellent inducers of lactase but cannot be broken down by the enzyme. This means that the cell is tricked into spending energy to make an enzyme it cannot use. The signal works, but it is a false alarm. Trickery in the opposite direction is also possible. There is an analogue of tryptophan, called 5-methyl tryptophan, that acts as a repressive signal, causing the cell to stop its production of tryptophan. But 5-methyl tryptophan cannot be incorporated into protein in place of the genuine amino acid. Without that essential amino acid the cell stops growing and dies of starvation. Thus the false signal in effect acts as an antibiotic.

If chemical signals control the pro-

duction of enzymes, may they not also control the more generalized activities of the cell, notably its self-replication? Jacob, Sydney Brenner and François Cuzin, working cooperatively at the Pasteur Institute and at the Laboratory of Molecular Biology at the University of Cambridge, recently discovered evidence of such a chemical control. They investigated the replication of the unique circular chromosome of *E. coli*. The synthesis of the deoxyribonucleic acid (DNA) of the chromosome, they found, is initiated by a signaling molecule that corresponds to the repressor of enzyme synthesis. The "initiator" has a positive effect rather than a repressive one. Like the repressor of enzyme synthesis, it is synthesized under the direction of a regulatory gene for replication. As the cell prepares for division, the initiator receives orders from the cell membrane and triggers the replication of its DNA by activating a genetic structure called the replicator (analogous to the "operator" of enzyme synthesis). Not much information has been gathered so far about the signal that prompts the initiator or about the

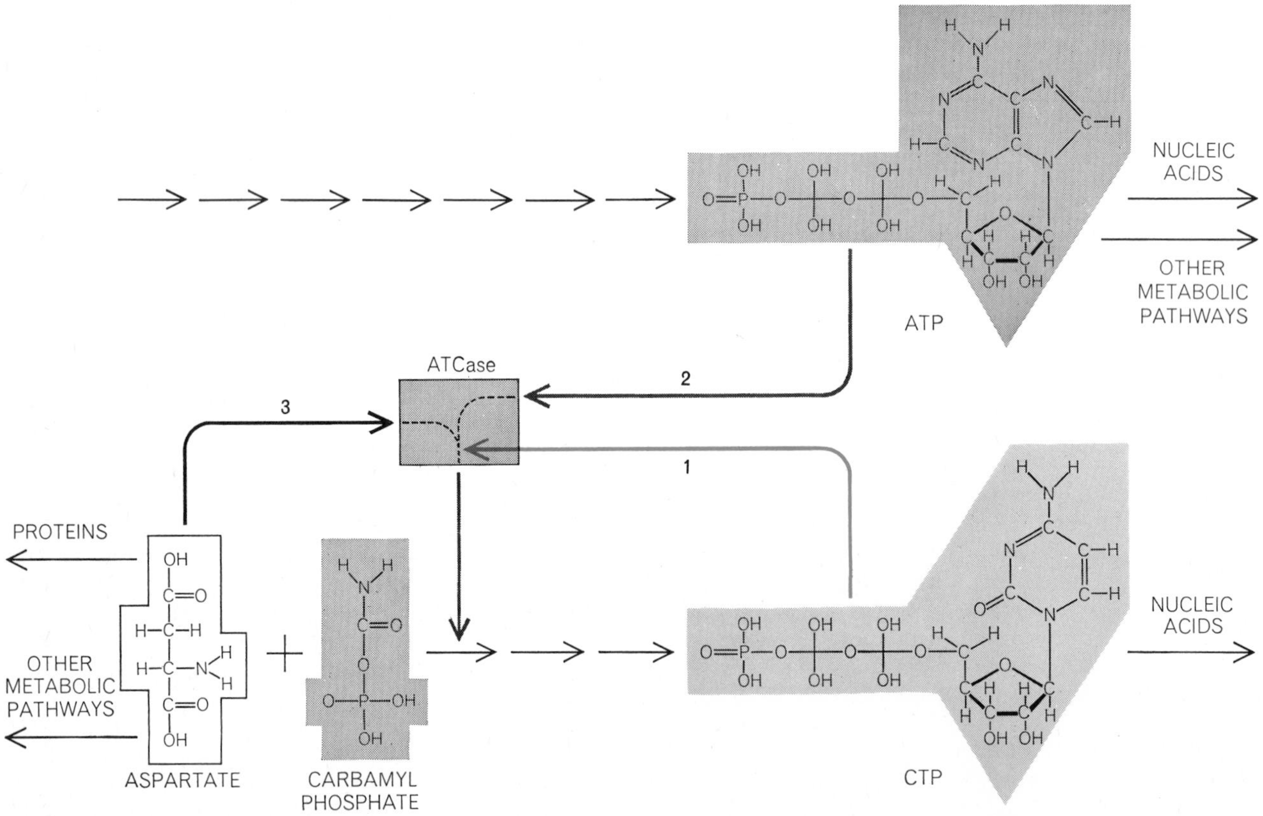

TWO NUCLEOTIDES, adenosine triphosphate (ATP) and cytidine triphosphate (CTP), are required by the cell in fixed proportions, so their production is regulated by interconnected feedback mechanisms operating on the first enzymes in the synthetic chains. In the case of CTP the enzyme is aspartate transcarbamylase (ATCase). It is inhibited by an excess of CTP (1), activated by an excess of ATP (2) and must also recognize and respond to the "cooperative" effects of aspartate, its substrate (3), which also plays a role in protein synthesis. Notice that ATP, CTP and aspartate have different shapes. How, then, can they all "fit" ATCase chemically?

details of the machinery it sets in motion, but it seems clear that cell division has its own system of chemical control and that it can adjust itself to the composition of the growth medium.

We have been considering the control of the synthesis of enzymes; now let us turn to the control of their activity. As I have mentioned, Umbarger and his colleagues found that the presence of L-isoleucine would not only cause *E. coli* to stop synthesizing the enzymes needed for its production but also inhibit the activity of the first enzyme in the chain leading to the formation of the amino acid. The phenomenon of control of enzyme activity had already been noted earlier in the 1950's by Aaron Novick and Leo Szilard of the University of Chicago. They had shown that an excess of tryptophan in the *E. coli* cell halted the cell's production of tryptophan immediately, which means that the signal inhibited the activity of enzymes already present in the cell. Umbarger went on to investigate the direct effect of L-isoleucine on the enzymes that synthesize it; these had been extracted from the cell. He demonstrated that L-isoleucine inhibited the first enzyme in the chain (L-threonine deaminase), and only the first. This action was extremely specific; no other amino acid—not even D-isoleucine, the mirror image of L-isoleucine—had any effect on the enzyme's activity.

One must pause to remark on the extraordinary economy and efficiency of this control system. As soon as the supply of L-isoleucine reaches an adequate level, the cell stops making it at once. The signal acts simply by turning off the activity of the first enzyme; that is enough to stop the whole production line. Most remarkable of all, once this first enzyme has been synthesized the control costs the cell no expenditure of energy whatever; this is shown by the fact that the amino acid will act to inhibit the enzyme outside the cell without any energy being supplied. A factory with control relays that require no energy for their operation would be the ultimate in industrial efficiency!

The L-isoleucine control system of *E. coli* is only one example of this type of regulation in the living cell. It has now been demonstrated that similar circuits control the cell's production of the other amino acids, vitamins and other major substances, including the purine and pyrimidine bases that are the precursors of DNA.

In all these cases the control is nega-

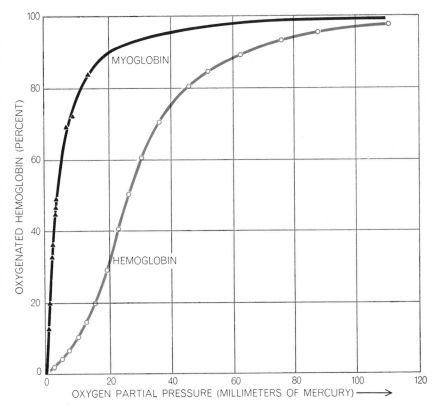

HEMOGLOBIN, like an enzyme, is a large molecule that binds a small one (oxygen) at specific sites. The curves show the rate of oxygen-binding by hemoglobin (*color*) and myoglobin (*black*), a related oxygen-carrier in muscle. The myoglobin curve is a hyperbola but the hemoglobin curve is S-shaped. Hemoglobin binds best at higher oxygen concentrations (in the lungs); the binding of a few oxygen molecules favors the binding of more.

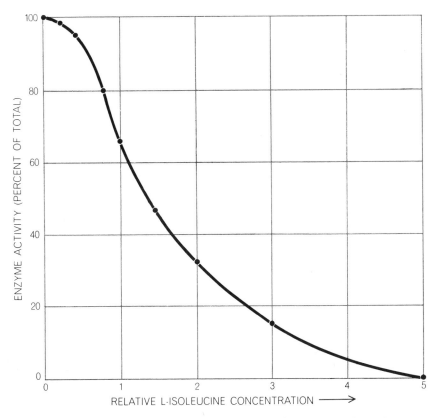

"COOPERATIVE EFFECT" occurs in regulatory enzymes as in hemoglobin. This curve shows the inhibition of L-threonine deaminase by L-isoleucine. The curve's S shape indicates that the effect of the regulatory signal is significant only above a threshold value.

a   *b*   *c*

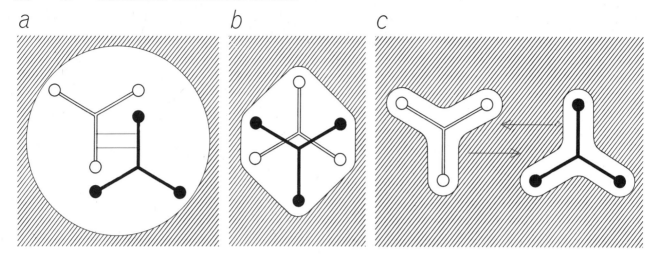

REGULATORY PROPERTY of an enzyme might be explained in three different ways. A regulatory signal (*open shape*) might combine with the substrate (*black shape*), participating directly in the chemical reaction it is controlling (*a*). But no such compounds have been found. A signal could simply get in the way of the substrate, excluding it from the enzyme's active site by "steric hindrance" (*b*). The different shapes of substrates and signals preclude this, and in any case steric hindrance could only account for enzyme inhibition, not activation. The only plausible hypothesis, confirmed by experiments with several enzymes, is that the signals and the substrate fit different sites on the enzyme and that the regulatory interactions of these sites are "allosteric," or indirect (*c*).

tive; that is, it involves the inhibition of enzymes. There are opposite situations, of course, in which the control system *activates* an enzyme when the circumstances call for it. An excellent example of such a positive control has to do with the cell's storage and use of energy.

Animal cells store reserve energy in the form of glycogen, or animal starch. Glycogen is synthesized from a precursor—glucose-6-phosphate—in three enzymatic steps. First glucose-6-phosphate is made into glucose-1-phosphate; then glucose-1-phosphate is made into uridine diphosphate D-glucose. Finally uridine diphosphate D-glucose is made into glycogen. When the cell has a good supply of energy, it produces considerable amounts of glucose-6-phosphate. This serves as a signal for stimulating the synthesis of glycogen. The signal works at the third step: the presence of a high level of glucose-6-phosphate strongly activates the enzyme that brings about the conversion of uridine diphosphate D-glucose into glycogen. On the other hand, when the supply of working energy in the cell falls to a low level, so that it must draw on the reserve stored in glycogen, it becomes necessary to activate an enzyme that splits the glycogen (the enzyme known as glycogen phosphorylase). One chemical signal known to be capable of activating this enzyme is adenosine monophosphate (AMP). AMP is a product of the splitting of adenosine triphosphate (ATP), the principal source of the cell's working energy, and an accumulation of AMP therefore indicates that the cell has used up its energy. The AMP signal activates the glycogen-splitting enzyme; the enzyme splits the glycogen molecule; the splitting releases energy, and the energy then is used to regenerate ATP.

The cell thus possesses mechanisms for two types of control of enzyme activity: negative (inhibited enzymes) and positive (activated enzymes). There are

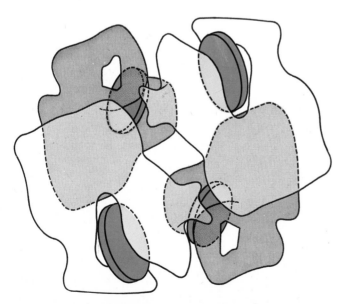

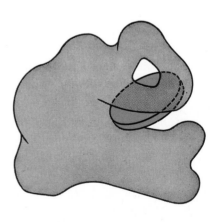

MOLECULE OF HEMOGLOBIN, shown (*left*) in very simplified form, has four heme groups (*color*), each of which is borne on a subunit, or chain, that is very similar to a myoglobin molecule (*right*). The heme groups of hemoglobin, each of which is a binding site for an oxygen molecule, are relatively far apart. Cooperative interactions among them must therefore be "allosteric."

situations in which both methods operate simultaneously. Consider, for example, the synthesis of a nucleic acid. It is assembled from purine and pyrimidine bases, combined in certain definite proportions. The purines and pyrimidines are synthesized on parallel production lines. For the sake of economy they should be produced roughly in the proportions in which they will be used.

This implies that the rate of production by each production line should feed back to control the output by the other. Such a system of mutual regulation must employ both negative and positive controls. Exactly this kind of system has been demonstrated in experiments with *E. coli* conducted by John C. Gerhart and Pardee at the University of California at Berkeley and at Princeton University. They showed that the output of the pyrimidine production line is controlled not only by its own end product (which inhibits the first enzyme in the synthetic sequence) but also by the end product of the purine production line, which counteracts the inhibition by the pyrimidine end product in vitro. Indeed, the purine end product can activate the pyrimidine production directly when no pyrimidine product is present! In short, the enzyme involved here is inhibited by one signal and activated by another.

Several enzymes involved in regulation have also been found to respond in this way to different signals. Moreover, this is not the only exceptional property of these enzymes. Let us now consider another property that will clarify the mechanism by which they are controlled.

A clue to this property seems to lie in the shape of the curve describing the rate at which the enzymes react with their substrates: the substances whose changes they catalyze. Ordinarily the rate of reaction of an enzyme increases as the concentration of substrate is increased. The increase is described by an experimental curve that fits a hyperbola. This kind of curve expresses the fact that the first step in the transformation of the substrate by the enzyme is the binding of the substrate to a specific attachment site on the enzyme.

When the concentration of substrate is increased, molecules of substrate tend to occupy more and more binding sites. Since the number of enzyme molecules is limited, at high concentrations of substrate nearly all the binding sites are occupied. At this point the rate of reaction levels off, hence the hyperbolic

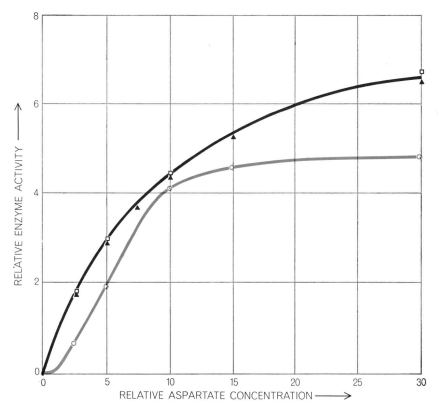

DESENSITIZATION of an enzyme affects all its regulatory properties. The substrate saturation curve of natural ATCase (*color*) is S-shaped as a result of the cooperative effect. If the enzyme is denatured by heating, the cooperative effect is lost (*black curve*). So is the effect of feedback inhibition by CTP, as shown by the fact that the curve is the same whether the enzyme is assayed without CTP (*triangles*) or with CTP added (*squares*).

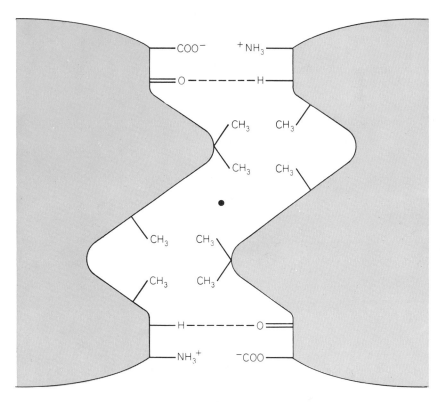

ALLOSTERIC PROTEINS are assumed by Monod, Jeffries Wyman and the author to be polymers, molecules composed of identical subunits, that have a definite axis of symmetry (*black dot*). A cross section through such a molecule (made up in this case of two subunits) shows how the symmetry results from the chemical bonds by which the units are associated.

shape of the curve. The regulatory enzymes, surprisingly, do not exactly follow this pattern: their reaction rate increases with the concentration of substrate but often the curve is sigmoid (S-shaped) rather than hyperbolic.

When one reflects on the saturation curve of the regulatory enzymes, one notes that it is strikingly like the curve describing the saturation of the hemoglobin of the blood with oxygen. There too the reaction rate traces a sigmoid curve; this remarkable property is related to hemoglobin's physiological function of carrying oxygen from the lungs to other tissues. In the lungs, where the oxygen pressure is high, the hemoglobin is readily charged with the gas; in the tissues, where the oxygen pressure is low, the hemoglobin readily discharges its oxygen. Consider now, however, the myoglobin of muscle tissue. It takes on oxygen, but its oxygenation follows a hyperbolic curve like the classical one for enzymes. A comparative chart shows that when the pressure of oxygen is increased, the amount of oxygen bound by hemoglobin increases faster than the amount bound by myoglobin [see top illustration, page 299]. It looks as if the first oxygen molecules picked up by the hemoglobin favor the binding of others—as if there is cooperation among the oxygen molecules in binding themselves to the carrier. Oxygen thus plays the role of a regulatory signal for its own binding.

Similarly, cooperation may be the key to the sigmoid pattern of binding activity in many of the regulatory enzymes. An example of such an enzyme is threonine deaminase. Here again physiological function is evident. The substrate of threonine deaminase is the amino acid threonine. If the amount of this amino acid falls to a very low level in the cell, the cell cannot synthesize proteins. In the absence of threonine, it would be a waste of energy to make isoleucine, the end product of the chain of which threonine deaminase is the first step; hence the economy-geared control system of the cell calls off the production of the second amino acid. In other words, threonine deaminase will not be active and isoleucine will not be produced unless at least threshold concentrations of threonine are present in the cell. In this situation threonine plays the role of regulatory signal for the reaction of which it is the specific substrate; it is an activator of its own transformation.

The most remarkable part of the story is that such cooperative effects are not restricted to the binding of substrate but also operate in the binding of more familiar regulatory signals: specific inhibitors or activators. Regulatory enzymes appear to be built in such a way that they not only recognize the configuration of specific substrates as signals but also gauge their response to whether or not the substrates and regulatory signals are present in certain threshold concentrations. (This is strongly reminiscent, of course, of electric relays—and, one may add, of nerve cells—which react only if the signal has a certain threshold strength.) The regulatory enzymes are thus capable of integrating several signals—both positive and negative—that modulate their activity.

We come now to the question: How do the regulatory relays work? The signals (either activators or inhibitors) are usually small molecules, and the receptor is a regulatory enzyme. In chemical terms, how does the enzyme translate and integrate the signals it receives? The answer to this question applies not only to regulatory enzymes but also to any other molecule that mediates a regulatory interaction. Since little is known about many of these molecules, the model I shall now describe is based on the experimental results obtained from regulatory enzymes. It seems legitimate, however, to extend the model to any category of regulatory molecule.

The question presents a biochemist with a difficult paradox. A molecule can "recognize" a message only in terms of geometry, that is, the shape or configuration of the molecule bearing the message. In this case the message is supposed to cause the enzyme to carry out (or refrain from carrying out) a certain reaction: conversion of a specific substrate into a specific product. Yet the molecule bearing the message often has no structural likeness to either the substrate or the product! How, then, can it promote or interfere with the enzyme's performance of its specific catalytic action on this substrate?

Considering several possible explana-

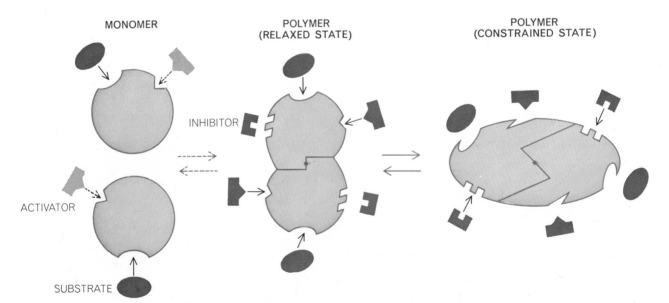

MONOMER   POLYMER (RELAXED STATE)   POLYMER (CONSTRAINED STATE)

INHIBITOR

ACTIVATOR

SUBSTRATE

REGULATORY CHANGES in an allosteric molecule are conceived of as arising from its shifting back and forth between two states. The polymeric molecule is made up of several monomers (two in this case), as shown at left. The polymer can exist in a "relaxed" state (middle) or a "constrained" state (right). In one condition it binds substrate and activators; in the other state it binds inhibitors. The binding of a signal tilts the balance toward one or the other state but the molecule's symmetry is preserved.

tions, Monod, Jacob and I have concluded that the only plausible one is that the signal and the substrate fit into separate binding sites on the enzyme and that the signal takes effect by an interaction between these sites [*see top illustration, page 300*]. There is strong experimental evidence in favor of this model. One of the most convincing lines of evidence is the recent discovery by Gerhart that the regulatory enzyme aspartate transcarbamylase has a binding site for its substrate on one subunit of the molecule and a site for an inhibitor of its activity on another subunit. When the subunits are split apart, one retains the ability to recognize the substrate, the other the ability to recognize the inhibitor.

We must now inquire into the nature of the interaction of these two categories of sites on the enzyme. How does the binding of a molecule at one site affect the binding of another molecule at the other site? The best clue to an understanding of the mechanism of the interaction seems to lie in a property of regulatory enzymes that I have already mentioned: the sigmoid curve describing their binding of substrate or of signal molecules, which indicates a cooperative effect among those molecules. Again it is instructive to consider the analogy of the binding of oxygen molecules by hemoglobin.

The hemoglobin molecule has four hemes that are well separated from one another; each is a binding site for an oxygen molecule. In view of the separation between the sites, their cooperation in binding oxygen must be "allosteric," or indirect. Myoglobin, which has only one binding site, binds oxygen hyperbolically (that is, without any control); hemoglobin, with its four sites, binds oxygen in a sigmoid pattern. It seems, therefore, that the key to hemoglobin's cooperative, controlled binding of oxygen lies in the molecule's four-part structure.

Now consider a regulatory enzyme. The binding of any particular molecule

EXPERIMENTAL DATA supporting the allosteric model come from X-ray diffraction maps of hemoglobin made by **M. F. Perutz** and his colleagues at the University of Cambridge. The contour lines based on electron densities suggest the shapes of the subunit chains of oxygenated hemoglobin (*top*), reduced hemoglobin (*middle*) and the two superposed (*bottom*). A conformational change of the kind proposed in the model on the opposite page is evident, as is preservation of the molecule's axis of symmetry.

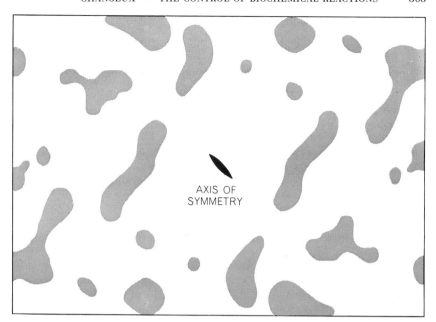

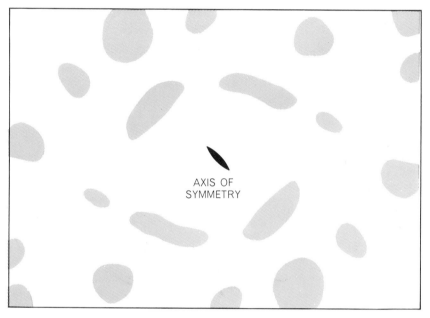

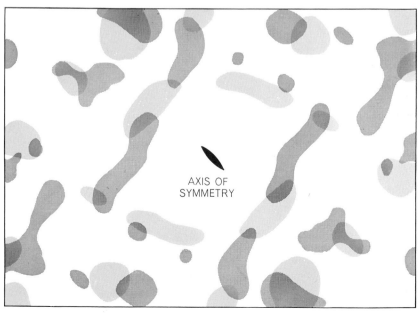

(substrate, inhibitor or activator) is sigmoid and therefore a cooperative affair; this implies that there is a set of reception sites for each specific molecule. There also appears to be interaction among the binding sites for different molecules, such as substrate and activator or substrate and inhibitor. Surprisingly the experimental evidence suggests that both types of allosteric interaction—that among the sites binding a particular molecule and that among the sites binding different molecules—may depend on one and the same mechanism, embodied in the structure of the enzyme molecule.

The most striking evidence comes from experiments in the alteration of the structure of regulatory enzyme molecules. Gerhart and Pardee at Berkeley and Princeton and I at the Pasteur Institute, working independently, have found that by changing the molecular structure of aspartate transcarbamylase or L-threonine deaminase (by means of heat, bacterial mutation or certain other procedures) it is possible to "desensitize" these regulatory enzymes so that they are no longer affected by a feedback inhibitor. They are still capable, however, of reacting with their respective substrates. The interesting point is that a change in the

enzyme's structure eliminates, along with the negative interaction of the feedback inhibitor and the substrate, all the cooperative interactions in the enzyme molecule. This applies particularly to the binding of the substrate, which changes from a sigmoid to a hyperbolic pattern.

What, then, is the crucial structural feature that accounts for the allosteric interactions within the enzyme molecule? Again hemoglobin offers a clue.

We have noted that the hemoglobin molecule is a four-part structure. It comprises four heme units, each of which is attached to a distinct chain of amino acid units. This molecule is thus made up of four subunits, each of which is so similar to a myoglobin molecule that hemoglobin can be considered essentially a combination of four myoglobin molecules. Hemoglobin displays cooperative interaction, whereas myoglobin does not; hence this property evidently is associated with its four-part structure. Now, experiments show that the binding of oxygen by hemoglobin is connected in some way with an adjustment in the bonding between the subunits making up the molecule [see the article "The Hemoglobin

Molecule," by M. F. Perutz, beginning on page 105]. The same turns out to be true of many of the regulatory enzymes; their binding of smaller molecules also depends on the adjustment of the bonds holding together their subunits.

On the strength of the experimental findings, Monod, Jeffries Wyman and I have proposed a model picturing the working of the regulatory enzyme system [see illustration on page 302]. It suggests that the enzyme molecule consists of a set of identical subunits, each subunit containing just one specific site for each of the molecules it may bind to itself, either substrate molecules or regulatory signals. Now, if a molecule is made up of a definite and limited number of subunits, the implication is that it has an axis of symmetry. Let us say that the enzyme molecule can switch back and forth between two states, and that in each state its symmetry is preserved. The two symmetrical states differ in the energy of bonding between the subunits: in the more relaxed state the enzyme molecule will preferentially bind activator and substrate; in the more constrained state it will bind inhibitor. Whichever compound it binds (substrate, inhibitor or activator) will tip the balance so that it then favors the binding of that category of small molecule. A change in the relative concentrations of substrate and signals may, depending on their molecular structure, tip the balance one way or the other. Thus the model indicates how the enzyme molecule's binding sites may interact, either cooperatively or antagonistically. It suggests that the enzyme may integrate different messages simply by adopting a characteristic state of spontaneous equilibrium between two states.

The major conclusion from the study of the regulatory enzymes is that their powers of control and regulation depend entirely on the form of their molecular structure. Built into that structure, as into a computer, is the capacity to recognize and integrate various signals. The enzyme molecule responds to the signals automatically with structural modifications that will determine the rate of production of the product in question. How did these biological "computers" come into being? Obviously they must owe their remarkable properties to nature's game of genetic mutation and selection, which in eons of time has refined their construction to a peak of exquisite efficiency.

MUTATIONS in the structural gene for L-threonine deaminase in *E. coli* affect the regulatory properties of the enzyme. Mutant enzymes respond differently to feedback inhibition.

30

GENETIC REPRESSORS

MARK PTASHNE AND WALTER GILBERT

June 1970

Genes do not operate continuously but are switched on and off. One control mechanism is repression. Now the first specific repressors have been isolated, confirming hypotheses put forward a decade ago

How are genes controlled? All cells must be able to turn their genes on and off. For example, a bacterial cell may need different enzymes in order to digest a new food offered by a new environment. As a simple virus goes through its life cycle its genes function sequentially, directing a series of timed events. As more complex organisms develop from the egg, their cells switch thousands of different genes on and off, and the switching continues throughout the organism's life cycle. This switching requires the action of many specific controls. During the past 10 years one mechanism of such control has been elucidated in molecular terms: the control of specific genes by molecules called repressors. Detailed understanding of control by repressors has come primarily through genetic and biochemical experiments with the bacterium *Escherichia coli* and certain viruses that infect it. In this article we shall first outline the current view of the action of repressors and then report some of the experimental evidence, involving the isolation of specific repressors and the description of how they work, that supports this picture.

A gene is a particular sequence of subunits called bases, ranged along a molecule of deoxyribonucleic acid (DNA), that is ultimately translated into a corresponding sequence of amino acids constituting a molecule of protein [*see illustration on next page*]. An enzyme, RNA polymerase, first transcribes the DNA sequence into a corresponding sequence of bases of ribonucleic acid (RNA), beginning at a specific site termed the promoter. This messenger RNA molecule then becomes attached to the particles called ribosomes, where its genetic information is translated into

protein. For certain genes this process may be controlled by the intervention of a protein, a repressor, the product of a separate control gene. The repressor binds, or attaches, directly to the DNA molecule at the beginning of the set of genes it controls, at a site called the operator, preventing the RNA polymerase from transcribing the gene into RNA and thus turning off the gene. Each set of independently regulated genes is controlled by a different repressor made by a different repressor gene.

The repressor determines when the gene turns on and off by functioning as an intermediate between the gene and an appropriate signal. Such a signal is often a small molecule that sticks to the repressor and alters or slightly distorts its shape. In some cases this change in shape renders the repressor inactive, that is, no longer able to bind to the operator, and so the gene is no longer repressed; the gene turns on when the small molecule, which here is called an inducer, is present. In other cases the complex of the repressor and the small molecule is the active form; the repressor is only able to bind to the operator when the small molecule (here called a corepressor) is present.

For example, consider the *E. coli* gene that codes for the enzyme beta-galactosidase. If the bacterium is to grow with lactose (milk sugar) as its source of carbon, this enzyme must be produced in order to split the sugar into its two components, galactose and glucose. In the absence of lactose the beta-galactosidase gene is not needed, and it is switched off by a specific repressor, the *lac* repressor. In the presence of lactose the required enzyme is synthesized because a breakdown product of the sugar acts as an inducer: it attaches directly to the repressor, causing the repressor to release

the DNA and allow the synthesis of the messenger RNA corresponding to the beta-galactosidase gene. When the lactose supply is exhausted, the repressor is freed of the sugar inducer and once again switches off the beta-galactosidase gene.

According to this picture the lactose repressor is a protein molecule that interacts (at two different sites) with two entirely different kinds of molecules. One site recognizes the unique sequence of DNA bases that constitutes the operator; the other binds the inducer (the lactose derivative). The binding of the inducer to the repressor induces a conformational change in the repressor that prevents its sticking to the operator [see the article "The Control of Biochemical Reactions," by Jean-Pierre Changeux, beginning on page 295].

Much of the genetic basis of the picture we have just described was presented in 1961 by François Jacob and Jacques Monod of the Pasteur Institute in Paris. One of their most illuminating examples was an analysis of the complete set of genes for the metabolism of lactose by *E. coli*, the *lac* system. Genetic and biochemical experiments showed that two kinds of genes are involved in the utilization of this sugar. There are three genes of one kind, one coding for beta-galactosidase (the *z* gene), one for a permease protein that concentrates lactose in the cell (the *y* gene) and one for a transacetylase enzyme with an unknown function (the *a* gene). Mutations in any one of these three genes change the structure of the corresponding enzyme but have no effect on the regulation of their synthesis. Mutations in the other kind of gene fail to change the structure of these enzymes but do affect their regulation. These mutations affecting regulation are located in

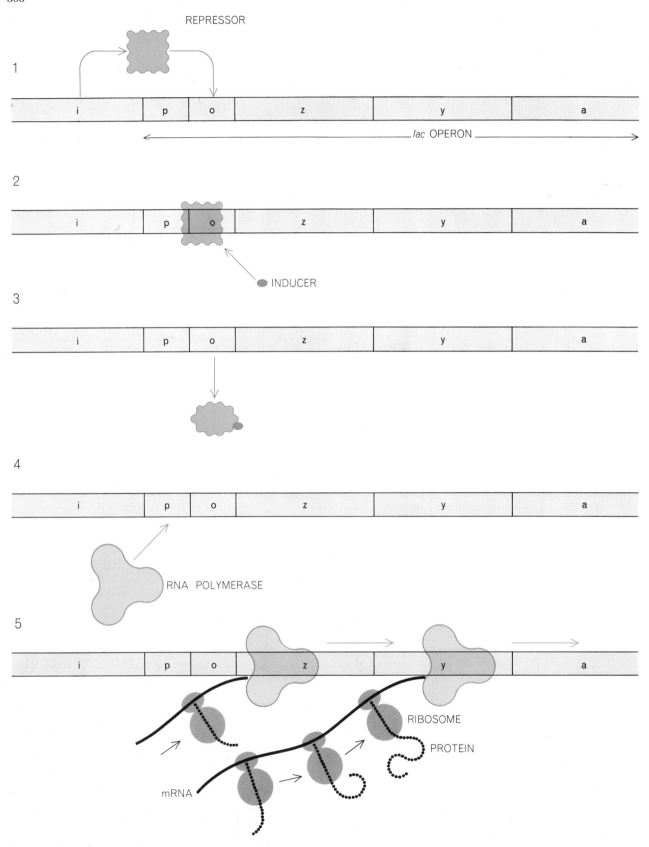

REPRESSOR FUNCTION is illustrated for the *lac* repressor, which controls genes that provide for the metabolism of lactose in the bacterium *Escherichia coli*. The repressor, produced by the *i* gene, binds to the DNA at the operator site (*o*), thus preventing the transcription of the *z*, *y* and *a* genes into RNA (*1, 2*). In the pres- ence of an inducer (a lactose derivative), which attaches to the repressor and changes its shape, the repressor releases the DNA (*3*), switching the gene on. Beginning at the promoter (*p*), RNA polymerase (*4*) transcribes the DNA into messenger RNA, which is translated into protein on particles termed ribosomes (*5*).

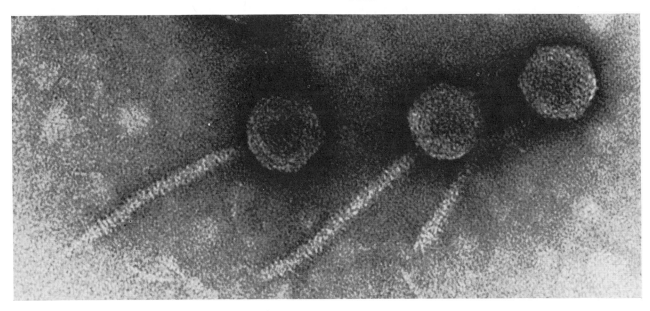

PHAGE LAMBDA, a virus that infects *E. coli,* is another organism whose genes are controlled by a repressor. The phage consists of a DNA molecule wrapped in a protein head to which a long tail is attached. In the electron micrograph, made by L. Kemp and A. F. Howatson of the University of Toronto, the lambda particles are enlarged 350,000 diameters and stained with uranyl acetate.

a single gene, the *i* gene. Typical *i* gene mutations studied by Jacob and Monod caused continual synthesis of *lac* enzymes, even in the absence of lactose.

What was the nature of this control? It might be positive, that is, the product of the *i* gene might keep the *lac* genes turned on. Or it might be negative: the *i* gene product might somehow turn the *lac* genes off. The actual nature of the control can be shown by selectively eliminating the *i* gene product from the cell by introducing a mutation that deletes the *i* gene. In such a mutant the *lac* enzymes are synthesized at the full rate whether or not the inducer is present. If the absence of the *i* gene results in continual synthesis, the *i* gene's control must be negative in character. This is confirmed if several different pieces of DNA carrying the *lac* genes are introduced into a cell, some with and some without the *i* gene. No *lac* enzymes are synthesized, indicating that the presence of any *i* gene is enough to turn off the *lac* genes on all the pieces of DNA, and therefore showing again that *i* gene control is negative. Jacob and Monod proposed the simplest possible mechanism based on this evidence: that the *i* gene product acts as a repressor—a substance that diffuses through the cell and shuts

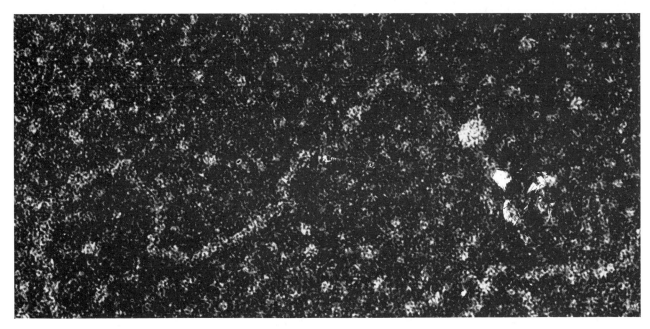

LAC REPRESSOR MOLECULE (*large white object*) sits on DNA in an electron micrograph made by Jack Griffiths of Cornell University. Purified *lac* repressor was put into a solution with DNA bearing the *lac* operon. When DNA molecules were caught on a carbon film and were shadowed by rotation in tungsten vapor so as to build contrast evenly on all sides, repressor molecules were seen bound to the DNA. The molecules have been enlarged 500,000 diameters. The granular background texture is due to the shadowing.

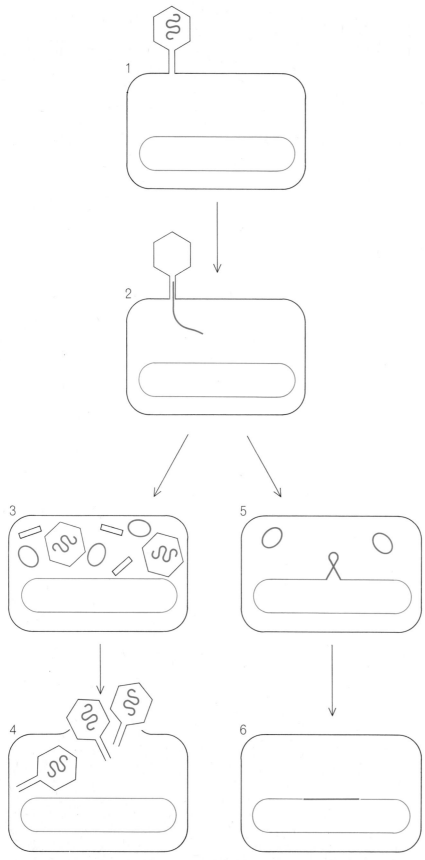

INFECTION can take either of two courses after the phage becomes attached to the bacterium (1) and injects its DNA (2). In lysis (*left*) the viral DNA (*thick colored line*) replicates and forms new particles (3), which burst the cell (4). In lysogeny (*right*) most of the viral genes are switched off. Instead of replicating, the viral DNA integrates into the bacterial DNA (*thin colored line*) by crossing over (5) and remains there, dormant (6).

off any *lac* genes that are present, not just the *lac* genes on the same piece of DNA.

Jacob and Monod went on to predict and find mutations that render the *lac* genes insensitive to the *lac* repressor—mutations that cause continual enzyme synthesis even in cells known to contain active repressor. These mutations, which are located on the bacterial DNA just before the beginning of the *z* gene, define the genetic determinant of the repressor's site of action. Note that one must distinguish between a genetically defined region of DNA, which can harbor mutations affecting a certain function, and the ultimate product of that region, which actually carries out the function. The genetic experiments could show only that mutations in an operator gene locus (*o*) render inactive some structure (the operator) that otherwise would interact with the repressor. The promoter, the operator locus and the three genes controlled by the *lac* repressor constitute what is called the *lac* operon.

Several other sets of genes controlled by specific repressors have been studied in *E. coli*. One that has played an important role in illuminating the mechanism of repression is not properly a group of bacterial genes at all but rather the genes of a bacterial virus, phage lambda, that infects *E. coli*. The phage particle consists of a single DNA molecule enclosed in a protein coat with a tail through which the DNA is injected into a bacterial cell. Once inside the cell the DNA can follow either of two courses [*see illustration at left*]. It can complete the ordinary course of infection, utilizing the cell's machinery to replicate its own virus DNA and wrap it in a protein coat, producing large numbers of progeny phage that lyse, or burst, the bacterium and go on to infect more cells. Alternatively, instead of completing this lytic cycle, the phage genes are switched off soon after they enter the cell. Then the dormant phage DNA becomes integrated into the host cell's chromosome and remains there inertly, replicating with the bacterial DNA and giving rise to a population of cells each of which contains a dormant phage chromosome, or prophage.

What shunts an ordinarily infectious phage into the dormant, or lysogenic, state? Genetic investigations conducted by Jacob, by A. Dale Kaiser of Stanford University and by others showed that the phage's ability to turn its genes off is conferred by a single phage gene designated C_I. Experiments and arguments strictly analogous to the ones that elucidated the *lac* system showed that the C_I gene makes a repressor, the lambda phage repressor, that switches off certain genes of the phage. Treatments (such as irradiation with ultraviolet) that turn the genes on again and restore the phage to the lytic state are believed to act by leading to the production of a small molecule–an inducer–that inactivates the phage repressor. Recent experiments have shown that the lambda repressor acts at two separate operator sites on the phage chromosome, controlling two different sets of genes that are read in opposite directions on the DNA [*see illustration below*]. The "early" genes directly controlled by the repressor include genes whose products are necessary to turn on the "late" phage genes. The mechanism of this positive control is not yet known.

The presentation of the repressor theory in 1961 caused great excitement among biologists because it proposed to explain gene control, which had been regarded as an extremely complicated matter, in terms of just three elements: the repressor, the operator and the inducer. The most important aspect of the model was the proposition that the *i* gene product or the C_I gene product functions directly to turn off its target genes without the intervention of other large molecules. A counterhypothesis would be, for example, that the *i* gene product is actually an enzyme that is involved in the formation of the ultimate repressor. Although the genetic evidence was entirely consistent with the simpler model, it could not exclude this more complicated possibility. Furthermore, the genetic analysis did not reveal the molecular nature of either the repressor or the operator. It was suggested that the repressor might consist at least partly of RNA, and in fact experimental evidence, now known to be incorrect, was presented to support this notion.

The question of the molecular nature of the operator was even more intriguing. One model had the repressor binding to the operator sequence on the DNA and directly blocking the transcription from DNA to RNA. It was possible, however, that the repressor had its effect at some other stage of the process. For example, if the operator were transcribed into RNA at the head of the messenger RNA molecule, the repressor might function by binding to this operator sequence on the messenger RNA and blocking the translation into protein. The results of experiments performed in many parts of the world in the five years following the introduction of the repressor theory made it clear that the only way to answer these questions was to isolate a genetically defined repressor and to show how it worked in the test tube.

In the summer of 1965 we began experiments at Harvard University to try to isolate specific repressors. One of us (Gilbert), in collaboration with Benno Müller-Hill, a postdoctoral fellow from Germany, attacked the *lac* repressor. The other (Ptashne) began a search for the lambda phage repressor aided by Nancy Hopkins, now a graduate student at Harvard. The actual experimental approaches used to isolate these two repressors were entirely different, although the same major problems were faced by both groups.

The problems were both scientific and psychological. We had no biochemical assay for repressors and could not devise a functional assay because we did not know how they would function. We suspected (correctly, as it turned out) that repressors are present only in extremely small quantities in the cell. (The average *E. coli* cell contains only about 10 or 20 molecules of the *lac* repressor, constituting only .002 percent of the cell's protein.) We did not even know whether to look for repressors among proteins, nucleic acids or other substances. Since there was no way to assay function, we had to surmise some other property that

would give us a foothold, but not until the experiments were completely successful could we know that our working hypothesis was relevant. In fact, we could not be sure that the experiments would have any outcome at all. It was possible that we would search but not find anything; such a failure would not prove the negative (that repressors did not exist) but would simply mean that the question was still open. Even if we could isolate the product of the repressor gene, we might still fail to understand how that product functioned.

Gilbert and Müller-Hill based their experiments on the simplest interpretation of the induction phenomenon of the *lac* operon: that a lactose-like inducer inactivates the repressor by binding directly to it. The inducer they used was isopropyl-thio-galactoside (IPTG), a substance that resembles the sugar inducer in molecular form and that is the most active experimental inducer known for the *lac* genes. They reasoned that they might detect the presence of repressor molecules in solution by the ability of these molecules to bind radioactively labeled IPTG molecules. Hoping to measure the extent of this binding (and thus establish a method of assay for the repressor), they resorted to the technique of equilibrium dialysis.

A concentrated extract of *E. coli* cells was placed in a bag made from a cellulose membrane. The membrane has pores that are small enough to prevent the passage of typical protein molecules but large enough to allow water and IPTG molecules to pass through freely. The bag was floated in water containing radioactively labeled IPTG molecules. Since the IPTG molecules could pass through the pores in the bag, the concentration of these molecules free in solution would soon become equal inside and outside the bag. If, however, there were large molecules inside the bag that would bind IPTG, then the total concentration (free plus bound) inside the bag

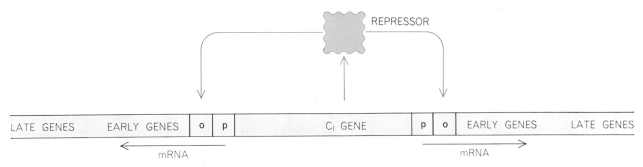

LYSOGENY of the lambda phage is brought about by a repressor, a product of the C_I gene on the viral chromosome mapped here.

The repressors bind to two different operators, blocking the transcription of two sets of "early" genes read in opposite directions.

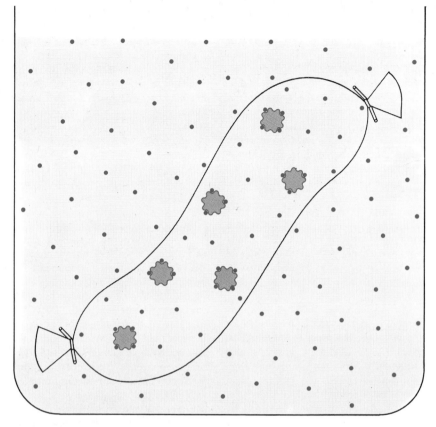

LAC REPRESSOR was identified by its ability to bind molecules of IPTG, an effective inducer. An *E. coli* extract was put inside a bag made of a semipermeable membrane. The bag was floated in a solution containing radioactive IPTG (*colored dots*). The IPTG diffused through the membrane, making the concentration of free IPTG the same inside and outside the bag. Repressor molecules (*light color*) in the bag bound some IPTG molecules, however. The excess of radioactivity inside the bag measured the extent of the binding and provided an assay for purifying the repressor: the protein that was binding IPTG.

would come to be greater than the concentration outside. Since the IPTG was radioactive, even a small difference in concentration would be revealed by comparing the radioactivity of samples taken from inside the bag and from the surrounding solution.

After nine months of false starts and failures, a way was finally found to prepare a highly concentrated extract of bacterial cells that would draw a tiny excess of radioactivity into the bag. That excess became the basis of the needed assay. After that it was a straightforward matter to fractionate the bacterial proteins and, by testing each fraction separately for the ability to bind IPTG, to purify the substance that could bind the inducer.

Was this stuff indeed the repressor? One critical test involved strains of cells that in genetic experiments made no repressor. These cells turned out not to provide protein that would bind IPTG in the dialysis experiment. An even more convincing proof was based on the existence of cells with special mutations in

the *i* gene that increase the affinity of the repressor for the inducer in the living cell. Extracts from these cells were shown to contain molecules that bound radioactive IPTG more tightly in the dialysis assay.

Meanwhile the search for the lambda repressor was proceeding along different lines. Ptashne's strategy was to attempt to create a situation in which the only protein synthesized by the cell would be the lambda repressor. Then radioactive amino acids, the protein precursors, could be incorporated in the repressor only, making the properties of the repressor molecule easy to study. To approach this ideal situation he tried first to stop all protein synthesis directed by the cellular DNA and second to stop the functioning of all the phage genes except the C_I gene itself. The first requirement was met by exposing *E. coli* cells to massive doses of ultraviolet radiation, which so damages the DNA that it cannot be copied into RNA, halting the further synthesis of bacterial proteins. The cells' protein-synthesizing machinery remains

functional, however, so that when phages inject undamaged DNA into irradiated cells, phage proteins can be synthesized. In order to meet the second requirement the experiment was done with *E. coli* cells that contained a special lambda prophage, one whose repressor is not inactivated by ultraviolet. After irradiation this repressor shuts off all the genes on the incoming phage except for the repressor gene itself.

If these stratagems worked perfectly, only the repressor would be synthesized in the irradiated and infected cells. In reality the procedures are quite imperfect, and so repressor synthesis accounts for only 5 to 10 percent of the total residual protein synthesis. An experiment using two different radioactive labels was therefore necessary in order to pick out the repressor from among the other labeled proteins [*see illustration on opposite page*].

The irradiated cells were divided into two portions. One was infected with lambda phages carrying an intact repressor gene and the other with phages bearing mutations that blocked the synthesis of the repressor. Both cultures were fed radioactive amino acids. The cells with an intact repressor gene got amino acids containing radioactive hydrogen, or tritium (H^3), whereas the cells that could not make repressor got amino acids containing radioactive carbon (C^{14}).

The beta particle emitted when the tritium nucleus decays has only a tenth the energy of the beta particle emitted by the carbon. This energy difference is easily detectable, so that proteins made from amino acids labeled with H^3 can be distinguished from those synthesized from amino acids labeled with C^{14}. Ptashne mixed the labeled cells, extracted their proteins and spread them out on a chromatographic column, which separates proteins from one another. In the pattern of radioactivity corresponding to the proteins synthesized in both kinds of cells, and thus labeled with both isotopes, one peak appeared that was labeled only with H^3—not with C^{14}. The H^3-labeled protein, made in one culture but not in the other, was presumably the repressor. This was confirmed in a second experiment when another phage, bearing a mutation in the C_I gene that modifies but does not eliminate the repressor, caused the synthesis of a protein with properties slightly different from those of the unmutated repressor.

Many biologists had expected that repressors would turn out to be similar to histone proteins, which are found in conjunction with DNA in the chromosomes of higher organisms. The histones are very small proteins characterized by a large excess of positive charges carried on their molecules at a neutral *p*H. Both the lambda and the *lac* repressor are weakly acidic, however, meaning that at neutral *p*H they have an overall negative charge. Furthermore, they are large proteins. The *lac* repressor has four identical subunits, each with a molecular weight of 38,000. Although the lambda repressor was originally isolated as a single polypeptide chain with a molecular weight of 28,000, we now know that the functional structure is a complex of four of these subunits. (Vincenzo Pirrotta, working in Ptashne's laboratory, has recently isolated another phage repressor that behaves like the lambda repressor, although it bears a slight overall positive charge at neutral *p*H.) Some biologists had thought repressors would contain RNA, which would allow them to recognize other nucleic acids by base pairing, but both the *lac* and the lambda repressor turned out to be quite ordinary proteins, with no striking properties other than their action.

Having isolated and purified these two products of control genes, we were still faced with fundamental questions. Were these substances in fact repressors in the sense that they would interact directly with their respective operators? And what, in fact, was the operator? One simple version of the Jacob-Monod theory was that repressors might bind directly to the operator region on a DNA molecule and so block the transcription into RNA. We verified this notion by examining directly the interaction between isolated repressor and purified DNA in the test tube. The first experiment involved mixing radioactive lambda repressor with DNA purified from lambda phage particles to see if the radioactive repressor would stick to the DNA and sediment with it in a centrifuge [*see il-*

LAMBDA REPRESSOR was identified as shown at left. *E. coli* cells were irradiated with ultraviolet to halt bacterial protein synthesis (*1*). The cells were divided (*2*) and half were infected with phages producing repressor (*left*), half with phages that could not produce repressor (*right*). The former were fed amino acids (*3*) labeled with radioactive hydrogen (H³), the latter amino acids labeled with radioactive carbon (C¹⁴). The cells were mixed, their proteins were extracted and separated by chromatography (*4*) and the protein fractions (*5*) were tested for radioactivity. One protein (*peak in colored curve*) appeared that was labeled only with H³. This was the repressor.

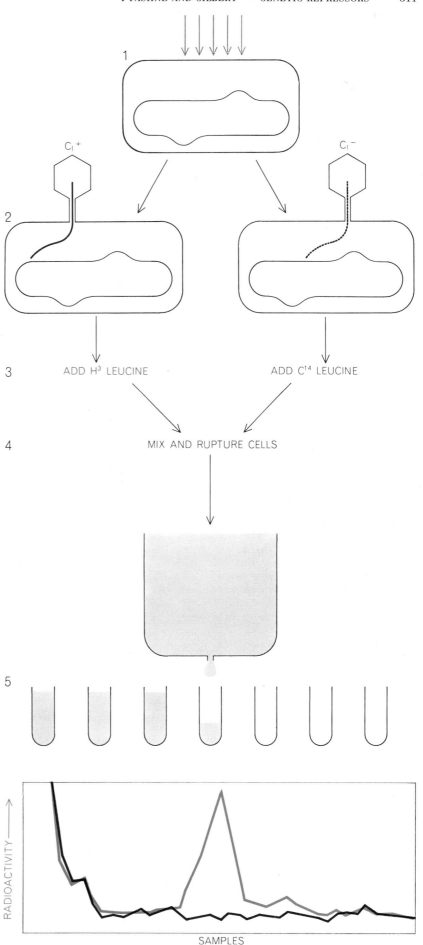

lustration at right]. A sample of the mixture was layered on top of a solution containing a gradient of sucrose and then spun at high speed in an ultracentrifuge. (The sucrose gradient, a solution that is denser at the bottom than at the top, simply prevents mixing and allows large molecules to move individually down the tube in the centrifugal field, which is several hundred thousand times normal gravity.) The DNA was much more massive than the repressor and would sediment faster. If the repressor was bound sufficiently tightly to the DNA, it would move at the high rate characteristic of the DNA; otherwise the repressor would remain near the top of the tube. When samples from different levels of the centrifuge tube were examined, it was clear that the radioactive lambda repressor did bind to lambda DNA. On the other hand, it did not bind to DNA isolated from a different phage.

Similarly, the *lac* repressor binds to DNA that contains the *lac* genes but not to DNA without *lac* genes. For these experiments the necessary radioactive repressor was made by purifying the *lac* repressor from bacterial cells that were grown on radioactive nutrients so that their proteins were labeled. The *lac* DNA was obtained not from *E. coli* but from a special phage that had been bred (from a phage very similar to phage lambda) by Jonathan R. Beckwith and Ethan Signer, who were then working at the Pasteur Institute. This phage carried the *lac* genes in place of some of the phage genes; thus it provided us with a conveniently small DNA molecule carrying the *lac* genes and in effect with a concentrated source of *lac* genes.

How specific is the repressor-DNA interaction? To verify the hypothesis we must know that the presumed repressor molecule interacts specifically with a unique site along the DNA. To show this we turned to mutant DNA's: molecules isolated from phages with mutations in their operators. The lambda repressor fails to stick to such mutant DNA's if both phage operators are damaged. If the *lac* operator is damaged (that is, if the DNA carries a mutation in the operator region), the *lac* repressor no longer binds. Such experiments show that these repressors do indeed interact with unique sites on the DNA molecules. If mutation alters even one base in the binding site, the protein molecule can no longer bind.

The action of the inducer of the *lac* operon can also be studied in the test tube. After radioactive *lac* repressor has

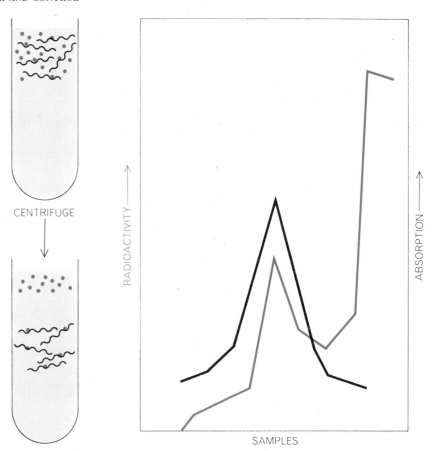

CENTRIFUGE

RADIOACTIVITY →

ABSORPTION →

SAMPLES

INTERACTION BETWEEN repressor and DNA was demonstrated by centrifugation. Lambda DNA and radioactive lambda repressor were layered on a sucrose gradient and spun in an ultracentrifuge. The DNA, being heavier, sedimented more rapidly than the free repressor. Much of the repressor, however, sedimented with the DNA, indicating that it had bound to the DNA. This was shown by the fact that a peak of radioactivity (*color*) coincided with the DNA peak (*black*). The remaining free repressor formed a second peak.

been bound to *lac* DNA, the inducer IPTG is added. The repressor then releases the DNA and, in the type of experiment we have described, no longer sediments with the DNA.

An ingenious alternative method of observing the interaction of the repressor with DNA was developed by Suzanne Bourgeois and Arthur D. Riggs of the Salk Institute for Biological Studies. They observed that when a solution of DNA molecules mixed with repressor was filtered through a cellulose nitrate filter, the repressor molecules would stick to the filter substance. If the molecule had previously been bound to DNA, the repressor molecule, in sticking to the filter, held the DNA on the filter. (The holes in the filter are gigantic compared with the repressor molecule, and they are too large to retain even the DNA alone.) This sticking property provides a simple assay of extraordinary sensitivity for the repressor-DNA interaction: one puts a radioactive label into the DNA and asks whether or not the label is

bound to the filter. The *lac* repressor must stick to the filter in an active form because, when the filter is washed with a solution of IPTG, the inducer interacts with the repressor and releases the DNA, leaving the repressor bound to the filter. Miss Bourgeois and Riggs have used this assay to study many of the quantitative aspects of the binding of the *lac* repressor to DNA.

The DNA binding experiments show directly that the operator, defined functionally as the physical target for the repressor, is a stretch of bases along a DNA molecule. We infer that when the repressor is bound to the operator, it prevents the RNA polymerase from initiating transcription. This has been confirmed recently for phage lambda: purified phage repressor directly blocks the copying of the two operons into RNA. Such a final demonstration has not yet been achieved for the *lac* repressor.

The mechanism by which the repressor recognizes the appropriate DNA sequence is not yet understood. It is clear that the interaction is with the native double strand of DNA, because if the strands are separated, or denatured, by heating, the repressor will not bind. Part of the interaction is presumably an attraction for the phosphates of the DNA backbone; although the repressors themselves have no strong overall charge, they have regions containing positively charged groups that would attract them to the negatively charged phosphates. In order to tell one sequence from another along the DNA molecule so as to bind at a specific site, however, the protein molecule must interact in some way with the bases in order to read the sequence information. It may be, for example, that the repressor forms hydrogen bonds with the outside of the bases, in the big groove of the DNA double helix, and in this way "sees" enough differences in the shapes of the bases to recognize the specific region.

How large is the operator? There is no direct evidence as yet, but a likely guess is that it is on the order of 12 bases long. (This is the shortest sequence that could be unique if the three million base pairs of the *E. coli* chromosome are considered as though arranged at random.) Such a sequence would be 40 angstroms long, a little more than one turn of the DNA helix and a short enough region to fit easily against a molecule of the repressor's size.

The gene for the *lac* repressor functions at a very low rate, producing only 10 molecules of repressor during the time between cell divisions. On the other hand, a very active gene in *E. coli*, such as the gene for beta-galactosidase, can make 10,000 molecules in each generation. Could one alter the rate at which the repressor gene functions? This was achieved by Müller-Hill, who found mutant forms of the *lac* repressor gene that make tenfold more repressor. Further mutations, found by Jeffrey Miller of the Harvard Medical School, raised this another five times, so that the final form of the gene works 50 times as rapidly as the unmutated form. We believe these mutations are changes in the promoter for the repressor gene that somehow make the promoter more active. In order to obtain still more repressor these superactive forms of the gene were introduced into the special *lac*-containing phage DNA we mentioned above. When this phage DNA multiplies inside a cell, many copies of the repressor gene are synthesized and there is a further twentyfold increase in repressor synthesis. The overall thousandfold increase means that the *lac* repressor is now one of the more easily obtainable proteins.

Control by repressors will probably turn out to be only one of many different mechanisms for gene control. For example, specific positive control elements are known to exist. The enzymes required to utilize the sugar arabinose, which have been studied by Ellis Englesberg of the University of California at Santa Barbara, have a single control gene, as lambda and the *lac* operon do, but the product of this gene seems to be required to turn the operon on rather than to turn it off. That is, if the control gene is deleted, the structural genes do not function. Such control could work through the attachment of arabinose to the product of the regulatory gene, with the resulting complex in turn binding to DNA and modifying the DNA structure in such a way as to permit the operon to be read. Confirmation of this notion awaits the isolation of the product of the control gene.

One specific molecular process of positive control is now known. Richard Burgess and Andrew Travers of Harvard and Ekkehard Bautz and John J. Dunn of Rutgers University have shown that RNA polymerase, which initiates the synthesis of RNA chains at the promoters, contains an easily dissociated subunit that is required for proper initiation. This subunit, the sigma factor, endows the enzyme to which it is complexed with the ability to read the correct promoters. Travers has shown that the *E. coli* phage T4 produces a new sigma factor that binds to the bacterial polymerase and enables it to read phage genes that the original enzyme-sigma complex cannot read. This change explains part of the timing of events after infection with T4. The first proteins made are synthesized under the direction of the bacterial sigma factor; among these proteins is a new sigma factor that directs the enzyme to read new promoters and make a new set of proteins. This control by changing sigma factors can regulate large blocks of genes. We imagine that in *E. coli* there are many classes of promoters and that each class is recognized by a different sigma factor, perhaps in conjunction with other large and small molecules.

Both the turning on and the turning off of specific genes depend ultimately on the same basic elements we have discussed here: the ability to recognize a specific sequence along the DNA molecule and to respond to molecular signals from the environment. The biochemical experiments with repressors demonstrate the first clear mechanism of gene control in molecular terms. Our detailed knowledge in this area has provided some tools with which to explore other mechanisms.

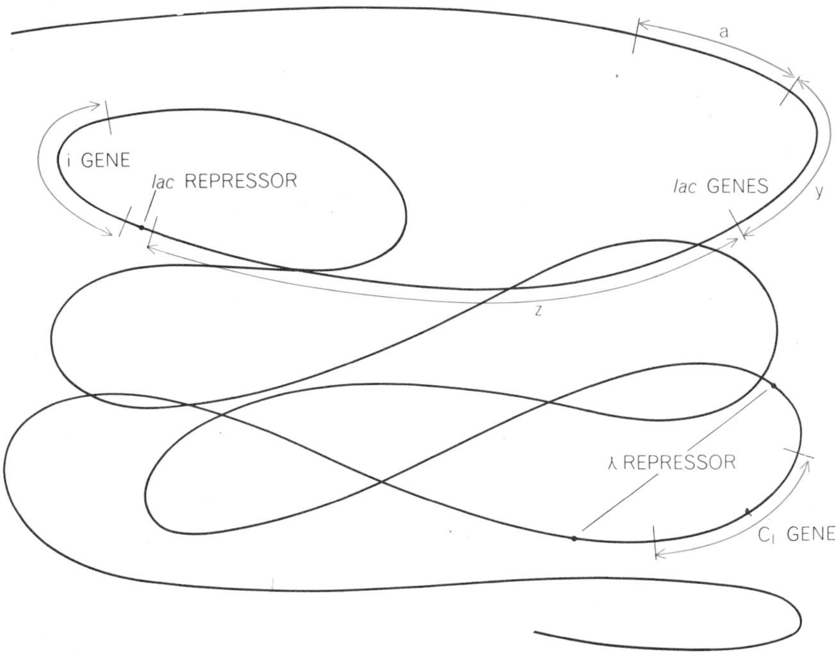

SPECIAL PHAGE CHROMOSOME was utilized in some experiments on the repressor-DNA interaction. The chromosome carries *lac* genes in addition to lambda genes. The diagram shows approximate relationships among some of the elements. The repressors are about twice as wide as the DNA molecule, which is some 8,000 times as long as it is wide.

31

CYCLIC AMP

IRA PASTAN
August 1972

This comparatively small molecule is a "second messenger" between a hormone and its effects within the cell. It operates in cells as diverse as bacteria and cancerous animal cells

The chemical reactions that proceed in the living cell are catalyzed by the large molecules called enzymes. If all the enzymes found within cells were working at top speed, the result would be chaos, and many mechanisms have evolved that control the speed at which these enzymes function. A small molecule that plays a key role in regulating the speed of chemical processes in organisms as distantly related as bacteria and man is cyclic-3'5'-adenosine monophosphate, more widely known as cyclic AMP. ("Cyclic" refers to the fact that the atoms in the single phosphate group of the molecule are arranged in a ring.) Among the many functions served

● CARBON   ● NITROGEN
○ OXYGEN   ○ HYDROGEN
     ● PHOSPHORUS

CYCLIC AMP is so named because the phosphate group in its molecule (*colored area*) forms a ring with the carbon atoms to which it is attached. Earl W. Sutherland, Jr., and his colleagues isolated substance in 1958.

by cyclic AMP in man and other animals is acting as a chemical messenger that regulates the enzymatic reactions within cells that store sugars and fats. Cyclic AMP has also been shown to control the activity of genes. Moreover, a precondition for one of the kinds of uncontrolled cell growth we call cancer appears to be an inadequate supply of cyclic AMP.

The first steps leading to the discovery of cyclic AMP were taken by Earl W. Sutherland, Jr., at Washington University some 25 years ago. For this work Sutherland was awarded the Nobel prize in physiology and medicine for 1971. Sutherland was trying to trace the sequence of events in a well-known physiological reaction whereby the hormone epinephrine (as adrenalin is generally known in the U.S.) causes an increase in the amount of glucose, or blood sugar, in the circulatory system. It is this reaction, usually a response to pain, anger or fear, that provides an animal with the energy either to fight or to flee. It is not a simple one-step process. Glycogen, a polymeric storage form of glucose, is held in reserve in the cells of the liver. What transforms the glycogen into glucose, which can then leave the liver and enter the bloodstream, is a series of steps involving intermediate substances. Sutherland measured the levels of these intermediates and concluded that only the initial step in the series (the transformation of glycogen into the intermediate sugar glucose-1-phosphate) was mediated by epinephrine. The transformation itself is actually catalyzed by an enzyme known as phosphorylase. Observing the activity of phosphorylase in cell-free extracts of liver tissue, Sutherland was able to enhance the enzyme's performance by first exposing to epinephrine the cells he used to make his extract.

Sutherland began to examine the properties of phosphorylase in more de-

tail. He found that the enzyme could exist in two forms: one that degraded glycogen rapidly and one that had no effect on it. The conversion of the enzyme from the active to the inactive form was catalyzed by a second enzyme, whose only action was to remove inorganic phosphate from the phosphorylase molecule. The conversion is worth noting; it is an important example of how the activity of an enzyme can be controlled by a relatively small change in its structure.

In collaboration with the first of a number of talented co-workers, Walter D. Wosilait, Sutherland next found still another liver enzyme, phosphorylase *b* kinase, which could restore the inactive form of phosphorylase to the active state. As one might expect, the reversal was accomplished by replacement of the missing phosphate; the donor of the phosphate was a close chemical relative of cyclic AMP, adenosine triphosphate (ATP). At about the same time Edwin G. Krebs and Edmond H. Fischer of the University of Washington detected a similar activating kinase in muscle tissue. ("Kinase" is the name reserved for transformations where ATP is the phosphate-donor.)

Having established the existence of two forms of phosphorylase, Sutherland and his colleagues concluded that the speed at which glycogen was broken down in the liver was a function of the amount of the enzyme present in its active form. Sutherland and Theodore W. Rall now made preparations of ruptured liver cells. When they added epinephrine to these broken cells, they found that in spite of the damage the hormone still increased the activity of the enzyme. The experiment, although simple, was extremely significant. Never before had a hormone been observed to function in a preparation that contained no intact cells. Once such a reaction can be shown to occur in a cell-free preparation the investigator can go on to test various cell components one at a time to determine just which ones are affected by the hormone.

This Sutherland and his co-workers proceeded to do. They knew that the phosphorylase was present in liver-cell cytoplasm: that part of the cell outside the nucleus and inside the cell membrane. When they added epinephrine to preparations composed of cytoplasm alone, however, there was no increase in enzyme activity. The absence of response suggested that the hormone exerted its effect on some other component of the liver cell. In due course they found that this component was the cell membrane.

Exactly what was the hormone doing to the cell membrane? In an effort to find out Sutherland employed a stratagem commonly used in biochemistry. He incubated a preparation of cell membrane (which itself contains no phosphorylase) with epinephrine. He then brought the mixture to the boiling point, expecting the heat to destroy the activity of the enzyme or enzymes in the membrane that were required for epinephrine action. When he added the now denatured mixture to a cell-free preparation that contained phosphorylase but no cell membrane, the activity of the phosphorylase was increased. The epinephrine had interacted with the cell membrane during the initial incubation period, evidently causing some enzyme in the membrane to produce a heat-stable factor that enhanced the activity of phosphorylase. Unfortunately the factor—whatever it was—was present in very small amounts and was therefore difficult to identify.

Sutherland eventually collected a large enough sample of the factor to determine that it belonged to the group of small molecules known as nucleotides. It did not, however, appear to be any of the known nucleotides. He wrote to Leon A. Heppel of the National Institutes of Health, who had developed many of the methods used in preparing and identifying a number of nucleotides, asking him for a quantity of an enzyme that breaks nucleotides into their component parts and thus facilitates their identification. This request set the stage for a remarkable coincidence.

It is Heppel's habit to let letters that do not require an immediate answer accumulate on his desk. He covers each few days' correspondence with a fresh sheet of wrapping paper, and every few months he clears his desk. Heppel immediately sent Sutherland the enzyme

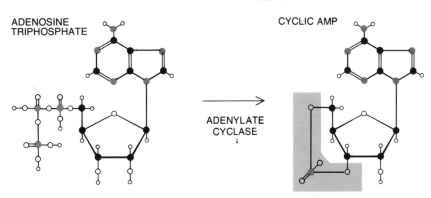

FORMATION OF CYCLIC AMP takes place when an enzyme in the cell membrane, adenylate cyclase, responds to the arrival of a hormone at the membrane. The enzyme transforms molecules of adenosine triphosphate, or ATP (left), within the cell into cyclic AMP (right).

but left the letter of request on his desk. By chance the stratum just below contained a chatty letter from a friend and former colleague, David Lipkin of Washington University, describing an experiment where ATP was treated with a solution of barium hydroxide. The result was the formation of an unusual nucleotide.

Heppel remembers coming into his office one Saturday to clear his desk. He found Sutherland's and Lipkin's letters in adjacent strata and consequently reread them together. It seemed likely to him that both men had isolated the same substance, and he proceeded to put them in touch. Lipkin's chemical synthesis readily produced the nucleotide in large quantities. This made it easy to establish that the synthetic substance was structurally identical with natural cyclic AMP. It also provided an abundant supply of synthetic cyclic AMP for experimental purposes.

Taking advantage of the demonstrated ability of cyclic AMP to increase phosphorylase activity in cell-free preparations, Sutherland and his co-workers were able to measure the amount of cyclic AMP present in a wide variety of cells. They found that it was 1,000 times less abundant than ATP, being present in cell water in a ratio of about one part per million in contrast to ATP's one part per 1,000. Although scanty in amount, cyclic AMP was present in virtually every organism they examined, from bacteria and brine-shrimp eggs to man. Among mammals it was present in almost every type of body cell. Sutherland's group went on to examine a number of tissues that are characterized by their secretion of various substances following stimulation by hormones. He discovered that the level of cyclic AMP in such cells rose soon after exposure to the hormone. Moreover, when the tissues

were exposed to nothing but cyclic AMP, or to derivatives of cyclic AMP that enter cells rapidly, they produced secretions just as readily.

The cyclic AMP molecule is formed from ATP by the action of a special enzyme: adenylate cyclase. The enzyme is located in the membrane of the cell wall. Normally its activity is low and the transformation of ATP into cyclic AMP takes place at a slow rate. Let us consider what happens, however, when a hormone enters the bloodstream. The hormone acts as a "first messenger." It travels to the target cell and then binds to specific receptor sites on the outside of the cell wall. Thyroid cells have receptors that "recognize" thyrotrophin, adrenal cells have receptors that recognize adrenocorticotrophin, and so forth. The binding of the hormone to the receptor site increases the activity of the adenylate cyclase in the cell membrane; just how this occurs has not yet been established. In any event, cyclic AMP is produced as a result, utilizing the abundant supply of ATP on the inner side of the cell membrane. The cyclic AMP is then free to diffuse throughout the cell, where it acts as a "second messenger," instructing the cell to respond in a characteristic way. For example, a thyroid cell responds to this second message by secreting more thyroxine, whereas an adrenal cell responds by producing and secreting steroid hormones. In the cells of the liver the instruction results in the conversion of glycogen into glucose.

Because cyclic AMP is such a powerful regulator of cell functions the cell must be able to control its level of concentration. In most cells control is accomplished by regulating the rate of synthesis of cyclic AMP and by the actions of one or more enzymes, known as phosphodiesterases, that degrade cyclic

AMP into an inert form of adenosine monophosphate. The deactivation results from a splitting of the ester bond that joins the phosphate to the 3′ carbon of the ribose ring. The quantity of the degradative enzymes in the cell is not kept constant; apparently more can be made whenever the level of cyclic AMP in the cell is elevated for more than a few minutes. The level of cyclic AMP is also controlled by diffusion through the cell wall; this is the mechanism operating in addition to enzyme degrada-

tion in bacteria and in the cells of some animal tissues.

Krebs, whose earlier work with Fischer had established the presence of active and inactive forms of phosphorylase in muscle tissue, was able in 1968 to specify the role played by cyclic AMP in activating the enzyme. Working with Donal Walsh, he found that the cyclic AMP binds to yet another enzyme, protein kinase, which is inactive until cyclic AMP is present. The activated kinase then performs the same function for a

related enzyme, phosphorylase kinase. It is this second enzyme that at last activates the phosphorylase. The result of the final activation is the breakdown of glycogen, the storage form of glucose, in a series of steps similar to those that proceed in liver cells.

Whenever glycogen is being degraded in order to satisfy the organism's need for glucose, it would be a waste of energy to continue the synthesis of additional glycogen. This waste is avoided. A specific enzyme mediates the syn-

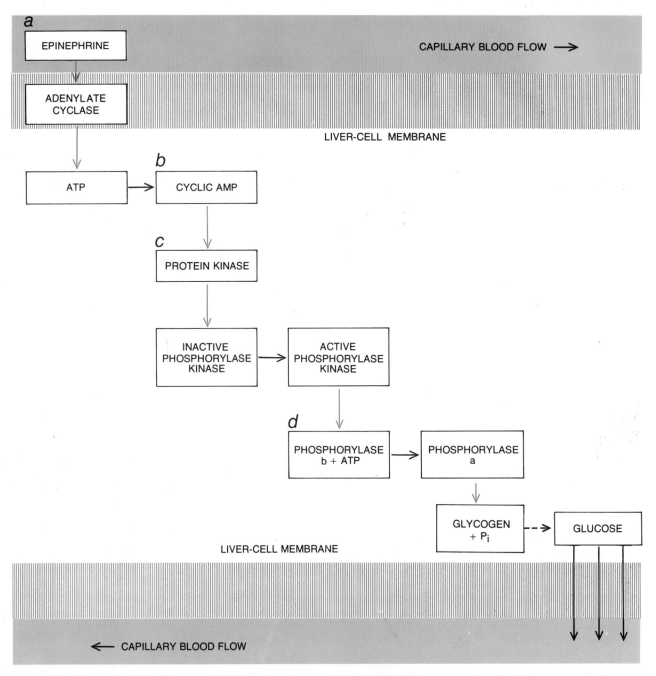

RELEASE OF BLOOD SUGAR by a glycogen storage cell in the liver is mediated by cyclic AMP. In the schematic diagram arrows in color symbolize actions and black arrows the results. "First messenger," epinephrine, arrives at the cell membrane and activates the enzyme, adenylate cyclase (a), causing it to convert some of the ATP present in the cytoplasm into cyclic AMP (b), the second messenger. The cyclic AMP then activates a protein kinase (c), which activates a second kinase. The second kinase (d) triggers a four-step sequence (not shown) that converts the glycogen into the assimilable sugar glucose, which then passes into the bloodstream.

thesis of glycogen. It is glycogen syn- thetase, and like phosphorylase and many other enzymes it has an active and an inactive form. At the time when some molecules of cyclic AMP are initiating the chain of events that leads to the breakdown of glycogen, other cyclic AMP molecules are at work converting glycogen synthetase from the active to the inactive form.

Just as cells that store glycogen have the task of supplying the fasting orga- nism with glucose, so the task of fat cells is to satisfy the organism's need for fatty acids. Fatty acids are present in the fat cell in the storage form triglyceride. The triglyceride can ocupy as much as 90 percent of the cell's volume. Here again it is cyclic AMP that initiates the break- down of the stored fat. In response to any of several hormonal stimuli the level of cyclic AMP in the fat cell begins to rise. As with muscle tissue, this activates a protein kinase. The kinase in turn ac- tivates a second enzyme, triglyceride li- pase. On being converted to the active form the second enzyme begins to de- grade the stored triglyceride into the re-

quired fatty acids. It should be noted here that protein kinases are present in the cells of many tissues other than fat and muscle; numerous other actions of cyclic AMP presumably also involve these ATP-powered enzymes.

In addition to such effects within cells cyclic AMP has been observed to stimulate the expression of genetic in- formation. How this stimulation is ac- complished in animal cells remains ob- scure. Almost all the detailed observa- tions of the regulation of gene activity involve a single microorganism: the com- mon intestinal bacterium *Escherichia coli.* There are many good reasons for molecular biologists who are engaged in genetic studies choosing to work with this simple organism. One of them is that a typical animal cell contains enough DNA to account for 10 million individual genes; *E. coli* has only enough DNA for about 10,000 genes.

Sutherland and another colleague, Richard Makman, established the pres- ence of cyclic AMP in *E. coli* cells in 1965. By then the substance was already

known to control a variety of cell proc- esses, and Robert Perlman and I at the National Institutes of Health guessed that it must also play an important role in the bacterium. But what role?

There were two clues. First, cultures of *E. coli* that are nourished exclusively with glucose show low levels of cyclic AMP. Second, such cultures can syn- thesize only very small amounts of a number of enzymes, including those needed to metabolize sugars other than glucose; this inhibition is called the "glu- cose effect." Putting these clues togeth- er, we reasoned that cyclic AMP was a chemical switch, so to speak; if it was present in *E. coli* in adequate quantities, it would activate the expression of those genes that are necessary for the synthesis of the missing enzymes. In order to test this speculation we added cyclic AMP to cultures of inhibited *E. coli.* The cells were then able to metabolize such sugars as maltose, lactose and arabinose in ad- dition to a variety of other nutrients.

Now, factors that operate at the level of the gene do so by stimulating the syn- thesis of the messenger RNA that in

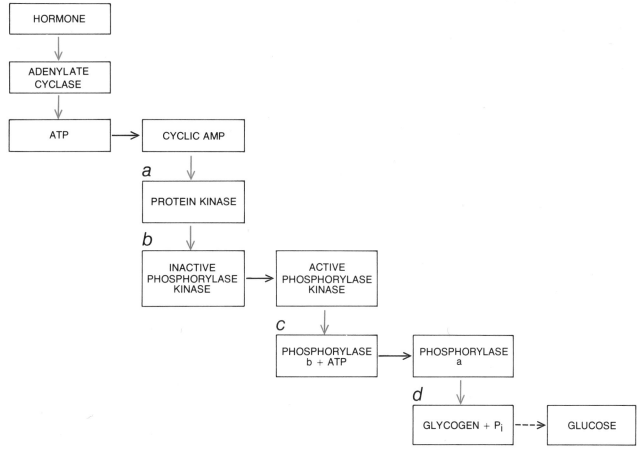

INITIAL SEQUENCE that triggers the transformation of glycogen into glucose in muscle tissue also involves cyclic AMP. First (a) the cyclic AMP that has been formed from ATP in the cell cytoplasm activates a protein kinase. The kinase in turn activates a second kinase (b) that is capable of transforming phosphorylase b into phosphorylase a. When this takes place (c), the transformed phos- phorylase then starts the sequence (d) that converts glycogen into glucose. Edwin Krebs and Edmond Fischer of the University of Washington discovered the phosphorylase alteration in muscle at the same time that Sutherland found the same process in the liver.

TISSUE		HORMONE	PRINCIPAL RESPONSE
FROG SKIN		MELANOCYTE-STIMULATING HORMONE	DARKENING
BONE		PARATHYROID HORMONE	CALCIUM RESORPTION
MUSCLE	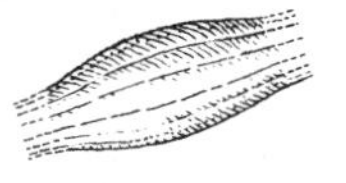	EPINEPHRINE	GLYCOGENOLYSIS
FAT	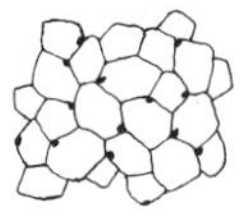	EPINEPHRINE	LIPOLYSIS
		ADRENOCORTICO-TROPHIC HORMONE	LIPOLYSIS
		GLUCAGON	LIPOLYSIS
BRAIN	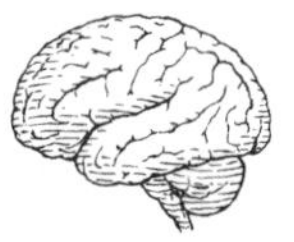	NOREPINEPHRINE	DISCHARGE OF PURKINJE CELLS
THYROID	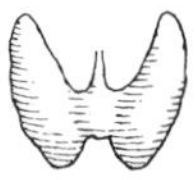	THYROID-STIMULATING HORMONE	THYROXIN SECRETION
HEART	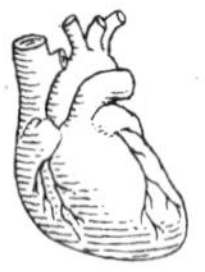	EPINEPHRINE	INCREASED CONTRACTILITY
LIVER	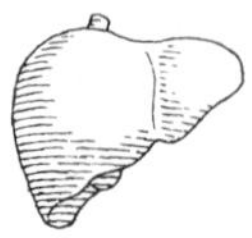	EPINEPHRINE	GLYCOGENOLYSIS
KIDNEY	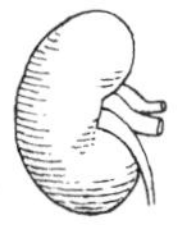	PARATHYROID HORMONE	PHOSPHATE EXCRETION
		VASOPRESSIN	WATER REABSORPTION
ADRENAL	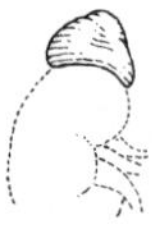	ADRENOCORTICO-TROPHIC HORMONE	HYDROCORTISONE SECRETION
OVARY	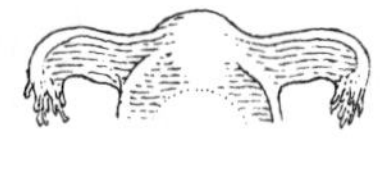	LUTEINIZING HORMONE	PROGESTERONE SECRETION

FOURTEEN EXAMPLES of hormonal activities that affect many different target tissues (*left*) have one factor in common: each causes an increase in the level of cyclic AMP in the tissue. It seems probable that all the responses are set in train by the sudden increase in level.

effect reproduces the information contained in the DNA of the gene. The messenger RNA then provides the ribosomes of the cell with the information they need to construct the appropriate enzyme. In our laboratory by this time we had isolated a mutant form of *E. coli* that lacked the cell-membrane enzyme required for the synthesis of cyclic AMP. Curiously, the mutant cells were viable; apparently the presence of cyclic AMP is not absolutely crucial to survival. For our next experiment we selected cultures of the mutant strain that contained a known amount of the two messenger RNA's needed for the metabolism of two sugars: lactose and galactose. When we added cyclic AMP to the mutant cultures, we found that the quantity of the two messenger RNA's was increased but that the expression of most of the other genes in the cells was not affected. In order to learn exactly how the increase took place we now needed to study the reaction in a cell-free preparation.

At Columbia University, Geoffrey L. Zubay and his co-workers were working with a complex cell-free preparation made from *E. coli*. When they added to the preparation DNA that was greatly enriched in the genes for lactose metabolism, small amounts of one of the enzymes of lactose metabolism, beta-galactosidase, were formed. The DNA was enriched because it was derived from a bacterial virus that had acquired the lactose genes from the *E. coli* host it had once lived in and now contained it permanently. The virus is a hybrid of two other viruses, designated λ and 80, and the DNA derived from it is λ80*lac* (for lactose) DNA.

When Zubay and his co-worker Donald Chambers added cyclic AMP to the cell-free preparation, synthesis of the enzyme for lactose metabolism was greatly stimulated. Soon thereafter my colleagues B. de Crombrugghe and H. Varmus showed that the synthesis of *lac* messenger RNA was also increased. John Parks in our laboratory, following Zubay's procedure, developed a cell-free *E. coli* preparation that responded to the addition of λ*gal* (for galactose) DNA by producing one of the enzymes of galactose metabolism. Like Zubay's preparation, Parks's synthesized much more of the enzyme when cyclic AMP was added.

One might now have expected that the same result could be achieved without even using the complex cell-free preparation. It would be necessary only to mix together in the test tube appropriate quantities of DNA rich in lactose genes and of the special enzyme that copies DNA into RNA, add some ATP and other nucleoside triphosphates as building blocks and cyclic AMP as mediator and the result would be the production of *lac* messenger RNA. The expectation proved to be false. Some RNA was formed in the test tube but none of it corresponded to the *lac* gene. Clearly some factor was missing. Fortunately there was a clue to its nature.

In our search for mutant strains of *E. coli* that could not make various enzymes known to be controlled by cyclic AMP we had found some mutants that produced cyclic AMP in abundance but were still unable to metabolize either lactose or several other sugars. Similar *E. coli* mutants had been isolated at the Harvard Medical School by Jonathan Beckwith and his colleagues. What was lacking in the mutants was a protein that has the ability to bind cyclic AMP. The protein, which is known either as catabolite gene activation protein (abbreviated CAP) or as cyclic AMP receptor protein (CRP), was difficult to purify but a pure form was finally prepared by my colleague Wayne B. Anderson.

De Crombrugghe added some of the pure protein to the test-tube mixture that had failed to yield *lac* messenger RNA. This time the effort was successful. Moreover, when CRP was added to a similar test-tube mixture that included λ*gal* DNA, *gal* messenger RNA was formed. These experiments, incidentally, were the first to achieve the transcription of a bacterial gene in a system containing only purified components. We could now proceed to examine in detail how the gene activity was controlled.

The initial step in the process involved the combination of cyclic AMP with CRP. The complex of the two substances was then bound to the DNA; once this took place the enzyme that copies DNA into RNA was enabled to bind to a specific site, called the "promoter," at the beginning of the *lac* gene in the bacterial DNA. After that the nucleoside triphosphates initiated the *lac* transcription process. We soon found that the transcription of the genes for galactose metabolism followed the same steps.

Of the 10,000 or so *E. coli* genes, perhaps some hundreds are regulated in this way. How the others are controlled remains a mystery. The regulation of even those genes that are controlled by cyclic AMP involves other substances as well. The *lac* transcription process provides an example. A specific protein, *lac* repressor, is bound to the DNA at a site (called the operator) at the beginning of

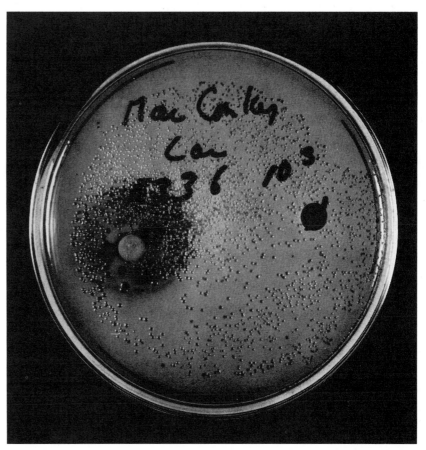

MUTANT STRAIN of the bacterium *Escherichia coli* lacks the enzyme that turns ATP into cyclic AMP. As the different reactions show, addition of cyclic AMP to one disk in the dish (*left*) enabled the colonies of bacteria nearby to metabolize certain sugars. The bacteria near the second disk (*right*), to which only 5' AMP was added, were not able to do this.

the *lac* gene. The operator site is near the site on the DNA where the complex of cyclic AMP and CRP binds. The cell cannot now produce *lac* messenger RNA even in the presence of CRP and cyclic AMP because the repressor prevents the copying enzyme from beginning the transcription [*see illustration on following page*]. Only when the repressor is removed by the action of a substance closely related to lactose can the transcription take place.

This pattern of events, where cyclic AMP is able to stimulate the transcription of many genes while at the same time individual repressors can prevent certain transcriptions, gives the *E. coli* cell the flexibility it needs for the efficient utilization of the foodstuffs in its environment. *E. coli* does not store large amounts of carbohydrate and must live on what is in its immediate vicinity. If the cell is exposed, say, both to glucose (which it can already metabolize) and to lactose (which it cannot), there is nothing to be gained by expending energy to produce the enzymes for lactose metabolism. As long as its supply of cyclic

AMP remains at a low level the *E. coli* cell conserves its capacity for protein synthesis.

That so many different cell functions should be under the control of a single substance is remarkable. Cyclic AMP also participates in the process of visual excitation and regulates the aggregation of certain social amoebae so that they can form complex reproductive structures [*see* "Hormones in Social Amoebae and Mammals," by John Tyler Bonner; SCIENTIFIC AMERICAN Offprint 1145]. A good example of the substance's versatility is provided by the contrast between the mechanisms of glycogen and lipid degradation on the one hand and the stimulation of gene transcription on the other. The degradation reactions involve enzymes and depend on protein phosphorylation. In *E. coli* gene transcription no phosphorylation is involved. It will be interesting to learn whether the *E. coli* mechanism is employed to control gene activity elsewhere in nature and whether cyclic AMP acts in still other ways. Most of the substance's actions in animal cells remain to be explored.

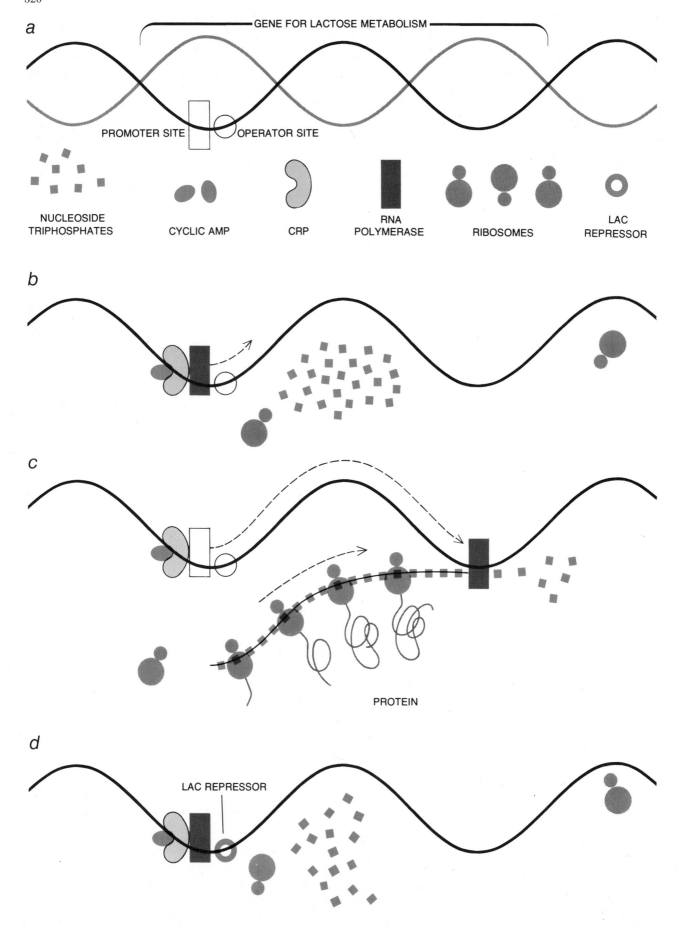

a

GENE FOR LACTOSE METABOLISM

PROMOTER SITE   OPERATOR SITE

NUCLEOSIDE
TRIPHOSPHATES   CYCLIC AMP   CRP   RNA
POLYMERASE   RIBOSOMES   LAC
REPRESSOR

b

c

PROTEIN

d

LAC REPRESSOR

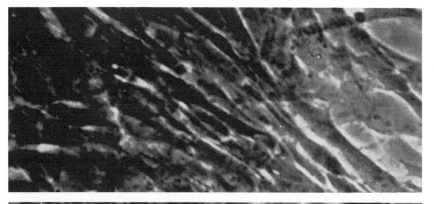

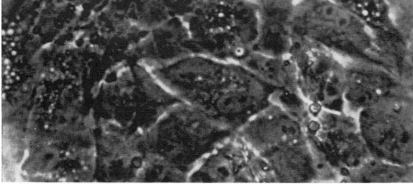

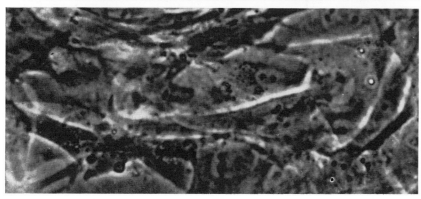

CHICK EMBRYO CELLS infected with a temperature-sensitive mutant strain of Rous sarcoma virus maintain a normal appearance (*top micrograph*) when they are cultured at a temperature of 40.5 degrees Celsius. When the temperature is reduced to 36 degrees, however, they quickly develop an abnormal appearance (*middle*). If cyclic AMP is added to the cells, their appearance remains normal (*bottom*) even after the temperature is reduced.

TRANSCRIPTION OF A GENE, the sequence of events shown schematically on the opposite page, occurred in a cell-free medium that was supplied (*a*) with RNA polymerase (*solid bar*), nucleoside triphosphates (*colored squares*), the protein CRP (*crescent shape*), which reversibly binds cyclic AMP (*colored ovals*), and quantities of DNA enriched with the gene for lactose metabolism (*bracketed area of helix*). Transcription begins when a combined unit of CRP and cyclic AMP activates a promoter site at the beginning of the *lac* operon (*b*); RNA polymerase now binds to the promoter site, ready to link the nucleoside triphosphates together in the correct sequence. As transcription proceeds (*c*), the RNA polymerase arranges the nucleoside triphosphates in the correct sequence. Ribosomes meanwhile begin the process of assembling the proteins that comprise the enzymes for lactose metabolism. If still another protein, *lac* repressor (*colored doughnut*), attaches itself to the operator site at the beginning of the *lac* operon (*d*), then the RNA polymerase can bind but cannot transcribe in spite of the presence of the CRP-cyclic AMP unit. This pattern, whereby cyclic AMP stimulates the transcription of many different operons, whereas the repressors are specific for a particular gene, makes for flexibility in the utilization of foodstuffs.

Abnormalities in the metabolism of cyclic AMP may explain the nature of certain diseases. For example, in cholera the bacteria responsible for the disease produce a toxin that stimulates the intestinal cells to secrete huge amounts of salt and water; it is the resulting dehydration, if it is unchecked, that is fatal [see "Cholera," by Norbert Hirschhorn and William B. Greenough III; SCIENTIFIC AMERICAN, August, 1971]. It appears that the toxin first stimulates the intestinal cells to accumulate excess cyclic AMP, whereupon the cyclic AMP instructs the cells to secrete the salty fluid.

In our own laboratory we are interested in understanding the difference between normal cells and cancer cells. Since cancer cells typically grow in an uncontrolled manner and lose their ability to carry out specialized functions, it seemed possible to us that some of their abnormal properties might be attributable to their inability to accumulate normal amounts of cyclic AMP. My colleague George Johnson and I have begun to investigate this possibility.

The cells that are commonly used for such studies are fibroblasts: cells that contribute to the formation of connective tissue. They are usually taken from the embryos of chickens, mice or other animals and allowed to grow in a nutrient fluid. After a short period of culture in a medium approximating blood serum in composition, the embryonic cells take on the appearance of the normal fibroblast cells in connective tissue and grow in bottles or dishes in a controlled manner. When the cultured cells are exposed to cancer-producing viruses or to chemical carcinogens, however, they begin to grow in an abnormal manner. They take on the appearance of tumor cells, and when they are injected into a suitable host, they usually produce tumors. The process of changing a normal cell to a cancer cell in culture is termed transformation. Among the properties of transformed cells are a change in appearance, accelerated growth, looser adherence to the container surface, alterations in the rate of production of specialized large molecules such as mucopolysaccharides and clumping on exposure to certain agglutinative plant proteins.

Now, cells that have been transformed and are then grown in the presence of cyclic AMP tend to return to normal. They grow more slowly, adhere more tenaciously to the container, synthesize certain large molecules at a faster rate and clump less when they are exposed to agglutinative plant protein. Their appearance also frequently returns toward normal. As far as we now know, the morphologic reversal occurs only in embryo cells and those derived from connective tissue; tumors from a few other kinds of cell do not show the same morphologic response.

In an event the reversal suggested to us that transformed cells might contain abnormally low levels of cyclic AMP. One of our co-workers, Jack Otten, investigated this possibility. He found that cells from chick embryos that had been transformed by exposure to Rous sarcoma virus contained much less cyclic AMP than normal chick-embryo cells. We could not be certain, however, which came first: the abnormal appearance of the transformed cells or the low level of cyclic AMP.

To settle the question we needed a way to transform cells very rapidly. We had at our disposal a mutant variety of the Rous sarcoma virus that made rapid cell transformation possible. For example, chick cells infected with the mutant virus (which had been isolated by our colleague John Bader of the National Cancer Institute) remain normal in appearance as long as the culture is kept at a temperature of 40.5 degrees Celsius. When the temperature is reduced to 36 degrees, however, the cells are rapidly transformed.

We grew cultures infected with this temperature-sensitive virus and kept them at the "normal" temperature. We incubated some of the cultures with a potent derivative of cyclic AMP and some without it. When we lowered the temperature of the cells incubated with cyclic AMP to the transformation level, they continued to look normal for some time [see illustration on preceding page]. The cultures without cyclic AMP, once the temperature was lowered, developed the characteristic transformed appearance within a few hours.

Otten measured the level of natural cyclic AMP in the readily transformed cells after their exposure to the lower temperature. He found that as soon as 20 minutes later the level had fallen greatly; this was well before the cells began to develop a transformed appearance.

This finding obviously leaves many questions unanswered. Is the phenomenon confined to tumors of connective tissue? How many of the abnormal properties of the transformed cells result from the low level of cyclic AMP? Which enzymes are responsible for lowering the level? Are there other tumor-forming viruses and chemical carcinogens that similarly lower the cell's supply of cyclic AMP? Not least, can the findings to date be exploited in a therapeutically useful manner? The search for answers continues in each of these areas.

DNA Transactions

32

THE RECOGNITION OF DNA IN BACTERIA

SALVADOR E. LURIA
January 1970

*Some bacteria have enzyme systems that scan invading
DNA molecules injected by viruses and break them
unless they are chemically marked at specific
recognition sites*

The genetic code closely resembles a universal language. As far as anyone knows every word in this language (that is, every triplet of nucleotides) means the same thing for all forms of life: it specifies a particular amino acid. By the same token, organization of a sequence of the words into a sentence, written in the form of a molecule of deoxyribonucleic acid (DNA), carries a similarly unvarying meaning: it specifies the construction of a particular protein. Hence the genetic script, like the script of a book printed in a universal language, is read in the same way by all organisms.

There are situations, however, where the organism examines the DNA script not as a linguist would but as a bibliophile would. A bibliophile may find in the structure of a book's script little signs or marks that identify the printer. It turns out that DNA often has just such identifying marks, and that these decisively affect the behavior of the organisms that recognize them. This unexpected finding has begun a fascinating chapter in the study of molecular biology and evolution.

The story begins with a curious discovery Mary L. Human and I made nearly 20 years ago in experiments with the bacteriophage, or bacterial virus, known as T2 [see "The T2 Mystery," by Salvador E. Luria; SCIENTIFIC AMERICAN Offprint 24]. Ordinarily T2 readily invades and multiplies in the bacterium *Escherichia coli*, but we found that when the virus was grown in a certain *E. coli* strain (called *B/4*), almost all the daughter phages that came out were altered in such a way that they could no longer multiply in the usual *E. coli* hosts of T2. It developed in further experiments that the altered T2 could multiply perfectly well in dysentery bacilli, and this breeding had the effect of transforming the

phage back to full ability to multiply in *E. coli* bacteria.

Since the change T2 had undergone was completely reversible, it was obviously not a genetic mutation; an alteration of a gene or genes would be expected to persist in the progeny that inherited it. What, then, could account for the modification of T2's character? The phenomenon we had observed was not a freak; this was soon shown in other experiments with phages conducted by Giuseppe Bertani in my laboratory at the University of Illinois and independently by Jean J. Weigle at the California Institute of Technology. They found, for example, that when a certain phage called lambda (λ) was grown in the cells of an *E. coli* strain called *C*, only one in 10,000 of the daughter λ phages coming out of *C* could reproduce in a different *E. coli* strain called *K*. The descendants of the few phage particles that did manage to grow in *K* were then able to multiply in *K* cells as well as in *C* cells. A single cycle of growth in *C* cells, however, would restore these phages to the original form: capable of growing in *C* cells but not in *K*. Similarly, it was found that certain other strains of *E. coli* bacteria could restrict the growth of specific phages or modify their form [see illustrations on pages 326 and 327]. A given phage could be modified to a series of different forms by shifting it from one host strain to another; each time the daughter phages became adapted to the new host, and the descendants could be returned to the original form by shifting them back to the original host.

The enigma of the changeable phages did not begin to clear up until 1961, when a Swiss investigator, Werner Arber of the University of Geneva, discovered a clue to the reason for the different reception of phages by different strains of

a bacterium. Arber and his colleague Daisy Roulland-Dussoix found that when phage lambda injects its genetic material into an "alien" bacterium (one that will restrict the phage's multiplication), the cell breaks up most of the phage's DNA molecules into small fragments. This happens, for example, when phages grown in *E. coli* of strain *K* invade bacterial cells of strain *B*. The broken DNA of course cannot reproduce new phage. A few of the DNA molecules injected into the *B* bacterium manage, however, to remain intact, and they multiply, producing about 100 daughter phages. These now have the ability to multiply in *B* cells. It turns out that two of them (that is, about 2 percent of modified phages) can also still multiply in *K* bacteria as well. Arber performed a pretty experiment that revealed why this is so [see top illustration on page 328].

He used tracer isotopes—heavy nitrogen (^{15}N) and heavy hydrogen (^{2}H, or deuterium)—to label the DNA of phages bred in *K* cells. This was done by growing the bacteria and phages in a medium containing the heavy isotopes. The "heavy" phages were then employed to infect *B* bacteria growing in a medium of ordinary (light) nitrogen and hydrogen. When the new crop of daughter phages emerged from the lysed bacteria, Arber separated the phages by weight by means of a centrifuge that layered the particles (in a cesium salt solution) according to their relative buoyancy, based on their density. Most of the daughter phages proved to be "light," indicating that they were composed of material newly synthesized in the infected bacteria. A few particles, however, contained DNA that was half-heavy; that is, their weight indicated that half of the DNA was made up of new material and the other half of material from the heavy parent phages that had infected the bac-

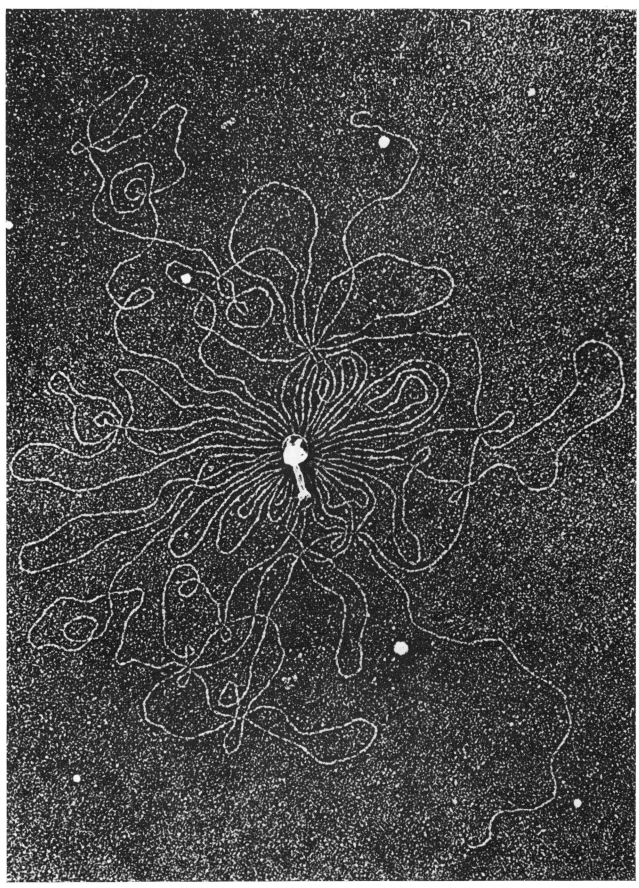

MOLECULE OF BACTERIOPHAGE DNA is shown here as a long, tangled thread after being released from the head of the T2 phage particle. The magnification is 100,000 diameters. The T2 bacteriophage is one of several "T even" phages that infect cells of *Escherichia coli*. When the phage DNA molecule is modified in a certain way, it can be "recognized" as an invader by the cell and destroyed. This electron micrograph was made by Albrecht K. Kleinschmidt of the New York University School of Medicine.

PROBABILITY OF SUCCESSFUL INFECTION IN

	E. COLI B/4	E. COLI B/4	S. DYSENTERIAE SH
T2•B	1	1	1
T2•SH	1	1	1
T2•B/4 = T*2	10^{-5}	10^{-5}	1
T2•B/4•SH	1	1	1

PROBABILITY OF SUCCESSFUL INFECTION IN

	E. COLI K	E. COLI B	E. COLI C
λ•K	1	10^{-4}	1
λ•B	4×10^{-4}	1	1
λ•C	4×10^{-4}	10^{-4}	1

MODIFICATION OF PHAGES can be demonstrated by growing phages in one strain of bacteria and observing how successfully the progeny grow when another strain serves as the host. In these tables the original host strain of T2 or lambda (λ) phages is indicated, in the first column, by a suffix such as *B* or *B/4*, which represents *E. coli* strains *B* or *B/4* respectively. The other columns indicate the probability of successful growth when the phage is obliged to grow again in the same strain or in another one. Fifteen years ago the author and Mary L. Human discovered that when phage T4 was grown in *E. coli B/4*, yielding T2 · *B/4* (originally called T*2), the phage was so altered that it would scarcely multiply in its usual hosts. It would grow freely, however, in dysentery bacilli (*Shigella dysenteriae Sh*). Daughter phages from this cycle, designated T2 · *B/4* · *Sh*, were completely normal.

teria. Since the phage DNA is a typical double-strand molecule, it could be deduced that in these phages one strand of their DNA was light and the other strand was heavy—contributed intact from the parent phage. Arber found that the phage particles containing this hybrid DNA retained the ability to multiply in *K* cells as well as in *B* cells. There was also a small fraction of fully heavy daughter phages (both DNA strands being heavy), and these too proved able to grow in both *K* and *B*.

The result of Arber's experiment made clear that the lambda phage's DNA strands carried some kind of marking that identified specific phages for the bacteria. In a phage generated in *K* cells the DNA somehow acquired a *K* marking, and a phage grown in *B* cells had its DNA marked *B*. When phage DNA that lacked the right marking was injected into a bacterium, the DNA was almost sure to be broken down. The experimental results also said something more: they said that the specific marking on the DNA molecule was attached by a stable, covalent chemical bond, since the mark was retained during the chemical events attending the construction of the daugh-

ter phage. Arber confirmed the stability of the bond by using highly purified phage DNA (instead of the intact phage) to infect bacteria; the DNA retained its specific marking even after going through the chemical purification treatment.

Arber's findings raised a number of interesting questions. What was the nature of the DNA markings? How did a bacterium recognize the absence of the right markings in phage DNA? What mechanism in the bacterium was responsible for destroying DNA with the "wrong" markings? A number of laboratories have looked into these questions and have worked out much of the answer. Actually two sets of answers have come out, one for phages of the lambda type (with *K*, *B* or *C* markings) and one for the T2 and other "T even" phages. Let us look first at the lambda markings.

It was known even before Arber's discoveries that every kind of DNA can occasionally have a seemingly spurious methyl group (CH_3) attached to some of the nucleotides. For example, the methyl group can be tacked on at a certain position in adenine; the base is then called

methylaminopurine, or MAP, instead of simply A. Similarly, the base cytosine (C) can be methylated, so that it becomes methylcytosine (MC). In each case the methyl group is added (with the help of a special enzyme) only after the DNA molecule has been built [*see bottom illustration, page 328*]. And it plays no observable part in the molecule's functioning: MAP behaves just like A, and MC just like C. In short, the methyl group does not alter the genetic spelling.

Could one now suppose the addition does change the style of the script? Is the methyl group perhaps a kind of serif tacked onto the letters? Arber tried the experiment of cutting down the amount of methylation of the phage DNA. He did this by growing *K* bacteria in a medium in which they were deprived of the amino acid methionine, the precursor of the substance that donates methyl groups to DNA. It turned out that phages grown in these bacteria generally did not acquire the *K* marking. Apparently they did not have the serifs (methyl groups) that would identify them as *K*.

Arber found further that the methyl groups serve as markers not just anywhere on the DNA molecule but only at certain strategic spots. This was clearly shown in experiments with a very tiny phage known as fd. Attached to each DNA molecule of this phage are a few MAP groups. Arber and his colleague Urs Kühnlein found that when the fd phage was grown in *K* bacteria, it had two fewer MAP groups than when it was grown in *B* bacteria. The *K*-marked phages failed to grow in *B* cells; these cells broke down the phage DNA.

This suggests that the fd phage's DNA has two sensitive sites, located on adenine bases. If the two sites are methylated (that is, marked MAP), the phage can grow equally well in both *K* and *B* bacteria. If, however, these sites are unmethylated, the *B* bacterium recognizes the phage as being the "wrong" kind and breaks down its DNA. Presumably the recognition is effected by a special enzyme that can split the DNA molecule. The *B* bacterium also has a methylating enzyme that can convert the two critical A bases to MAP (that is, attach a serif) and thus make the phage DNA acceptable so that it multiplies in *B* cells [*see top illustration on page 329*].

This interpretation has now been confirmed by various experiments. Some of the specific enzymes that break the vulnerable DNA sites and some of those that can methylate them have been isolated and identified. For example, Arber's group in Geneva and Matthew S. Meselson and Robert T.-Y. Yuan at

Harvard University have identified the enzymes that break the DNA molecules marked *K* and those that break the DNA molecules marked *B*. As expected from Arber's vulnerable-site hypothesis, it turns out that in each case the enzyme makes just a few breaks in the DNA molecule. Similarly, Arber and his colleagues have isolated a marking enzyme: from bacteria of the *B* strain they extracted the enzyme that can transfer methyl groups from a suitable methyl-donor substance (S-adenosyl methionine) to the *K*-marked (unmethylated) DNA of fd phages.

Thus we see that the bacteria possess a well-defined system for marking and recognizing phages. Shifting from our printing metaphor, we might say that each strain of *E. coli* bacteria stamps its own trademark on the DNA of the phages it produces, just as a factory brands its commercial product.

Work in several laboratories has elicited further details of the bacteria's branding and recognition system. It has been learned that the system in *E. coli* involves three closely linked genes. One directs the synthesis of the DNA-break-ing enzyme, another controls the synthesis of the methylating enzyme, and the third gene generates a mechanism that is responsible for the recognition of the critical DNA sites that are to be either broken or methylated. It seems likely that this mechanism, or component, is a protein chain that associates itself with both the breaking and the methylating enzymes, and that can recognize the specific sequence of nucleotides that represents the critical DNA sites [*see bottom illustration, page 329*].

The marking and restricting system is widespread among the *E. coli* and related strains of bacteria; it is possessed not only by the *K* and *B* strains but also by others (except for the *C* strain and certain others that have lost the recognition system, perhaps through mutation). Also Arber and his co-workers found that DNA incompatibility is not confined to the case of invasion by a phage. It also applies to exchanges of DNA among the bacteria themselves. When a female *E. coli* cell mates with a male cell carrying the wrong DNA brand (for example, if the female is *B* and the male *K*, or vice versa), the female on receiving the male's DNA will break it down, just as if it belonged to a phage.

The answer to the T2 mystery turned out to be a different story. The investigation of this problem, carried out in our laboratory at the Massachusetts Institute of Technology by my colleagues Toshio Fukasawa, Costa P. Georgopoulos, Stanley Hattman and Helen R. Revel, demonstrated that the DNA in the T2 phage is not branded in the same way as that in the lambda and fd phages.

The DNA of T2 (and of its even-numbered relatives T4 and T6) carries an oddity: the base cytosine always has a hydroxymethyl group (CH_2OH) attached to it, so that the nucleotide contains hydroxymethylcytosine (HMC) instead of cytosine as its base. The discovery of the HMC base by Gerard R. Wyatt of Yale University and Seymour S. Cohen of the University of Pennsylvania played a major role in phage biochemistry. In most of the HMC nucleotides of these phages one or more sugar molecules (glucose) normally are attached to the hydroxymethyl group [*see illustration on page 330*]. What our laboratory team discovered was that when an *E. coli* bacterium of the *B/4* strain

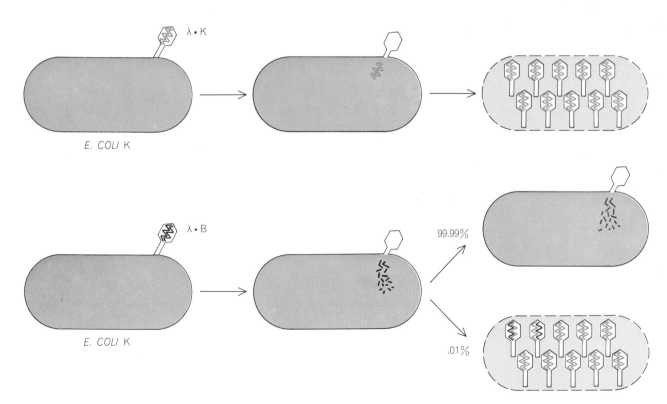

GROWTH RESTRICTION is demonstrated by cells of *E. coli* strain *K*. No restriction occurs when *K* cells are infected by phage λ · *K* (*top*). The deoxyribonucleic acid (DNA) of the phage (*colored zigzag shape*) enters the cell and exploits the cell's chemical machinery to produce about 100 new phage particles, which are released when the cell lyses, or dissolves. Restriction occurs (*bottom*) when *K* cells are infected by λ · *B* particles: lambda phages previously grown on *E. coli* strain *B*. The DNA of the phage (*black zigzag*) enters the cell but is broken down. In about one cell in 10,000, however, the phage DNA manages to multiply and give rise to phage progeny. About 2 percent of the progeny can grow in both *B* and *K* cells because their DNA is a hybrid molecule consisting of one strand of DNA (*black*) from the original λ · *B* and one strand (*color*) newly made inside the *K* cell. The remaining progeny are now modified so that they will grow normally in *K* cells but with a probability of only 10^{-4} in *B* cells.

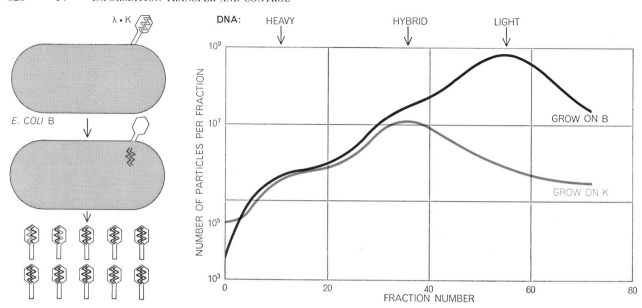

LABELING OF DNA demonstrates that phage particles possessing the ability to grow on both *B* and *K* strains of *E. coli* are predominantly those with hybrid DNA molecules. The experiment was conducted at the University of Geneva by Werner Arber and Daisy Roulland-Dussoix. They infected cells of strain *B* made up of atoms of ordinary weight with phage $\lambda \cdot K$ that had been grown on *E. coli K* cells incorporating heavy hydrogen (deuterium) and heavy nitrogen (^{15}N). Thus the DNA molecules in particles of $\lambda \cdot K$ were labeled with heavy atoms. The progeny produced in the light *B* cells were spun in a centrifuge containing a gradient of cesium chloride. The phage particles become distributed in the centrifuge tube according to their weight: heaviest particles at the bottom, lightest at the top. The phage particles in different fractions are then tested for their ability to grow on *B* and *K* cells, with the result plotted at the right. The particles containing hybrid DNA, about 2 percent of the total, grow in both kinds of cell. A small fraction of phage inherits two heavy strands of DNA from its $\lambda \cdot K$ parent and can also grow in both *B* and *K*. For this experiment Arber used a special strain of *E. coli B* that can impress the *B* modification on the phage DNA but lacks the *B* restriction function.

modifies, or brands, a T2 phage, the way it does so is to fail to attach the glucose to the HMC at the sensitive sites in the DNA molecule. In short, it changes the DNA from "sweet" to "sour"!

The soured phage can no longer multiply in *E. coli* cells; the cells detect the absence of glucose at certain critical sites and break down the DNA. Desugared T2 can, however, grow in the cells of dysentery bacilli, and these cells mark the phage DNA by attaching glucose at the necessary sites, thus restoring the phage's ability to grow in *E. coli*.

The special strains of *E. coli* that transform the phage from sweet to sour lack a substance (uridine diphosphoglucose, or UDPG) that is needed to donate glucose to the phage's DNA. This is one way for a T2 phage to acquire the sour branding. The brand can also be put on

"MARKED" BASES found in phage DNA that has been modified are 5-methylcytosine (MC) and 6-methylaminopurine (MAP). They are formed from the standard bases cytosine and adenine by the addition of a methyl group (CH_3), supplied by S-adenosyl methionine. Adenine (A) and cytosine (C) supply two of the four "letters" that spell out the genetic message in DNA molecules.

it, however, by a genetic mutation in the phage itself, that is, by an alteration of the phage genes that normally are responsible for production of the enzymes that attach glucose to the HMC [*see illustration on page 331*]. The *E. coli* bacteria, for their part, possess special genes that endow them with the ability to recognize and destroy sour DNA. These genes produce two enzymes, apparently different, that recognize and attack specific unsweetened sites in the DNA molecule. Experiments have shown that mutation of these two genes eliminates the rejection mechanism, so that unsweetened phages can grow in the mutated *E. coli* cells.

The two enzymes responsible for distinguishing between sweetened and unsweetened DNA have not yet been isolated. All we know about them so far (from genetic analysis) is that they are not the same as the enzymes that detect the difference between the methylated and unmethylated brands on the DNA of lambda and fd phages and on the DNA of different strains of *E. coli* bacteria.

How does the enzyme, in each case, recognize the brand? By what mechanism does the enzyme (protein) molecule find and identify the significant chemical markings on the DNA molecule? This is an intriguing question that applies to many other phenomena in molecular biology. For example, the enzymes that bring about the replication of DNA and those that build RNA molecules using DNA as a template must find the right place on the DNA at which to start the construction of the new molecule. How do they recognize these starting points?

We must suppose that the identifying marker on the DNA molecule in each case is a certain sequence of nucleotides, and that the enzyme establishes recognition by temporary attachment to this sequence—by fit and "feel," so to speak. There are good reasons to believe the identifying sequences are generally rather short; the length of a globular protein would not span many nucleotides in their linear array in the DNA molecule. Arber has suggested that the identifying sequences in branded DNA are probably no more than six to eight nucleotides long. It may soon be possible to work out the exact sequence of bases that constitutes the recognition site for at least one of the restrictive enzymes. This would be a big step forward in the study of protein-DNA interaction.

What kind of search do the enzymes carry on? Are the sensitive sites recognized and attacked only when enzymes

PHAGE	CHARACTERISTIC	PROBABILITY OF GROWTH IN *E. COLI* B
fd • K	2 MAP PER DNA MOLECULE	.001
fd • B	4 MAP PER DNA MOLECULE	1.0
fd m1 • K	3 MAP PER DNA MOLECULE	.03
fd m2 • K	2 MAP PER DNA MOLECULE	1.0

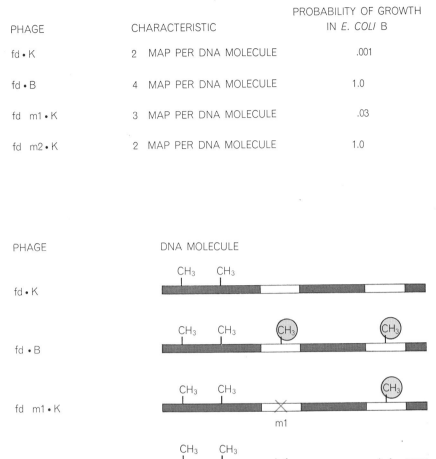

PHAGE	DNA MOLECULE
fd • K	
fd • B	
fd m1 • K	
fd m1 • m2 • K	

VULNERABLE SITES in the DNA molecule of the phage fd are represented by the white regions in the lower part of the illustration. These sites, or sequences of bases, are broken with a probability of 97 percent by the "restricting" enzyme of *E. coli* B unless an A (adenine) base in that sequence has been converted to MAP by the modifying (that is, methylating) enzyme present in B cells. There are also two irrelevant CH_3 groups elsewhere in the fd DNA molecule. The table (*top*) shows for various strains of fd the probability of growth in *E. coli* B. Strains of fd · K with certain mutations, m1 and m2 (*indicated by* X *marks*), are insensitive to the restricting enzyme in *E. coli* B even if they are not methylated.

BACTERIAL GENE	FUNCTION	MUTANTS
r	RESTRICTION MECHANISM ("NUCLEASE" COMPONENT)	r⁻: RESTRICTING FUNCTION LOST
m	MODIFICATION MECHANISM ("METHYLASE" COMPONENT)	m⁻: MODIFYING FUNCTION LOST
s	SPECIFICITY (SITE-RECOGNITION COMPONENT)	s⁻: BOTH RESTRICTING AND MODIFYING FUNCTIONS LOST

GENETIC CONTROL OF RECOGNITION SYSTEM seems to involve three closely linked genes in *E. coli*. Gene *r* makes the restricting enzyme that breaks the phage DNA molecule unless it carries the proper MAP marks. Gene *m* makes the methylating enzymes that provide the marks. Gene *s* makes a component responsible for specific recognition of sites where the marks may or may not be present. Presumably this recognition component is part of both enzymes: the one that can break DNA and the one capable of methylating it.

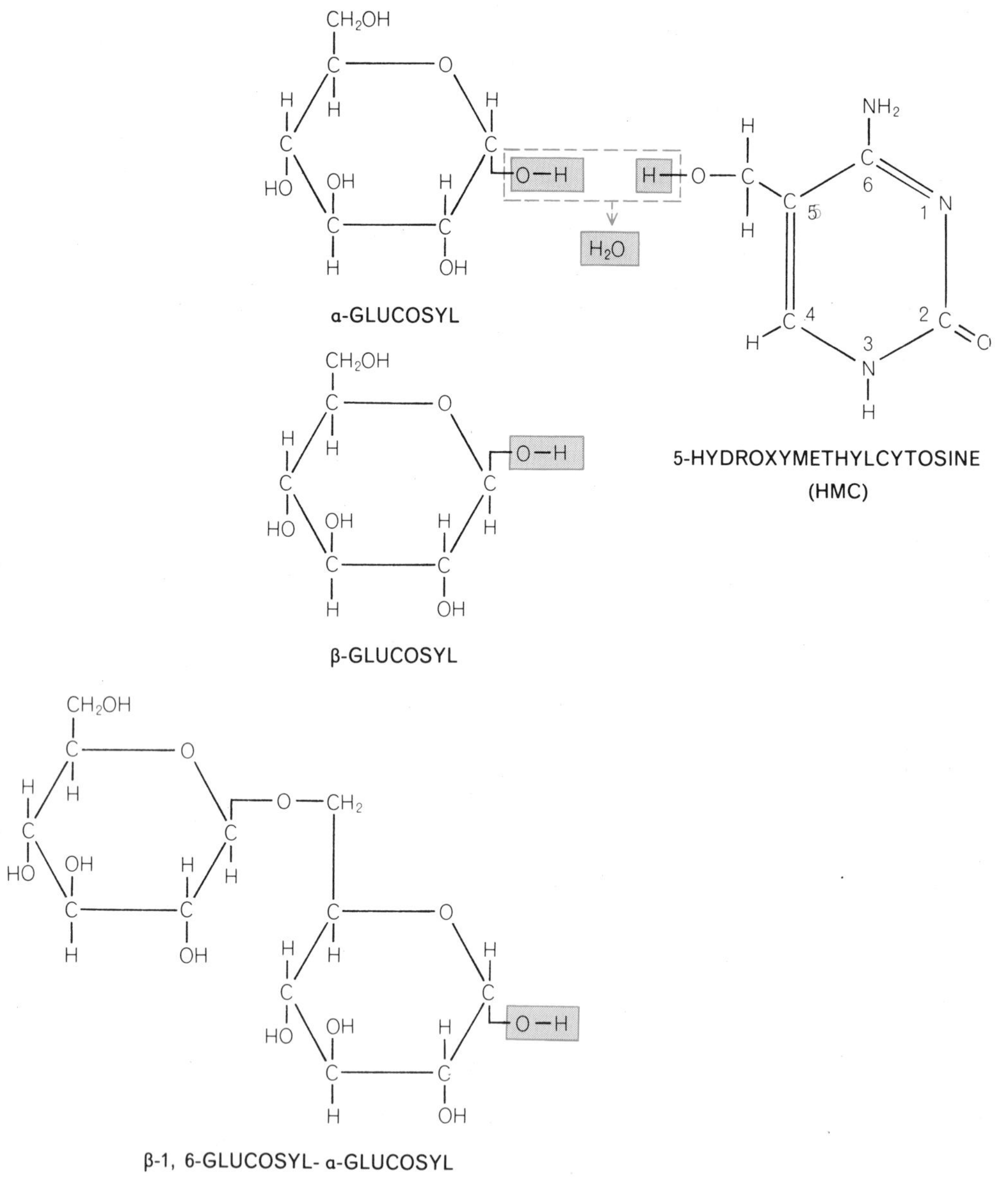

α-GLUCOSYL

β-GLUCOSYL

5-HYDROXYMETHYLCYTOSINE
(HMC)

β-1, 6-GLUCOSYL- α-GLUCOSYL

"SWEET" AND "SOUR" LABELS provide another recognition system that enables *E. coli* cells to discriminate among invading molecules of phage DNA. In this system, which applies to the phages T2, T4 and T6, the base cytosine (C) is replaced by 5-hydroxymethylcytosine (HMC), a base formed when a hydroxymethyl group (CH₂OH) replaces hydrogen on the No. 5 carbon atom of cytosine. The normal phage has 70 percent or more of its HMC bases linked to one of the three sugar structures shown at the left. Such DNA molecules can be regarded as sweet. When the sugar units are missing, the DNA is sour and is readily broken down by *E. coli* cells.

happen to fall on them directly? Or do the enzymes browse along the phage DNA molecules, possibly as the latter nose into the bacterium, until they find the telltale markers? A finding by Georgopoulos and Revel in our laboratory suggests that the second alternative may be correct. They found that even a trace of "sweetness" (that is, glucosylation of comparatively few of the HMC bases, such as occurs in certain mutant T2 phages) was sufficient to protect the DNA of these phages from an enzyme that attacked only one particular site on the molecule and then only when it was unsweetened. It seems reasonable to as-

sume that the small amount of glucose in these phages attaches itself to HMC groups at random, and therefore that the HMC groups at the particular site this enzyme attacks may often be unsweetened. We can guess the reason this site usually escapes attack is that the enzyme explores the DNA molecule and stops its exploration or falls off as soon as it encounters any glucose.

Another puzzling question is posed by the finding that even if only one (either one) of the two strands of a phage's DNA carries the protective marking, an enzyme will not break the molecule. This indicates that in the case of a vul-

nerable molecule the enzyme scans both strands separately before breaking the molecule. How does it manage to scan the strands individually?

The entire set of phenomena I have described in this article raises a more general question. What is the evolutionary significance of the DNA-branding system? What function does it serve for the bacteria? One obvious suggestion is that the system gives a bacterium a defense against certain phages. Evolutionary development could, however, have provided the bacteria with more effective defense mechanisms, such as eliminating the surface sites on the bacterial cell to which phages attach themselves. Furthermore, the bacteria that can recognize and destroy branded DNA have no defense against phages that are not branded. It is hard to believe the branding system evolved simply to protect bacteria of one strain against phages coming from another strain. A more plausible speculation is that the system serves primarily to prevent "undesirable" mixing of the bacterial genes (DNA) between the strains of bacteria themselves. Presumably in nature the *E. coli* bacteria do mate or otherwise exchange genetic material. If this is so, the ability of a *B*-strain bacterium to reject *K*-marked DNA (and vice versa) is an effective device for keeping each strain "pure." Thus the branding-and-rejection system facilitates the evolution of bacterial strains in diverging directions, in the same way that isolation mechanisms in cross-fertilization play a role in the evolution of plant and animal species.

The case of sweet and sour DNA seems to tell a clearer and more dramatic story. The T2 phages have the ability to break down the DNA of the bacteria they attack, which contains cytosine instead of HMC. Hence the hydroxymethylation of cytosine in the phages' own DNA may have been an important step in the evolution of their ability to multiply in the bacteria. We can surmise, then, that in reply some strains of *E. coli* by natural selection evolved the ability to break down phage DNA marked with

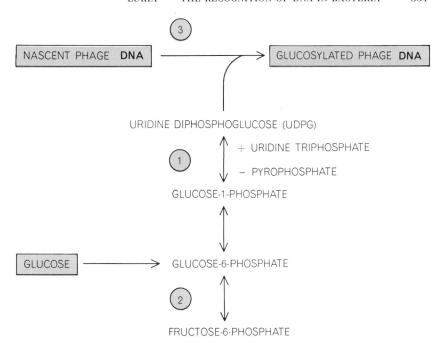

SYNTHESIS OF SWEET DNA requires a series of biosynthetic steps culminating in the addition of glucose units to the nascent DNA of the T-even phages. Mutants of *E. coli* can have blockages in the enzyme systems that carry out either or both of the first two numbered steps. Mutants with a blockage in enzyme No. 1 produce T-even phage with unglucosylated, or sour, DNA (T* phage). Mutants with a block in enzyme No. 2 can make the glucose donor substance, uridine diphosphoglucose, only if glucose is present in the growth medium. The last step, No. 3, is carried out by enzymes not normally present in *E. coli* cells; they are made from instructions coded in the DNA molecules of T-even phages.

HMC. The phage in turn evolved the capacity to tack glucose onto HMC to protect the vulnerable sites of its DNA. Here, then, the evidence of the branded DNA may be unfolding for us a scene in the dynamic acting out of evolution at the most elementary level: the chemistry of the genetic material itself.

33

THE REPAIR OF DNA

PHILIP C. HANAWALT AND ROBERT H. HAYNES
February 1967

*The two-strand molecule that incorporates the genetic
information of the living cell is subject to damage.
Experiments with bacteria reveal that the cell has a
remarkable ability to repair such damage*

One of the most impressive achievements of modern industry is its ability to mass-produce units that are virtually identical. This ability is based not solely on the inherent precision of the production facilities. It also involves intensive application of quality-control procedures for the correction of manufacturing errors, since even the best assembly lines can introduce faulty parts at an unacceptable rate. In addition industry provides replacement parts for the repair of a product that is subsequently damaged by exposure to the hazards of its natural environment. Recent studies have demonstrated that living organisms employ analogous processes for repairing defective parts in their genetic material: deoxyribonucleic acid (DNA). This giant molecule must be replicated with extraordinary fidelity if the organism is to survive and make successful copies of itself. Thus the existence of quality-control mechanisms in living cells may account in large part for the fact that "like produces like" over many generations.

Until recently it had been thought that if the DNA in a living cell were damaged or altered, for example by ionizing radiation, the cell might give rise either to mutant "daughter" cells or to no daughter cells at all. Now it appears that many cells are equipped to deal with some of the most serious hazards the environment can present. In this article we shall describe the experimental results that have given rise to this important new concept.

The instructions for the production of new cells are encoded in the sequences of molecular subunits called bases that are strung together along a backbone of phosphate and sugar groups to form the chainlike molecules of DNA. A sequence of a few thousand bases constitutes a single gene, and each DNA molecule comprises several thousand genes. Before a cell can divide and give rise to two daughter cells, the DNA molecule (or molecules) in the parent cell must be duplicated so that each daughter cell can be supplied with a complete set of genes. On the basis of experiments made with the "chemostat"—a device for maintaining a constant number of bacteria in a steady state of growth—Aaron Novick and the late Leo Szilard estimated that bacterial genes may be duplicated as many as 100 million times before there is a 50 percent chance that even one gene will be altered. This is a remarkable record for any process, and it seems unlikely that it could be achieved without the help of an error-correcting mechanism.

The ability of cells to repair defects in their DNA may well have been a significant factor in biological evolution. On the one hand, repair would be advantageous in enabling a species to maintain its genetic stability in an environment that caused mutations at a high rate. On the other hand, without mutations there would be no evolution, mutations being the changes that allow variation among the individuals of a population. The individuals whose characteristics are best adapted to their environment will leave more offspring than those that are less well adapted. Presumably even the efficiency of genetic repair mechanisms may be subject to selection by evolution. If the repair mechanism were too efficient, it might reduce the natural mutation frequency to such a low level that a population could become trapped in an evolutionary dead end.

Although the error-correcting mechanism cannot yet be described in detail, one can see in the molecular architecture of DNA certain features that should facilitate both recognition of damage and repair of damage. The genetic material of all cells consists of two complementary strands of DNA linked side by side by hydrogen bonds to form a double helix [*see upper illustration, page 334*]. Normally DNA contains four chemically distinct bases: two purines (adenine and guanine) and two pyrimidines (thymine and cytosine). The two strands of DNA are complementary because adenine in one strand is always hydrogen-bonded to thymine in the other, and guanine is similarly paired with cytosine [*see lower illustration on page 334*]. Thus the sequence of bases that constitute the code letters of the cell's genetic message is supplied in redundant form. Redundancy is a familiar stratagem to designers of error-detecting and error-correcting codes. If a portion of one strand of the DNA helix were damaged, the information in that portion could be retrieved from the complementary strand. That is, the cell could use the undamaged strand of DNA as a template for the reconstruction of a damaged segment in the complementary strand. Recent experimental evidence indicates that this is precisely what happens in many species of bacteria, particularly those that are known to be highly resistant to radiation.

The ability to recover from injury is a characteristic feature of living organisms. There is a fundamental difficulty, however, in detecting repair processes in bacteria. For example, when a population of bacteria is exposed to a dose of ultraviolet radiation or X rays, there is no way to determine in advance what proportion of the population will die. How can one tell whether the observed mortality accurately reflects all the damage sustained by the irradiated cells or whether some of the damaged cells have repaired themselves? Fortunately it is possible to turn the repair mechanism on or off at will.

A striking example can be found in the process called photoreactivation [*see bottom illustration on page 336*]. Although hints of its existence can be traced back to 1904, photoreactivation was not adequately appreciated until Albert Kelner rediscovered the effect in 1948 at the Carnegie Institution of Washington's Department of Genetics in Cold Spring Harbor, N.Y. Kelner was puzzled to find that the number of soil organisms (actinomycetes) that survived large doses of ultraviolet radiation could be increased by a factor of several hundred thousand if the irradiated bacteria were subsequently exposed to an intense source of visible light. He concluded that ultraviolet radiation had its principal effect on the nucleic acid of the cell, but he had no inkling what the effect was. In an article published before the genetic significance of DNA was generally appreciated, Kelner wrote: "Perhaps the real stumbling block [to understanding photoreactivation] is that we do not yet understand at all well the biological role of that omnipresent and important substance—nucleic acid" [see "Revival by Light," by Albert Kelner; SCIENTIFIC AMERICAN, May, 1951].

It is now known that the germicidal action of ultraviolet radiation arises chiefly from the formation of two unwanted chemical bonds between pyrimidine bases that are adjacent to each other on one strand of the DNA molecule. Two molecules bonded in this way are called dimers; of the three possible types of pyrimidine dimer in DNA, the thymine dimer is the one that forms most readily [*see upper illustration on page 335*]. It is therefore not surprising that a given dose of ultraviolet radiation will create more dimers in DNA molecules that contain a high proportion of thymine bases than in DNA molecules with fewer such bases. Consequently bacteria whose DNA is rich in thymine tend to be more sensitive to ultraviolet radiation than those whose DNA is not.

Richard B. Setlow, his wife Jane K. Setlow and their co-workers at the Oak Ridge National Laboratory have shown that pyrimidine dimers block normal replication of DNA and that bacteria with even a few such defects are unable to divide and form colonies [see "Ultraviolet Radiation and Nucleic Acid," by R. A. Deering; SCIENTIFIC AMERICAN Offprint 143]. In the normal replication of DNA each parental DNA strand serves as a template for the synthesis of a complementary daughter strand. This mode of replication is termed semiconservative because the parental strands separate in the course of DNA synthesis; each daughter cell receives a "hybrid" DNA molecule that consists of one parental strand and one newly synthesized complementary strand. The effect of a pyrimidine dimer on DNA replication may be analogous to the effect on a zipper of fusing two adjacent teeth.

Claud S. Rupert and his associates at

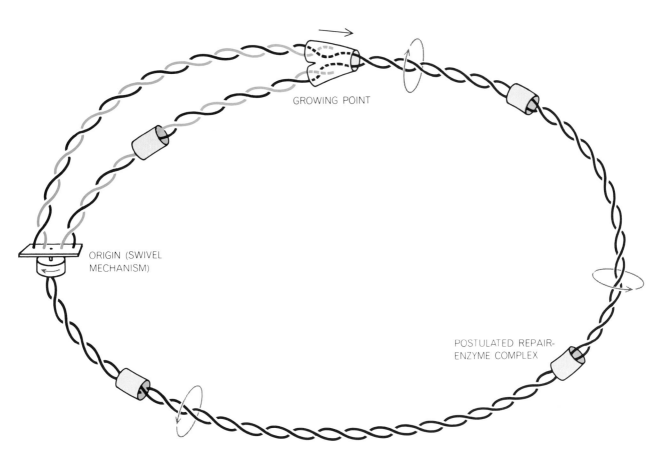

GROWING POINT

ORIGIN (SWIVEL MECHANISM)

POSTULATED REPAIR-ENZYME COMPLEX

REPLICATION OF BACTERIAL CHROMOSOME, a ring-shaped molecule of deoxyribonucleic acid (DNA), has now been shown to take two forms: normal replication and repair replication. In the former process the two strands that constitute the double helix of DNA are unwound and a daughter strand (*color*) is synthesized against each of them. In this way the genetic "message" is transmitted from generation to generation. The pairing of complementary subunits that underlies this process is illustrated on the next page. In repair replication, defects that arise in individual strands of DNA are removed and replaced by good segments. It is hypothesized that "repair complexes," composed of enzymes, are responsible for the quality control of the DNA structure. Although this diagram shows the swivel mechanism for unwinding the parent strands to be at the origin, it may in fact be located at the growing point.

Johns Hopkins University have shown that photoreactivation involves the action of an enzyme that is selectively bound to DNA that has been irradiated with ultraviolet. When this enzyme is activated by visible light (which simply serves as a source of energy), it cleaves the pyrimidine dimers, thereby restoring the two bases to their original form. Photoreactivation is thus a repair process that can be turned on or off merely by flicking a light switch.

Let us now consider another kind of repair mechanism in which light plays no role and that is therefore termed dark reactivation. This type of repair process can be turned off genetically, by finding mutant strains of bacteria that lack the repair capabilities of the original radiation-resistant strain. The "B/r" strain of the bacterium *Escherichia coli*,

first isolated in 1946 by Evelyn Witkin of Columbia University, is an example of a microorganism that is particularly resistant to radiation. The first radiation-sensitive mutants of this strain, known as B_{s-1}, were discovered in 1958 by Ruth Hill, also of Columbia.

Not long after the discovery of the B_{s-1} strain a number of people suggested that its sensitivity to radiation might be due to the malfunction of a particular enzyme system that enabled resistant bacteria such as B/r to repair DNA that had been damaged by radiation. This was a reasonable suggestion in view of the steadily accumulating evidence that DNA is the principal target for many kinds of radiobiological damage. Experiments conducted by Howard I. Adler at Oak Ridge and by Paul Howard-Flanders at Yale University lent further support to this hypothesis. It had been

known for some years that bacteria can exchange genes by direct transfer through a primitive form of sexual mating [see "Viruses and Genes," by François Jacob and Elie L. Wollman; SCIENTIFIC AMERICAN Offprint 89]. Howard-Flanders and his co-workers found that bacteria of a certain radiation-resistant strain of *E. coli* (strain *K*-12) have at least three genes that can be transferred by bacterial mating to radiation-sensitive cells, thereby making them radiation-resistant. Since genes direct the synthesis of all enzymes in the living cell, these experiments supported the hypothesis that B_{s-1} and other radiation-sensitive bacteria lack one or more enzymes needed for the repair of radiation-damaged DNA.

The question now arises: Do the enzymes involved in dark reactivation operate in the same way as the enzyme that

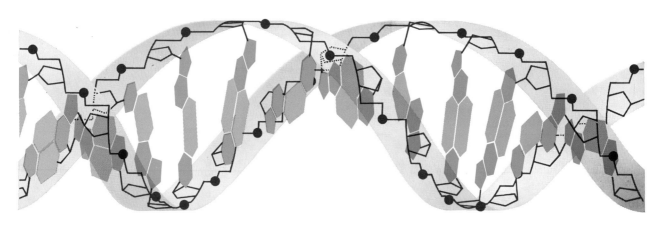

DNA MOLECULE is a double helix that carries the genetic message in redundant form. The backbone of each helix consists of repeating units of deoxyribose sugar (*pentagons*) and phosphate (*black dots*). The backbones are linked by hydrogen bonds between pairs of four kinds of base: adenine, guanine, thymine and cytosine. The bases are the "letters" in which the genetic message is written. Because adenine invariably pairs with thymine and guanine with cytosine, the two strands carry equivalent information.

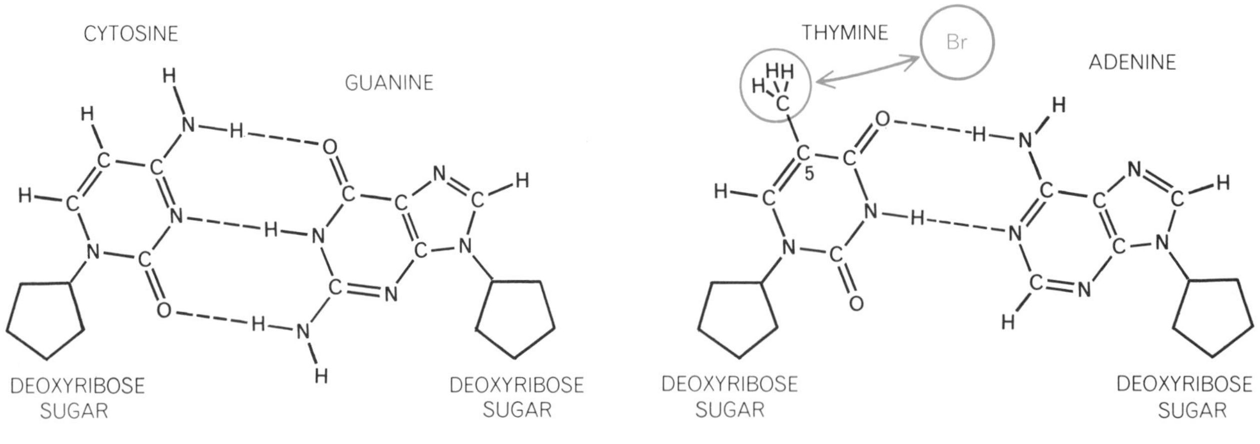

CYTOSINE

GUANINE

THYMINE   Br

ADENINE

DEOXYRIBOSE SUGAR   DEOXYRIBOSE SUGAR   DEOXYRIBOSE SUGAR   DEOXYRIBOSE SUGAR

DNA BASES are held together in pairs by hydrogen bonds. The cytosine-guanine pair (*left*) involve three hydrogen bonds, the thymine-adenine pair (*right*) two bonds. If the CH_3 group in thymine is replaced by an atom of bromine (*Br*), the resulting molecule is called 5-bromouracil. Thymine and 5-bromouracil are so similar that bacteria will incorporate either in synthesizing DNA. Because the bromine compound is so much heavier than thymine its presence can be detected by its effect on the weight of the DNA.

is known to split pyrimidine dimers in the photoreactivation process? Another possibility is that the resistant cells might somehow bypass the dimers during replication of DNA and leave them permanently present, although harmless, in their descendant molecules.

The actual mechanism is even more elegant than either of these possibilities; it exploits the redundancy inherent in the genetic message. The radiation-resistant strains of bacteria possess several enzymes that operate sequentially in removing the dimers and replacing the defective bases with the proper complements of the bases in the adjacent "good" strand. We shall recount the two key observations that substantiate this postulated repair scheme.

The excision of dimers was first demonstrated by Richard Setlow and William L. Carrier at Oak Ridge and was soon confirmed by Richard P. Boyce and Howard-Flanders at Yale. In their studies cultures of ultraviolet-resistant and ultraviolet-sensitive bacteria were grown separately in the presence of radioactive thymine, which was thereupon incorporated into the newly synthesized DNA. The cells were then exposed to ultraviolet radiation. After about 30 minutes they were broken open so that the fate of the labeled thymine could be traced. In the ultraviolet-sensitive strains all the thymine that had been incorporated into DNA was associated with the intact DNA molecules. Therefore any thymine dimers formed by ultraviolet radiation remained within the DNA. In the ultraviolet-resistant strains, however, dimers originally formed in the DNA were found to be associated with small molecular fragments consisting of no more than three bases each. (Thymine dimers can easily be distinguished from the individual bases or combinations of bases by paper chromatography, the technique by which substances are separated by their characteristic rate of travel along a piece of paper that has been wetted with a solvent.) These experiments provided strong evidence that dark repair of ultraviolet-damaged DNA does not involve the splitting of dimers in place, as it does in photoreactivation, but does depend on their actual removal from the DNA molecule.

Direct evidence for the repair step was not long in coming. At Stanford University one of us (Hanawalt), together with a graduate student, David Pettijohn, had been studying the replication of DNA after ultraviolet irradiation of a radiation-resistant strain of E. coli. In

EFFECT OF ULTRAVIOLET RADIATION on DNA is to fuse adjacent pyrimidine units: thymine or cytosine. The commonest linkage involves two units of thymine, which are coupled by the opening of double bonds. The resulting structure is known as a dimer.

EFFECT OF NITROGEN MUSTARD, a compound related to mustard gas, is to cross-link units of guanine within the DNA molecule. Unless repaired, the structural defects caused by nitrogen mustard and ultraviolet radiation can prevent the normal replication of DNA.

these experiments we used as a tracer a chemical analogue of thymine called 5-bromouracil. This compound is so similar to the natural base thymine that a bacterium cannot easily tell the two apart [see right half of lower illustration on opposite page]. When 5-bromouracil is substituted for thymine in the growth medium of certain strains of bacteria that are unable to synthesize thymine, it is incorporated into the newly replicated DNA. The fate of 5-bromouracil can be traced because the bromine atom in it is more than five times heavier than the methyl (CH_3) group in normal thymine that it replaces. Therefore DNA fragments containing 5-bromouracil are denser than normal fragments containing thymine. The density difference can be detected by density-gradient centrifugation, a technique introduced in 1957 by Matthew S. Meselson, Franklin W. Stahl and Jerome Vinograd at

the California Institute of Technology.

In density-gradient centrifugation DNA fragments are suspended in a solution of the heavy salt cesium chloride and are spun in a high-speed centrifuge for several days. When equilibrium is reached, the density of the solution varies from 1.5 grams per milliliter at the top of the tube to two grams per milliliter at the bottom. If normal DNA from E. coli is also present, it will eventually concentrate in a band corresponding to a density of 1.71 grams per milliliter. A DNA containing 5-bromouracil instead of thymine has a density of 1.8 and so will form a band closer to the bottom of the tube.

The entire genetic message of E. coli is contained in a single two-strand molecule of DNA whose length is nearly 1,000 times as long as the cell itself. This long molecule must be coiled up like a

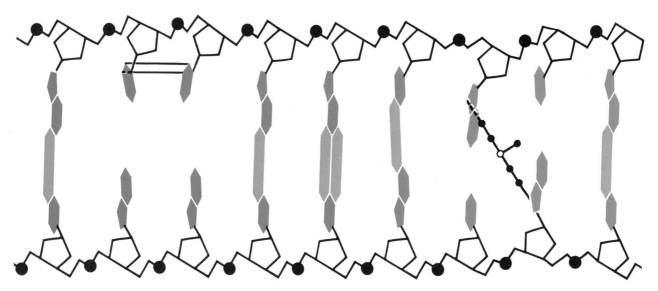

DEFECTS IN DNA probably distort the symmetry of its helical structure. To make the distortions more apparent this diagram shows the DNA in flattened form. A thymine dimer appears in the left half of the structure; a guanine-guanine cross-link is shown in the right half. The authors believe the repair complex recognizes the distortions rather than the actual defects in the bases.

skein of yarn to be accommodated within the cell [see the article "The Bacterial Chromosome," by John Cairns, beginning on page 151]. Such a molecule is extremely sensitive to fluid shearing forces; in the course of being extracted from the cell it is usually broken into several hundred pieces.

If 5-bromouracil is added to a culture of growing bacteria for a few minutes (a small fraction of one generation), the DNA fragments isolated from the cells fall into several categories of differing density, each of which forms a distinct band in a cesium chloride density gradient. The lightest band will consist of unlabeled fragments: regions of the DNA molecule that were not replicated during the period when 5-bromouracil was present. A distinctly heavier band will contain fragments from regions that have undergone replication during the labeling period. This is the band containing hybrid DNA: molecules made up of one old strand containing thymine and one new strand containing 5-bromouracil in place of thymine. If synthesis proceeds until the chromosome has completed one cycle of replication and has started on the next cycle, some DNA fragments will have 5-bromouracil in both strands and therefore will form a band still heavier than the band containing the hybrid fragments. Finally, one fragment from each chromosome will include the "growing point" where the new strands are being synthesized on the pattern of the old ones, and thus will consist of a mixture of replicated and unreplicated DNA. This fragment, which is presumably shaped like a Y, will show up in the density gradient at a position

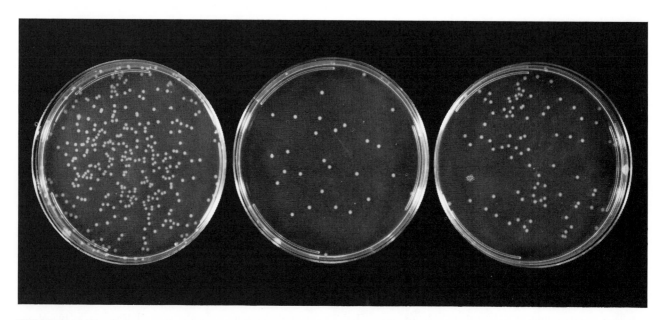

PHOTOREACTIVATION, a type of DNA repair process, is demonstrated in this photograph of three bacterial culture dishes. The dish at the left is a control: it contains 368 colonies of B/r strain of Escherichia coli. The middle dish contains bacteria exposed to ultraviolet radiation; only 35 cells have survived to form colonies. The bacteria in the dish at right were exposed to visible light following ultraviolet irradiation; it contains 93 colonies. Thus exposure to visible light increased the survival rate nearly threefold.

intermediate between the unlabeled DNA and the hybrid fragments containing 5-bromouracil.

When we used this technique to study DNA replication in bacteria exposed to ultraviolet radiation, we observed a pattern quite different from the one expected for normal replication. The DNA fragments containing the 5-bromouracil appeared in the gradient at the same position as normal fragments containing thymine! There could be no doubt of this because in these experiments we used 5-bromouracil labeled with tritium (the radioactive isotope of hydrogen) and thymine labeled with carbon 14 [*see illustration on next page*].

This pattern, which at first seems puzzling, is just the one to be expected if many thymine dimers—created at random throughout the DNA by ultraviolet radiation—had been excised and if 5-bromouracil had been substituted for thymine in the repaired regions. As a result many DNA fragments would contain 5-bromouracil, but no one fragment would contain enough 5-bromouracil to affect its density appreciably.

How can we be sure that the density distribution of 5-bromouracil observed in the foregoing experiment arises from "repair replication" rather than from normal replication? A variety of tests confirmed the repair interpretation. By using enzymes to break down the DNA molecule and separating the bases by paper chromatography we verified that the radioactive label was still in 5-bromouracil and had not been transferred to some other base. Various physical studies showed that the 5-bromouracil had been incorporated into extremely short segments that were distributed randomly throughout both DNA strands. This mode of DNA replication was not observed in the B_{s-1} strain of *E. coli*, the radiation-sensitive mutant that cannot excise pyrimidine dimers and therefore could not be expected to perform repair replication. Moreover, repair replication was not observed in the radiation-resistant bacteria in which visible light had triggered the splitting of pyrimidine dimers by photoreactivation; this indicates that repair replication is not necessary if the dimers are otherwise repaired *in situ*.

Finally it was shown that DNA repaired by dimer excision and strand reconstruction could ultimately replicate in the normal semiconservative fashion. This is rather compelling evidence for the idea that biologically functional DNA results from repair replication and that the process is not some aberrant

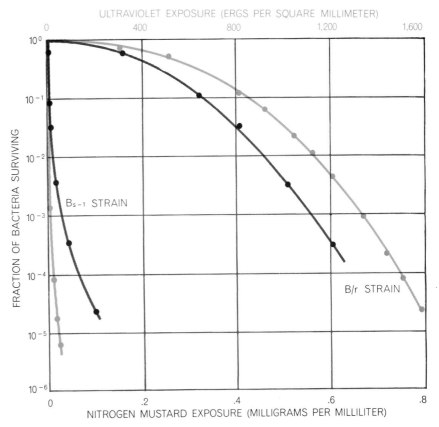

RESISTANCE TO LETHAL AGENTS is demonstrated by certain strains of *E. coli* but not by others. The *B/r* strain, for example, shows a high tolerance to doses of ultraviolet radiation (*colored curves*) and nitrogen mustard (*black curves*) that kill a large percentage of the sensitive B_{s-1} strain. This result suggests that the DNA repair mechanism of the *B/r* strain is effective in removing guanine-guanine cross-links produced by nitrogen mustard as well as in removing thymine dimers formed by exposure to ultraviolet radiation.

form of synthesis with no biological importance.

How can one visualize the detailed sequence of events that must be involved in this type of repair? Two models have been suggested, and the present experimental data seem to be equally compatible with each. The two models are distinguished colloquially by the terms "cut and patch" and "patch and cut" [*see illustrations on page 339*]. The former refers to the model originally proposed by Richard Setlow, Howard-Flanders and others. The latter refers to a model that took form during a discussion at a recent conference on DNA repair mechanisms held in Chicago.

The cut-and-patch scheme postulates an enzyme that excises a short, single-strand segment of the damaged DNA. The resulting gap is enlarged by further enzyme attack and then the missing bases are replaced by repair replication in the genetically correct sequence according to the rules that govern the pairing of bases.

In the patch-and-cut scheme the proc-

ess is assumed to be initiated by a single incision that cuts the strand of DNA near the defective bases. Repair replication begins immediately at this point and is accompanied by a "peeling back" of the defective strand as the new bases are inserted. This patch-and-cut scheme is attractive because it could conceivably be carried out by a single enzyme complex or particle that moves in one direction along the DNA molecule, repairing defects as it goes. Furthermore, it does not involve the introduction of long, vulnerable single-strand regions into the DNA molecule while the repair is taking place. Both models are undoubtedly oversimplifications of the actual molecular events inside the living cell, but they have great intuitive appeal and are helpful in planning further studies of the DNA repair process.

Repair replication would be of interest only to radiation specialists if it were not for the evidence that DNA structural defects other than pyrimidine dimers can be repaired and that similar repair

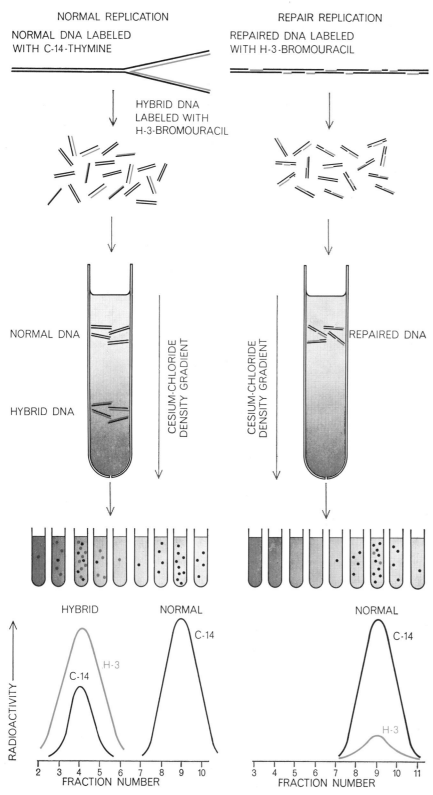

NORMAL REPLICATION

NORMAL DNA LABELED
WITH C-14-THYMINE

HYBRID DNA
LABELED WITH
H-3-BROMOURACIL

REPAIR REPLICATION

REPAIRED DNA LABELED
WITH H-3-BROMOURACIL

NORMAL DNA

HYBRID DNA

CESIUM-CHLORIDE
DENSITY GRADIENT

REPAIRED DNA

CESIUM-CHLORIDE
DENSITY GRADIENT

HYBRID   NORMAL

RADIOACTIVITY

C-14   H-3   C-14

FRACTION NUMBER
2 3 4 5 6 7 8 9 10

NORMAL

C-14   H-3

FRACTION NUMBER
3 4 5 6 7 8 9 10 11

TWO KINDS OF REPLICATION can be demonstrated by growing bacteria first in a culture containing thymine labeled with radioactive carbon 14 and then in a culture containing 5-bromouracil labeled with radioactive hydrogen 3. In normal replication (*left*), also known as semiconservative replication, daughter strands of "hybrid" DNA incorporate the 5-bromouracil (*color*). Because 5-bromouracil is much heavier than the thymine for which it substitutes, fragments of hybrid DNA form a separate heavier layer when they have been centrifuged and have reached equilibrium in a density gradient of cesium chloride. When the radioactivity in the various fractions is analyzed (*bottom left*), carbon 14 appears in two peaks but hydrogen 3 occurs in only one peak. If the experiment is repeated with DNA fragments that have undergone repair replication (*right*), they all appear to be of normal density. This implies that relatively little 5-bromouracil has been incorporated and also that the repaired segments are randomly scattered throughout the DNA molecule.

phenomena occur in organisms other than *E. coli*. We shall review some of the evidence indicating that repair replication is of general biological significance.

Just as strains of *E. coli* vary considerably in their sensitivity to ultraviolet radiation, so they vary considerably in their sensitivity to other mutagenic agents. One such agent is nitrogen mustard, so named because it is chemically related to the mustard gas used in World War I. It was the first chemical agent known to be capable of producing mutations and chromosome breaks in fruit flies and other organisms. Its biological action arises primarily from its ability to react with neighboring guanine bases in DNA, thereby producing guanine-guanine cross-links [*see lower illustration on page 335*].

If one compares the survival curves of different strains of bacteria treated with nitrogen mustard with survival curves for bacteria subjected to ultraviolet radiation, one finds that the curves are almost identical [*see illustration on preceding page*]. This similarity led us to suggest that it is not the altered bases themselves that are "recognized" by the repair enzymes but rather the associated distortions, or kinks, that the alteration of the bases produces in the backbone of the DNA molecule. On this hypothesis one would predict that a wide variety of chemically different structural defects in DNA might be repaired by a common mechanism.

A substantial amount of biochemical evidence has now accumulated in support of this idea. We have established, for example, that repair replication of DNA takes place in *E. coli* that have been treated with nitrogen mustard. Others have found evidence that defects produced by agents as diverse as X rays, the chemical mutagen nitrosoguanidine and the antibiotic mitomycin *C* can all be repaired in radiation-resistant strains of *E. coli*. Walter Doerfler and David Hogness of Stanford have even found evidence that simple mispairing of bases between two strands of DNA can be corrected.

Finally, it now seems that certain steps in repair replication may also be involved in such phenomena as genetic recombination and the reading of the DNA code in preparation for protein synthesis. Evidence for these exciting possibilities has begun to appear in the work of Howard-Flanders, Meselson (who is now at Harvard University), Alvin J. Clark, of the University of California at Berkeley, Crellin Pauling of the University of California at Riverside and other investigators.

Repair replication has also been observed in a number of bacterial species other than *E. coli*. For example, Douglas Smith, a graduate student at Stanford, has demonstrated the repair of DNA in the pleuropneumonia-like organisms, which are probably the smallest living cells. These organisms, which are even smaller than some viruses, are thought to possess only the minimum number of structures needed for self-replication and independent existence [see "The Smallest Living Cells," by Harold J. Morowitz and Mark E. Tourtellotte; SCIENTIFIC AMERICAN Offprint 1005]. This suggests that repair replication may be a fundamental requirement for the evolution of free-living organisms.

In view of the impressive versatility of the repair replication process it is natural to ask if there is any type of DNA damage that cannot be mended by the cell. The evidence so far is limited and indirect, but William Rodger Inch, working at the Lawrence Radiation Laboratory of the University of California, has found that the *B/r* strain of *E. coli* is unable to repair all the damage caused when it is exposed to certain energetic beams of atomic nuclei produced by the heavy-ion linear accelerator (HILAC). Considering the extensive damage that must be done to cells by a beam of such intensely ionizing radiation, the result is not too surprising.

The discovery that cells have the facility to repair defects in DNA is a recent one. It is already apparent, however, that the process of repair replication could have broad significance for biology and medicine. Many questions remain to be investigated: What is the structure of the various repair enzymes? Are they organized into particulate units within the cell? What range of DNA defects can be recognized and repaired? Does DNA repair, as we now understand it, take place in the cells of mammals, or do even more complicated processes underlie the recovery phenomena that are observed after these higher types of cells are exposed to radiation? Might it be possible to increase the radiosensitivity of tumors by inhibiting the DNA repair mechanisms that may operate in cancer cells? If so, the idea could be of great practical value in the treatment of cancer.

These and many related questions are now being investigated in many laboratories around the world. Once again it has been demonstrated that the study of what may appear to be rather obscure properties of the simplest forms of life can yield rich dividends of much intellectual and practical value.

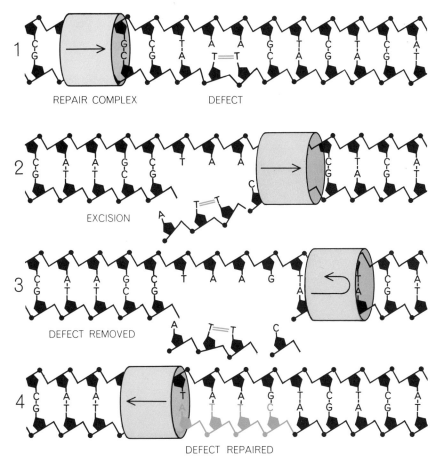

"CUT AND PATCH" repair mechanism was the first one proposed to explain how bacterial cells might remove thymine dimers (*1*) and similar defects from a DNA molecule. The hypothetical repair complex severs the defective strand (*2*) and removes the defective region (*3*). Retracing its path, it inserts new bases according to the rules of base pairing (*4*).

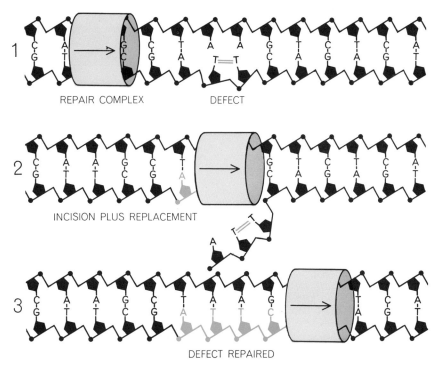

"PATCH AND CUT" mechanism has been proposed as an alternative to the cut-and-patch scheme. On the new model the repair complex inserts new bases as it removes defective ones.

34

CELLULAR FACTORS IN GENETIC TRANSFORMATION

ALEXANDER TOMASZ

January 1969

In transformation certain bacteria change their hereditary makeup by absorbing DNA molecules from their environment. The ability to do this is induced by a protein-like factor synthesized by the cell

The phenomenon of transformation in bacteria was first observed in 1928, led to the identification of DNA as the genetic material in 1944 and has since been recognized as a significant form of genetic intervention: a means whereby bacterial cells can acquire new genes (and thus new traits) with a frequency many orders of magnitude higher than if such changes occurred only through random mutation. Transformation is easy to demonstrate, as is now done routinely in college and high school laboratories, and yet its actual mechanism has not been well understood. The importance of learning more about the mechanism is obvious, since the invasion of cells by extraneous genetic material is not restricted to the world of bacteria. Such events are the essence of all viral infections and may be responsible for the induction of some forms of cancer.

In bacterial transformation a bit of DNA penetrates the boundary of a bacterial cell and becomes incorporated into the cell's genetic apparatus. How does it get in? In particular, why are only certain species of bacteria transformable, and then only under certain conditions and at certain times? Our group at Rockefeller University has been investigating such questions for several years, and we are beginning to learn something about a mechanism by which the cell apparently controls its own transformability.

Transformation was first noted as a bizarre phenomenon peculiar to the pneumonia bacterium *Diplococcus pneumoniae*. In 1928 the British bacteriologist Fred Griffith found that if heat-killed pneumococci of a virulent strain were injected into mice along with live pneumococci of a normally nonvirulent strain, the mice died. Moreover, large numbers of live, virulent bacteria

could be isolated from their bodies. It was not that the heat-killed virulent bacteria had come to life; rather, Griffith showed, some substance from the killed bacteria had "transformed" the normally harmless nonvirulent cells into virulent ones. It was not until 1944 that Oswald T. Avery, Colin M. MacLeod and Maclyn McCarty of the Rockefeller Institute for Medical Research identified that substance as deoxyribonucleic acid, which by that time was known to be associated with chromosomes in the cells of higher organisms.

Avery and his colleagues utilized transformation to develop experimental techniques with which to investigate the role of DNA in heredity [see "Transformed Bacteria," by Rollin D. Hotchkiss and Esther Weiss; SCIENTIFIC AMERICAN Offprint 18]. For example, one can extract the DNA from disrupted cells of a mutant pneumococcus strain that can survive and multiply in the presence of streptomycin and then add the DNA to a culture of a more typical strain that is susceptible to the antibiotic. Sometimes nothing happens. Sometimes, however a piece of the giant DNA molecule makes its way into the cells of the streptomycin-sensitive strain [see *illustration on following page*]. Within five minutes after the contact is established a segment of DNA has become an integral part of the recipient cell's chromosome. If the segment includes the sequence of DNA units that encodes the property of resistance to streptomycin, the recipient cell is soon able to survive and multiply in the presence of streptomycin [see *illustration on page 343*]. Even more important, the blueprint for streptomycin resistance becomes a permanent, heritable property of the cell, which replicates it and eventually passes it on to its progeny.

Through this process new, heritable

properties can be introduced into pneumococcus populations with startling frequency. By treating about a million cells of a streptomycin-sensitive strain for 20 minutes with a few billionths of a gram of DNA from resistant cells, one can render several hundred thousand cells permanently and heritably resistant to the antibiotic; without the added DNA one could not expect to find even a single resistant cell among a million such pneumococci. Clearly transformation is a powerful method of introducing changes into the hereditary makeup of a living bacterial cell. It appears that bacteria may practice this kind of genetic engineering on their own. If one allows cultures of two pneumococcal strains that differ in a number of properties to grow in a common medium, the bacteria engage spontaneously in bilateral exchanges of genes: cells of one strain are transformed by DNA liberated into the medium by the other strain. (Whether this DNA is excreted by living cells or originates only from the hulks of a few dead cells we still do not know.)

Of the many species of bacteria that have been tested, only about half a dozen have been found that can absorb DNA and undergo genetic transformation. What special properties make them susceptible? These properties must somehow account for the ability of DNA to penetrate the tough multilayered bacterial surface, which is not normally permeable to a molecule of even moderate size. Yet DNA segments with molecular weights on the order of a million to 100 million are rapidly absorbed in transformation. Moreover, this DNA is a bare molecule. In the other known forms of genetic transfer in bacteria the transfer is mediated by outside agents. In conjugation, or sexual mating, it is the "male" bacterium that actively mediates

342

DNA MOLECULE entering a pneumonia bacterium (*Diplococcus pneumoniae*) is seen in electron micrographs made by the author and arranged here vertically to present the full length of the DNA. The enlargement is about 70,000 diameters. The bacterium, one of two, is at the bottom right, shadowed with uranium, which visualizes the DNA as a fine line extending to the left and across the upper micrographs. The DNA, about seven microns long, encompasses about 21,000 base pairs, or enough to code for 10 to 14 proteins.

the transfer of DNA into a "female" of the same species; in transduction by a phage, or bacterial virus, the contractile injection mechanism of the virus helps to introduce a small amount of DNA from one bacterial cell into another cell along with its own viral genetic material. In transformation, on the other hand, a sizable segment of DNA enters a cell without any outside help, and so the entire apparatus for the uptake must reside in the recipient cell. One of the most interesting aspects of genetic transformation is therefore the cellular one: how do the bacteria transport the giant DNA molecules through the multiple layers of their surface?

There is evidence that the ability to absorb DNA is correlated with the chemical composition of the bacterial surface. Frank Young of the Scripps Clinic and Research Foundation in La Jolla, Calif., has noted that the cell wall of transformable strains of the bacterium *Bacillus subtilis* is richer in the amino sugar galactosamine than the wall of strains that cannot absorb DNA. In our laboratory we were able to affect the transformability of pneumococci by changing the composition of the cell wall. One of the components of the cell surface is a giant molecule of which one building block is choline. When, by suitable means, we forced the bacteria to replace the choline with such close structural analogues as ethanolamine or monomethyl-amino ethanolamine, the ability to absorb DNA was completely inhibited, although the cells survived and multiplied.

I mentioned earlier that even potentially transformable cells are "competent" to participate in transformation only at a certain time, and our major line of investigation has been directed toward clarifying this observation. If one allows a culture to multiply from a relatively small concentration—say 10,000 or 100,000 cells per cubic centimeter of growth medium—practically none of the cells are capable of undergoing transformation for a long period during which they grow and divide normally. Then, as the concentration reaches a million or 10 million cells per cubic centimeter, the property of competence appears abruptly and spreads with explosive speed to practically all the bacteria. With 10 million bacteria per cubic centimeter present, the number of competent cells can increase more than a millionfold in 10 minutes. (Under the same conditions a single division, or doubling, of pneumococci would require more than an hour.) We found that it was specifically the degree of concentra-

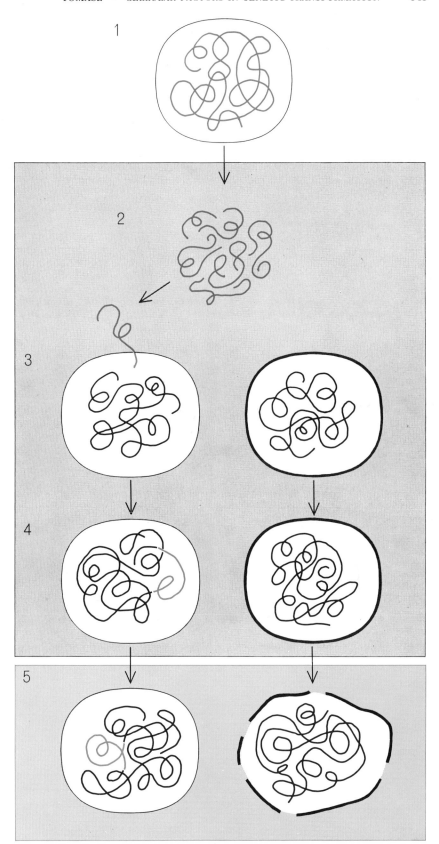

TRANSFORMATION EXPERIMENT begins with the extraction of DNA from a strain of pneumococci resistant to streptomycin (1). Placed in a medium with a strain susceptible to streptomycin (2), some of the DNA is absorbed by "competent" cells (*left*), but it cannot be absorbed by incompetent cells (*right*). The absorbed DNA is integrated into the chromosome of competent cells (4), transforming them so they (and their progeny) survive and can multiply in a streptomycin environment (*color*), whereas incompetent cells cannot (5).

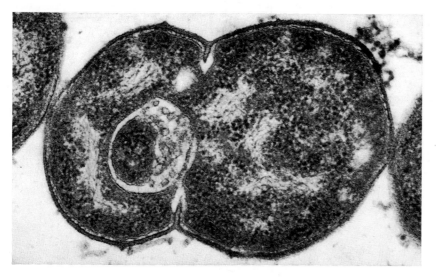

DIVIDING PNEUMOCOCCUS in thin section is enlarged 125,000 diameters in an electron micrograph made by James D. Jamieson of Rockefeller University (from a paper by the author, Jamieson and Elena Ottolenghi, *Journal of Cell Biology*, Vol. 22, page 453, 1964).

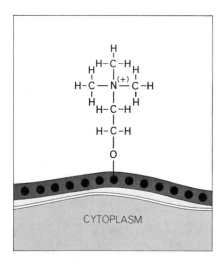

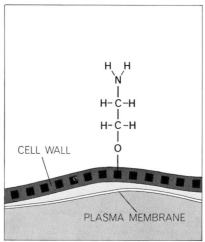

CELL WALL of pneumococcus is a network of macromolecules: glycopeptides, polysaccharides, proteins. Their exact structure is not known but one important component is the amino alcohol choline (*left*). The cells can be led to synthesize a wall in which these choline molecules are replaced by molecules of a structural analogue, ethanolamine (*right*). This modifies the cell wall so that the cell cannot become competent for transformation.

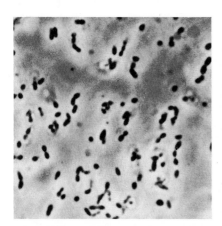

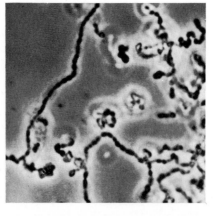

GROWTH PATTERN of pneumococci is altered by the surface change that occurs with substitution of ethanolamine. In the mediums used here, pneumococci usually grow alone, in pairs or in short chains (*left*). After substitution they tend to grow in long chains (*right*).

tion that governs the appearance of competence, not the rate of growth and not the fact that a culture may be nearing the stationary phase of growth during which the concentration levels off.

After the competence of the culture has reached its peak, we found, it soon starts to decline, and eventually it falls to an undetectable level. In short, it seemed that for some mysterious reason the ability to undergo transformation is expressed only during a brief phase of the culture cycle—and then it is expressed in a synchronous manner in practically all the cells present. This sudden and synchronous emergence of competence suggested that some kind of induction process might be operating to bring it on. Could it be that competent cells have a way of forcing incompetent ones to become competent ahead of time? Such a process might be demonstrated if incompetent cells were put into a common environment with bacteria that had already reached the competent state.

In designing an experiment we took advantage of the fact that the fewer cells one inoculates into a culture medium, the longer it will take for the culture to reach the competent phase. Indeed, given the original concentration and controlled conditions, one can predict the time of maximum competence fairly accurately. We took two strains of pneumococcus, distinguishable from each other because each was resistant to a different antibiotic, and inoculated them as follows. We put a high concentration of the first strain into one test tube and a low concentration of the second into another test tube; these were the controls. We also mixed the two strains in a single test tube, each at the same initial cell concentration as in the control tubes [*see illustration on page 346*]. Since the time of peak competence depends on initial concentration, we could predict that of the two control cultures the first strain, with the high concentration, would arrive at its peak of competence much earlier than the low-concentration second strain. The question was: What would happen in the mixed culture? Would the early development of competence in the first strain of cells influence the time-course of competence in the second strain?

We incubated the cultures and allowed them to grow. From time to time we monitored the ability of cells of each strain to react with a transforming DNA carrying genes for resistance to a third antibiotic, streptomycin. Because of their different double-resistance patterns on mediums containing streptomycin and one of the other antibiotics, we could selectively count the transformed cells of both test strains even when they came from the mixed culture.

The results were striking. The time-course of the appearance of competence was as predicted for the two strains grown separately. It was the same for the high-concentration first strain whether it was grown separately or in mixed culture. Although the low-concentration second strain became competent at the predicted slow rate in the control tube, however, the same strain grown at the same concentration in a mixed culture became competent quickly; it precisely copied the time-course of the first strain! The answer to our experimental question was therefore affirmative: Pneumococci in the competent state are apparently capable of transferring competence to cells that are as yet incompetent.

The problem now was to identify the agent responsible for this induction effect. First we considered the possibility that a competent cell had to come into physical contact with an incompetent one in order to induce competence. To test this possibility we repeated the experiment in a modified form. We used a U-shaped tube separated into two compartments by a filter membrane that had pores too small to pass cells but that was quite permeable to the growth medium. We put a culture that was already competent in the left arm of the U-tube and an incompetent culture in the right arm [see top illustration, page 347]. We then immersed the tube in a warm bath and pumped the medium slowly back and forth between the two chambers, at intervals analyzing the two separated strains for competence in the same manner as before. Once again the result was positive: the competent state was induced rapidly in spite of the fact that physical contact between the two strains of cells was precluded. Apparently, then, we were dealing with an inducer substance that was produced and secreted into the medium by the competent cells and was able to move through the filter.

The next problem was to isolate this inducer substance, which we have provisionally called "activator." Setting about this task, we removed competent pneumococci from their culture medium by centrifugation. Somewhat to our surprise we found that very little, if any, activator remained behind in the cell-free medium, indicating that most of the activator was bound to the competent bacteria. This seemed to be in conflict

with the result of the U-tube experiment, in which some of the activator had clearly been free in the medium. The two findings could be reconciled if the activator were loosely bound to the surface of the cells and could be detached rather easily, perhaps by rubbing against the filter membrane in the U-tube. Apparently that is the case. When we stirred the cells or exposed them to a few seconds of ultrasonic irradiation, substantial quantities of activator were detached and appeared in the medium. These experiments indicated that the activator was a surface component of competent cells and should not be hard to purify. We found that this could best be done by washing competent cells free of growth medium in a centrifuge, resuspending them in a salt solution and then heating the suspension to detach the activator from the cell debris and kill the cells. Repeated extraction of heat-killed cells with salt solution finally yielded highly purified preparations of the activator, which we then set out to characterize.

Chromatography revealed that the activator is a large molecule (molecular weight about 10,000) with a positive charge. It must be a protein or at least

contain protein that is essential to its activity, since it is completely inactivated by small amounts of protein-digesting enzymes; treatment with enzymes that attack nucleic acids, polysaccharides, glycopeptides and phospholipids have no effect on it. We are still working toward a fuller biochemical description of the molecule.

These procedures have required the development of a more precise test for the presence and potency of activator than we had used to determine peak competence. The cells exposed to an activator preparation must be completely incompetent and not in the process of developing spontaneous competence, and yet be sensitive to activator. We accomplish this by growing pneumococci in a slightly acid medium (pH 6.8), in which activator does not function, and then transferring them at the desired concentration into a more alkaline medium (pH 7.6), where they immediately respond to any activator present. Next the length of the activation process must be accurately established in each experiment. We do this by transferring the cells being tested, after the desired exposure, to test tubes containing both transforming DNA and a proteolytic en-

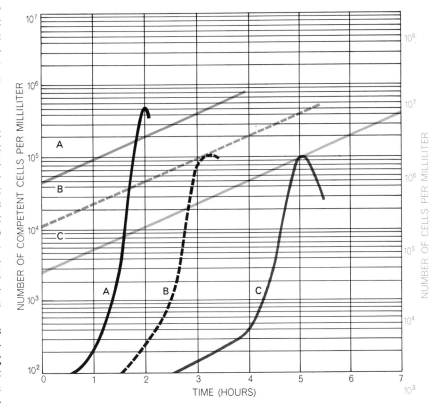

RELATION of competence to cell concentration is illustrated. Three cultures of pneumococci (A, B, C) differ in their initial cell concentrations. The time required to reach peak competence (black curves, scale at left) varies approximately inversely with concentration (colored curves, scale at right): the lower the concentration, the later the peak.

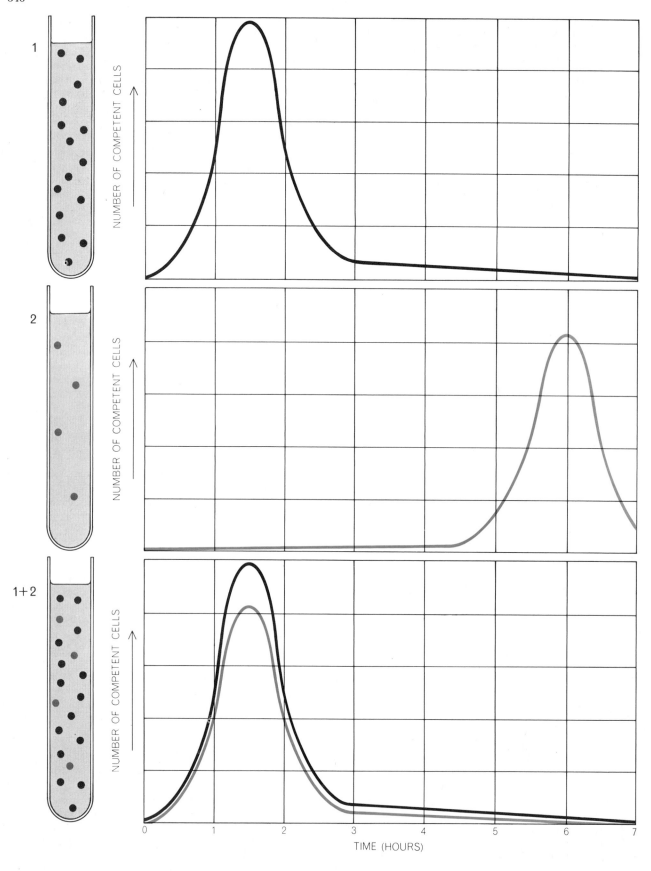

INDUCTION OF COMPETENCE was demonstrated by growing two strains of pneumococci at different concentrations in control cultures (*top and middle*) and in a mixed culture (*bottom*). The graphs show, for each strain, the number of cells that were found to be competent for transformation. In the high-concentration cul- ture (*1*) the degree of competence increases rapidly to a peak (*top*). The low-concentration culture (*2*), grown alone, takes a longer time to reach its peak of competence (*middle*). Yet when the two strains are grown in the same tube, cells of the second strain "copy" the competence curve of the first strain (*bottom*).

zyme that destroys any activator. Finally we add the enzyme DNA-ase, which destroys any DNA remaining free in the medium and thus ends the process of transformation. All of this makes for a procedure in which clear beginning and end points are established for the two phases of our assay procedure, first a "reaction" of bacteria with activator and then a "reaction" of activated cells with transforming DNA: (1) incompetent cells + activator → competent cells; (2) competent cells + transforming DNA → transformed cells. And so the induction of competence, which we originally discovered as a property of live competent cells, can now be more accurately studied as a reaction between bacterial cells and molecules of a purified activator substance. A few hundredths of a microgram of the activator can induce competence in several hundred thousand incompetent pneumococci in 10 minutes.

The existence of an activator explains how and why pneumococci express their competence in synchrony. Competence is apparently "contagious": it is a physiological state that can be provoked in the rest of the culture through the mediation of the activator. What controls the production of the activator in the first place? It is clear that the lack of competence in dilute cultures is caused by the lack of sufficient quantities of activator supplied by the cells themselves; when we supply activator, even a dilute culture becomes competent. In other words, the production (or accumulation) of this substance by the cells themselves seems to be a function of their concentration, but just how the production is triggered at a certain cell concentration we still do not know. Nor do we know what the actual change is that the activator causes in bacteria to enable them to absorb DNA molecules from the environment. At this point we can only say that the change probably involves some modification of the cell surface. This is indicated by the fact that the activator molecules seem to be located on the surface and by the importance of the choline-containing surface macromolecule, which I mentioned earlier. In addition, Sam M. Beiser and his colleagues at the Columbia University College of Physicians and Surgeons have by immunological methods detected some change in the surface of pneumococci while they are in their competent state.

The existence of an activator for transformation is not uniquely a property of the pneumococcus. Working at the University of Warsaw, Roman Pakula, now of the University of Toronto, independently isolated a similar competence-inducing substance that is apparently released into the growth medium by streptococci. The pneumococcal and the streptococcal activators are species-specific: neither induces competence in cells of the other species.

Work is now in progress in a number of laboratories, in Canada, Czechoslovakia and France as well as in the U.S. and Poland, aimed at learning more about the mechanisms that somehow open and close the gates of cells to the entry of foreign genetic material. This work could eventually lead to better understanding of viral infection and could even contribute to the possibility of deliberate genetic intervention in higher organisms. The hope is that once again, as so many times in the past, the bacteria can provide important clues to understanding how our own cells function.

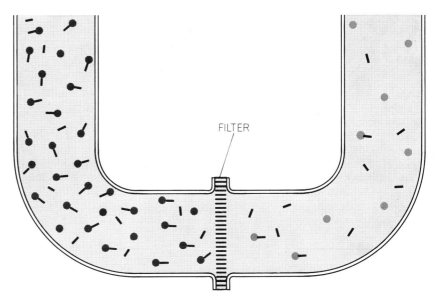

EXTRACELLULAR NATURE of the inducing agent was demonstrated by separating a competent culture (*left*) from an incompetent one (*right*) by a filter that was impermeable to the cells. When medium was pumped through the filter, competence was induced in the incompetent cells, indicating that "activator" molecules (*rods*) were crossing the filter.

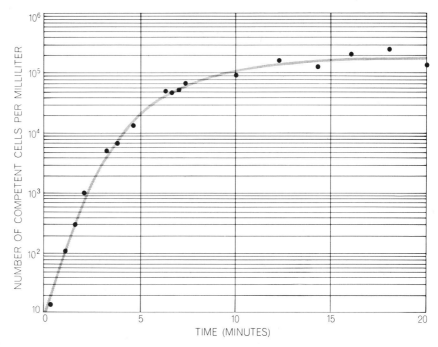

PURIFIED ACTIVATOR has the same inducing effect as competent cells have. Incompetent bacteria exposed to a small amount of activator develop competence rapidly, as shown here. In the absence of activator the same cells would remain incompetent for several hours.

35

REPEATED SEGMENTS OF DNA

ROY J. BRITTEN AND DAVID E. KOHNE
April 1970

In cells of higher organisms a significant fraction of the genetic material appears in as many as a million identical or very similar copies. The origin and function of repetitive DNA remain unknown

DNA was conclusively identified as the genetic material in 1944 and its structure was established in 1953. In the next few years it became clear how that structure is critical to DNA's replication, its transcription into RNA and its subsequent translation into particular chains of amino acids to form enzymes and other proteins. Everything that had been learned about DNA by the early 1960's emphasized the significance of the precise and unique sequence of nitrogenous bases that constitutes a gene and suggested that DNA consisted of chains of different genes. It was therefore a major surprise when it was discovered in 1964 that much of the DNA in the cells of the mouse consists of multiple copies of the same or very similar base sequences.

Repeated DNA has now been identified in all higher species that have been examined. It constitutes as little as 20 percent of all the DNA in the cell nuclei of some species and as much as 80 percent in others. The precision of repetition is often imperfect, so that members of a "family" of repeated DNA are usually closely related rather than identical. The number of related base sequences in a family ranges from 50 to two million. The wide distribution of repeated DNA, its persistence through millions of years of evolution and the fact that at least some of it is "expressed," or transcribed into RNA, all suggest that it is important to cell function and the survival of the organism. Yet it is still not known if the repeated segments are genes or parts of genes or if they carry out some role other than the specification of gene products. At this stage it seems likely that the most significant phenomenon may not be at the level of gene repetition; the repeated DNA may have an organizational or regulatory function. More insight into the evolutionary history and the present role of repeated DNA will surely lead to a new understanding of evolution and of the regulation of genetic functions in the living cell.

Before telling the story of the discovery of repeated DNA by our group in the Department of Terrestrial Magnetism of the Carnegie Institution of Washington, it will be well to briefly review the structure and behavior of DNA. Most DNA consists of sequences of only four nitrogenous bases: adenine (*A*), thymine (*T*), guanine (*G*) and cytosine (*C*). Together these bases form the genetic alphabet, and long ordered sequences of them contain in coded form much, if not all, of the information present in the genes. The DNA molecule in its native state (as it is found in the cell) is made up of two strands wound helically around each other. The bases face inward and each is specifically bonded to a complementary base on the other strand; *T* is always linked with *A* and *G* is always linked with *C*. These complementary bases have an affinity for each other such that, when they are paired, they contribute to the stability of the entire double-strand molecule.

In native DNA every base is paired with its complement and the molecule is quite stable. In isolated DNA the bonds between the bases can be broken by heating, and the two strands of the DNA can be completely separated from each other. This dissociation is usually accomplished simply by boiling the DNA solution for a few minutes. Left in solution, the single strands collide with complementary partners and, if the conditions are right, double-strand helixes are formed again. The requirements for such reassociation of DNA are surprisingly simple: the salt concentration must be fairly high, the temperature should be about 25 degrees Celsius below the dissociation temperature and enough time must be allowed for effective collisions to occur between complementary strands.

The extent of reassociation is the critical observation in most experiments with repeated DNA. Reassociation can be measured in a variety of ways, each depending on some easily detected difference between single-strand and double-strand DNA. Some time ago it was discovered that single-strand DNA absorbs more ultraviolet radiation than

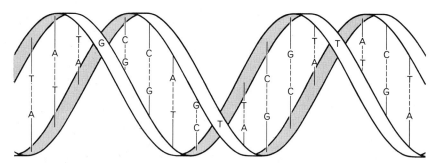

DNA in its native state is a double helix composed of two single strands. Each strand has a backbone of sugar and phosphate groups and a series of nitrogenous bases: adenine (*A*), thymine (*T*), guanine (*G*) and cytosine (*C*). The bases project inward and are linked by hydrogen bonds, which hold the two strands together like the rungs of a ladder; *A* always bonds with *T* and *G* bonds with *C*. The sequence of bases encodes genetic information.

double-strand DNA does; with a spectrophotometer one can measure changes in ultraviolet absorption and thus in the amount of reassociation. In another method single-strand DNA labeled with radioactive atoms is incubated over other strands fixed in agar, and the quantity of radioactive DNA that is reassociated, and thereby bound to the agar, is measured. Our own preferred method for the study of repeated DNA utilizes hydroxyapatite, a crystalline form of calcium phosphate that was originally used for the fractionation of proteins. A number of years ago it was discovered that single-strand and double-strand DNA have different affinities for hydroxyapatite crystals. We have helped to bring this technique to the point where the solution of DNA used for reassociation can simply be passed through the hydroxyapatite. The single-strand DNA passes through the column of crystals; the double-strand DNA is absorbed on the hydroxyapatite and can be removed for assay. The solution and its temperature (about 60 degrees C.) are such that the amount of single-strand DNA that is absorbed is low but the double-strand DNA is efficiently bound. The temperature is only about 25 degrees C. below the dissociation temperature, which prevents the formation of strand pairs held together by very short or very imprecise complementary regions.

When DNA strands derived from different but related sources, such as two related animals, are reassociated, their bases will not be perfectly complementary. The best means now known for recognizing and measuring the extent of differences between similar DNA sequences is to form hybrid-strand pairs between the two DNA's and measure the stability of the resulting helixes. The presence of a few mismatched bases reduces the stability of an otherwise complementary strand pair and leads to its dissociation at a lower temperature. The best current estimate is that for every 1.5 percent of the base pairs that are mismatched the dissociation temperature is reduced one degree. In order to determine the stability of reassociated DNA one need only absorb hybrid-strand pairs on hydroxyapatite and wash the column free of single-strand DNA at 60 degrees. Then the temperature is raised in steps and the DNA that dissociates at each temperature is washed out and assayed.

In 1962 Ellis T. Bolton and Brian J. McCarthy of our biophysics group at the Carnegie Institution had devised the DNA-agar method for measuring the relatedness between different DNA's

REASSOCIATION of strands of DNA is the experimental procedure that gives information about repeated DNA. The schematic drawings show the process of dissociation and reassociation for a hypothetical genome, or gene complement, composed of six copies of a repeated DNA sequence (*color*) and four different (*black*) sequences (*a*). Heating the DNA above the dissociation temperature breaks the hydrogen bonds between complementary base pairs, thus separating the strands (*b*). After the temperature is lowered complementary regions of DNA strands that collide may match up, forming reassociated double strands (*c*). Collision of complementary regions is more likely in the case of repeated DNA than for single-copy DNA. Speed of reassociation is proportional to degree of repetition.

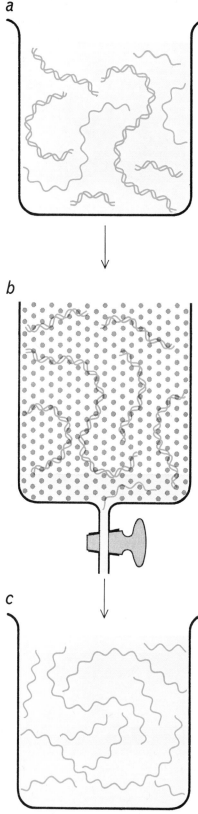

a

b

c

EXTENT OF REASSOCIATION is measured by separating double-strand DNA from single-strand DNA. A solution containing the two (*a*) is passed through a column of hydroxyapatite crystals (*b*). The double strands bind to the hydroxyapatite, whereas the single strands pass through the column (*c*). The double strands may then be collected by raising the column's temperature.

and RNA's and with it had measured the relations among various species of bacteria. With Bill H. Hoyer they undertook to apply the same method to the relations among vertebrate species. The first animal DNA they tried to reassociate was from mouse cells, and they succeeded. They went on to conduct a fascinating series of experiments on the relations between the genes of higher species and the expression of those genes in various tissues.

The trouble was that, according to what was then known about DNA, their experiments should not have worked. The cells of vertebrate animals contain much more DNA than bacteria do. As a result it had been expected that the concentration of each particular DNA sequence from vertebrate cells would be very small, and that after dissociation the complementary strands would rarely collide with each other. Yet the dissociated strands found their partners faster in the vertebrate DNA than they did in the bacterial DNA. Why? The question was set aside for future explanation, but as time passed it increasingly disturbed one of us (Britten). Had the reaction been accelerated by some unknown catalytic action of the agar or by the immobilization of the DNA? An experiment ruled out those possibilities: hybridization proceeded rapidly in free solution. By midsummer of 1964 only one explanation seemed possible and the basic hypothesis was proposed. Certain sequences in the DNA were reiterated again and again. Reiteration, or repetition, would make the concentration of some DNA base sequences much higher than had been expected and would thus account for the rapid reassociation.

On the basis of evidence that these related sequences were not identical with one another, the early hypothesis further suggested that some kind of multiplication process had occurred for selected segments of the DNA, forming large families of related sequences that became integrated into the genetic material and were then passed on to descendants by the usual hereditary mechanisms. It was further supposed that sequence changes, or mutations, altered various members of the families of repeated sequences, so that precise relationship was slowly lost among the member sequences. An essential part of the hypothesis was that new events of multiplication must have occurred at intervals throughout the course of evolution, producing new populations of repeated DNA segments. The evidence obtained since that time has supported this set of hypotheses.

The first direct evidence for repetition came out of an improbable piece of good fortune, again involving the mouse. Michael Waring arrived from the University of Cambridge as a postdoctoral fellow and chose to join in the search for actual pieces of repeated DNA. One of us (Britten) and Waring decided to extract DNA from mouse tissue because that tissue was convenient and available. We were quite unaware that mouse DNA has a singular family of repeated sequences. It is the most obvious one known even today; it makes up a tenth of the total mouse DNA and consists of about a million copies of a short sequence of some 300 base pairs. We had not expected anything so outlandish, but we immediately recognized the highly repetitive component. Because of its special qualities (it is lighter than most mouse DNA and forms a distinct "satellite" band when it is centrifuged in a cesium chloride density gradient), it could be purified in a few weeks and was shown to be an example of repetitive DNA. Soon afterward Ann McClaren and P. M. B. Walker of the University of Edinburgh, who were also working with mouse DNA, observed that 10 percent of it was bound to hydroxyapatite when it was expected to be single-strand and therefore not to bind. They did not imagine that reassociation could be as rapid as their results suggested and they identified their DNA as a stable fraction whose strands either had not separated or had folded back on themselves instantly. Walker's group at Edinburgh continued to investigate rapidly reassociating DNA in rodents. And since 1965 the authors of this article have collaborated on an extensive series of experiments to characterize the repeated sequences in the DNA of many different species.

Let us examine in more detail the paradox of the mouse-DNA reassociation, which led to the discovery of repetitive DNA. As we have noted, the reason animal DNA was expected to reassociate slowly is the large amount of DNA per animal cell [*see illustration on opposite page*]. If all the base sequences in the DNA of an animal were different from one another, then in a reassociation reaction each sequence would in effect be diluted by all the others. The dilution would lead to a reduced rate of reassociation, and so the rate of reassociation of the DNA from an animal with a large genome size, or DNA content per nucleus, should be less than the rate for DNA from one with a small genome.

This is exactly what we observe if we put aside the effects of the DNA present in multiple copies. In fact, the time required for the reassociation of unrepeated DNA to proceed halfway to completion is proportional to the genome size [*see illustration on next page*]. The reason is that complementary single strands must collide with each other before they can reassociate. Although most collisions are ineffective, occasionally complementary regions of the two strands are matched up in such a way that a short double-strand region results. If the neighboring regions are also complementary, then the double-strand region will grow into a stable helix; if they are not, the short region is likely to be unstable and therefore to dissociate. Under most practical conditions the rate of collisions controls the rate at which reassociation occurs. Reassociation between DNA strands therefore is normally, in the terminology of chemical kinetics, a second-order reaction. When its time course is plotted on a logarithmic scale, it takes the characteristic form of an S-shaped curve.

Measurement of the rate of reassociation is the best way to identify repetitive DNA. Because the rate of reassociation depends on a number of conditions, the DNA of a standard organism, the bacterium *Escherichia coli*, is taken as a reference. The length of the DNA of the *E. coli* chromosome has been well measured and its genome size is known. Virtually all the nucleotide sequences of its DNA occur only once, that is, it has little repetition. The time course of the reassociation of *E. coli* DNA [*see illustration on page 353*] reveals the S-shaped curve expected for a single-component reaction, and the precision of agreement with the predicted curve sets a low limit for the presence of repeated DNA. (The agreement also shows that such variables as fragment size have negligible effects and that the hydroxyapatite method adequately separates double-strand DNA from single-strand.)

The time required for half-reassociation is proportional to the DNA content per cell only if each sequence of bases occurs once in the cell's DNA. If a stretch of DNA sequence were to occur twice (if a gene were duplicated, for example), then the concentration of complementary fragments derived from the duplicate region would be twice as great as the concentration of the rest of the DNA. As a result these fragments would take only half as long to reassociate. If there were more copies, the period required for reassociation would be proportionately reduced. This simple rule apparently continues to hold even when as many as a million copies are present,

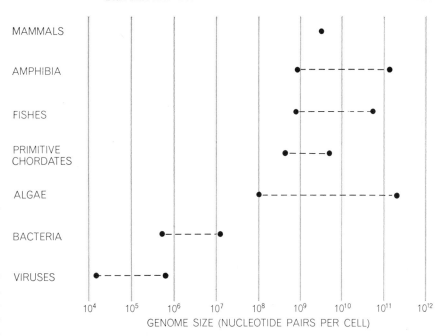

GENOME SIZE, or the amount of DNA per cell, increases with increasing evolutionary complexity. For each group of organisms shown here the minimum and maximum genome size recorded in various species are plotted, expressed as number of nucleotide pairs per haploid cell (a germ cell in higher forms). Mammals all have about the same genome size.

as is the case for the most rapidly associating family in mouse DNA. With this rule we can evaluate the number of copies from measurements of the rate of reassociation.

The DNA that has been most thoroughly studied is the DNA of the calf (because calf thymus glands, which are rich in DNA, are readily available from meat-packers). When its reassociation curve is plotted in the same way as the *E. coli* curve, there are some obvious differences. The reaction starts much sooner and ends much later; it is evidently divided into two quite separate regions. The later region of the curve has been shown to be due to the reassociation of DNA present in single copies. In the earlier part of the curve the DNA on the average reassociates 100,000 times faster, indicating that for each of the segments reassociating in this early region there must be about 100,000 other segments in the genome that are similar enough to form complementary pairs of strands. More detailed measurements show there are two major families of repeated sequences whose reactions overlap. One is present in about a million copies and amounts to only a few percent of the DNA. The other repeated component, accounting for nearly 40 percent of the total DNA of the calf, is present in 66,000 copies, not all of them exactly alike.

The total quantity of DNA in the form of repetitive sequences is quite large:

about 45 percent for calf DNA, according to the hydroxyapatite measurements. Actually this is a somewhat arbitrary figure. It depends on the degree of sequence relationship—the proportion of matching bases in each pair of strands—that is recognized in the particular measurement. We do not yet know how much of the DNA would have to be included if all degrees of sequence relationship greater than chance expectation were taken into account. In several species—Pacific salmon, wheat, onion and the salamander *Amphiuma*—more than 80 percent of the DNA shows repetition that is recognizable under the conditions in which calf shows only 45 percent repetition. Certainly the repeated sequences are not a minor part of the DNA. Why is there so much and what role does it play in the operations of the cell? Before considering the evidence bearing on these unsolved questions we will summarize what is known about the distribution of repeated DNA in various organisms, the frequency of repetition, the apparent evolutionary history of repeated DNA and the distribution of the material in the cell.

In the past few years the DNA of more than 60 plant and animal species has been examined. Significant quantities of repeated sequences have been found in the following groups: primates, other mammals, birds, amphibians, bony and cartilaginous fishes, amphioxus, echinoderms, brachiopods, insects, other arthropods, mollusks, coelenterates, sponges, protozoans and plants ranging from algae to wheat. Small quantities may be present in the nuclei of fungi, but here a clear separation from DNA outside the nucleus has not yet been made. Bacteria contain a number of copies of the genes that specify ribosomal RNA, but no other nuclear repeated DNA has been observed in the few species tested. (All species appear to have a few copies of certain special genes such as the genes for ribosomal or transfer RNA, and so one can say that repeated DNA probably occurs universally. The large quantities and high frequencies we are describing, however, appear to occur in species higher on the scale of complexity than fungi.)

The number of DNA segments from one cell that have similar base sequences is the repetition frequency for that set. One can regard the set of different repetition frequencies for each species as a spectrum [see illustration on page 354]. These spectra are somewhat tentative in detail, since measurements with adequate technique have just been undertaken; the overall patterns are nonetheless correct, and some of the spectra, such as the spectrum for calf DNA, have been fairly well measured. The repetition spectra show few common features. Calf cells have little or no DNA with frequencies between 10 and 10,000. Toad and snail cells have no families with more than 10,000 members, and their dominant components are in the broad frequency range that is not present in calf DNA. It is too early in the exploration to discern the regularities in frequency spectra that probably exist. For example, it appears that components in the range from 100,000 to a million copies may exist only in the DNA of mammals; each of the mammalian species examined so far has such highly repetitive components, which do not appear even in other vertebrates. Many more measurements are required to establish such generalizations.

It seems likely that repeated DNA originated in past events of DNA replication, each of which produced a family made up of identical copies of some pre-existing DNA sequence. Then in the period of time since the original production of the copies a number of things must have happened. Segments were translocated and scattered into various parts of the genome; subunits were substituted and deleted. The traces of such events can be found in the present populations, and the result is that in most cases the members of a family of repeated sequences differ somewhat from one another.

Estimates of the degree of sequence difference can be made from the thermal stability of reassociated DNA, since the imperfectly complementary strand pairs have reduced stability compared with perfectly matched strands [see top illustration, page 355]. Typically when the entire population of repeated DNA in one organism is examined, a wide range of thermal stability is seen. Some sequences are quite closely related to others, even to the extent of being perfect copies, whereas other sequences differ so much that the strand pairs barely hold together under mild conditions. In the most divergent examples changes appear to have taken place in as many as half of the nucleotides. We believe all degrees of divergence are present, as would be expected if the processes of production and divergence had been active throughout the evolution of higher species.

When did the events that produced these enormous families occur? Some of them happened a very long time ago, perhaps several hundred million years ago. It is possible to date reasonably well the time in evolutionary history when there lived the last common ancestor of two related but now distinct modern species. This time may be considered, if enough fossils have been discovered and examined, to be the time of divergence of the ancestral lines of species that led to the present-day organisms. Any feature, molecular or morphological, that is now exhibited by both species probably originated before the time of divergence. We assume that any family of repetitive DNA held in common by two species probably originated before their ancestral lines diverged. (There are risks in this assumption where gene flow between species is possible, as it is in plants.) The relation between the fraction of the repeated DNA held in common between species and the time since their species lines diverged can be established [see bottom illustration on page 355]. Assuming, as seems reasonable, that in each species line there is production of new families of repeated DNA as preexisting families are slowly lost, the curve can be considered a "decay-of-relatedness curve," by analogy

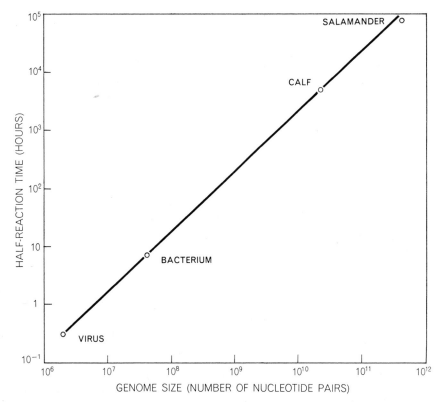

TIME FOR REASSOCIATION of single-copy DNA varies with the size of the genome and therefore increases with the complexity of the organism. Here the time for half-reassociation is plotted against genome size for a virus (T4) and a bacterium (*Escherichia coli*) and for the nonrepetitive DNA of the calf and a salamander, *Amphiuma*. The times shown are for a particular set of conditions; the time can be accelerated by increasing the concentrations of DNA and salt, but for large genomes it is difficult to measure a complete reaction.

with radioactive decay. For the vertebrates shown in the illustration the median age of the repeated families seems to be about 100 million years—the time in which half of the repeated DNA is replaced.

Some of the more abstract but nevertheless fascinating questions about repeated DNA sequences have to do with their organization in the cell. How long are the repeated elements? Do they occur together in the genome? Are they organized in special places in the DNA in keeping with some role in chromosome structure, or are they scattered throughout the DNA?

Several different classes of experiments yield information about the arrangement of the repeated DNA in the genome. One case—perhaps an atypical case—is the mouse satellite DNA, which has recently been studied in much detail by Walker and by William G. Flamm, now at the National Institute of Environmental Health Science, and by others. The sequences of this family are of fairly recent evolutionary origin, having appeared since the ancestral lines leading to the house (or laboratory) mouse and the rat diverged 10 or 20 million years ago. They have been shown to be arrayed in clusters containing as many as 100 copies, and according to recent measurements the cluster's appear on many of the chromosomes of the mouse. This family is somewhat exceptional, since it is a very short repeated element present in very many copies and is found in only the one species.

Such detailed evidence is not yet available for any other family of repeated sequences, but there is evidence from a number of species that the repeated sequences are intimately scattered through the DNA. If DNA is prepared in fragments about 20,000 bases long, each fragment is almost certain to include a repeated sequence, and it usually includes more than one. The extent of scattering is known in a little more detail in calf DNA, where three-quarters of the fragments broken down to about 4,000 bases contain segments of repeated DNA.

One approach to the difficult problem of the function of repeated DNA is to assay the extent to which it is expressed. All the genes of an organism's genome are present in all its cells, and yet many of the genes carry out no function in particular tissues or at particular times. Such genes are said to be unexpressed, whereas those that are active are said to be ex-

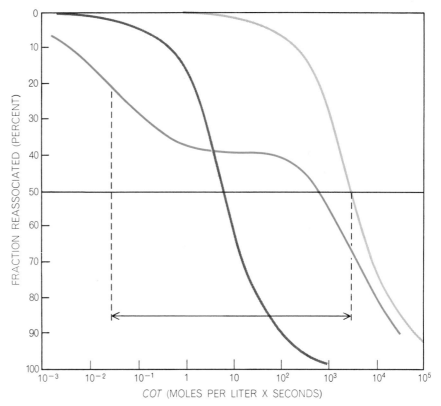

TIME COURSE of reassociation is shown for bacterial and for calf DNA. The amount of reassociation is plotted against the *COT*, a coined word for the product of DNA concentration and time. The time course for bacterial (*E. coli*) DNA, almost all of which is single-copy DNA, plots as an *S*-shaped curve (*gray*). If calf DNA were also single-copy DNA, it would plot as a similar curve displaced to the right (*light color*). The actual curve for calf DNA (*dark color*) has a different shape, however. The shape of the curve indicates that the reaction takes place in two different stages, one early and one late; the midpoints of the two stages (*broken vertical lines*) are separated by a factor of 100,000. The early stage represents the reassociation of repeated DNA and the later stage that of single-copy DNA.

pressed. The function of many genes is to specify the synthesis of proteins, and the mechanism of these genes' expression is now quite well understood. The first step in the process is the transcription of the base sequence of the DNA into "messenger" RNA by the pairing of complementary RNA bases with the DNA as template. When the messenger RNA reaches the site of protein synthesis, its base sequence supplies information that is translated according to the genetic code to determine the amino acid sequence of the protein. Our interest here is only in the first stage, the synthesis of RNA on the DNA template, as an indication of gene expression. Clearly one can say that a gene is expressed if RNA complementary to one of its DNA strands is synthesized. Therefore a good test for genetic activity is to search for an RNA complementary to the DNA of the gene. In most cases this would be difficult, since preparations of the DNA of particular genes are not ordinarily available. In the case of repeated DNA taken as a whole, however, it is quite easy. The high

frequency of repetition increases the rate of RNA-DNA hybridization just as it increases the rate of DNA reassociation. When proper precautions are taken, RNA that hybridizes rapidly with DNA can be assumed to be RNA that was synthesized on a repeated DNA sequence.

Hybridization experiments have been conducted in a number of laboratories in attempts to learn the mechanism of gene expression and its control. We now know that the limitations of the available techniques restricted the observations to rapidly hybridizing sequences—the sequences involving repeated DNA. The observations therefore made available a large body of measurements of the expression of repeated DNA sequences in a variety of cell types from a number of organisms. These measurements yield two principal results: first, repeated DNA sequences are expressed in every higher organism and cell type that has been examined; second, different families of repeated DNA are expressed in different tissues.

A particularly significant measurement of differences in patterns of expression can be made with a technique called competition. For this purpose RNA from one tissue is mixed at a high concentration with radioactively labeled RNA from a second tissue in the presence of DNA. To the extent that there are similarities between the RNA populations, the unlabeled RNA "occupies" the DNA and prevents the labeled RNA from hybridizing. In this way RNA populations from different tissues and different stages of embryonic development have been compared. In a number of instances the populations of RNA molecules are entirely dissimilar, indicating that the DNA sequences expressed are also dissimilar, that is, they are members of different families of repeated sequences. Why should a particular family of DNA sequences be expressed in an early embryo and not be expressed at a later stage of development? The answer to this question would probably go a long way toward revealing the role of repeated DNA in the molecular machinery of the cell.

The fact that repetitive DNA is expressed (transcribed into RNA) leaves no doubt in our minds about the significance of some of the repeated sequences. That brings us up against the problem of the hundreds of thousands and millions of copies present in the mammals. What role could such very large numbers of copies have? If these frequencies were observed even throughout the vertebrates, one might consider a structural or organizing role for the highly repetitive DNA, but it is unlikely that 1,000 times more copies would be required for such a purpose in the mouse than in, say, the toad. Mammalian chromosomes do not seem to be that much more complex. They have comparable amounts of DNA per cell; in fact, some amphibian cells contain a great deal more DNA than mammalian cells.

Taking the evidence all in all, the large size of some families of repeated DNA is probably best explained if the production of a family of repeated sequences is taken to be an analogue of a more familiar kind of mutation. Repeated sequences could be produced in large numbers "by accident," that is, by an event not directly related to their ultimate function. In the long run, of course, their fate would presumably be determined by natural selection, but for some time a large number of unexpressed

copies might remain in the genome. In other words, we are led to the somewhat unorthodox view that a significant part of the DNA in the cell may not have a current genetic function.

We have called the appearance of a family of repeated sequences "saltatory replication" to imply a sudden event of multiplication. The evidence indicates that on an evolutionary time scale the appearance of a family is sudden, although we cannot tell whether the events occur within the lifetime of an organism or are spread over a few million years. Several steps are obviously required. First a segment of DNA must be multiplied; then the copies must be transmitted by heredity. The set of copies must become intimately associated with the chromosomes and ultimately must be linked directly into the principal DNA strands. If that much is accomplished in a small population within the species, this DNA must be transmitted throughout the species over a large number of succeeding generations, presumably by natural selection. Such a process would be inexplicable in neo-Darwinian terms unless some genetic advantage were associated with the family of repeated DNA sequences. We are unable

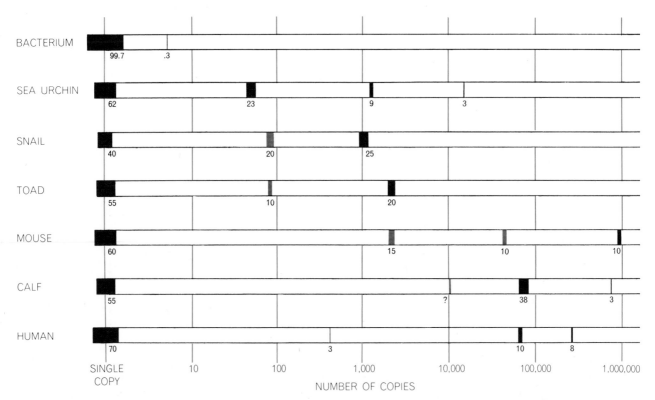

PATTERNS OF REPETITION of DNA sequences are shown in the form of spectra for a number of organisms. Each spectral band represents a repetitive class of DNA that has been identified to date. The width of the band indicates what percent (*given below the band*) of the total DNA is represented by that class. The position of the band indicates the frequency, or number of copies. (*Gray bands are those whose precise frequencies are not known.*) For example, the most repetitive mouse component amounts to 10 percent of its DNA and has about a million copies. Snail and toad data are from Eric H. Davidson of Rockefeller University.

to specify what the advantage might be unless an actual function were already being carried out by the repetitive DNA at the earliest stages of the process.

The time period would be greatly compressed if the saltatory replication were due to a virus infection that swept through a large population. A model for such an event might be the behavior of bacteria that survive a viral infection because they have previously accepted the virus genome into an integrated, or lysogenic, state within their own chromosome and by so doing have achieved immunity to the virus. Cellular resistance to virus infection by this kind of mechanism has not been clearly demonstrated in animals. The transformation of animal cells by viruses, however, is well known, and in many cases it may be due to the integration of the viral genome with the host chromosomes. Even so, there is no reason to expect that an enormous number of copies would be involved. It is therefore merely possible—not likely—that an infection of some kind is responsible for the transmission of a family of repeated sequences to all the surviving members of a species.

Obviously we do not understand how a family of repeated DNA originates, but such families exist, and they must have originated somehow. What can be said of their subsequent history? We know that some of the sequences have become functional because RNA complementary to them is produced. We do not know when in their history the useful attributes were developed, how many of the members of a family of sequences are utilized or what function they carry out. It is clear, however, that the families of repeated sequences evolve and that the sequences change slowly. It is even possible to estimate the rate of change. Estimating the "decay of relatedness" is one approach, and more elegant methods are now available. Fairly good estimates can be made of the actual rate of base substitution in DNA by measuring the stability of strand pairs hybridized between chosen species. Such measurements have only been started and no broad generalizations can yet be drawn, but we should soon know whether or not natural selection operates to conserve the sequences of some of the members of a family while allowing others to change at a more rapid rate. We now know a little about the history of repetitive sequences and we possess techniques to learn a great deal more, but a comprehensive set of measurements is needed that will require much work at many laboratories.

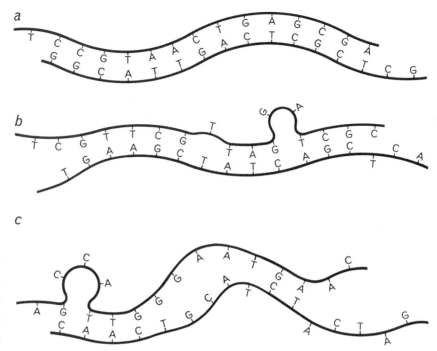

STABILITY of reassociated DNA depends on how well the two strands are hydrogen-bonded, which depends in turn on the extent to which the two strands are composed of complementary base sequences. In native DNA or in two strands from identical copies of a repetitive DNA all the bases match and the strand pair is quite stable (a). If some bases do not match, the stability is reduced (b), and if many are mismatched, the stability is poor (c). The stability of DNA can be determined by measuring its dissociation temperature.

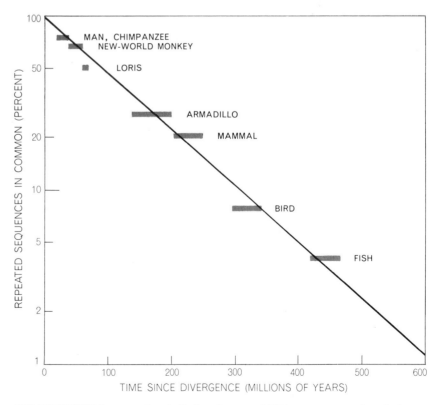

RELATIONSHIP between the similarity of repeated-DNA sequences and evolutionary history is shown, based on measurements made by Bill H. Hoyer, Brian J. McCarthy and Ellis T. Bolton of the Carnegie Institution. Vertical scale shows the fraction of repeated sequences held in common by the DNA of the rhesus monkey and that of other animals. The time of divergence of species lines leading to the rhesus and to other modern species is plotted on the horizontal scale. ("Mammal" refers to nonprimates other than armadillo.)

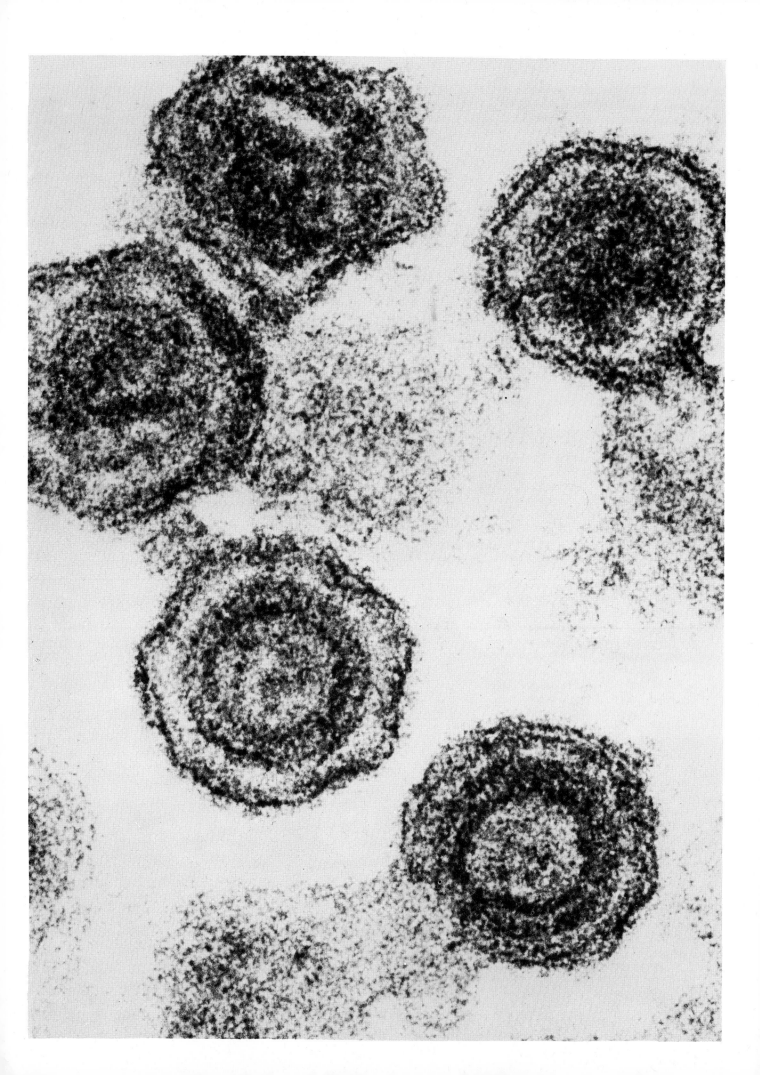

36

RNA-DIRECTED DNA SYNTHESIS

HOWARD M. TEMIN

January 1972

The discovery that in certain cancer-causing animal viruses genetic information flows "in reverse"—from RNA to DNA—has important implications for studies of cancer in humans

A major goal of present-day biology is to learn how information is coded in molecular structures and how it is transmitted from molecule to molecule in biological systems. Discovery of the rules governing this transmission is an integral part of understanding

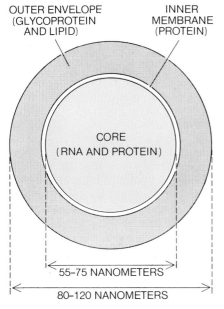

OUTER ENVELOPE (GLYCOPROTEIN AND LIPID)

INNER MEMBRANE (PROTEIN)

CORE (RNA AND PROTEIN)

55–75 NANOMETERS

80–120 NANOMETERS

VIRIONS, or individual particles, of an "RNA-DNA virus," an animal-tumor virus that transfers genetic information from RNA (ribonucleic acid) to DNA (deoxyribonucleic acid) in addition to the normal modes of information transfer used by cells and other viruses, are enlarged about 700,000 diameters in the electron micrograph on the opposite page. The particular RNA-DNA virions shown in thin section in the micrograph cause leukemia in mice; they are similar in structure and function to the Rous sarcoma virions discussed in this article. The electron micrograph was made by N. Sarkar of the Institute for Medical Research in Camden, N.J. A diagram of structure of a virion of this type is given above.

how embryonic cells differentiate into the hundreds of distinct types of cell observed in plants and animals and how normal healthy cells become cancerous.

It has now been known for nearly 20 years that the genetic information in all living cells is encoded in molecules of deoxyribonucleic acid (DNA) consisting of two long strands of DNA wound in a double helix. The genetic information for each organism is written in a four-letter alphabet, the "letters" being the four different chemical units called bases. In the normal cell short passages of the genetic message (individual genes) are transcribed from DNA into the closely related single-strand molecule ribonucleic acid (RNA). A length of RNA representing a gene is then translated into a particular protein, a molecule constructed with a 20-letter alphabet, the 20 amino acids. When a cell divides, the information contained in each of the two strands of DNA is replicated, thereby equipping the daughter cell with the full genetic blueprint of the parent.

Francis Crick, one of the codiscoverers of the helical structure of DNA, originally proposed that information can be transferred from nucleic acid to nucleic acid and from nucleic acid to protein, but that "once information has passed into protein it cannot get out again," that is, information cannot be transferred from protein to protein or from protein to nucleic acid. These concepts were simplified into what came to be known as the "central dogma" of molecular biology, which held that information is sequentially transferred from DNA to RNA to protein [*see illustration on next page*]. Although Crick's original formulation contained no proscription against a "reverse" flow of information from RNA to DNA, organisms seemed to have

no need for such a flow, and many molecular biologists came to believe that if it were discovered, it would violate the central dogma.

I shall describe experiments that originally hinted at a flow of information from RNA to DNA and that since have provided strong evidence that the "reverse" flow of information not only takes place but also accounts for the puzzling behavior of a sizable group of animal viruses whose genetic information is encoded in RNA rather than in DNA. Many of these viruses also produce cancer in animals. Although they have not yet been linked to cancer in man, their ability to transmit information from RNA to DNA inside the living cell makes it attractive to unify two hypotheses of the cause of human cancer that had previously seemed separate: the genetic hypothesis and the viral hypothesis.

There are two broad classes of viruses: viruses whose genome, or complete set of genes, consists of DNA and viruses whose genome consists of RNA. In the cells that they infect the DNA viruses replicate their DNA into new DNA and transmit information from DNA to RNA and thence into protein. Most RNA viruses, such as the viruses that cause poliomyelitis, the common cold and influenza, replicate RNA directly into new copies of RNA and translate information from RNA into protein; no DNA is directly involved in their replication.

In the past few years it has become apparent that a group of viruses, variously called the RNA tumor viruses, the leukoviruses or the rousviruses (after their discoverer, Peyton Rous), replicate by another mode of information transfer. The rousviruses use information transfer from RNA to DNA in addition to the modes of information transfer (DNA to DNA, DNA to RNA and RNA to protein) that are found in cells and in DNA viruses. The rousviruses do not transfer information from RNA to RNA, as other RNA viruses do. The existence of the RNA-to-DNA mode of information transfer in the replication of rousviruses has led some to suggest that there should be three major classes of viruses: DNA viruses, RNA viruses and RNA-DNA viruses [see top illustration on opposite page].

The prototype RNA tumor virus, the Rous chicken sarcoma virus, was discovered by Rous 61 years ago at the Rockefeller Institute for Medical Research. An RNA tumor virus had actually been found earlier by V. Ellerman and O. Bang of Copenhagen, but their virus was little studied because it caused leukemia in chickens and was harder to work with than Rous's virus. Rous was studying a transplantable tumor of the barred Plymouth Rock hen. Originally he observed that he could transfer the tumor by the transfer of cells. In 1911 he found that the tumor could also be transferred by means of fluid from which the cells had been filtered. Demonstration that a disease can be transmitted by a cell-free filtrate is commonly accepted as evidence that it is caused by a virus. Descendants of the virus originally discovered by Rous are still being worked on in laboratories all over the world. At the time, however, Rous's discovery was met with disbelief, and after 10 years Rous himself stopped working with the tumor. It was not until nearly 30 years later, when Ludwik Gross of the Veterans Administration Hospital in the Bronx discovered that RNA tumor viruses cause leukemia in mice, that the study of rousviruses became popular.

It is now known that viruses in the same group as the virus originally discovered by Rous, or closely related to it, can cause tumors not only in chickens and mice but also in rats, hamsters, monkeys and many other species of animals. Moreover, viruses of the same group have been isolated from nonmammalian species, including snakes. As yet no bona fide human rousvirus has been discovered. It also appears that some members of this group, for example some of the "associated viruses," do not produce cancer.

In the 1950's, with the beginning of the application of cell-culture methods to animal virology, a tissue-culture assay for the Rous sarcoma virus was developed, first by Robert A. Manaker and Vincent Groupé at Rutgers University and subsequently by Harry Rubin and me at the California Institute of Technology. The assay involves adding suspensions of the virus to sparse cultures of cells taken from the body wall of chicken embryos. The Rous sarcoma virus infects some cells and transforms them into tumor cells. The transformed tumor cells differ in morphology and in growth properties from normal cells and therefore create a focus of altered cells. Assays of the same type have been developed for infections that the Rous sarcoma virus causes in cells taken from turkeys, ducks, quail and rats. Similar assays have also been developed for other transforming rousviruses.

The number of foci of transformed cells is proportional to the number of infectious units of the virus added to the cell culture and provides a rapid and reproducible assay for the Rous sarcoma virus. The use of this assay led to the discovery that the Rous sarcoma virus differs from the other viruses that had been studied up to that time in the way

it interacts with the cell. The replication of most viruses is incompatible with cell division; in other words, the virus causes the infected cells to die. Chicken cells infected with the Rous sarcoma virus not only survive but also continue to divide and produce new virus particles [see middle illustration on opposite page]. When the Rous sarcoma virus infects rat cells, there is a slightly different interaction of the cell and the virus. The rat cells are transformed into cancer cells, which divide, but the transformed cells do not produce the Rous sarcoma virus even though the genome (DNA) of the virus can be shown to be present. Production of the Rous sarcoma virus can be induced if the transformed rat cells are fused with normal chicken cells.

In the early 1960's the antibiotic actinomycin D was found to be very useful in unraveling the flow of genetic information in cells infected with RNA viruses. The antibiotic inhibits the synthesis of RNA made on a DNA template but not the synthesis of RNA made on an RNA template. The antibiotic therefore stops all RNA synthesis in cells infected by RNA viruses except for RNA specifically related to the viral genome. With this new tool it became easy to determine which RNA's were specific for the viruses.

When I added actinomycin D to cultures of cells producing Rous sarcoma virus, however, I found that the antibiotic inhibited the production of all RNA. One would have expected the replication of RNA on the template of an RNA viral genome to continue without hindrance [see bottom illustration on opposite page]. This result was the first direct evidence that the molecular biology of the replication of Rous sarcoma virus was different from that of other RNA viruses. Since that observation was made the inhibition of the replication of rousviruses by actinomycin D has been recognized as one of their defining characteristics. The actinomycin D experiments suggested to me that the Rous sarcoma virus might replicate through a DNA intermediate. This hypothesis is called the DNA provirus hypothesis.

Further experiments, carried out by me and by John P. Bader at the National Cancer Institute, demonstrated that if one inhibits the synthesis of DNA in cells immediately after they have been inoculated with Rous sarcoma virus, one can protect the cells from infection. Here the inhibitors were amethopterin, fluorodeoxyuridine and cytosine arabinoside. These experiments appeared to support the idea that infection requires the synthesis of new viral DNA pro-

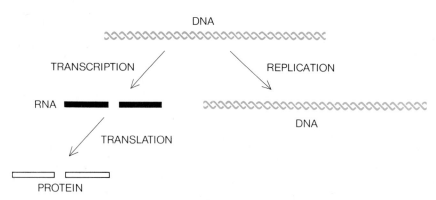

"CENTRAL DOGMA" of molecular biology, originally formulated by Francis Crick, states that within an organism genetic information can be transferred from DNA to DNA or from DNA to RNA to protein, but that it cannot be transferred from protein to protein or from protein to either DNA or RNA. Although a "reverse" flow of genetic information from RNA to DNA was not proscribed in Crick's original formulation, many molecular biologists came to believe that if such a flow were ever discovered, it would violate the central dogma.

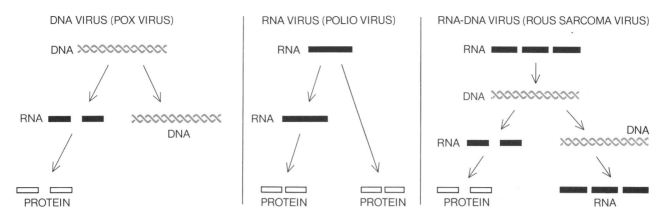

VIRUSES CAN BE GROUPED into three major classes: DNA viruses (*left*), whose genome, or complete set of genes, consists of DNA; RNA viruses (*middle*), whose genome consists of RNA, and RNA-DNA viruses (*right*), the most recently discovered group, whose genome consists alternately of RNA and DNA. A prototype virus in each major class is indicated in parentheses next to the class name. The diagrams illustrate the mode of information transfer that characterizes the replication of viruses in each class.

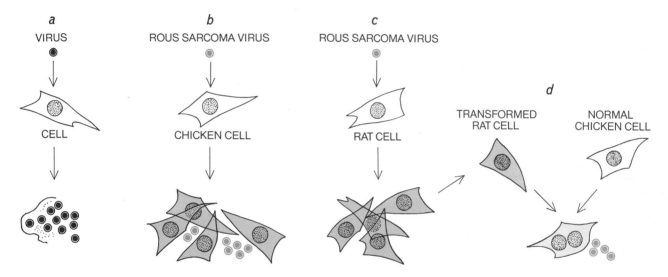

VIRUS-CELL INTERACTION usually leads to the death of the infected cell (*a*), since the replication of most viruses is incompatible with cell division. The Rous sarcoma virus, however, interacts with cells in a different way. Chicken cells infected with the Rous sarcoma virus (*b*) not only survive but also are transformed into cancer cells, which continue to divide and produce new virions. Rat cells infected with the Rous sarcoma virus (*c*) are transformed into cancer cells, which divide but do not produce new virions. By fusing the transformed rat cells with normal chicken cells the production of Rous sarcoma virions can be induced (*d*).

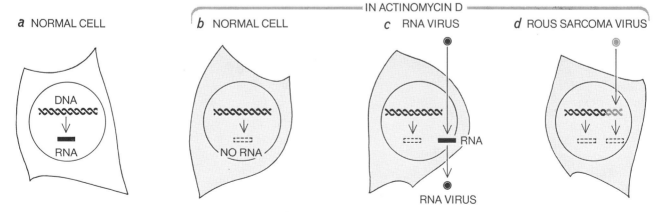

SYNTHESIS OF RNA on a DNA template in normal cells (*a*) is inhibited by the addition of the antibiotic actinomycin D (*b*). Since the antibiotic does not affect the synthesis of RNA made on an RNA template, however, it does not stop RNA synthesis specifically related to the viral genome in cells infected by most RNA viruses (*c*). The finding that actinomycin D inhibited the production of *all* RNA in cells producing Rous sarcoma virus (*d*) was the first direct evidence that the molecular biology of the replication of Rous sarcoma virus was different from that of other RNA viruses. The actinomycin D experiments led the author to propose the DNA provirus hypothesis, which holds that rousviruses such as the Rous sarcoma virus replicate through a DNA intermediate.

duced on an RNA template. This interpretation was not unequivocal, however, because successful production of Rous sarcoma virus requires that the cells divide normally after infection. Therefore the inhibition of DNA synthesis after infection could inhibit production of Rous sarcoma virus not only by blocking possible new viral DNA synthesis but also by preventing normal cell division.

To get around this problem I introduced the idea of infecting cultures of stationary, or nondividing, cells with Rous sarcoma virus. Cells in culture usually require specific factors in blood serum to support their multiplication. If the serum is removed from the medium of the cell cultures, the cells stop dividing. If they are then exposed to Rous sarcoma virus, they become infected but there is no virus production or morphological transformation until serum is added back and the cells divide once again. When such stationary cells are exposed to inhibitors of DNA synthesis, the cells are not killed because they are not making DNA. When the stationary cells are exposed simultaneously to Rous sarcoma virus and to inhibitors of DNA synthesis, the cells are not killed but neither are they infected [*see illustration at right*].

If one now removes the inhibitor of DNA synthesis and adds serum, enabling the cells to divide once more, one finds that the cells remain free of infection. They do not become transformed and they do not produce virus. These experiments supported the hypothesis that after cells are infected by the Rous sarcoma virus new viral DNA is synthesized at a time different from the cell's normal synthesis of DNA. The new viral DNA is evidently synthesized on a template of viral RNA.

A further extension of this approach to understanding the replication of Rous sarcoma virus was carried out by one of my students, David E. Boettiger, and independently by Piero Balduzzi and Herbert R. Morgan at the University of Rochester School of Medicine and Dentistry. It had been found by others that if 5-bromodeoxyuridine, an analogue of the DNA constituent thymidine, is incorporated into DNA, the DNA becomes sensitized so that it can be inactivated by light. Under the same conditions normal thymidine-containing DNA is not affected by light. Boettiger therefore exposed stationary cells to Rous sarcoma virus in the presence of bromodeoxyuridine and then exposed the cells to light. Although the cells were not killed, the treatment prevented their

being infected by the virus. When serum was again added to enable the cells to divide, they did not become transformed and did not produce virus [*see illustration on page 362*].

In a related experiment Boettiger showed that the rate of inactivation of the infection by Rous sarcoma virus was dependent on the number of viruses infecting a cell. As he raised the number of viruses infecting each cell, he found that the infection became increasingly resistant to inactivation by light. We interpreted these experiments as showing that each infecting virus makes a new specific DNA, and that the more viruses that infect a cell, the more molecules of new viral DNA that are produced. The experiment seemed to effectively rule out the alternative hypothesis, which was that the infecting virus provokes a new synthesis of some preexisting cellular DNA.

Unfortunately no one has yet been able to unequivocally demonstrate the existence of newly synthesized viral DNA in cells infected with the Rous sarcoma virus. The available techniques are evidently too crude to detect the tiny

amounts of new viral DNA expected to be present. Certain results have been reported, however, with transformed cells. One approach has been to bring DNA from infected cells together with labeled viral RNA to see if single strands of the two molecules would coalesce into a double-strand hybrid molecule. Such hybrids are readily created when the base sequences in the DNA are complementary to the base sequences in the RNA, indicating that both carry the same genetic message and hence that each could arise from the transcription of the other.

The hybridization experiments reported thus far have aroused a great deal of controversy. Although some experiments, notably those of Marcel A. Baluda and Debi P. Nayak of the University of California at Los Angeles, have seemed to demonstrate the presence in infected cells of DNA complementary to viral RNA, the results have not been universally accepted. The finding of an intermediate viral DNA is an essential link in the chain of evidence that is still needed to establish firmly the DNA provirus hypothesis.

Meanwhile strong support for the hy-

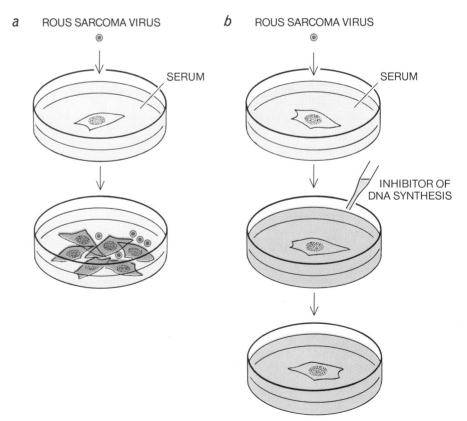

EXPERIMENTS carried out by the author and by John P. Bader at the National Cancer Institute supported the hypothesis that the infection of cells with Rous sarcoma virus requires the synthesis of new viral DNA produced on an RNA template. When the virus is added to cultures of normally dividing cells (*a*), the cells are transformed into cancer cells, which divide and produce new Rous sarcoma virus. By adding a substance that inhibits the synthesis of DNA in the cells immediately after they have been inoculated with Rous sarpothesis has come from experiments of a different kind. In 1969 Satoshi Mizutani, who had written his doctoral thesis on bacterial viruses, came to my laboratory for postdoctoral training. We decided to ask the question: What is the origin of the enzyme (a protein) responsible for forming proviral DNA using the viral RNA as template? When Mizutani exposed stationary cells to Rous sarcoma virus in the presence of inhibitors of protein synthesis, he found that the cells still became infected. We interpreted this experiment to mean that the enzyme that synthesizes DNA from the viral RNA template is already in existence before the infection.

Somewhat earlier other workers had fractionated virions—the actual virus particles as distinct from the forms assumed by the virus inside cells—and had found RNA polymerases, enzymes that catalyze the synthesis of RNA from its building blocks: four different ribonucleoside triphosphates. In 1967 Joseph Kates and B. R. McAuslan of Princeton University and William Munyon, E. Paoletti and J. T. Grace, Jr., of the Roswell Park Institute had found RNA polymer-

ases in a poxvirus, a large DNA virus. Other workers had found another RNA polymerase in a reovirus, a double-strand RNA virus. Therefore we decided to look in the virions of Rous sarcoma virus for a DNA polymerase capable of using the viral RNA as a template. After several months of preliminary experiments we succeeded in showing the existence of a DNA polymerase in purified virions of Rous sarcoma virus.

Before discussing this result I should digress briefly to describe the structure of the Rous sarcoma virus [*see illustration on page 356*]. The virion of the Rous sarcoma virus has a diameter of about 100 nanometers, which makes it larger than the particles of the viruses that cause poliomyelitis and smaller than the particles of the viruses that cause smallpox. The virion of the Rous sarcoma virus consists of a lipid-containing envelope (derived by budding from the cell membrane), an inner membrane and a nucleoid, or core, that contains the viral RNA and certain proteins.

In order to demonstrate that the Rous sarcoma virus contains a polymerase capable of producing DNA on an RNA

template, we first treated the virion with a detergent to disrupt its lipid-containing envelope. We then added to the disrupted virus the four deoxyribonucleoside triphosphates that are the building blocks of DNA. One of the deoxyribonucleoside triphosphates was radioactively labeled.

When the mixture was incubated at 40 degrees Celsius, it incorporated the radioactive label into an acid-insoluble substance that met the usual tests for DNA. The substance was stable in the presence of alkali and the enzyme ribonuclease, treatments that are known to destroy RNA, whereas it was attacked and fragmented by an enzyme that destroys DNA. When we repeated the experiment with disrupted virions pretreated with ribonuclease, an enzyme that destroys RNA, little or no DNA was produced, indicating that intact viral RNA was needed as the template for the synthesis of DNA [*see top illustration on page 363*].

After we had announced these results at the Tenth International Cancer Congress in Houston in May, 1970, we learned that David Baltimore of the

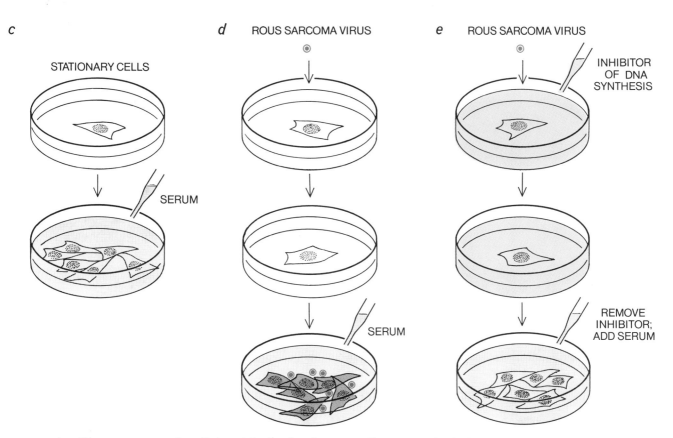

coma virus (*b*), one can protect the cells from infection. In subsequent experiments by the author cultures of stationary, or nondividing, cells were used; when blood serum is added to such cultures (*c*), they divide normally. If such stationary cells are first exposed to Rous sarcoma virus (*d*), however, they become infected but there is no virus production or morphological transformation until serum is added back and the cells divide once again. When the stationary cells are exposed simultaneously to Rous sarcoma virus and to an inhibitor of DNA synthesis (*e*), the cells are not killed but neither are they infected; when the inhibitor of DNA synthesis is removed and serum is added, cells divide normally, are not infected, do not become transformed and do not produce virus.

Massachusetts Institute of Technology had independently made similar observations with the virion of a mouse leukemia virus. The two papers describing these findings were published together in the June 27, 1970, issue of *Nature*, the British scientific weekly. The two publications stimulated an enormous amount of work whose peak is not yet in sight.

In our early papers we called the new viral enzyme RNA-dependent DNA polymerase because the template was RNA and the product was DNA. Subsequently we and others found that the enzyme could also use DNA as a template for DNA synthesis. We therefore decided to change the word "dependent" to "directed," so that we now refer to the enzyme as RNA-directed DNA polymerase. The revised name makes no statement about the origin of the enzyme or its relation to other DNA polymerases. Independently *Nature* began referring to the enzyme as "reverse

transcriptase," a name that I do not like because of its ambiguity but that has gained wide currency.

All the later studies confirm the original finding that the virions of RNA tumor viruses contain a DNA polymerase system that is activated by treating the virion with a detergent and that is sensitive to ribonuclease. Moreover, the virion enzyme functions only as a DNA polymerase; it will not act as an RNA polymerase. As I have mentioned, however, other unrelated RNA viruses do contain an RNA polymerase.

If the DNA produced by the RNA-directed DNA polymerase is isolated free of protein, the size of its molecule can be estimated by spinning it at high speed in a sucrose gradient in an ultracentrifuge. The molecule is surprisingly small: less than a tenth as long as one would expect a copy of the complete viral RNA to be. The reason for the small size is still elusive. If the isolated DNA product is centrifuged in a cesium sulfate density gradient, which separates RNA from

DNA on the basis of their different densities, one finds that the product has the density of DNA [*see bottom illustration on opposite page*]. Further characterization, for example by treatment with enzymes that specifically attack either single- or double-strand DNA, shows that the product of the DNA polymerase system is a double strand. From such studies one can conclude that the DNA polymerase system of the virion makes short pieces of double-strand DNA.

Many workers have demonstrated that the DNA product of the RNA-directed DNA polymerase system has a base sequence complementary to the viral RNA [*see top illustration on page 364*]. This conclusion is drawn from annealing, or molecular hybridization, experiments. Labeled DNA from the virion polymerase reaction is treated so that the strands of the DNA dissociate. The single-strand DNA is added to unlabeled viral RNA, and the mixture is incubated so that complementary strands can form a hybrid combination. The mixture is then centrifuged in a cesium sulfate density gradient. About half of the product DNA forms a band at a density characteristic of RNA or of hybrid RNA-DNA molecules rather than at a density characteristic of DNA. The test is quite specific and indicates that the DNA polymerase of the virion copies the sequence of the bases of the viral RNA into DNA. This experiment, however, still does not demonstrate that such a copying process takes place in cells infected by Rous sarcoma virus.

The viral DNA polymerase was shown to be present in the core of the virion by the following experiment carried out by George Todaro's group at the National Cancer Institute and by John M. Coffin in my laboratory at the University of Wisconsin. Rousvirus virions were treated with a detergent to disrupt the envelope. Then the disrupted virus was centrifuged in a sucrose density gradient. Most of the viral RNA, about 20 percent of the protein and most of the RNA-directed DNA polymerase activity were found to sediment together in "cores," a term given to structures that are denser than whole virions [*see bottom illustration on page 364*]. Further studies showed that with more extensive disruption of the virion the viral DNA polymerase can be freed from the viral RNA and then purified. The purified enzyme is capable of directing the synthesis of DNA on a variety of templates: synthetic and natural DNA, RNA and RNA-DNA hybrids.

The general conclusion from studies in a number of laboratories is that the

FURTHER EXPERIMENTS, carried out by one of the author's students, David E. Boettiger, and independently by Piero Balduzzi and Herbert R. Morgan at the University of Rochester, involved exposing stationary cells to Rous sarcoma virus in the presence of 5-bromodeoxyuridine, an analogue of the DNA constituent thymidine that, when incorporated into DNA, sensitizes the DNA to inactivation by light. As a control some of the treated cells were first not exposed to light (*left*); after serum was added to these cells to enable them to divide they were transformed into cancer cells and began to produce virus. When another culture of treated cells was exposed to light (*right*), the cells were not killed, but the treatment prevented their infection by the virus. When serum was again added to enable these cells to divide, they did not become transformed and did not produce virus.

rousvirus DNA polymerase closely resembles the other DNA polymerases described above that are present in more familiar biological systems and that catalyze the synthesis of DNA on a DNA template. In other words, it is not a unique property of the rousvirus DNA polymerase to be able to use RNA as a template for DNA synthesis. (This was first proposed several years ago by Sylvia Lee Huang and Liebe F. Cavalieri of the Sloan-Kettering Institute.) What is unique so far is the apparent biological role of RNA-directed DNA synthesis in the replication of rousviruses.

Further work in my laboratory has shown that preparations of purified virions of the Rous sarcoma virus contain other enzymes related to DNA replication. The most unusual of them is an enzyme that is named polynucleotide ligase, which repairs breaks in DNA molecules. It is an attractive hypothesis that the function of the ligase is to join the viral DNA to the chromosomal DNA of the host cell, thus integrating the viral genome with the cell genome. After this integration the genetic information of the virus would be replicated with that of the host and passed from the parent cell to the daughter cell. The Rous sarcoma virus virion also contains many other enzymes whose role is completely unknown. We do not know whether they participate in the life cycle of the virus or whether they are merely accidental contaminants picked up in the formation of the virion.

After the first discovery of a DNA polymerase in the virions of RNA tumor viruses, a great many other RNA viruses were examined to see if they contain a similar DNA polymerase system. First it was found that all the viruses previously classified in the RNA tumor virus group contain such an enzyme system. This group of RNA viruses includes both the rousviruses that cause tumors and those that do not cause tumors. Even more interesting, it was found that two types of virus that had not been classified in the same group with RNA tumor viruses also contain a DNA polymerase system. One of these viruses is Visna virus, which causes a slowly developing neurological disease in sheep. After the demonstration of a DNA polymerase in virions of the Visna virus, Kenneth Kaname Takemoto and L. B. Stone at the National Institutes of Health showed that the same virus could cause cancerous transformation of mouse cells in culture. Therefore Visna virus can now be considered a transforming rousvirus. The other type of virus that

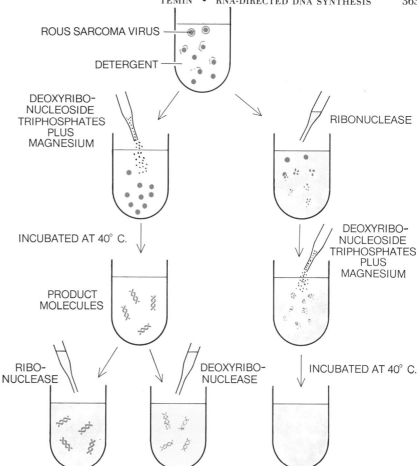

EXISTENCE OF A POLYMERASE capable of producing DNA on an RNA template in RNA tumor viruses was demonstrated by the author and his colleague Satoshi Mizutani (and also independently by David Baltimore of the Massachusetts Institute of Technology). In the experiment conducted by Mizutani and the author purified virions of Rous sarcoma virus were first treated with a detergent to disrupt their lipid-containing envelope. Four deoxyribonucleoside triphosphates, the "building blocks" of DNA, were then added to the disrupted virions. When the mixture was incubated, it incorporated the radioactive label associated with one of the building blocks into an acid-insoluble substance that was stable in the presence of ribonuclease (an enzyme known to destroy RNA), whereas it was fragmented by deoxyribonuclease (an enzyme that destroys DNA). When the experiment was repeated with disrupted virions pretreated with ribonuclease, little or no DNA was produced, indicating that intact viral RNA was needed as template for synthesis of DNA.

has been found to have a DNA polymerase system is the "foamy," or syncytium-forming, viruses. These viruses, isolated from monkeys and cats, have not been connected with any particular disease but are common contaminants of cell cultures. They have not yet been shown to cause tumors or cancerous transformation.

The DNA polymerase present in RNA tumor viruses may not only explain how these viruses produce stable cancerous transformations in the cells they infect but also account for some viral latency,

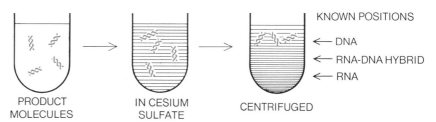

CENTRIFUGATION of the isolated DNA product of the RNA-directed DNA polymerase system in a cesium sulfate density gradient (which separates RNA from DNA on the basis of their different densities) resulted in the finding that the product has the density of DNA. In combination with other findings this result led to the conclusion that the DNA polymerase system of the Rous sarcoma virus virion makes short pieces of double-strand DNA.

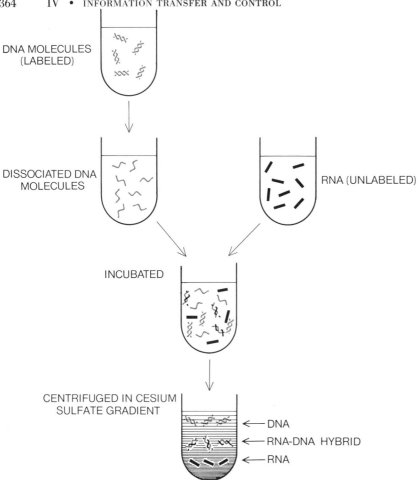

DNA MOLECULES (LABELED)

DISSOCIATED DNA MOLECULES

RNA (UNLABELED)

INCUBATED

CENTRIFUGED IN CESIUM SULFATE GRADIENT

← DNA
← RNA-DNA HYBRID
← RNA

MOLECULAR-HYBRIDIZATION EXPERIMENTS demonstrated that the DNA product of the RNA-directed DNA polymerase system within the virion copies the sequence of bases of the viral RNA into DNA. Labeled DNA from the virion polymerase reaction was first treated so that strands of the DNA dissociated. The single-strand DNA was then added to unlabeled viral RNA, and the mixture was incubated at high temperature so that complementary strands could form a hybrid combination. When the resulting "annealed" mixture was centrifuged in a cesium sulfate density gradient, about half of the product DNA was observed to form a band at a density characteristic of hybrid RNA-DNA molecules.

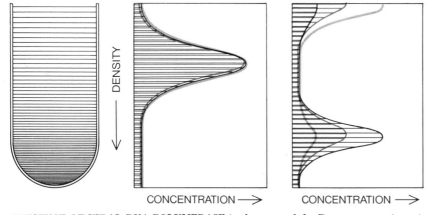

DENSITY

CONCENTRATION →   CONCENTRATION →

PRESENCE OF VIRAL DNA POLYMERASE in the cores of the Rous sarcoma virus virions was demonstrated by John M. Coffin in the author's laboratory. The curves at center show the density distribution of various radioactively labeled constituents of the whole virions as determined by centrifugation in a sucrose density gradient (*left*). The curves at right show the density distribution of the same constituents determined by centrifugation after the virions were treated with a detergent to disrupt their envelopes. Most of the viral RNA (*black curves*), about 20 percent of the protein (*gray curves*) and most of the RNA-directed DNA polymerase activity (*colored curves*) of the disrupted virions were found to sediment together at a higher density than the corresponding constituents of the whole virions, indicating that these constituents are concentrated in the cores of the virions.

the phenomenon in which a virus disappears after infecting an organism only to reappear months or years later. Once an RNA virus has transferred its genetic information to DNA, it would be able to remain latent in a cell and be replicated by the cellular enzyme systems that replicate and repair the cell DNA. After some later activation the virus could appear again as infectious virus particles [*see top illustration on opposite page*].

About a year ago considerable public excitement was generated by the reported discovery of "RNA-dependent DNA polymerases" in human tumor cells. The general conclusion I would draw now from most of this work has been stated above: All DNA polymerases are capable, under the appropriate conditions, of transcribing information from RNA into DNA. At present we lack generally accepted criteria for determining whether or not such syntheses have any biological role or any relation to rousviruses.

In my laboratory we have taken a slightly different approach to the question of RNA-directed DNA synthesis in cells. We have used detergent activation and ribonuclease sensitivity as criteria in a broad search for DNA polymerase systems in a variety of animal cells. That is, we have looked in cells for a DNA polymerase system similar to viral "cores." Coffin has found such a DNA polymerase system in normal, uninfected rat embryo cells. So far we do not know the full significance of this discovery, but it suggests that ribonuclease-sensitive DNA polymerase systems are present in cells other than tumor cells or virus-infected cells.

For many years I have favored the idea that RNA-directed DNA synthesis may be important in normal cellular processes, particularly those involved in the embryonic differentiation of cells. This idea has been expanded in the form of the protovirus hypothesis [*see bottom illustration on opposite page*]. The general idea is that in normal cells there are regions of DNA that serve as templates for the synthesis of RNA, and that this RNA serves in turn as a template for the synthesis of DNA that subsequently becomes integrated with the cellular DNA. By this means certain regions of DNA can be amplified. With additional processes that introduce changes in the DNA, the DNA of different cells can be made different. This difference might serve as a means of distinguishing different cells.

What, then, are the general implications of this work for the prevention or treatment of human cancer? We can

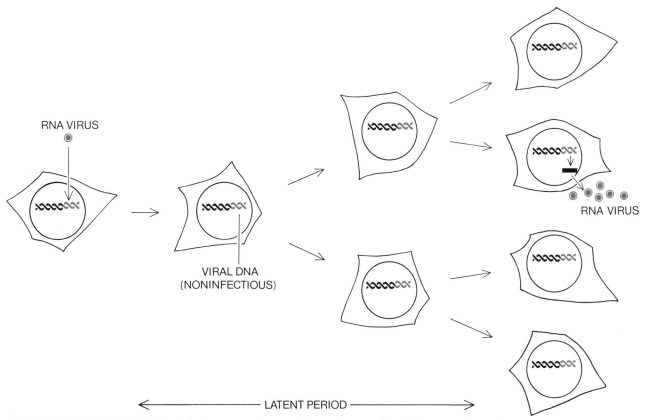

LATENCY OF RNA VIRUSES after infecting an organism may be attributable to the DNA polymerase system present in the cores of such viruses. After transferring its genetic information to DNA in the cell nucleus (*left*), the RNA virus would disappear, remaining latent in the cell by virtue of its replication by the cellular enzyme systems that replicate and repair the cell's DNA (*left center, right center*). Months or years later some form of activation could then cause the infectious RNA virions to appear again (*upper right*).

conclude only that some biological systems utilize a previously undescribed mode of information transfer: from RNA to DNA. It is an interesting coincidence that this new mode of information transfer was first discovered in tumor-causing viruses. We cannot say, however, that RNA-directed DNA synthesis is an exclusive property of such viruses. What the discovery of RNA-directed DNA synthesis does mean is that we now have some simple biochemical tests to determine whether or not newly discovered human viruses are members of the same group as the RNA viruses that produce tumors and cancerous transformations in animal cells, and to look for information related to these viruses in human cancers. We cannot now say that inhibitors of RNA-directed DNA synthesis would have any effect on human cancer. In rousvirus-induced tumors in animals the synthesis of new viral DNA appears to be important only at the initial stage of cancerous transformation, not thereafter.

Probably the most important implication of this discovery for the understanding of cancer in man has been the removal of the dichotomy between viral and genetic theories of the origin of cancer. At a time when genes were thought to consist of DNA alterable only by mutation, and when most of the known cancer-causing animal viruses were of the RNA type, it was hard to imagine common features of genetic and viral theories. Now that we have uncovered evidence that cancer-causing RNA viruses can produce a DNA transcript of the viral RNA, one can readily formulate hypotheses in which elements related to viral RNA are attached to the genome of the cell and transmitted genetically to become activated at some future time and cause "spontaneous" cancer. Experiments designed to test this idea are now in progress in a number of laboratories around the world.

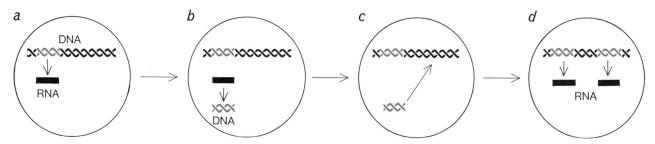

PROTOVIRUS HYPOTHESIS, put forward by the author, embodies the idea that RNA-directed DNA synthesis may be important in normal cellular processes. According to this view, there are regions of DNA in normal cells that serve as templates for the synthesis of RNA (*a*). This RNA serves in turn as a template for the synthesis of DNA (*b*), which later becomes integrated with the cellular DNA (*c*). The amplification of certain regions of DNA resulting from the repetition of the process (*d*) may, in conjunction with additional processes that introduce changes in the DNA, play an important role in the embryonic differentiation of cells.

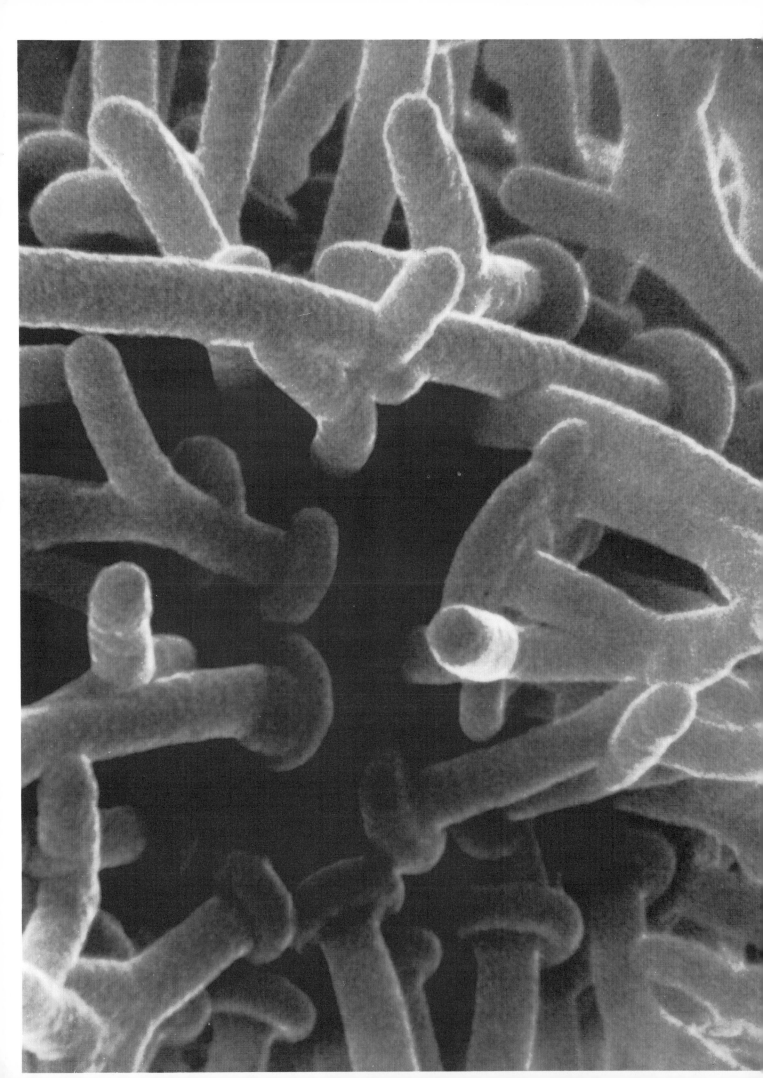

37

CELLULAR COMMUNICATION

GUNTHER S. STENT
September 1972

*Cells communicate by means of hormones and nerve
fibers. Such communication and all other forms of
communication are founded on the information
incorporated in the molecules of nucleic acid*

The capacity to communicate is a fundamental feature of living cells. As John R. Pierce notes in the introduction to this issue, the types of information that are the subject of cellular communication can be grouped into three general classes: genetic, metabolic and nervous. The genetic information of an organism is embodied in the precise sequence of the four kinds of nucleotide base—adenine, guanine, thymine and cytosine—in the DNA molecules of the nuclei of its cellular constituency. The meaning of that information is the specification of the precise sequence of the 20 amino acids in a myriad of different kinds of protein molecule. It is the ensemble of cellular proteins that functions to make the cell what it is: an engine built of highly specific structural members and enzymes that carries out a complex network of catalytically facilitated metabolic reactions. In the course of cell reproduction the parental DNA molecules are replicated and each of the two daughter cells is endowed with a complete store of the genetic information of the mother cell.

In addition to this "vertical" inheritance, in which each cell receives its genetic information from only one parent cell, there is a "horizontal" mode of inheritance, in which cells communicate and thus exchange their genetic information to give rise to offspring of mixed parentage. There are at least four different mechanisms by which such communication of genetic information is re-

alized. The most primitive is genetic transformation, in which a donor cell simply releases some of its DNA molecules into its surroundings. One of the released donor DNA molecules is then taken up by a recipient cell, which incorporates the molecule into its own genetic structures. The second mechanism is genetic transduction. On the infection of a host cell with a virus one or another of the host-cell DNA molecules is incorporated into the shell of one of the progeny virus particles to which the infection gives rise. After its release from the host cell the bastard virus particle infects a recipient cell and thereby transfers to the new host the donor-cell DNA molecule it brought along for the ride. The third mechanism is genetic conjugation. Here two cells meet and establish a conjugal tube, or thin bridge, between them. In the conjugal act the donor member of the cell pair mobilizes a part of its complement of DNA and passes it through the tube to the recipient member.

The natural occurrence of these three mechanisms has so far been demonstrated only for bacteria, the lowest form of cellular life. Higher forms of life, from fungi and protozoa up to man, employ the fourth and most elaborate mechanism of genetic communication, namely sex. Here two gametes, cell types specialized for just such intercourse, meet, fuse and pool their entire complement of DNA. The process gives rise to offspring to which both parents have com-

municated an equal amount of genetic information. This result is to be contrasted with what happens in the first three processes, where the offspring come into possession of only a small fraction of the total donor-cell DNA and are endowed mainly with the genetic information of the recipient cell.

The biological function of genetic communication is to increase the evolutionary plasticity of the species. The ultimate source of the genetic diversity on which natural selection, the motor that drives evolution, feeds are rare changes, or mutations, in the nucleotide base sequence and hence the genetic information contained in the DNA. Thus in the purely vertical mode of inheritance genetic diversity among the members of a cell population all descended from a single ancestral mother cell could be built up very slowly by the accumulation of mutations in the individual lines of descent. In the horizontal mode of inheritance and its intercellular communication, however, there develops quite quickly a rich individual genetic diversity as mutations that have arisen in different lines of descent are continually combined and recombined among the offspring of the interbreeding population. Thus in anticipation of any environmental changes that may affect its fitness the population presents for natural selection a spectrum of diverse types among which one or another may possess a greater fitness for the future state of the species.

Metabolic information is embodied in the quality and concentration of large and small molecules that participate in the chemical processes by which cells reproduce, develop and maintain their living state. In contrast to the communication of genetic information,

TERMINAL ENDS OF NERVE-CELL AXONS in the nervous system of the "sea hare" *Aplysia* are the button-like objects in the scanning electron micrograph on the opposite page. Information is communicated from cell to cell at the junctions where the axons make synaptic contact. Micrograph, which enlarges structures 30,000 diameters, is by Edwin R. Lewis, Thomas E. Everhart and Yehoshua Y. Zeevi of University of California at Berkeley.

whose utility pertains to time periods that are long compared with the lifespan of the members of populations, the communication of metabolic information is of importance to organized societies of cells over much shorter intervals. Here the mechanism of communication gen-

erally consists in the release by a secretory cell of one of its constituent molecules. The released molecule diffuses through the space occupied by the cell society, and on encountering a target cell it intervenes in some highly specific manner in the metabolism of that cell.

Such messenger molecules of metabolic information are generally called hormones.

The biological function of metabolic communication is mainly twofold. In the first instance hormones control the orderly development of multicellular animals

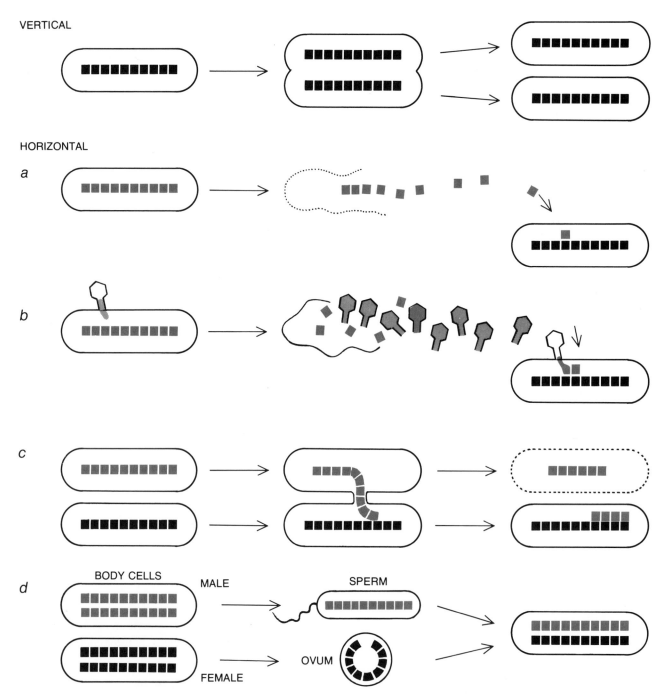

TYPES OF GENETIC COMMUNICATION can be classified as vertical or horizontal. In cellular reproduction (top) the communication is vertical; the replication of the parental genes before cell division provides both offspring with complete and identical sets of genetic information. The three simpler kinds of horizontal communication are found in bacteria. In transformation (a) the donor cell releases the DNA molecules embodying its genetic information and some of the molecules are taken up by the recipient cell. In transduction (b), which may occur when a virus infects a cell and new viruses are formed, one of the new viruses may incorporate a

DNA molecule into its shell; if the new virus now infects another cell, the donor-cell molecule and the information it embodies are transferred to the recipient cell. In conjugation (c) donor and recipient cells come into physical contact; a bridge is formed between them and some donor DNA molecules are transferred to the recipient. The fourth and most elaborate kind of horizontal genetic communication (d) is sex. Parental organisms give rise to male (color) and female (black) gametes: specialized cells that contain only half of the information an offspring requires. When these fuse, each parent gives the offspring an equal amount of information.

and plants. These organisms represent societies made up of millions or billions of cells that come into being by a series of successive cycles of cell growth and division from a single fused pair of gametes, for example from the sperm-fertilized egg. This process of cell multiplication is accompanied by the process of cell differentiation, in which each cell acquires the particular molecular ensemble that enables it to play its destined specialized role in the life of the organism. In many cases cells receive from hormones their instructions concerning how and when to differentiate in the developmental sequence that leads to the adult organism. The female sex hormone estrogen, which belongs to the steroid class of molecules, is a well-known example of such a developmental messenger. Estrogen is released by secretory cells in the ovary of the female animal, particularly at the onset of puberty. Estrogen reaches target cells in almost all tissues and induces in these cells the metabolic reactions that eventually lead to the development of the secondary sexual characteristics of the body.

In the second instance hormones serve in the homeostatic processes by means of which all organisms minimize for their internal environment the consequences of changes in the external environment. The protein hormone insulin, well known through its connection with diabetes, is an example of such a homeostatic chemical messenger. The rate of release of insulin by secretory cells of the pancreas is accelerated in response to high glucose concentrations in the blood. On reaching target cells in liver and muscles insulin signals to these organs to remove glucose from the blood and either to store it in the form of glycogen or to burn it. Once, thanks to this increase in removal rates, the glucose blood concentration has returned to its normal level, the release of insulin by the pancreas is slowed down and so is the removal of glucose from the blood by liver and muscles. Thus insulin makes possible the maintenance of a relatively constant blood-sugar concentration in the face of great fluctuations of the animal's rate of sugar intake.

Nervous information is embodied in the activity of a special cell type possessed by all multicellular animals, the nerve cell or neuron. Although the physiological time spans over which the communication of metabolic information is relevant are much shorter than the evolutionary periods for which the communication of genetic information is intended, these physiological time spans still extend over hours, days or weeks. In

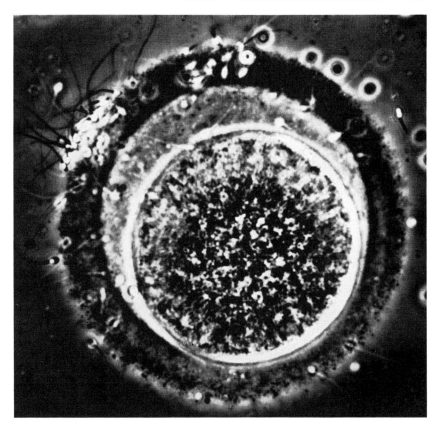

HUMAN OVUM AND SPERM CELLS at the moment of fertilization exemplify the union of gametes and the pooling of an equal amount of genetic information from each parent. Of the numerous sperm cells surrounding the ovum only one will actually deliver the male parental genes. This micrograph, which enlarges the cells some 400 diameters, was made by Landrum B. Shettles of the Columbia University College of Physicians and Surgeons.

order to stay alive, however, most animals must respond to certain events in their environment within time spans of seconds or even milliseconds. And since the diffusion of a molecule such as glucose through the space occupied by the cell society that makes up even so small an animal as a fly requires a few hours, animals must have communication channels that are faster than those provided by hormones. These channels are provided by neurons, and the biological function of the communication of nervous information they perform is to generate the rapid stimulus-response reactions that comprise the animal's behavior.

Neurons are endowed with two singular features that make them particularly suitable for this purpose. First, unlike most other cell types, they possess long and thin extensions: axons. With their axons neurons reach and come into contact with other neurons at distant sites and thereby form an interconnected network extending over the entire animal body. Second, unlike most other cell types, neurons give rise to electrical signals in response to physical or chemical stimuli. They conduct these signals along their axons and transmit them to

other neurons with which they are in contact. The interconnected network of neurons and its traffic of electrical signals forms the nervous system. It is, of course, the nervous system that is both the source and the destination of all the information with which the diverse communication systems under discussion in this issue are concerned.

Like Roman Gaul, the nervous system is divisible into three parts: (1) an input, or sensory, part that informs the animal about its condition with respect to the state of its external and internal environment; (2) an output, or effector, part that produces motion by commanding muscle contraction, and (3) an internuncial part (from the Latin *nuncius*, meaning messenger) that connects the sensory and effector parts. The most elaborate portion of the internuncial part, concentrated in the head of those animals that have heads, is the brain.

The processing of data by the internuncial part consists in the main in making an abstraction of the vast amount of data continuously gathered by the sensory part. This abstraction is the result of a selective destruction of portions of

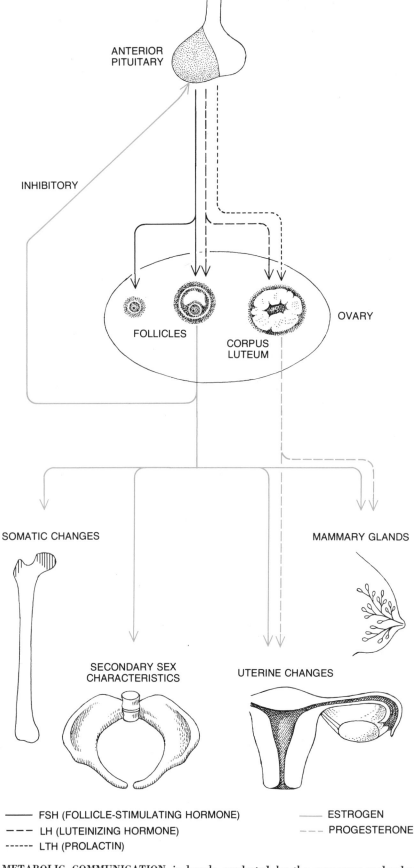

ANTERIOR PITUITARY

INHIBITORY

OVARY

FOLLICLES

CORPUS LUTEUM

SOMATIC CHANGES

MAMMARY GLANDS

SECONDARY SEX CHARACTERISTICS

UTERINE CHANGES

——— FSH (FOLLICLE-STIMULATING HORMONE)
– – – LH (LUTEINIZING HORMONE)
------ LTH (PROLACTIN)

——— ESTROGEN
- - - PROGESTERONE

METABOLIC COMMUNICATION is largely conducted by the messenger molecules known as hormones. One basic role of these chemical messengers is mediation of cell differentiation. Illustrated here is the action of hormones from the anterior pituitary that induce maturation in the human female. The ovarian hormones that are secreted following stimulus by pituitary hormones affect cells in bone, uterine and mammary tissues.

the input data in order to transform these data into manageable categories that are meaningful to the animal. It should be noted that the particular command pattern issued to the muscles by the internuncial part depends not only on here-and-now sensory inputs but also on the history of past inputs. Stated more plainly, neurons can learn from experience. Until not so long ago attempts to fathom how the nervous system actually manages to abstract sensory data and learn from experience were confined mainly to philosophical speculations, psychological formalisms or biochemical naïvetés. In recent years neurophysiologists, however, have made some important experimental findings that have provided for a beginning of a scientific approach to these deep problems. Here I can do no more than describe briefly one example of these recent advances and sketch some of the insights to which it has led.

Before discussing these advances we must give brief consideration to how electrical signals arise and travel in the nervous system. Neurons, like nearly all other cells, maintain a difference in electric potential of about a tenth of a volt across their cell membranes. This potential difference arises from the unequal distribution of the three most abundant inorganic ions of living tissue, sodium (Na^+), potassium (K^+) and chlorine (Cl^-), between the inside of the cell and the outside, and from the low and unequal specific permeability of the cell membrane to the diffusion of these ions. In response to physical or chemical stimulation the cell membrane of a neuron may increase or decrease one or another of these specific ion permeabilities, which usually results in a shift in the electric potential across the membrane. One of the most important of these changes in ion permeability is responsible for the action potential, or nerve impulse. Here there is a rather large transient change in the membrane potential lasting for only one or two thousandths of a second once a prior shift in the potential has exceeded a certain much lower threshold value. Thanks mainly to its capacity for generating such impulses, the neuron (a very poor conductor of electric current compared with an insulated copper wire) can carry electrical signals throughout the body of an animal whose dimensions are of the order of inches or feet. The transient change in membrane potential set off by the impulse is propagated with undiminished intensity along the thin axons. Thus the basic element of signaling in the nervous system is the nerve impulse, and the information transmitted by an axon is encoded in the frequency with which impulses propagate along it.

Neurophysiologists have developed methods by which it is possible to listen to the impulse traffic in a single neuron of the nervous system. For this purpose a recording electrode with a very fine tip (less than a ten-thousandth of an inch in diameter) is inserted into the nervous tissue and brought very close to the surface of a neuron. A neutral electrode is placed at a remote site on the animal's body. Each impulse that arises in the neuron then gives rise to a transient difference in potential between the recording electrode and the neutral electrode. With suitable electronic hardware this transient potential difference can be displayed as a blip on an oscilloscope screen or made audible as a click in a loudspeaker.

The point at which two neurons come into functional contact is called a synapse. Here the impulse signals arriving at the axon terminal of the presynaptic neuron are transferred to the postsynaptic neuron that is to receive them. The transfer is mediated not by direct electrical conduction but by the diffusion of a chemical molecule, the transmitter, across the narrow gap that separates the presynaptic axon terminal from the membrane of the postsynaptic cell. That is to say, the arrival of each impulse at the presynaptic axon terminal causes the release there of a small quantity of transmitter, which reaches the postsynaptic membrane and induces a transient change in its ion permeability. Depending on the chemical identity of the transmitter and the nature of its interaction with the postsynaptic membrane, the permeability change may have one of two diametrically opposite results. On the one hand it may increase the chance that there will arise an impulse in the postsynaptic cell. In that case the synapse is said to be excitatory. On the other hand it may reduce that chance, in which case the synapse is said to be inhibitory. Most neurons of the internuncial part receive synaptic contacts from not just one but many different presynaptic neurons, some axon terminals providing excitatory inputs and others inhibitory ones. Hence the frequency with which impulses arise in any postsynaptic neuron reflects an ongoing process of summation, more exactly a temporal integration, of the ensemble of its synaptic inputs.

We are now ready to proceed to our example of an important advance in the understanding of the internuncial ner-

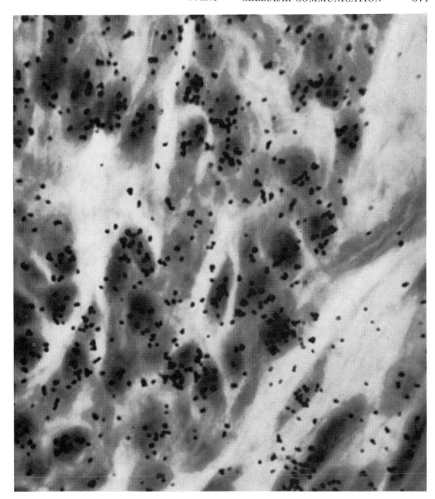

HORMONE AND TARGET CELLS appear in a radioautograph of muscle tissue from a rat uterus. Estradiol, a component of the hormone estrogen that stimulates the growth of uterine muscle, was labeled with tritium; the tissue section was prepared one hour after the hormone was injected. Black dots show concentrations of the hormone. Most are grouped in or near the nuclei of muscle cells. Micrograph was made by W. E. Stumpf, M. Sar and R. N. Prasad of the Laboratories for Reproductive Biology of the University of North Carolina.

vous system, the analysis of the visual pathway in the brain of higher mammals. It is along this pathway that the visual image formed on the retina by light rays entering the eye is transformed into a visual percept, on the basis of which appropriate commands to the muscles are issued. The visual pathway begins at the mosaic of approximately 100 million primary light-receptor cells of the retina. They transform the light image into a spatial pattern of electrical signals, much as a television camera does. Still within the retina, however, the axons of the primary light-receptor cells make synapses with neurons already belonging to the internuncial part of the nervous system. After one or two further synaptic transfers within the retina the signals emanating from the primary light-receptor cells eventually converge on about a million retinal ganglion cells. These ganglion cells send their axons into the optic nerve, which con-

nects the eye with the brain. Thus it is as impulse traffic in ganglion-cell axons that the visual input leaves the eye.

In 1953 Stephen W. Kuffler, who was then working at Johns Hopkins University, discovered that what the impulse traffic in ganglion-cell axons carries to the brain is not raw sensory data but an abstracted version of the primary visual input. This discovery emerged from Kuffler's efforts to ascertain the ganglion-cell receptive field, or that territory of the retinal receptor-cell mosaic whose interaction with incident light influences the impulse activity of individual ganglion cells. For this purpose Kuffler inserted a recording electrode into the immediate vicinity of a ganglion cell in a cat's retina. At the very outset of the study Kuffler made a somewhat unexpected finding, namely that even in the dark, retinal ganglion cells produce impulses at a fairly steady rate (20 to 30 times per second) and that illuminating

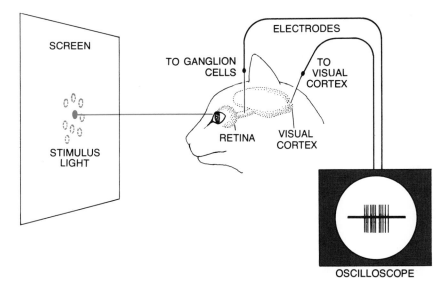

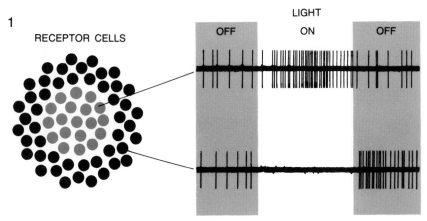

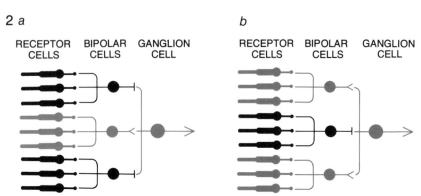

NERVE COMMUNICATION depends for its effectiveness on a process of abstraction that has been demonstrated in studies of visual perception in cats. Use of the experimental situation illustrated at top has revealed the selective destruction of sensory inputs at three successively higher levels of organization within the visual system. At the first of these levels impulses from ganglion cells in the retina, displayed on an oscilloscope, indicate that the cells receive signals from circular arrays of receptor cells. The center and the peripheral cells of the array react differently when stimulated by a spot of light. If the array is an "on center" one (1, color), the number of impulses produced by the ganglion cell will increase when light falls on the center but diminish when light falls on the periphery; the reaction is reversed if the array is an "off center" one. An oversimplified diagram of the neuronal circuits involved (2) suggests that in on-center arrays the central receptor cells (a, color) feed an excitatory synapse on the ganglion cell while the peripheral cells feed inhibitory synapses. Off-center arrays (b) would have the reverse circuitry. Thus the ganglion cell does not report the data on levels of illumination that are collected by the individual receptor cells but sends to the next level of abstraction a summary of the contrast between two concentric regions of the receptive field. The successive stages in this process of selective destruction of sensory inputs are illustrated on the opposite page and on the next two pages.

the entire retina with diffuse light does not have any dramatic effect on that impulse rate. This finding suggested paradoxically that light does not affect the output activity of the retina. Kuffler then, however, projected a tiny spot of light into the cat's eye and moved the image of the spot over various areas of the retina. In this way he found that the impulse activity of an individual ganglion cell does change when the light spot falls on a small circular territory surrounding the retinal position of the ganglion cell. That territory is the receptive field of the cell.

On mapping the receptive fields of many individual ganglion cells Kuffler discovered that every field can be subdivided into two concentric regions: an "on" region, in which incident light increases the impulse rate of the ganglion cell, and an "off" region, in which incident light decreases the impulse rate. Furthermore, he found that the structure of the receptive fields divides retinal ganglion cells into two classes: on-center cells, whose receptive field consists of a circular central "on" region and a surrounding circular "off" region, and off-center cells, whose receptive field consists of a circular central "off" region and a surrounding circular "on" region. In both the on-center and the off-center cells the net impulse activity arising from partial illumination of the receptive field is the result of an algebraic summation: two spots shining on different points of the "on" region give rise to a more vigorous response than either spot alone, whereas one spot shining on the "on" and the other on the "off" region give rise to a weaker response than either spot alone. Uniform illumination of the entire receptive field, the condition that exists under diffuse illumination of the retina, gives rise to virtually no response because of the mutual cancellation of the antagonistic responses from "on" and "off" regions.

It could be concluded, therefore, that the function of retinal ganglion cells is not so much to report to the brain the intensity of light registered by the primary receptor cells of a particular territory of the retina as it is to report the degree of light and dark contrast that exists between the two concentric regions of its receptive field. As can be readily appreciated, such contrast information is essential for the recognition of shapes and forms in the animal's visual field, which is what the eyes are mainly for. Thus we encounter the first example in this discussion of how the nervous system abstracts information by selective destruction of information. The absolute light-intensity data gathered by the primary light-receptor cells are selectively destroyed in the algebraic summation process of "on" and "off" responses and are thereby transformed into the perceptually more meaningful relative-contrast data.

When one thinks about the neuronal circuits that might be responsible for this retinal abstraction process, the first possibility that comes to mind is that they embody the antagonistic function of excitatory and inhibitory synaptic inputs to the same postsynaptic neuron. One might suppose that to produce an on-center receptive field the axon terminals of primary receptor cells from the central "on" territory simply make excitatory synapses with their retinal ganglion cell and primary cells from the peripheral "off" territory make inhibitory synapses [*see illustration on opposite page*]. Detailed analyses of the anatomy and physiology of retinal neurons conducted in recent years have shown that on the one hand the real situation is much more complicated than this simple picture, but that on the other hand the actual neuronal circuits do involve matching of excitatory and inhibitory synapses in the pathways leading from the primary light receptor of antagonistic receptive-field regions to the ganglion cell.

In the late 1950's David H. Hubel and Torsten N. Wiesel, two associates of Kuffler's (who was by then, and still is, at the Harvard Medical School), began to extend these studies on the structure and character of visual receptive fields to the next-highest stage of information processing [see "The Visual Cortex of the Brain," by David H. Hubel; SCIENTIFIC AMERICAN Offprint 168]. For this purpose they examined the further fate of the impulse signals conducted away from the eye by the million or so retinal ganglion-cell axons in the optic nerve to the brain. The optic nerves from the two eyes meet near the center of the head at the optic chiasm. In animals such as cats and men there is a partial crossover of the optic nerves at the optic chiasm; some retinal ganglion-cell axons cross over to the opposite brain hemisphere and some do not. This partial crossover provides the right hemisphere of the brain with the binocular input that the retinas of both eyes receive from the left half of the animal's visual field, and vice versa. Behind the optic chiasm the output of the retinal ganglion cells passes through a way station in the midbrain that for the purposes of this discussion can be consid-

ered a simple neuron-to-neuron replay and finally reaches the cerebral cortex at the back of the head. The cortical destination of the visual input is designated as the visual cortex. Here the incoming axons make synaptic contact with the nerve cells of the cortex. The first cortical cells with which the axon projecting from the eye comes in contact in turn send their axons to other cells in the visual cortex for further processing of the visual input. From that point one must still find the trail that eventually leads to the motor centers of the brain, where, if the visual stimulus is to elicit a behavioral act, commands must be issued to the muscles.

Hubel and Wiesel's procedure was to insert a recording electrode into the visual cortex of a cat and to observe the impulse activity of individual cortical neurons in response to various light stimuli projected on a screen in front of the cat's eyes. In this way they found that

these higher-order neurons of the visual pathway also respond only to stimuli falling on a limited retinal territory of light-receptor cells. The character of the receptive fields of cortical neurons turned out, however, to be dramatically different from that of the retinal ganglion cells. Instead of having circular receptive fields with concentric "on" and "off" regions, the cortical neurons were found to respond to straight edges of light-dark contrast, such as bright bars on a dark background. Furthermore, for the straight edge to produce its optimum response it must be in a particular orientation in the receptive field. A bright bar projected vertically on the screen that produces a vigorous response in a particular cortical cell will no longer elicit the response as soon as its projection is tilted slightly away from the vertical.

In their first studies Hubel and Wiesel found two different classes of cells in the

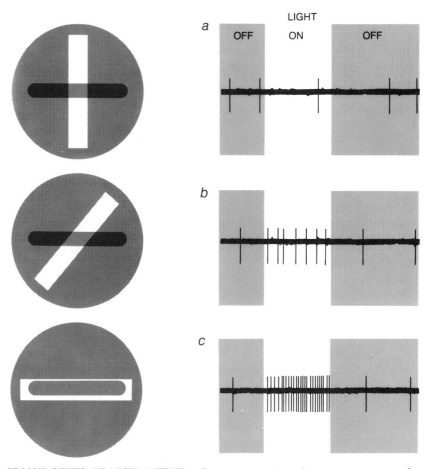

SECOND LEVEL OF ABSTRACTION in the processing of visual inputs occurs among the "simple" cells of the visual cortex, which receive summarized data from many ganglion cells. Cortical-cell impulses increase only when the many retinal receptor cells providing input data are stimulated by light-dark contrast in the form of a straight edge, such as a bar of light shown on a dark background. Moreover, the simple cortical cells fail to react if the edge is not precisely oriented, as the difference in oscilloscope impulse frequencies (a–c) indicates. In effect, by means of selective destruction of input data, the simple cortical cells transform the ganglion-cell information about light-dark contrast at various points in the visual field into information about light-dark contrasts along straight-line sets of points.

visual cortex: simple cells and complex cells. The response of simple cells demands that the straight-edge stimulus must have not only a given orientation but also a precise position in the receptive field. The stimulus requirements of complex cells are less demanding, however, in that their response is sustained on a parallel displacement (but not a tilt) of the straight-edge stimuli within the receptive field. Thus the process of abstraction of the visual input begun in the retina is continued at higher levels in the visual cortex. The simple cells, which are evidently the next abstraction stage, transform the data supplied by the retinal ganglion cells concerning the light-dark contrast at individual points of the visual field into information concerning the contrast present at particular straight-line sets of points. This transformation is achieved by the selective destruction of the information concerning just how much contrast exists at just which point of the straight-line set. The complex cells carry out the next stage of abstraction. They transform the contrast data concerning particular straight-line sets of visual-field points into information concerning the contrast present at parallel sets of straight-line point sets. In other words, here there is a selective destruction of the information concerning just how much contrast exists at each member of a set of parallel straight lines.

The neuronal circuits responsible for these next stages of abstraction of the visual input can now be fathomed. Let us consider first the simple cell of the visual cortex that responds best to a bright bar on a dark background projected in a particular orientation and position on the retinal receptor-cell mosaic. Here we may visualize the simple cell being so connected to the output of

the retina that it receives synaptic inputs from axons reporting the impulse activity of a set of on-center retinal ganglion cells with receptive fields arranged in a straight line. Therefore a bright bar falling on all the central "on" regions of this row of receptive fields but on none of the peripheral "off" regions will activate the entire set of retinal ganglion cells and provide maximal excitation for the simple cortical cell. If the retinal projection of the bar is slightly displaced or tilted, however, some light will also strike the peripheral "off" regions and the excitation provided for the simple cell is diminished.

We next consider the response of a complex cortical cell to a bright bar of a particular orientation in any one of several parallel positions in the receptive field. This response can be easily explained on the basis that the complex cell receives its synaptic inputs from the axons of a set of simple cortical cells. All the simple cells of this set would have receptive fields that respond optimally to a bright bar projected in the same field orientation, but they differ in the field position of their optimal response. A suitably oriented bright bar projected anywhere in the complex receptive field will always activate one of the component simple cells and so also the complex cell.

In their later work Hubel and Wiesel were able to identify cells in the visual cortex whose optimal stimuli reflect even higher levels of abstraction than parallel straight lines, such as straight-line ends and corners. It is not so clear at present how far this process of abstraction by convergence of communication channels ought to be imagined as going. In particular, should one think that there exists for every pattern of whose specific recognition an animal is capable at least

one particular cell in the cerebral cortex that responds with impulse activity when that pattern appears in the visual field? In view of the vast number of such patterns we recognize in a lifetime, that might seem somewhat improbable. So far, however, no other plausible explanation of perception capable of advancing neurophysiological research appears to have been put forward.

Admittedly, ever since the discipline of neurophysiology was founded more than a century ago, there have been adherents of a "holistic" theory of the brain. This theory envisions specific functions of the brain, including perception, depending not on the activity of particular localized cells or centers but flowing instead from general and widely distributed activity patterns. With the discovery of functionally specialized brain loci such as the visual cortex the holistic theory has had to retreat from its original extreme position, yet it may still hold in some more limited way. Quite aside from being hard to fathom, however, the theory seems to be better suited to inspiring experiments that show the defects of the localization concept than to explaining how the brain might actually work.

In any case the findings on the nature of nervous communication described here have some important general implications, in that they lend physiological support to the latter-day philosophical view that has come to be known as "structuralism." In recent years the structuralist view emerged more or less simultaneously, independently and in different guises in diverse fields of study, for example in analytical psychology, cognitive psychology, linguistics and anthropology. The names most often associated with each of these developments are those of Carl Jung, Wolfgang Köhler, Noam Chomsky and Claude Lévi-Strauss. The emergence of structuralism represents the overthrow of "positivism" (and its psychological counterpart "behaviorism") that held sway since the late 19th century and marks a return to Immanuel Kant's late-18th-century critique of pure reason. Structuralism admits, as positivism does not, the existence of innate ideas, or of knowledge without learning. Furthermore, structuralism recognizes that information about the world enters the mind not as raw data but as highly abstract structures that are the result of a preconscious set of step-by-step transformations of the sensory input. Each transformation step involves the selective destruction of information,

a   *b*

THIRD LEVEL OF ABSTRACTION occurs when data from the simple cortical cells reach the complex cortical cells. These higher nerve cells increase their impulses only when the retinal stimulus consists of a bar of light in motion across the visual field. Moreover, only the vertical movement of a horizontal bar through the field (*a*) induces the complex cells of the cortex to respond; a vertical bar moved horizontally (*b*) goes unnoticed. A diagram of the circuitry responsible for this type of abstraction appears on the opposite page.

according to a program that preexists in the brain. Any set of primary sense data becomes meaningful only after a series of such operations performed on it has transformed the data set into a pattern that matches a preexisting mental structure. These conclusions of structuralist philosophy were reached entirely from the study of human behavior without recourse to physiological observations. As the experimental work discussed in this article shows, however, the manner in which sensory input into the retina is processed along the visual pathway corresponds exactly to the structuralist tenets.

It should be mentioned in this connection that studies on the visual cortex of monkeys have led to results entirely analogous to those obtained with cats, namely that there are simple and complex cells responding to parallel straight-line patterns in the visual field. It is therefore reasonable to expect that the organization of the human visual cortex follows the same general plan. That is to say, our own visual perception of the outer world is filtered through a stage in which data are processed in terms of straight parallel lines, thanks to the way in which the input channels coming from the primary light-receptors of the retina are hooked up to the brain. This fact cannot fail to have profound psychological consequences; evidently a geometry based on straight parallel lines, and hence by extension on plane surfaces, is most immediately compatible with our mental equipment. It need not have been this way, since (at least from the neurophysiological point of view) the retinal ganglion cells could just as well have been connected to the higher cells in the visual cortex in such a way that their concentric on-center and off-center receptive fields form arcs rather than straight lines. If evolution had given rise to that other circuitry, curved rather than plane surfaces would have been our primary spatial concept. Thus neurobiology has now shown why it is human—and all too human—to hold Euclidean geometry and its nonintersecting coplanar parallel lines to be a self-evident truth. Non-Euclidean geometries of convex or concave surfaces, although our brain is evidently capable of conceiving them, are more alien to our built-in spatial-perception processes. Apparently a beginning has now been made in providing, in terms of cellular communication, an explanation for one of the deepest of all philosophical problems: the relation between reality and the mind.

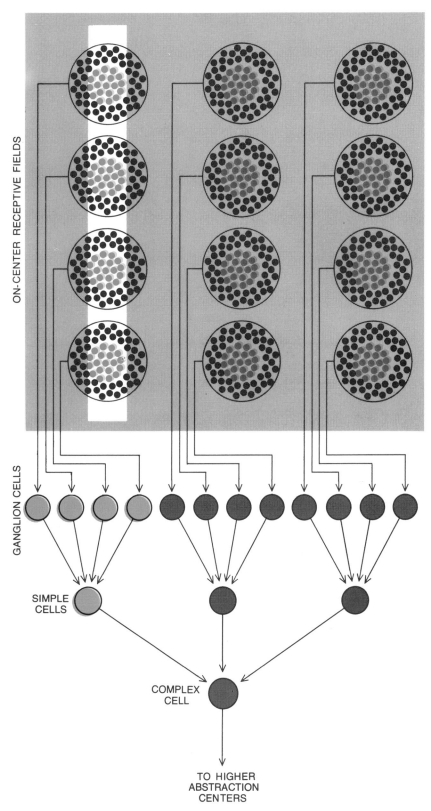

CIRCUITRY OF ABSTRACTION inferred to exist within a cat's visual cortex is illustrated schematically. It begins with a rectangular array of receptive fields. The array is comprised of three parallel vertical rows and each row contains four receptive fields. The axons from the retinal ganglion cells that correspond to each row converge on one of three simple cortical cells; the axons from the simple cells then converge on one complex cortical cell. A bar of light (left) falling on a row of receptive fields excites the four ganglion cells corresponding to that row and also the simple cortical cell connected to them (color).

CONCLUDING ESSAY

38

PREMATURITY AND UNIQUENESS IN SCIENTIFIC DISCOVERY

GUNTHER S. STENT

December 1972

A molecular geneticist reflects on two general historical questions: (1) What does it mean to say a discovery is "ahead of its time"? (2) Are scientific creations any less unique than artistic creations?

The fantastically rapid progress of molecular genetics in the past 25 years now obliges merely middle-aged participants in its early development to look back on their early work from a depth of historical perspective that for scientific specialties flowering in earlier times came only after all the witnesses of the first blossoming were long dead. It is as if the late-18th-century colleagues of Joseph Priestley and Antoine Lavoisier had still been active in chemical research and teaching in the 1930's, after atomic structure and the nature of the chemical bond had been revealed. This somewhat depressing personal vantage provides a singular opportunity to assay the evolution of a scientific field. In reflecting on the history of molecular genetics from the viewpoint of my own experience I have found that two of its most famous incidents—Oswald Avery's identification of DNA as the active principle in bacterial transformation and hence as genetic material, and James Watson and Francis Crick's discovery of the DNA double helix—illuminate two general problems of cultural history. The case of Avery throws light on the question of whether it is meaningful or merely tautologous to say that a discovery is "ahead of its time," or premature. And the case of Watson and Crick can be used, and in fact has been used, to discuss the question of whether there is anything unique in a scientific discovery, in view of the likelihood that if Dr. *A* had not discovered Fact *X* today, Dr. *B* would have discovered it tomorrow.

Five years ago I published a brief retrospective essay on molecular genetics, with particular emphasis on its origins. In that historical account I mentioned neither Avery's name nor DNA-mediated bacterial transformation. My essay elicited a letter to the editor by a microbiologist, who complained: "It is a sad and surprising omission that... Stent makes no mention of the definitive proof of DNA as the basic hereditary substance by O. T. Avery, C. M. MacLeod and Maclyn McCarty. The growth of [molecular genetics] rests upon this experimental proof.... I am old enough to remember the excitement and enthusiasm induced by the publication of the paper by Avery, MacLeod and McCarty. Avery, an effective bacteriologist, was a quiet, self-effacing, non-disputatious gentleman. These characteristics of personality should not [cause] the general scientific public... to let his name go unrecognized."

I was taken aback by this letter and replied that I should indeed have mentioned Avery's 1944 proof that DNA is the hereditary substance. I went on to say, however, that in my opinion it is not true that the growth of molecular genetics rests on Avery's proof. For many years that proof actually had little impact on geneticists. The reason for the delay was not that Avery's work was unknown to or mistrusted by geneticists but that it was "premature."

My *prima facie* reason for saying Avery's discovery was premature is that it was not appreciated in its day. By lack of appreciation I do not mean that Avery's discovery went unnoticed, or even that it was not considered important. What I do mean is that geneticists did not seem to be able to do much with it or build on it. That is, in its day Avery's discovery had virtually no effect on the general discourse of genetics.

This statement can be readily supported by an examination of the scientific literature. For example, a convincing demonstration of the lack of appreciation of Avery's discovery is provided by the 1950 golden jubilee of genetics symposium "Genetics in the 20th Century." In the proceedings of that symposium some of the most eminent geneticists published essays that surveyed the progress of the first 50 years of genetics and assessed its status at that time. Only one of the 26 essayists saw fit to make more than a passing reference to Avery's discovery, then six years old. He was a colleague of Avery's at the Rockefeller Institute, and he expressed some doubt that the active transforming principle was really pure DNA. The then leading philosopher of the gene, H. J. Muller of Indiana University, contributed an essay on the nature of the gene that mentions neither Avery nor DNA.

So why was Avery's discovery not appreciated in its day? Because it was "premature." But is this really an explanation or is it merely an empty tautology? In other words, is there a way of providing a criterion of the prematurity of a discovery other than its failure to make an impact? Yes, there is such a criterion: A discovery is premature if its implications cannot be connected by a series of simple logical steps to canonical, or generally accepted, knowledge.

Why could Avery's discovery not be connected with canonical knowledge? Ever since DNA had been discovered in the cell nucleus by Friedrich Miescher in 1869 it had been suspected of exerting some function in hereditary processes. This suspicion became stronger in the 1920's, when it was found that DNA is a major component of the chromosomes. The then current view of the molecular nature of DNA, however, made it well-nigh inconceivable that DNA could be the carrier of hereditary information. First, until well into the 1930's DNA was generally thought to be merely a tetranucleotide composed of one unit each of adenylic, guanylic, thymidylic and cytidylic acids. Second, even when it was finally realized by the early 1940's that the molecular weight of DNA is actually much higher than the tetranucleotide hypothesis required, it was still widely believed the tetranucleotide was the basic repeating unit of the large DNA polymer in which the four units mentioned recur in regular sequence. DNA was therefore viewed as a uniform macromolecule that, like other monotonous polymers such as starch or cellulose, is always the same, no matter what its biological source. The ubiquitous presence of DNA in the chromosomes was therefore generally explained in purely physiological or structural terms. It was usually to the chromosomal protein that the informational role of the genes had been assigned, since the great differences in the specificity of structure that exist between various proteins in the same or-

PICASSO'S "LES DESMOISELLES D'AVIGNON," painted in Paris in 1907, is often cited by art historians as the first major Cubist painting and a milestone in the development of modern art. It is reproduced here as an archetype of the proposition that works of artistic creation are unique (in the sense that if Picasso had not existed, it would never have been painted), whereas works of scientific creation are inevitable (in the sense that if Dr. *A* had not discovered Fact *X* today, Dr. *B* would discover it tomorrow). The validity of the proposition is disputed by the author. The painting is in the collection of the Museum of Modern Art in New York.

ganism, or between similar proteins in different organisms, had been appreciated since the beginning of the century. The conceptual difficulty of assigning the genetic role to DNA had not escaped Avery. In the conclusion of his paper he stated: "If the results of the present study of the transforming principle are confirmed, then nucleic acids must be regarded as possessing biological specificity the chemical basis of which is as yet undetermined."

By 1950, however, the tetranucleotide hypothesis had been overthrown, thanks largely to the work of Erwin Chargaff of the Columbia University College of Physicians and Surgeons. He showed that, contrary to the demands of that hypothesis, the four nucleotides are not necessarily present in DNA in equal proportions. He found, furthermore, that the exact nucleotide composition of DNA differs according to its biological source, suggesting that DNA might not be a monotonous polymer after all. And so when two years later, in 1952, Alfred Hershey and Martha Chase of the Carnegie Institution's laboratory in Cold Spring Harbor, N.Y., showed that on infection of the host bacterium by a bacterial virus at least 80 percent of the viral DNA enters the cell and at least 80 percent of the viral protein remains outside, it was possible to connect their conclusion that DNA is the genetic material with canonical knowledge. Avery's "as yet undetermined chemical basis of the biological specificity of nucleic acids" could now be seen as the precise sequence of the four nucleotides along the polynucleotide chain. The general impact of the Hershey-Chase experiment was immediate and dramatic. DNA was suddenly in and protein was out, as far as thinking about the nature of the gene was concerned. Within a few months there arose the first speculations about the genetic code, and Watson and Crick were inspired to set out to discover the structure of DNA.

Of course, Avery's discovery is only one of many premature discoveries in the history of science. I have presented it here for consideration mainly because of my own failure to appreciate it when I joined Max Delbrück's bacterial virus group at the California Institute of Technology in 1948. Since then I have often wondered what my later career would have been like if I had only been astute enough to appreciate Avery's discovery and infer from it four years before Hershey and Chase that DNA must also be the genetic material of our own experimental organism.

Probably the most famous case of prematurity in the history of biology is associated with the name of Gregor Mendel, whose discovery of the gene in 1865 had to wait 35 years before it was "rediscovered" at the turn of the century. Mendel's discovery made no immediate impact, it can be argued, because the concept of discrete hereditary units could not be connected with canonical knowledge of anatomy and physiology in the middle of the 19th century. Furthermore, the statistical methodology by means of which Mendel interpreted the results of his pea-breeding experiments was entirely foreign to the way of thinking of contemporary biologists. By the end of the 19th century, however, chromosomes and the chromosome-dividing processes of mitosis and meiosis had been discovered and Mendel's results could now be accounted for in terms of structures visible in the microscope. Moreover, by then the application of statistics to biology had become commonplace. Nonetheless, in some respects Avery's discovery is a more dramatic example of prematurity than Mendel's. Whereas Mendel's discovery seems hardly to have been mentioned by anyone until its rediscovery, Avery's discovery was widely discussed and yet it could not be appreciated for eight years.

Cases of delayed appreciation of a discovery exist also in the physical sciences. One example (as well as an explanation of its circumstances in terms of the concept to which I refer here as prematurity) has been provided by Michael Polanyi on the basis of his own experience. In the years 1914–1916 Polanyi published a theory of the adsorption of gases on solids which assumed that the force attracting a gas molecule to a solid surface depends only on the position of the molecule, and not on the presence of other molecules, in the force field. In spite of the fact that Polanyi was able to provide strong experimental evidence in favor of his theory, it was generally rejected. Not only was the theory rejected, it was also considered so ridiculous by the leading authorities of the time that Polanyi believes continued defense of his theory would have ended his professional career if he had not managed to publish work on more palatable ideas. The reason for the general rejection of Polanyi's adsorption theory was that at the very time he put it forward the role of electrical forces in the architecture of matter had just been discovered. Hence there seemed to be no doubt that the adsorption of gases must also involve an elec-

trical attraction between the gas molecules and the solid surface. That point of view, however, was irreconcilable with Polanyi's basic assumption of the mutual independence of individual gas molecules in the adsorption process. It was only in the 1930's, after a new theory of cohesive molecular forces based on quantum-mechanical resonance rather than on electrostatic attraction had been developed, that it became conceivable gas molecules could behave in the way Polanyi's experiments indicated they were actually behaving. Meanwhile Polanyi's theory had been consigned so authoritatively to the ashcan of crackpot ideas that it was rediscovered only in the 1950's.

Still, can the notion of prematurity be said to be a useful historical concept? First of all, is prematurity the only possible explanation for the lack of contemporary appreciation of a discovery? Evidently not. For example, my microbiologist critic suggested that it was the "quiet, self-effacing, non-disputatious" personality of Avery that was the cause of the failure of his contribution to be recognized. Furthermore, in an essay on the history of DNA research Chargaff supports the idea that personal modesty and aversion to self-advertisement account for the lack of contemporary scientific appreciation. He attributes the 75-year lag between Miescher's discovery of DNA and the general appreciation of its importance to Miescher's being "one of the quiet in the land," who lived when "the giant publicity machines, which today accompany even the smallest move on the chess-board of nature with enormous fanfares, were not yet in place." Indeed, the 35-year hiatus

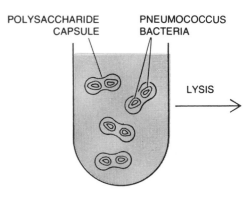

EXPERIMENT OF 1944 with which Oswald Avery correctly identified the chemical nature of the genetic material is regarded by the author as a classic example of a premature scientific discovery. The virulent normal, or S-type, pneumococcus, a bacteriin the appreciation of Mendel's discovery is often attributed to Mendel's having been a modest monk living in an out-of-the-way Moravian monastery. Hence the notion of prematurity provides an alternative to the invocation—in my opinion an inappropriate one for the instances mentioned here—of the lack of publicity as an explanation for delayed appreciation.

More important, does the prematurity concept pertain only to retrospective judgments made with the wisdom of hindsight? No, I think it can be used also to judge the present. Some recent discoveries are still premature at this very time. One example of here-and-now prematurity is the alleged finding that experiential information received by an animal can be stored in nucleic acids or other macromolecules.

Some 10 years ago there began to appear reports by experimental psychologists purporting to have shown that the engram, or memory trace, of a task learned by a trained animal can be transferred to a naïve animal by injecting or feeding the recipient with an extract made from the tissues of the donor. At that time the central lesson of molecular genetics—that nucleic acids and proteins are informational macromolecules—had just gained wide currency, and the facile equation of nervous information with genetic information soon led to the proposal that macromolecules—DNA, RNA or protein—store memory. As it happens, the experiments on which the macromolecular theory of memory is based have been difficult to repeat, and the results claimed for them may indeed not be true at all. It is nonetheless significant that few neurophysiologists have even bothered to check

these experiments, even though it is common knowledge that the possibility of chemical memory transfer would constitute a fact of capital importance. The lack of interest of neurophysiologists in the macromolecular theory of memory can be accounted for by recognizing that the theory, whether true or false, is clearly premature. There is no chain of reasonable inferences by means of which our present, albeit highly imperfect, view of the functional organization of the brain can be reconciled with the possibility of its acquiring, storing and retrieving nervous information by encoding such information in molecules of nucleic acid or protein. Accordingly for the community of neurophysiologists there is no point in devoting time to checking on experiments whose results, even if they were true as alleged, could not be connected with canonical knowledge.

The concept of here-and-now prematurity can be applied also to the troublesome subject of ESP, or extrasensory perception. In the summer of 1948 I happened to hear a heated argument at Cold Spring Harbor between two future mandarins of molecular biology, Salvador Luria of Indiana University and R. E. Roberts of the Carnegie Institution's laboratory in Washington. Roberts was then interested in ESP, and he felt it had not been given fair consideration by the scientific community. As I recall, he thought that one might be able to set up experiments with molecular beams that could provide more definitive data on the possibility of mind-induced departures from random distributions than J. B. Rhine's then much discussed card-guessing procedures. Luria declared that not only was he not

interested in Roberts' proposed experiments but also in his opinion it was unworthy of anyone claiming to be a scientist even to discuss such rubbish. How could an intelligent fellow such as Roberts entertain the possibility of phenomena totally irreconcilable with the most elementary physical laws? Moreover, a phenomenon that is manifest only to specially endowed subjects, as claimed by "parapsychologists" to be the case for ESP, is outside the proper realm of science, which must deal with phenomena accessible to every observer. Roberts replied that far from him being unscientific, it was Luria whose bigoted attitude toward the unknown was unworthy of a true scientist. The fact that not everyone has ESP only means that it is an elusive phenomenon, similar to musical genius. And just because a phenomenon cannot be reconciled with what we now know, we need not shut our eyes to it. On the contrary, it is the duty of the scientist to try to devise experiments designed to probe its truth or falsity.

It seemed to me then that both Luria and Roberts were right, and in the intervening years I often thought about this puzzling disagreement, unable to resolve it in my own mind. Finally six years ago I read a review of a book on ESP by my Berkeley colleague C. West Churchman, and I began to see my way toward a resolution. Churchman stated that there are three different possible scientific approaches to ESP. The first of these is that the truth or falsity of ESP, like the truth or falsity of the existence of God or of the immortality of the soul, is totally independent of either the methods or the findings of empirical science. Thus the problem of ESP is de-

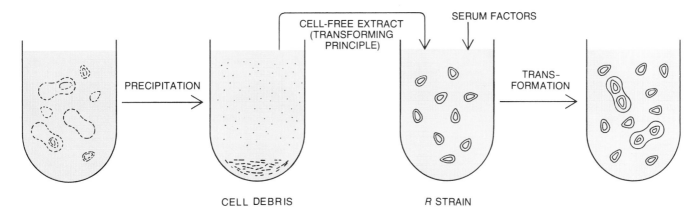

CELL DEBRIS   R STRAIN

um that causes pneumonia in mammals, is enclosed in a smooth (hence S) polysaccharide capsule that protects the bacterium from the ordinary defense mechanisms of the infected animal. The avirulent mutant, or R-type (R for rough), strain has lost the genetic capacity to form this protective capsule and hence is comparatively harmless. When a "transforming principle" extracted from normal

S donor bacteria was added to mutant R recipient bacteria, some of the mutants were found to regain the genetic capacity to form the capsule and thus were transformed back into the normal, virulent S type. Avery purified the transforming principle and succeeded in showing that it is DNA. The significance of Avery's discovery was not appreciated by molecular geneticists until 1952.

fined out of existence. I imagine that this was more or less Luria's position.

Churchman's second approach is to reformulate the ESP phenomenon in terms of currently acceptable scientific notions, such as unconscious perception or conscious fraud. Hence, rather than defining ESP out of existence, it is trivialized. The second approach probably would have been acceptable to Luria too, but not to Roberts.

The third approach is to take the proposition of ESP literally and to attempt to examine in all seriousness the evidence for its validity. That was more or less Roberts' position. As Churchman points out, however, this approach is not likely to lead to satisfactory results. Parapsychologists can maintain with

some justice that the existence of ESP has already been proved to the hilt, since no other set of hypotheses in psychology has received the degree of critical scrutiny that has been given to ESP experiments. Moreover, many other phenomena have been accepted on much less statistical evidence than what is offered for ESP. The reason Churchman advances for the futility of a strictly evidential approach to ESP is that in the absence of a hypothesis of how ESP could work it is not possible to decide whether any set of relevant observations can be accounted for only by ESP to the exclusion of alternative explanations.

After reading Churchman's review I realized that Roberts would have been ill-advised to proceed with his ESP ex-

periments, not because, as Luria had claimed, they would not be "science" but because any positive evidence he might have found in favor of ESP would have been, and would still be, premature. That is, until it is possible to connect ESP with canonical knowledge of, say, electromagnetic radiation and neurophysiology no demonstration of its occurrence could be appreciated.

Is the lack of appreciation of premature discoveries merely attributable to the intellectual shortcoming or innate conservatism of scientists who, if they were only more perceptive or more open-minded, would give immediate recognition to any well-documented scientific proposition? Polanyi is not of that opinion. Reflecting on the cruel fate of his theory half a century after first advancing it, he declared: "This miscarriage of the scientific method could not have been avoided.... There must be at all times a predominantly accepted scientific view of the nature of things, in the light of which research is jointly conducted by members of the community of scientists. A strong presumption that any evidence which contradicts this view is invalid must prevail. Such evidence has to be disregarded, even if it cannot be accounted for, in the hope that it will eventually turn out to be false or irrelevant."

That is a view of the operation of science rather different from the one commonly held, under which acceptance of authority is seen as something to be avoided at all costs. The good scientist is seen as an unprejudiced man with an open mind who is ready to embrace any new idea supported by the facts. The history of science shows, however, that its practitioners do not appear to act according to that popular view.

Five years ago Chargaff wrote one of the many reviews of *The Double Helix*, Watson's autobiographical account of his and Crick's discovery of the structure of DNA. In his review Chargaff observes that scientific autobiography is "a most awkward literary genre." Most such works, he says, "give the impression of having been written for the remainder tables of bookstores, reaching them almost before they are published." The reasons for this, according to Chargaff, are not far to seek: scientists "lead monotonous and uneventful lives and... besides often do not know how to write." Moreover, "there may also be profounder reasons for the general triteness of scientific autobiographies. *Timon of Athens* could not have been written, 'Les Desmoiselles d'Avignon' not have

OLD VIEW of the chemical structure of DNA, widely held until well into the 1930's, saw the molecule as being merely a tetranucleotide composed of one unit each of adenylic, guanylic, thymidylic and cytidylic acids. This hypothesis demanded that the molecular weight of DNA be little more than 1,000 and that the four nucleotide bases (adenine, guanine, thymine and cytosine) occur in exactly equal proportions. Even when it was finally realized in the 1940's that the molecular weight of DNA is much higher (in the millions or billions), it was still widely believed that the tetranucleotide was the basic repeating unit of the large DNA polymer. The mistaken belief in this uniform macromolecular structure proved to be an obstacle to the eventual acceptance of the idea that DNA is the genetic material.

GUANINE   CYTOSINE   ADENINE   THYMINE   THYMINE   CYTOSINE   ADENINE   GUANINE

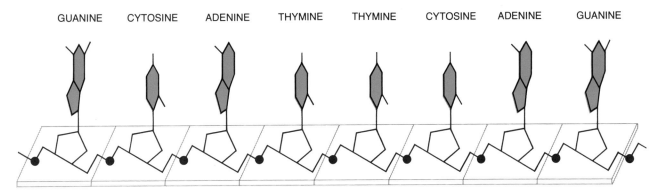

PRESENT VIEW of the chemical structure of DNA sees the molecule as a long chain in which the four nucleotide bases can be arranged in any arbitrary order. Although the proportion of adenine is always equal to that of thymine and the proportion of guanine is always equal to that of cytosine, the ratio of adenine-thymine to guanine-cytosine can vary over a large range, depending on the biological source of the DNA. With the elaboration of this single-strand structure it became possible to envision that genetic information is encoded in the DNA molecule as a specific sequence of the four nucleotide bases (*see illustration on next page*).

been painted, had Shakespeare and Picasso not existed. But of how many scientific achievements can this be claimed? One could almost say that, with very few exceptions, it is not the men that make science, it is science that makes the men. What *A* does today, *B* or *C* or *D* could surely do tomorrow."

On reading this passage, I found myself in full agreement on the general lack of literary skills among men of science. I was surprised, however, to find an eminent scientist embracing historicism (the theory championed by Hegel and Marx holding that history is determined by immutable forces rather than by human agency) as an explanation for the evolution of science while at the same time professing belief in the libertarian "great man" view of history for the evolution of art. Since it had not occurred to me that anyone could hold such contradictory, and to me obviously false, views concerning these two most important domains of human creation, I began to ask scientific friends and colleagues whether they too, by any chance, thought there was an important qualitative difference between the achievements of art and of science, namely that the former are unique and the latter inevitable. To my even greater surprise, I found that most of them seemed to agree with Chargaff. Yes, they said, it is quite true that we would not have had *Timon of Athens* or "Les Desmoiselles d'Avignon" if Shakespeare and Picasso had not existed, but if Watson and Crick had not existed, we would have had the DNA double helix anyway. Therefore, contrary to my first impression, it does not seem to be all that obvious that this proposition has little philosophical or historical merit. Hence I shall now attempt to show that there is no such profound difference between

the arts and sciences in regard to the uniqueness of their creations.

Before discussing the proposition of differential uniqueness of creation it is necessary to make an explicit statement of the meaning of "art" and of "science." My understanding of these terms is based on the view that both the arts and the sciences are activities that endeavor to discover and communicate truths about the world. The domain to which the artist addresses himself is the inner, subjective world of the emotions. Artistic statements therefore pertain mainly to relations between private events of affective significance. The domain of the scientist, in contrast, is the outer, objective world of physical phenomena. Scientific statements therefore pertain mainly to relations between or among public events. Thus the transmission of information and the perception of meaning in that information constitute the central content of both the arts and the sciences. A creative act on the part of either an artist or a scientist would mean his formulation of a new meaningful statement about the world, an addition to the accumulated capital of what is sometimes called "our cultural heritage." Let us therefore examine the proposition that only Shakespeare could have formulated the semantic structures represented by *Timon*, whereas people other than Watson and Crick might have made the communication represented by their paper, "A Structure for Deoxyribonucleic Acid," published in *Nature* in the spring of 1953.

First, it is evident that the exact word sequence that Watson and Crick published in *Nature* would not have been written if the authors had not existed, any more than the exact word sequence of *Timon* would have been written without Shakespeare, at least not until the

fabulous monkey typists complete their random work at the British Museum. And so both creations are from that point of view unique. We are not really concerned, however, with the exact word sequence. We are concerned with the content. Thus we admit that people other than Watson and Crick would eventually have described a satisfactory molecular structure for DNA. But then the character of *Timon* and the story of his trials and tribulations not only might have been written without Shakespeare but also were written without him. Shakespeare merely reworked the story of *Timon* he had read in William Painter's collection of classic tales, *The Palace of Pleasure*, published 40 years earlier, and Painter in turn had used as his sources Plutarch and Lucian. But then we do not really care about Timon's story; what counts are the deep insights into human emotions that Shakespeare provides in his play. He shows us here how a man may make his response to the injuries of life, how he may turn from lighthearted benevolence to passionate hatred toward his fellow men. Can one be sure, however, that *Timon* is unique from this bare-bones standpoint of the work's artistic essence? No, because who is to say that if Shakespeare had not existed no other dramatist would have provided for us the same insights? Another dramatist would surely have used an entirely different story (as Shakespeare himself did in his much more successful *King Lear*) to treat the same theme and he might have succeeded in pulling it off. The reason no one seems to have done it since is that Shakespeare had already done it in 1607, just as no one discovered the structure of DNA after Watson and Crick had already discovered it in 1953.

Hence we are finally reduced to as-

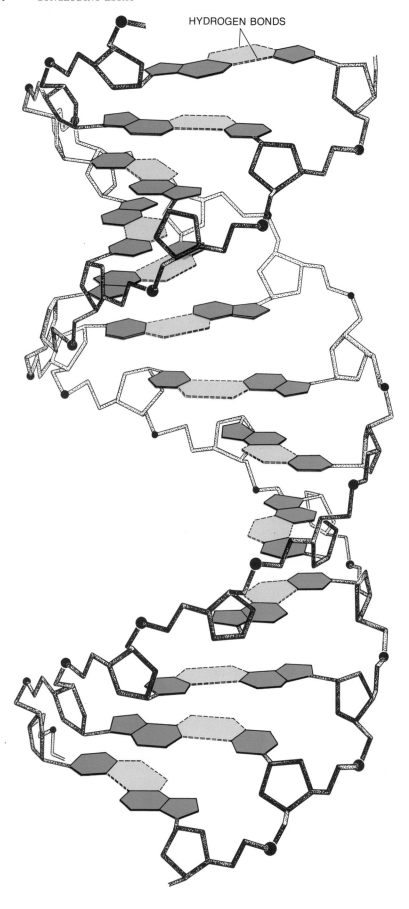

HYDROGEN BONDS

WATSON-CRICK MODEL of the structure of DNA, the discovery of which was announced in 1953, can now be described adequately as a double-strand self-complementary helix.

serting that *Timon* is uniquely Shakespeare's, because no other dramatist, although he might have brought us more or less the same insights, would have done it in quite the same exquisite way as Shakespeare. But here we must not shortchange Watson and Crick and take for granted that those other people who eventually would have found the structure of DNA would have found it in just the same way and produced the same revolutionary effect on contemporary biology. On the basis of my acquaintance with the personalities then engaged in trying to uncover the structure of DNA, I believe that if Watson and Crick had not existed, the insights they provided in one single package would have come out much more gradually over a period of many months or years. Dr. *B* might have seen that DNA is a double-strand helix, and Dr. *C* might later have recognized the hydrogen bonding between the strands. Dr. *D* later yet might have proposed a complementary purine-pyrimidine bonding, with Dr. *E* in a subsequent paper proposing the specific adenine-thymine and guanine-cytosine nucleotide pairs. Finally, we might have had to wait for Dr. *G* to propose the replication mechanism of DNA based on the complementary nature of the two strands. All the while Drs. *H, I, J, K* and *L* would have been confusing the issue by publishing incorrect structures and proposals. Thus I fully agree with the judgment offered by Sir Peter Medawar in his review of *The Double Helix:* "The great thing about [Watson and Crick's] discovery was its completeness, its air of finality. If Watson and Crick had been seen groping toward an answer, if they had published a partly right solution and had been obliged to follow it up with corrections and glosses, some of them made by other people; if the solution had come out piecemeal instead of in a blaze of understanding; then it would still have been a great episode in biological history; but something more in the common run of things; something splendidly well done, but not in the grand romantic manner."

Why is it that so many scientists apparently fail to see that it can be said of both art and science that whereas "what *A* does today, *B* or *C* or *D* could surely do tomorrow," *B* or *C* or *D* might nevertheless not do it as well as *A*, in the same "grand romantic manner." I think a variety of reasons can be put forward to account for this strange myopia. The first of them is simply that most scientists are not familiar with the working methods of artists. They tend to picture the artist's act of creation in the terms of Hollywood: Cornel Wilde in the role of the one and only Frédéric Chopin gazing fondly at Merle Oberon as his muse and mistress George Sand and then sitting down at the Pleyel pianoforte to compose his "Preludes." As scientists know full well, science is done quite differently: Dozens of stereotyped and ambitious researchers are slaving away in as many identical laboratories, all trying to make similar discoveries, all using more or less the same knowledge and techniques, some of them succeeding and some not. Artists, on the other hand, tend to conceive of the scientific act of creation in equally unrealistic terms: Paul Muni in the role of the one and only Louis Pasteur, who while burning the midnight oil in his laboratory has the inspiration to take some bottles from the shelf, mix their contents and thus discover the vaccine for rabies. Artists, in turn, know that art is done quite differently: Dozens of stereotyped and ambitious writers,

painters and composers are slaving away in as many identical garrets, all trying to produce similar works, all using more or less the same knowledge and techniques, some succeeding and some not.

A second reason is that the belief in the inevitability of scientific discoveries appears to derive support from the often-told tales of famous cases in the history of science where the same discovery was made independently two or more times by different people. For instance, the independent invention of the calculus by Leibniz and Newton or the independent recognition of the role of natural selection in evolution by Wallace and Darwin. As the study of such "multiple discoveries" by Robert Merton of Columbia University has shown, however, on detailed examination they are rarely, if ever, identical. The reason they are said to be multiple is simply that in spite of their differences one can recognize a semantic overlap between them that is transformable into a congruent set of ideas.

The third, and somewhat more profound, reason is that whereas the cumulative character of scientific creation is at once apparent to every scientist, the similarly cumulative character of artistic creation is not. For instance, it is obvious that no present-day working geneticist has any need to read the original papers of Mendel, because they have been completely superseded by the work of the past century. Mendel's papers contain no useful information that cannot be better obtained from any modern textbook or the current genetical literature. In contrast, the modern writer, composer or painter still needs to read, listen or look at the original works of Shakespeare, Bach or Leonardo, which, so it is thought, have not been superseded at all. In spite of the seeming truth of this proposition, it must be said that art is no less cumulative than science, in that artists no more work in a traditionless vacuum than scientists do. Artists also build on the work of their predecessors; they start with and later

SCIENTISTS' MISCONCEPTION of the working methods of artists is idealized in this scene from the 1945 Columbia Pictures production *A Song to Remember*. Frédéric Chopin (played by Cornel Wilde), after gazing fondly at his muse George Sand (Merle Oberon), sits down at the Pleyel pianoforte and composes his "Preludes." Science, as any scientist knows, is done quite differently.

improve on the styles and insights that have been handed down to them from their teachers, just as scientists do. To stay with our main example, Shakespeare's *Timon* has its roots in the works of Aeschylus, Sophocles and Euripides. It was those authors of Greek antiquity who discovered tragedy as a vehicle for communicating deep insights into affects, and Shakespeare, drawing on many earlier sources, finally developed that Greek discovery to its ultimate height. To some limited extent, therefore, the plays of the Greek dramatists have been superseded by Shakespeare's. Why, then, have Shakespeare's plays not been superseded by the work of later, lesser dramatists?

Here we finally do encounter an important difference between the creations of art and of science, namely the feasibility of paraphrase. The semantic content of an artistic work—a play, a cantata or a painting—is critically dependent on the exact manner of its realization; that is, the greater an artistic work is, the

more likely it is that any omissions or changes from the original detract from its content. In other words, to paraphrase a great work of art—for instance to rewrite *Timon*—without loss of artistic quality requires a genius equal to the genius of the original creator. Such a successful paraphrase would, in fact, constitute a great work of art in its own right. The semantic content of a great scientific paper, on the other hand, although its impact at the time of publication may also be critically dependent on the exact manner in which it is presented, can later be paraphrased without serious loss of semantic content by lesser scientists. Thus the simple statement "DNA is a double-strand, self-complementary helix" now suffices to communicate the essence of Watson and Crick's great discovery, whereas "A man responds to the injuries of life by turning from lighthearted benevolence to passionate hatred toward his fellow men" is merely a platitude and not a paraphrase of *Timon*. It took the writing

of *King Lear* to paraphrase (and improve on) *Timon,* and indeed the former has superseded the latter in the Shakespearean dramatic repertoire.

The fourth, and probably deepest, reason for the apparent prevalence among scientists of the proposition that artistic creations are unique and scientific creations are not can be attributed to a contradictory epistemological attitude toward the events in the outer and the inner world. The outer world, which science tries to fathom, is often viewed from the standpoint of materialism, according to which events and the relations between them have an existence independent of the human mind. Hence the outer world and its scientific laws are simply there, and it is the job of the scientist to find them. Thus going after scientific discoveries is like picking wild strawberries in a public park: the berries A does not find today B or C or D will surely find tomorrow. At the same time, many scientists view the inner world, which art tries to fathom,

ARTISTS' MISCONCEPTION of the scientific act of creation is equally unrealistic. In this scene from the 1935 Warner Brothers film *The Story of Louis Pasteur* the great scientist (played by Paul Muni) has the sudden inspiration to discover the vaccine for rabies. Art, as any artist knows, is done quite differently. Both photographs are from the Museum of Modern Art Film Stills Archive.

from the standpoint of idealism, according to which events and relations between them have no reality other than their reflection in human thought. Hence there is nothing to be found in the inner world, and artistic creations are cut simply from whole cloth. Here *B* or *C* or *D* could not possibly find tomorrow what *A* found today, because what *A* found had never been there. It is not altogether surprising, of course, to find this split epistemological attitude toward the two worlds, since of these two antithetical traditions in Western philosophical thought, materialism is obviously an unsatisfactory approach to art and idealism an unsatisfactory approach to science.

It is only in the past 20 years or so, more or less contemporaneously with the growth of molecular biology, that a resolution of the age-old epistemological conflict of materialism v. idealism was found in the form of what has come to be known as structuralism. Structuralism emerged simultaneously, independently and in different guises in several diverse fields of study, for example in psychology, linguistics, anthropology and biology.

Both materialism and idealism take it for granted that all the information gathered by our senses actually reaches our mind; materialism envisions that thanks to this information reality is mirrored in the mind, whereas idealism envisions that thanks to this information reality is constructed by the mind. Structuralism, on the other hand, has provided the insight that knowledge about the world enters the mind not as raw data but in already highly abstracted form, namely as structures. In the preconscious process of converting the primary data of our experience step by step into structures, information is necessarily lost, because the creation of structures, or the recognition of patterns, is nothing else than the selective destruction of information. Thus since the mind does not gain access to the full set of data about the world, it can neither mirror nor construct reality. Instead for the mind reality is a set of structural transforms of primary data taken from the world. This transformation process is hierarchical, in that "stronger" structures are formed from "weaker" structures through selective destruction of information. Any set of primary data becomes meaningful only after a series of such operations has so transformed it that it has become congruent with a stronger structure preexisting in the mind. Neurophysiological studies carried out in recent years on the process of visual perception in higher mammals have not only shown directly that the brain actually operates according to the tenets of structuralism but also offer an easily understood illustration of those tenets.

Finally, we may consider the relevance of structuralist philosophy for the two problems in the history of science under discussion here. As far as prematurity of discovery is concerned, structuralism provides us with an understanding of why a discovery cannot be appreciated until it can be connected logically to contemporary canonical knowledge. In the parlance of structuralism, canonical knowledge is simply the set of preexisting "strong" structures with which primary scientific data are made congruent in the mental-abstraction process. Hence data that cannot be transformed into a structure congruent with canonical knowledge are a dead end; in the last analysis they remain meaningless. That is, they remain meaningless until a way has been shown to transform them into a structure that is congruent with the canon.

As far as uniqueness of discovery is concerned, structuralism leads to the recognition that every creative act in the arts and sciences is both commonplace and unique. On the one hand, it is commonplace in the sense that there is an innate, or genetically determined, correspondence in the transformational operations that different individuals perform on the same primary data. With reference to science, cognitive psychology has taught that different individuals recognize the same "chairness" of a chair because they all make a given set of sense impressions from the outer world congruent with the same *Gestalt*, or mental structure. With reference to art, analytic psychology has taught that there is a sameness in the subconscious life of different individuals because an innate human archetype causes them to make the same structural transformations of the events of the inner world. And with reference to both art and science structural linguistics has taught that communication between different individuals is possible only because an innate human grammar causes them to transform a given set of semantic symbols into the same syntactic structure. On the other hand, every creative act is unique in the sense that no two individuals are quite the same and hence never perform exactly the same transformational operations on a given set of primary data. Although all creative acts in both art and science are therefore both commonplace and unique, some may nonetheless be more unique than others.

BIBLIOGRAPHIES

The articles in this volume were accompanied by the following bibliographies when they appeared in the SCIENTIFIC AMERICAN. Philip C. Hanawalt and Robert H. Haynes have provided suggestions for further reading.

I ENERGY FLOW IN LIVING SYSTEMS

1. The Origin of Life

THE ORIGIN OF LIFE. A. I. Oparin. Dover Publications, Inc., 1953.

A PRODUCTION OF AMINO ACIDS UNDER POSSIBLE PRIMITIVE EARTH CONDITIONS. Stanley L. Miller in *Science*, Vol. 117, No. 3046, page 528; May 15, 1953.

TIME'S ARROW AND EVOLUTION. Harold F. Blum. Princeton University Press, 1951.

2. How Light Interacts with Living Matter

CHEMISTRY AND BIOCHEMISTRY OF PLANT PIGMENTS. Edited by T. W. Goodwin. Academic Press, 1965.

PHOTORECEPTORS. *Cold Spring Harbor Symposia on Quantitative Biology: Vol. XXX.* Cold Spring Harbor Laboratory of Quantitative Biology, 1965.

PLANT BIOCHEMISTRY. Edited by James Bonner and J. E. Varner. Academic Press, 1966.

3. The Mechanism of Photosynthesis

PHOTOPHOSPHORYLATION AND THE CHEMI-OSMOTIC HYPOTHESIS. André T. Jagendorf and E. Uribe in *Brookhaven Symposia in Biology*, Vol. 19, pages 215–245; 1966.

ELECTRON TRANSPORT PATHWAYS IN PHOTOSYNTHESIS. Geoffrey Hind and John M. Olson in *Annual Review of Plant Physiology: Vol. XIX*, edited by Leonard Machlis. Annual Reviews, Inc., 1968.

HAEM-PROTEINS IN PHOTOSYNTHESIS. D. S. Bendall and R. Hill in *Annual Review of Plant Physiology: Vol. XIX*, edited by Leonard Machlis. Annual Reviews, Inc., 1968.

4. The Membrane of the Mitochondrion

THE ENERGY-LINKED FUNCTION OF MITOCHONDRIA. Edited by Britton Chance. Academic Press Inc., 1963.

MITOCHONDRIAL OXIDATIONS AND ENERGY COUPLING. Maynard E. Pullman and Gottfried Schatz in *Annual Review of Biochemistry*, Vol. 36, pages 539–610; 1967.

PARTIAL RESOLUTION OF THE ENZYMES CATALYZING OXIDATIVE PHOSPHORYLATION, X: CORRELATION OF MORPHOLOGY AND FUNCTION IN SUBMITO-CHONDRIAL PARTICLES. Yasuo Kagawa and Efraim Racker with Rolf E. Hauser in *The Journal of Biological Chemistry*, Vol. 241, No. 10, pages 2475–2484; May 25, 1966.

SYMPOSIUM ON ENERGY COUPLING IN ELECTRON TRANSPORT. Albert L. Lehninger, Efraim Racker, Britton Chance, Chuan-Pu Lee, Leena Mela, S. N. Grazen, S. Estrada-O, Henry A. Lardy, A. T. Jagendorf and P. Mitchell in *Federation Proceedings*, Vol. 26, No. 5, pages 1333–1334; September–October, 1967.

SUGGESTIONS FOR FURTHER READING

Clayton, Roderick K. LIGHT AND LIVING MATTER: A GUIDE TO THE STUDY OF PHOTOBIOLOGY, *Volume 1: THE PHYSICAL PART, Volume II: THE BIOLOGICAL PART.* New York: McGraw Hill, 1971.

Kenyon, Dean H. and Gary Steinman BIOCHEMICAL PREDESTINATION. New York: McGraw Hill, 1969.

Lehninger, A. L. BIOENERGETICS, 2nd ed., Menlo Park, California: W. A. Benjamin, Inc., 1971.

Morowitz, H. J. ENERGY FLOW IN BIOLOGY. New York: Academic Press, 1968.

Oparin, A. I. LIFE, ITS NATURE, ORIGIN AND DEVELOPMENT. New York: Academic Press, 1962.

Orgel, L. E. THE ORIGINS OF LIFE. New York: John Wiley and Sons, Inc., 1973.

II MOLECULAR ARCHITECTURE

THE STRUCTURES AND FUNCTIONAL ROLES OF PROTEINS

5. Proteins

THE NATIVE AND DENATURED STATES OF SOLUBLE COLLAGEN. Helga Boedtker and Paul Doty in *Journal of the American Chemical Society*, Vol. 78, No. 17, pages 4267–4280; September 5, 1956.

THE OPTICAL ROTARY DISPERSION OF POLYPEPTIDES AND PROTEINS IN RELATION TO CONFIGURATION. Jen Tsi Yang and Paul Doty in *Journal of the American Chemical Society*, Vol. 79, No. 4, pages 761–775; February 27, 1957.

POLYPEPTIDES, VIII: MOLECULAR CONFIGURATIONS OF POLY-L-GLUTAMIC ACID IN WATER-DIOXANE SOLUTION. Paul Doty, A. Wada, Jen Tsi Yang and E. R. Blout in *Journal of Polymer Science*, Vol. 23, No. 104, pages 851–861; February, 1957.

SYNTHETIC POLYPEPTIDES: PREPARATION, STRUCTURE, AND PROPERTIES. C. H. Bamford, A. Elliott and W. E. Hanby. Academic Press Inc., 1956.

6. Collagen

THE CHARGE PROFILE OF THE TROPOCOLLAGEN MACROMOLECULE AND THE PACKING ARRANGEMENT IN NATIVE TYPE COLLAGEN FIBRILS. Alan J. Hodge and Francis O. Schmitt in *Proceedings of the National Academy of Sciences*, Vol. 46, No. 2, pages 186–197; February, 1960.

THE CHEMISTRY AND REACTIVITY OF COLLAGEN. Karl Helmer Gustavson. Academic Press, Inc., 1956.

THE STRUCTURE OF COLLAGEN FIBRILS. Richard S. Bear in *Advances in Protein Chemistry*, Vol. 7, pages 69–160; 1952.

7. The Insulin Molecule

THE AMINO-ACID SEQUENCE IN THE GLYCYL CHAIN OF INSULIN. F. Sanger and E. O. P. Thompson in *The Biochemical Journal*, Vol. 53, No. 3, pages 353–374; February, 1953.

THE CHEMISTRY OF INSULIN. F. Sanger in *Annual Reports on the Chemical Society*, Vol. 45, pages 283–292; 1949.

THE PRINCIPLES OF CHROMATOGRAPHY. A. J. P. Martin in *Endeavour*, Vol. 6, No. 21, pages 21–28; January, 1947.

8. The Structure and History of an Ancient Protein

THE STRUCTURE AND ACTION OF PROTEINS. Richard E. Dickerson and Irving Geis. Harper & Row, Publishers, 1969.

THE STRUCTURE OF CYTOCHROME *c* AND THE RATES OF MOLECULAR EVOLUTION. Richard E. Dickerson in *Journal of Molecular Evolution*, Vol. 1, No. 1, pages 26–45; 1971.

FERRICYTOCHROME *c*: I, GENERAL FEATURES OF THE HORSE AND BONITO PROTEINS AT 2.8 A RESOLUTION. Richard E. Dickerson, Tsunehiro Takano, David Eisenberg, Olga B. Kallai, Lalli Samson, Angela Cooper and E. Margoliash in *The Journal of Biological Chemistry*, Vol. 246, No. 5, pages 1511–1535; March 10, 1971.

CONFORMATIONAL CHANGES UPON REDUCTION OF CYTOCHROME *c*. T. Takano, R. Swanson, O. B. Kallai and R. E. Dickerson in *Cold Spring Harbor Symposium on Quantitative Biology,* in press.

9. The Structure and Function of Antibodies

COLD SPRING HARBOR SYMPOSIA ON QUANTITATIVE BIOLOGY, VOL XXXII: ANTIBODIES. Cold Spring Harbor Laboratory of Quantitative Biology, 1967.

NOBEL SYMPOSIUM 3: GAMMA GLOBULINS—STRUCTURE AND CONTROL OF BIOSYNTHESIS. Edited by Johan Killander. Interscience Publishers, 1968.

THE ANTIBODY PROBLEM. Gerald M. Edelman and W. Einar Gall in *Annual Review of Biochemistry: Vol. XXXVIII*, edited by Esmond E. Snell. Annual Reviews, Inc., 1969.

THE COVALENT STRUCTURE OF AN ENTIRE γG IMMUNOGLOBULIN MOLECULE. Gerald M. Edelman, Bruce A. Cunningham, W. Einar Gall, Paul D. Gottlieb, Urs Rutishauser and Myron J. Waxdal in *Proceedings of the National Academy of Sciences*, Vol. 63, No. 1, pages 78–85; May 15, 1969.

10. The Hemoglobin Molecule

THE CHEMISTRY AND FUNCTION OF PROTEINS. Felix Haurowitz. Academic Press, 1963.

RELATION BETWEEN STRUCTURE AND SEQUENCE OF HAEMOGLOBIN. M. F. Perutz in *Nature*, Vol. 194, No. 4832, pages 914–918; June, 1962.

STRUCTURE OF HAEMOGLOBIN: A THREE-DIMENSIONAL FOURIER SYNTHESIS OF REDUCED HUMAN HAEMOGLOBIN AT 5.5 Å RESOLUTION. Hilary Muirhead and M. F. Perutz in *Nature*, Vol. 199, No. 4894, pages 633–639; August, 1963.

11. The Three-Dimensional Structure of an Enzyme Molecule

BIOSYNTHESIS OF MACROMOLECULES. Vernon M. Ingram. W. A. Benjamin, Inc., 1965.

INTRODUCTION TO MOLECULAR BIOLOGY. G. H. Haggis, D. Michie, A. R. Muir, K. B. Roberts and P. M. B. Walker. John Wiley & Sons, Inc., 1964.

THE MOLECULAR BIOLOGY OF THE GENE. J. D. Watson. W. A. Benjamin, Inc., 1965.

PROTEIN AND NUCLEIC ACIDS: STRUCTURE AND FUNCTION. M. F. Perutz. American Elsevier Publishing Company, Inc., 1962.

STRUCTURE OF HEN EGG-WHITE LYSOZYME: A THREE-DIMENSIONAL FOURIER SYNTHESIS AT 2 A. RESOLUTION. C. C. F. Blake, D. F. Koenig, G. A. Mair, A. C. T. North, D. C. Phillips and V. R. Sarma in *Nature*, Vol. 206, No. 4986, pages 757–763; May 22, 1965.

STORAGE, REPLICATION, AND EXPRESSION OF GENETIC INFORMATION

12. The Discovery of DNA

THE CELL IN DEVELOPMENT AND HEREDITY. Edmund B. Wilson. The Macmillan Company, 1928.

CELL, NUCLEUS, AND INHERITANCE: AN HISTORICAL STUDY. William Coleman in *Proceedings of the American Philosophical Society*, Vol. 109, No. 3, pages 124–158; June, 1965.

DIE HISTOCHEMISCHEN UND PHYSIOLOGISCHEN ARBEITEN. Friedrich Miescher. Verlag von F. C. W. Vogel, 1897.

ZELLSUBSTANZ, KERN UND ZELLTHEILUNG. Walther Flemming. Vogel Verlag Leipzig, 1882.

13. The Synthesis of DNA

ENZYMATIC SYNTHESIS OF DNA. Arthur Kornberg. John Wiley & Sons, Inc., 1961.

ENZYMATIC SYNTHESIS OF DEOXYRIBONUCLEIC ACID, VIII: FREQUENCIES OF NEAREST NEIGHBOR BASE SEQUENCES IN DEOXYRIBONUCLEIC ACID. John Josse, A. D. Kaiser and Arthur Kornberg in *The Journal of Biological Chemistry*, Vol. 236, No. 3, pages 864–875; March, 1961.

ENZYMATIC SYNTHESIS OF DNA, XXIV: SYNTHESIS OF INFECTIOUS PHAGE ϕX174 DNA. Mehran Goulian, Arthur Kornberg and Robert L. Sinsheimer in *Proceedings of the National Academy of Sciences*, Vol. 58, No. 6, pages 2321–2328; December 15, 1967.

14. The Bacterial Chromosome

THE BACTERIAL CHROMOSOME AND ITS MANNER OF REPLICATION AS SEEN BY AUTORADIOGRAPHY. John Cairns, in *Journal of Molecular Biology*, Vol. 6, No. 3, pages 208–213; March, 1963.

COLD SPRING HARBOR SYMPOSIA ON QUANTITATIVE BIOLOGY, VOLUME XXVIII: SYNTHESIS AND STRUCTURE OF MACROMOLECULES. Cold Spring Harbor Laboratory of Quantitative Biology, 1963.

MOLECULAR BIOLOGY OF THE GENE. James D. Watson. W. A. Benjamin, Inc., 1965. See pages 255–296.

15. The Fine Structure of the Gene

THE ELEMENTARY UNITS OF HEREDITY. Seymour Benzer in *The Chemical Basis of Heredity*, edited by William D. McElroy and Bentley Glass, pages 70–93. The Johns Hopkins Press, 1957.

GENETIC RECOMBINATION BETWEEN HOST-RANGE AND PLAQUE-TYPE MUTANTS OF BACTERIOPHAGE IN SINGLE BACTERIAL CELLS. A. D. Hershey and Raquel Rotman in *Genetics*, Vol. 34, No. 1, pages 44–71; January, 1949.

INDUCTION OF SPECIFIC MUTATIONS WITH 5-BROMOURACIL. Seymour Benzer and Ernst Freese in *Proceedings of the National Academy of Sciences*, Vol. 44, No. 2, pages 112–119; February, 1958.

THE STRUCTURE OF THE HEREDITARY MATERIAL. F. H. C. Crick in *Scientific American*, Vol. 191, No. 4, pages 54–61; October, 1954.

ON THE TOPOGRAPHY OF THE GENETIC FINE STRUCTURE. Seymour Benzer in *Proceedings of the National Academy of Sciences*, Vol. 47, No. 3, pages 403–415; March, 1961.

16. The Nucleotide Sequence of a Nucleic Acid

ISOLATION OF LARGE OLIGONUCLEOTIDE FRAGMENTS FROM THE ALANINE RNA. Jean Apgar, George A. Everett and Robert W. Holley in *Proceedings of the National Academy of Sciences*, Vol. 53, No. 3, pages 546–548; March, 1965.

LABORATORY EXTRACTION AND COUNTERCURRENT DISTRIBUTION. Lyman C. Craig and David Craig in *Technique of Organic Chemistry, Volume III, Part I: Separation and Purification*, edited by Arnold Weissberger. Interscience Publishers, Inc., 1956. See pages 149–332.

SPECIFIC CLEAVAGE OF THE YEAST ALANINE RNA INTO TWO LARGE FRAGMENTS. John Robert Penswick and Robert W. Holley in *Proceedings of the National Academy of Sciences*, Vol. 53, No. 3, pages 543–546; March, 1965.

STRUCTURE OF A RIBONUCLEIC ACID. Robert W. Holley, Jean Apgar, George A. Everett, James T. Madison, Mark Marquisee, Susan H. Merrill, John Robert Penswick and Ada Zamir in *Science*, Vol. 147, No. 3664, pages 1462–1465; March 19, 1965.

17. The Genetic Code

THE FINE STRUCTURE OF THE GENE. Seymour Benzer in *Scientific American*, Vol. 206, No. 1, pages 70–84; January, 1962.

GENERAL NATURE OF THE GENETIC CODE FOR PROTEINS. F. H. C. Crick, Leslie Barnett, S. Brenner and R. J. Watts-Tobin in *Nature*, Vol. 192, No. 4809, pages 1227–1232; December 30, 1961.

MESSENGER RNA. Jerard Hurwitz and J. J. Furth in *Scientific American*, Vol. 206, No. 2, pages 41–49; February, 1962.

THE NUCLEIC ACIDS: Vol. III. Edited by Erwin Chargaff and J. N. Davidson, Academic Press Inc., 1960.

18. The Genetic Code: III

THE GENETIC CODE, VOL. XXXI: 1966 COLD SPRING HARBOR SYMPOSIA ON QUANTITATIVE BIOLOGY. Cold Spring Harbor Laboratory of Quantitative Biology, in press.

MOLECULAR BIOLOGY OF THE GENE. James D. Watson. W. A. Benjamin, Inc., 1965.

RNA CODEWORDS AND PROTEIN SYNTHESIS, VII: ON THE GENERAL NATURE OF THE RNA CODE. M. Nirenberg, P. Leder, M. Bernfield, R. Brimacombe, J. Trupin, F. Rottman and C. O'Neal in *Proceedings of the National Academy of Sciences*, Vol. 53, No. 5, pages 1161–1168; May, 1965.

STUDIES ON POLYNUCLEOTIDES, LVI: FURTHER SYNTHESES, IN VITRO, OF COPOLYPEPTIDES CONTAINING TWO AMINO ACIDS IN ALTERNATING SEQUENCE DEPENDENT UPON DNA-LIKE POLYMERS CONTAINING TWO NUCLEOTIDES IN ALTERNATING SEQUENCE. D. S. Jones, S. Nishimura and H. G. Khorana in *Journal of Molecular Biology*, Vol. 16, No. 2, pages 454–472; April, 1966.

SUGGESTIONS FOR FURTHER READINGS

Cold Spring Harbor Symposium on Quantitative Biology, Volume XXXIII, "Replication of DNA in Microorganisms." Cold Spring Harbor, New York, 1968.

Cold Spring Harbor Symposium on Quantitative Biology, Volume XXXVI, "Structure and Function of Proteins at the Three-dimensional Level." Cold Spring Harbor, New York, 1972.

Dickerson, R. E. and Irving Geis THE STRUCTURE AND ACTION OF PROTEINS. New York: Harper and Row, 1969.

Lehninger, A. L. BIOCHEMISTRY. New York: Worth Publisher, Inc., 1970.

Rich, Alexander and Norman Davidson, Eds. STRUCTURAL CHEMISTRY AND MOLECULAR BIOLOGY, *A volume dedicated to Linus Pauling by his students, colleagues, and friends.* San Francisco: W. H. Freeman and Company, 1968.

Watson, J. D. THE DOUBLE HELIX. New York: Atheneum, 1968.

III   MACROMOLECULAR COMPLEXES AND THEIR ASSEMBLY

FUNCTIONAL ASSOCIATIONS OF NUCLEIC ACIDS AND PROTEINS

19. Ribosomes

RECONSTITUTION OF FUNCTIONALLY ACTIVE RIBOSOMES FROM INACTIVE SUBPARTICLES AND PROTEINS. Keiichi Hosokawa, Robert K. Fujimura and Masayasu Nomura in *Proceedings of the National Academy of Sciences*, Vol. 55, No. 1, pages 198–204; January, 1966.

STRUCTURE AND FUNCTION OF E. COLI RIBOSOMES, V: RECONSTITUTION OF FUNCTIONALLY ACTIVE 30S RIBOSOMAL PARTICLES FROM RNA AND PROTEINS. P. Traub and M. Nomura in *Proceedings of the National Academy of Sciences*, Vol. 59, No. 3, pages 777–784; March, 1968.

STRUCTURE AND FUNCTION OF ESCHERICHIA COLI RIBOSOMES, VI: MECHANISM OF ASSEMBLY OF 30 S RIBOSOMES.

20. The Multiplication of Bacterial Viruses

BACTERIAL VIRUSES AND SEX. Max and Mary Bruce Delbrück in *Scientific American*; November, 1948.

21. The Structure of Viruses

THE FINE STRUCTURE OF POLYOMA VIRUS. P. Wildy, M. G. P. Stoker, I. A. Macpherson, and R. W. Horne in *Virology*, Vol. 11, No. 2, pages 444–457; June, 1960.

A HELICAL STRUCTURE IN MUMPS, NEWCASTLE DISEASE AND SENDAI VIRUSES. R. W. Horne and A. P. Waterson in *Journal of Molecular Biology*, Vol. 2, No. 1, pages 75–77; April, 1960.

THE ICOSAHEDRAL FORM OF AN ADENOVIRUS. R. W. Horne, S. Brenner, A. P. Waterson and P. Wildy in *Journal of Molecular Biology*, Vol. 1, No. 1, pages 84–86; April, 1959.

THE MORPHOLOGY OF HERPES VIRUS. P. Wildy, W. C. Russell and R. W. Horne in *Virology*, Vol. 12, No. 2, pages 204–222; October, 1960.

THE STRUCTURE AND COMPOSITION OF THE MYXOVIRUSES. R. W. Horne, A. P. Waterson, P. Wildy and A. E. Farnham in *Virology*, Vol. 11, No. 1, pages 79–98; May, 1960.

SYMMETRY IN VIRUS ARCHITECTURE. R. W. Horne and P. Wildy in *Virology*, Vol. 15, No. 3, pages 348–373; November, 1961.

22. Building a Bacterial Virus

CONDITIONAL MUTATIONS IN BACTERIOPHAGE T4. R. S. Edgar and R. H. Epstein in *Genetics Today*, edited by S. J. Geerts. Pergamon Press, 1963.

GENE ACTION IN THE CONTROL OF BACTERIOPHAGE T4 MORPHOGENESIS. W. B. Wood in *Proceedings of the Thomas Hunt Morgan Centennial Symposium.* University of Kentucky, in press.

SOME STEPS IN THE MORPHOGENESIS OF BACTERIOPHAGE T4. R. S. Edgar and I. Lielausis in *Journal of Molecular Biology*, in press.

FUNCTIONAL ASSOCIATIONS OF LIPIDS, CARBOHYDRATES, AND PROTEINS

23. The Receptor Site for A Bacterial Virus

STUDIES ON THE CHEMICAL BASIS OF THE PHAGE CONVERSION OF O-ANTIGENS IN THE E-GROUP SALMONELLAE. P. W. Robbins and T. Uchida in *Biochemistry*, Vol. I, No. 2, pages 323–335; March, 1962.

MECHANISM OF ϵ^{15} CONVERSION STUDIED WITH A BACTERIAL MUTANT. Richard Losick and P. W. Robbins in *Journal of Molecular Biology*, Vol. 30, No. 3, pages 445–455; December 28, 1967.

MECHANISM OF ϵ^{15} CONVERSION STUDIED WITH BACTERIOPHAGE MUTANTS. D. Bray and P. W. Robbins in *Journal of Molecular Biology*, Vol. 30, No. 3, pages 457–475; December 28, 1967.

BIOSYNTHESIS OF CELL WALL LIPOPOLYSACCHARIDE IN GRAM-NEGATIVE ENTERIC BACTERIA. Hiroshi Nikaido in *Advances in Enzymology and Related Areas of Molecular Biology: Vol. XXXI*, edited by F. F. Nord. Interscience Publishers, 1968.

ISOLATION OF A TRYPSIN-SENSITIVE INHIBITOR OF O-ANTIGEN SYNTHESIS INVOLVED IN LYSOGENIC CONVERSION BY BACTERIOPHAGE ϵ^{15}. Richard Losick in *Journal of Molecular Biology*, Vol. 42, No. 2, pages 237–246; June 14, 1969.

24. The Bacterial Cell Wall

THE BACTERIAL CELL WALL. Milton R. J. Salton. Elsevier Publishing Company, 1964.

BACTERIAL CELL WALL SYNTHESIS AND STRUCTURE IN RELATION TO THE MECHANISM OF ACTION OF PENICILLINS AND OTHER ANTIBACTERIAL AGENTS. Jack L. Strominger and Donald J. Tipper in *The American Journal of Medicine*, Vol. 39, No. 5, pages 708–721; November, 1965.

ISOLATION AND STUDY OF THE CHEMICAL STRUCTURE OF LOW MOLECULAR WEIGHT GLYCOPEPTIDES FROM MICROCOCCUS LYSODEIKTICUS CELL WALLS. David Mirelman and Nathan Sharon in *The Journal of Biological Chemistry*, Vol. 242, No. 15, pages 3414–3427; August 10, 1967.

25. The Structure of Cell Membranes

MEMBRANES OF MITOCHONDRIA AND CHLOROPLASTS. Edited by Efraim Racker. Van Nostrand Reinhold Company, 1969.

STRUCTURE AND FUNCTION OF BIOLOGICAL MEMBRANES. Edited by Lawrence I. Rothfield. Academic Press, 1971.

MEMBRANE MOLECULAR BIOLOGY. Edited by C. F. Fox and A. Keith. Sinauer Associates, Stamford, Conn., in press.

SUGGESTIONS FOR FURTHER READINGS

Fox, C. Fred, Ed., MEMBRANE RESEARCH. New York: Academic Press, 1972.

Goodheart, C. R. AN INTRODUCTION TO VIROLOGY. New York: Saunders, 1969.

Hershey, A. D., Ed. THE BACTERIOPHAGE LAMDA. Cold Spring Harbor, New York: CSH Lab. of Quantitative Biology, 1971.

Rothfield, L. I., Ed. STRUCTURE AND FUNCTION OF BIOLOGICAL MEMBRANES. New York: Academic Press, 1971.

IV INFORMATION TRANSFER AND CONTROL

PROTEIN SYNTHESIS AND REGULATION

26. The Visualization of Genes in Action

PORTRAIT OF A GENE. Oscar L. Miller, Jr., and Barbara R. Beatty in *Journal of Cellular Physiology*, Vol. 74, No. 2, Part 2, Supplement 1, pages 225–232; October, 1969.

ELECTRON MICROSCOPIC VISUALIZATION OF TRANSCRIPTION. O. L. Miller, Jr., Barbara R. Beatty, Barbara A. Hamkalo and C. A. Thomas, Jr., in *Cold Spring Harbor Symposia on Quantitative Biology*, Vol. 35, pages 505–512; 1970.

VISUALIZATION OF BACTERIAL GENES IN ACTION. O. L. Miller, Jr., Barbara A. Hamkalo and C. A. Thomas, Jr., in *Science*, Vol. 169, No. 3943, pages 392–395; July 24, 1970.

MORPHOLOGICAL STUDIES OF TRANSCRIPTION. O. L. Miller, Jr., and Aimée H. Bakken in *Acta Endocrinologica*, Supplementum 168, pages 155–173; 1972.

VISUALIZATION OF RNA SYNTHESIS ON CHROMOSOMES. O. L. Miller, Jr., and Barbara A. Hamkalo in *The International Review of Cytology*, Vol. 33, pages 1–25; 1972.

27. Gene Structure and Protein Structure

CO-LINEARITY OF β-GALACTOSIDASE WITH ITS GENE BY IMMUNOLOGICAL DETECTION OF INCOMPLETE POLYPEPTIDE CHAINS. Audree V. Fowler and Irving Zabin in *Science*, Vol. 154, No. 3752, pages 1027–1029; November 25, 1966.

CO-LINEARITY OF THE GENE WITH THE POLYPEPTIDE CHAIN. A. S. Sarabhai, A. O. W. Stretton, S. Brenner and A. Bolle in *Nature*, Vol. 201, No. 4914, pages 13–17; January 4, 1964.

THE COMPLETE AMINO ACID SEQUENCE OF THE TRYPTOPHAN SYNTHETASE A PROTEIN (α SUBUNIT) AND ITS COLINEAR RELATIONSHIP WITH THE GENETIC MAP OF THE A GENE. Charles Yanofsky, Gabriel R. Drapeau, John R. Guest and Bruce C. Carlton in *Proceedings of the National Academy of Sciences*, Vol. 57, No. 2, pages 296–298; February, 1967.

MUTATIONALLY INDUCED AMINO ACID SUBSTITUTIONS IN A TRYPTIC PEPTIDE OF THE TRYPTOPHAN SYNTHETASE A PROTEIN. John R. Guest and Charles Yanofsky in *Journal of Biological Chemistry*, Vol. 240, No. 2, pages 679–689; February, 1965.

ON THE COLINEARITY OF GENE STRUCTURE AND PROTEIN STRUCTURE. C. Yanofsky, B. C. Carlton, J. R. Guest, D. R. Helinski and U. Henning in *Proceedings of the National Academy of Sciences*, Vol. 51, No. 2, pages 266–272; February, 1964.

28. How Proteins Start

A GTP REQUIREMENT FOR BINDING INITIATOR tRNA TO RIBOSOMES. John S. Anderson, Mark S. Bretscher, Brian F. C. Clark and Kjeld A. Marcker in *Nature*, Vol. 215, No. 5100, pages 490–492; July 29, 1967.

N-FORMYL-METHIONYL-S-RNA. K. Marcker and F. Sanger in *Journal of Molecular Biology*, Vol. 8, No. 6, pages 835–840; June, 1964.

THE ROLE OF N-FORMYL-METHIONYL-SRNA IN PROTEIN BIOSYNTHESIS. B. F. C. Clark and K. A. Marcker in *Journal of Molecular Biology*, Vol. 17, No. 2, pages 394–406; June, 1966.

STUDIES ON POLYNUCLEOTIDES, LXVII: INITIATION OF PROTEIN SYNTHESIS IN VITRO AS STUDIED BY USING RIBOPOLYNUCLEOTIDES WITH REPEATING NUCLEOTIDE SEQUENCES AS MESSENGERS. H. P. Ghosh, D. Söll and H. G. Khorana in *Journal of Molecular Biology*, Vol. 25, No. 2, pages 275–298; April 28, 1967.

29. The Control of Biochemical Reactions

ALLOSTERIC PROTEINS AND CELLULAR CONTROL SYSTEMS. Jacques Monod, Jean-Pierre Changeux and François Jacob in *Journal of Molecular Biology*, Vol. 6, No. 4, pages 306–329; April, 1963.

GENETIC REGULATORY MECHANISMS IN THE SYNTHESIS OF PROTEINS. François Jacob and Jacques Monod in *Journal of Molecular Biology*, Vol. 3, No. 3, pages 318–356; June, 1961.

ON THE REGULATION OF DNA REPLICATION IN BACTERIA. François Jacob, Sydney Brenner and François Cuzin in *Cold Spring Harbor Symposia on Quantitative Biology, Vol. XXVIII.* 1963.

A PLAUSIBLE MODEL OF ALLOSTERIC TRANSITION. Jacques Monod, Jeffries Wyman and Jean-Pierre Changeux in *Journal of Molecular Biology*, in press.

30. Genetic Responses

ISOLATION OF THE LAC REPRESSOR. Walter Gilbert and Benno Müller-Hill in *Proceedings of the National Academy of Sciences*, Vol. 56, No. 6, pages 1891–1898; December, 1966.

ISOLATION OF THE λ PHAGE REPRESSOR. Mark Ptashne in *Proceedings of the National Academy of Sciences*, Vol. 57, No. 2, pages 306–313; February, 1967.

SPECIFIC BINDING OF THE λ PHAGE REPRESSOR TO λ DNA. Mark Ptashne in *Nature*, Vol. 214, No. 5085, pages 232–234; April 15, 1967.

THE LAC OPERATOR IS DNA. Walter Gilbert and Benno Müller-Hill in *Proceedings of the National Academy of Sciences*, Vol. 58, No. 6, pages 2415–2421; December, 1967.

31. Cyclic AMP

CYCLIC ADENOSINE MONOPHOSPHATE IN BACTERIA. Ira Pastan and Robert Perlman in *Science*, Vol. 169, No. 3943, pages 339–344; July 24, 1970.

CYCLIC AMP. G. Alan Robison, Reginald W. Butcher and Earl W. Sutherland Academic Press, 1971.

CYCLIC AMP AND CELL FUNCTION. Edited by G. Alan Robison, Gabriel G. Nahas and Lubos Triner in *Annals of the New York Academy of Sciences*, Vol. 185; December 3, 1971.

DNA TRANSACTIONS

32. The Recognition of DNA in Bacteria

HOST-INDUCED MODIFICATION OF T-EVEN PHAGES DUE TO DEFECTIVE GLUCOSYLATION OF THEIR DNA. Stanley Hattman and Toshio Fukasawa in *Proceedings of the National Academy of Sciences of the United States of America*, Vol. 50, No. 2, pages 297–300; August 15, 1963.

GENERAL VIROLOGY. S. E. Luria and James E. Darnell, Jr. John Wiley & Sons, Inc., 1967.

RESTRICTION OF NONGLUCOSYLATED T-EVEN BACTERIOPHAGE: PROPERTIES OF PERMISSIVE MUTANTS OF ESCHERICHIA COLI B AND K12. Helen R. Revel in *Virology*, Vol. 31, No. 4, pages 688–701; April, 1967.

DNA MODIFICATION AND RESTRICTION. Werner Arber and Stuart Linn in *Annual Review of Biochemistry*, Vol. 38, pages 467–500; 1969.

RESTRICTION OF NONGLUCOSYLATED T-EVEN BACTERIOPHAGES BY PROPHAGE P1. Helen R. Revel and C. P. Georgopoulos in *Virology*, Vol. 39, No. 1, pages 1–17; September, 1969.

33. The Repair of DNA

THE DISAPPEARANCE OF THYMINE DIMERS FROM DNA: AN ERROR-CORRECTING MECHANISM. R. B. Setlow and W. L. Carrier in *Proceedings of the National Academy of Sciences*, Vol. 51, No. 2, pages 226–231; February, 1964.

EVIDENCE FOR REPAIR REPLICATION OF ULTRAVIOLET-DAMAGED DNA IN BACTERIA. David Pettijohn and Philip C. Hanawalt in *Journal of Molecular Biology*, Vol. 9, No. 2, pages 395–410; August, 1964.

A GENETIC LOCUS IN E. COLI K 12 THAT CONTROLS THE REACTIVATION OF UV-PHOTOPRODUCTS ASSOCIATED WITH THYMINE IN DNA. P. Howard-Flanders, Richard P. Boyce, Eva Simson and Lee Theriot in *Proceedings of the National Academy of Sciences*, Vol. 48, No. 12, pages 2109–2115; December 15, 1962.

STRUCTURAL DEFECTS IN DNA AND THEIR REPAIR IN MICROORGANISMS. Radiation Research, Supplement 6, edited by Robert H. Haynes, Sheldon Wolff and James E. Till. Academic Press, in press.

34. Cellular Factors in Genetic Transformation

THE EFFECT OF ENVIRONMENTAL FACTORS ON TRANSFORMABILITY OF A STREPTOCOCCUS. Roman Pakula, Janina Cybulska and Wlodzimierz Walczak in *Acta Microbiologica Polonica*, Vol. 12, pages 245–257; 1963.

TRANSFORMATION. Pierre Schaeffer in *The Bacteria, a Treatise on Structure and Function: Vol. V, Heredity,* edited by I. C. Gunsalus and Roger Y. Stanier. Academic Press, 1964.

REGULATION OF THE TRANSFORMABILITY OF PNEUMOCOCCAL CULTURES BY MACROMOLECULAR CELL PRODUCTS. Alexander Tomasz and Rollin D. Hotchkiss in *Proceedings of the National Academy of Sciences*, Vol. 51, No. 3, pages 480–487; March 15, 1964.

BIOLOGICAL CONSEQUENCES OF THE REPLACEMENT OF CHOLINE BY ETHANOLAMINE IN THE CELL WALL OF PNEUMOCOCCUS: CHAIN FORMATION, LOSS OF TRANSFORMABILITY AND LOSS OF AUTOLYSIS. Alexander Tomasz in *Proceedings of the National Academy of Sciences*, Vol. 59, No. 1, pages 86–93; January 15, 1968.

35. Repeated Segments of DNA

A MOLECULAR APPROACH IN THE SYSTEMATICS OF HIGHER ORGANISMS. B. H. Hoyer, B. J. McCarthy and E. T. Bolton in *Science*, Vol. 144, No. 3621, pages 959–967; May 22, 1964.

NUCLEOTIDE SEQUENCE REPETITION: A RAPIDLY REASSOCIATING FRACTION OF MOUSE DNA. Michael Waring and Roy J. Britten in *Science*, Vol. 154, No. 3750, pages 791–794; November 11, 1966.

REPEATED SEQUENCES IN DNA. R. J. Britten and D. E. Kohne in *Science*, Vol. 161, No. 3841, pages 529–540; August 9, 1968.

HIGHLY REPETITIVE DNA IN RODENTS. P. M. B. Walker, W. G. Flamm and A. McClaren in *Handbook of Molecular Cytology*, edited by A. Lima-De-Faria. North-Holland Publishing, 1969.

36. RNA-directed DNA Synthesis

ONCOGENIC RIBOVIRUSES. Frank Fenner in *The Biology of Animal Viruses: Vol. II*. Academic Press, 1968.

CENTRAL DOGMA OF MOLECULAR BIOLOGY. Francis Crick in *Nature*, Vol. 227, No. 5258, pages 561–563; August 8, 1970.

MECHANISM OF CELL TRANSFORMATION BY RNA TUMOR VIRUSES. Howard M. Temin in *Annual Review of Microbiology: Vol. XXV*. Annual Reviews, Inc., 1971.

THE PROTOVIRUS HYPOTHESIS: SPECULATIONS ON THE SIGNIFICANCE OF RNA-DIRECTED DNA SYNTHESIS FOR NORMAL DEVELOPMENT AND FOR CARCINOGENESIS. Howard M. Temin in *Journal of the National Cancer Institute*, Vol. 46, No. 2, pages III–VII; February, 1971.

37. Cellular Communication

STRUCTURALISM. Jean Piaget. Basic Books, Inc., 1970.

PAPERS IN CELLULAR NEUROPHYSIOLOGY. Edited by I. Cooke and M. Lipkin, Jr. Holt, Rinehart and Winston, 1972.

38. Prematurity and Uniqueness in Scientific Discovery

THE POTENTIAL THEORY OF ADSORPTION. Michael Polanyi in *Science*, Vol. 141, No. 3585, pages 1010–1013; September 13, 1963.

PERCEPTION AND DECEPTION. C. W. Churchman in *Science*, Vol. 153, No. 3740, pages 1088–1090; September 2, 1966.

MOLECULAR GENETICS: AN INTRODUCTORY NARRATIVE. Gunther S. Stent. W. H. Freeman and Company, 1971.

SUGGESTIONS FOR FURTHER READINGS

Beckwith, J. R., and D. Zipser, Eds. THE LACTOSE OPERON. Cold Spring Harbor, New York: CSH Lab. of Quantitative Biology, 1970.

Davidson, E. H. GENE ACTIVITY IN EARLY DEVELOPMENT. New York: Academic Press, 1968.

Hayes, W. THE GENETICS OF BACTERIA AND THEIR VIRUSES, 2nd ed., New York: John Wiley and Sons, Inc., 1968.

Kenney, F. T., B. A. Hamkalo, G. Favelukes, and J. T. August, Eds. GENE EXPRESSION AND ITS REGULATION. New York: Plenum Press, 1973.

Smith, K. C. and P. C. Hanawalt. MOLECULAR PHOTOBIOLOGY: INACTIVATION AND RECOVERY. New York: Academic Press, 1969.

Stent, G. S. MOLECULAR GENETICS. San Francisco: W. H. Freeman and Company, 1969.

Watson, J. D. MOLECULAR BIOLOGY OF THE GENE. 2nd ed., New York: W. A. Benjamin, 1970.

INDEX OF SUBJECTS

In this index 23f means separate references on pp. 23 and 24; 23ff means separate references on pp. 23, 24 and 25; 23–25 means a continuous discussion. *Passim* meaning "here and there", is used for a cluster of references in close but not consecutive sequence (for example, 22, 23, 24, 26, 27, 30 would be written as 22–30 *passim*).

INDEX OF NAMES